SECOND EDITION

RUBBER COMPOUNDING

Chemistry and Applications

SECOND EDITION

RUBBER COMPOUNDING

Chemistry and Applications

Edited by
BRENDAN RODGERS

CRC Press
Taylor & Francis Group
Boca Raton London New York

CRC Press is an imprint of the
Taylor & Francis Group, an **informa** business

CRC Press
Taylor & Francis Group
6000 Broken Sound Parkway NW, Suite 300
Boca Raton, FL 33487-2742

First issued in paperback 2020

© 2016 by Taylor & Francis Group, LLC
CRC Press is an imprint of Taylor & Francis Group, an Informa business

No claim to original U.S. Government works

ISBN-13: 978-1-4822-3549-4 (hbk)
ISBN-13: 978-0-367-78340-2 (pbk)

Library of Congress Cataloging-in-Publication Data

Rubber compounding : chemistry and applications / editor: Brendan Rodgers. -- Second edition.
    pages cm
Includes bibliographical references and index.
ISBN 978-1-4822-3549-4 (hardcover : alk. paper) 1. Rubber--Mixing. 2. Elastomers. 3. Polymers. 4. Rubber chemistry. I. Rodgers, Brendan, editor.

TS1890.R77 2015
547'.8426--dc23   2015026058

Visit the Taylor & Francis Web site at
http://www.taylorandfrancis.com

and the CRC Press Web site at
http://www.crcpress.com

Contents

Preface...vii
Editor ix
Contributors xi

Chapter 1 Natural Rubber and Other Naturally Occurring
Compounding Materials..1

Brendan Rodgers

Chapter 2 General-Purpose Elastomers...33

Howard Colvin

Chapter 3 Special-Purpose Elastomers...83

Sudhin Datta and Syamal S. Tallury

Chapter 4 Butyl Rubbers..109

Walter H. Waddell and Andy H. Tsou

Chapter 5 Thermoplastic Elastomers: Fundamentals 139

Tonson Abraham

Chapter 6 Carbon Black..209

Wesley A. Wampler, Leszek Nikiel, and Erika N. Evans

Chapter 7 Silica and Silanes 251

*Anke Blume, Louis Gatti, Hans-Detlef Luginsland,
Dominik Maschke, Ralph Moser, J.C. Nian, Caren Röben,
and André Wehmeier*

Chapter 8 General Compounding 333

Harry G. Moneypenny, Karl-Hans Menting, and F. Michael Gragg

Chapter 9 Resins 379

James E. Duddey

Chapter 10 Antioxidants and Other Protectant Systems 419

Sung W. Hong

Chapter 11 Vulcanization 461

Frederick Ignatz-Hoover and Byron H. To

Chapter 12 Recycling of Rubber 523

Brendan Rodgers and Bernard D'Cruz

Chapter 13 Industrial Rubber Products 537

George Burrowes

Chapter 14 Tire Technology 579

Brendan Rodgers and Bernard D'Cruz

Index 599

Preface

Rubber compounding describes the science of elastomer chemistry and the modification of elastomers and elastomer blends by addition of other materials to meet a set of required mechanical properties. It is therefore among the most complex disciplines in that the materials scientist requires a thorough understanding of materials physics, organic and polymer chemistry, inorganic chemistry, thermodynamics, and reaction kinetics.

The rubber industry has changed since the publication of the book's last edition. Asia and particularly China and India have emerged as major tire and industrial rubber products manufacturing centers. Other developing countries are actively pursuing development of rubber industries because of the technology and skills the industry brings. Every investor in the industry can acquire the same equipment, design the same types of factories, and build similar tire constructions and designs. However, materials science and technology cannot be replicated and, along with uniformity, has become the most important factor in differentiating the quality of tires and industrial rubber products from different manufacturers. Tire uniformity, tire and industrial rubber products durability, and performance parameters such as tire rolling resistance or conveyor belt energy efficiency are all dependent on materials quality and compound technologies. The new global regulatory environment has set new standards in tire performance, which, again, depend to a very large extent on materials technology.

The tire industry has evolved from production of bias to tubeless radial constructions and now ultralow-profile designs. The service lives of tires and industrial products such as automobile engine hoses have improved dramatically. None of these improvements would have been possible without an emphasis on the understanding of the chemistry of raw materials and compounds. As was mentioned in the first edition of this book, the advances in materials technologies over the last number of years have included

1. Commercialization of functionalized and coupled, solution-polymerized polymers, leading to improvements in tire wear resistance and traction performance
2. Development of the silica tread compound for high-performance tires, leading to a significant improvement or reduction in tire rolling resistance and in turn vehicle fuel economy
3. Thermoplastic elastomers
4. Hybrid filler systems and nanocomposite technologies, allowing improvement in tire air pressure retention
5. Reversion-resistant vulcanization systems, leading to improved tire durability

6. Halobutyl polymers which were the foundation for the development of the tubeless radial tire

7. A new emphasis on recycling and renewable sources for raw materials

As was the case in the first edition, the philosophy behind this work continues to emphasize more the chemistry of the materials used in building a compound formulation for a tire or engineered product and also now elaborates on product technologies. The depth to which subjects are presented is not at an introductory level nor is it an advanced treatise. Rather, it is intended as a tool for the industrial compounder, teacher, or other academic scientist to provide basic information on materials used in the rubber industry. In addition, it continues to address a gap in the body of literature available to the chemists in industry and academia. The text has been redesigned to add sections on recycling, expanded discussion on tire technology and industrial rubber products adding more information on hydraulic hose and conveyor belting.

Fred Barlow mentioned in the first edition that no comprehensive review of a subject such as this could be written by one individual. The compilation of this work thus depended on many contributors, and I express my thanks to the authors who participated in the project. All are recognized authorities in their field, and this is reflected in the quality of their contribution. I also express my thanks to the ExxonMobil Chemical Company for permission to pursue this project; to Dr. James P Stokes, Polymers Technology Manager at ExxonMobil Chemical Company, for his support; and most importantly to my wife, Elizabeth, for her encouragement.

Brendan Rodgers
ExxonMobil Chemical Company
ExxonMobil
Baytown, Texas

Editor

Brendan Rodgers was senior engineer at ExxonMobil's Baytown Technology and Engineering Center in Baytown, Texas, and is now technology advisor at the ExxonMobil Chemical Company, Shanghai Technical Center, China. Previously, he was as an engineer with the Goodyear Tire and Rubber Company with assignments in Europe and the United States. He earned his BS from the University of Ulster, Northern Ireland; his MS in polymer technology; and PhD in chemical engineering from Queen's University, Belfast, Northern Ireland.

Contributors

Tonson Abraham
ExxonMobil Chemical Company
Akron, Ohio

Anke Blume
Evonik Industries AG
Wesseling, Germany

George Burrowes
Veyance Technologies, Inc.
Fairlawn, Ohio

Howard Colvin
Cooper Tire & Rubber Company
Findlay, Ohio

Sudhin Datta
ExxonMobil Chemical Company
Baytown, Texas

Bernard D'Cruz
ExxonMobil Chemical Company
Baytown, Texas

James E. Duddey (Retired)
Akron, Ohio

Erika N. Evans
Sid Richardson Carbon & Energy
  Company
Fort Worth, Texas

Louis Gatti
Evonik Industries AG
Wesseling, Germany

F. Michael Gragg
ExxonMobil Lubricants & Petroleum
  Specialties Company
Fairfax, Virginia

Sung W. Hong
Crompton Corporation
Uniroyal Chemical
Naugatuck, Connecticut

Frederick Ignatz-Hoover
Eastman Chemical Company
Akron, Ohio

Hans-Detlef Luginsland
Evonik Industries AG
Wesseling, Germany

Dominik Maschke
Evonik Industries AG
Wesseling, Germany

Karl-Hans Menting
Schill + Seilacher "Struktol" GmbH
Hamburg, Germany

Harry G. Moneypenny
Moneypenny Tire & Rubber
  Consultants
Den Haag, the Netherlands

Ralph Moser
Evonik Industries AG
Wesseling, Germany

J.C. Nian
Evonik Industries AG
Wesseling, Germany

Leszek Nikiel
Sid Richardson Carbon & Energy
   Company
Fort Worth, Texas

Wolfgang Pille-Wolf
Arizona Chemical B.V.
Almere, the Netherlands

Caren Röben
Evonik Industries AG
Wesseling, Germany

Brendan Rodgers
ExxonMobil Chemical Company
Baytown, Texas

Syamal S. Tallury
ExxonMobil Chemical Company
Baytown, Texas

Byron H. To (Retired)
Flexsys America L.P.
Baltimore, Maryland

Andy H. Tsou
ExxonMobil Chemical Company
Baytown, Texas

Walter H. Waddell
ExxonMobil Chemical Company
Baytown, Texas

Wesley A. Wampler
Sid Richardson Carbon & Energy
   Company
Fort Worth, Texas

André Wehmeier
Evonik Industries AG
Wesseling, Germany

1 Natural Rubber and Other Naturally Occurring Compounding Materials

Brendan Rodgers

CONTENTS

I.   Introduction 1
II. Natural Rubber 2
    A. Chemistry of Natural Rubber...2
    B. Production of Natural Rubber...6
    C. Natural Rubber Products and Grades 7
        1. Sheet Rubber 8
        2. Crepe Rubber...9
        3. Technical Classification of Visually Inspected Rubbers.................12
        4. TSR...12
    D. Viscosity and Viscosity Stabilization of Natural Rubber 15
    E. Special-Purpose Natural Rubbers..20
    F. Quality...23
        1. Consistency and Uniformity...23
        2. Packaging 23
        3. Contamination 23
        4. Fatty Acids 24
    G. Use of Natural Rubber in Tires...25
III. Other Naturally Occurring Materials..27
IV. Summary 29
References...29

I. INTRODUCTION

The nature of the tire and rubber industry has changed over the last 20–30 years in that, like all other industries, it has come to recognize the value of using renewable sources of raw materials, recycling materials whenever possible, and examining the potential of reclaiming used materials for fresh applications. Renewable raw materials range from natural rubber (NR), more of which is used than any other elastomer, naturally occurring process aids such as pine tars and resins, and novel biological materials such as silica derived from the ash of burned rice husks. Naturally occurring materials include inorganic fillers such as calcium carbonate, which is distinct from naturally occurring organic material, whose total supply may be restricted. However, NR is by far the most important of these materials both in terms of the quantity used and also in meeting the performance target of tires and other industrial rubber products.

II. NATURAL RUBBER

Of the range of elastomers available to technologists, NR is among the most important, because it is the building block of most rubber compounds used in products today. In the previous edition of this text, Barlow [1] presented a good introductory discussion of this strategic raw material. Roberts [2] edited a very thorough review of NR covering topics ranging from basic chemistry and physics to production and applications. NR, which is a truly renewable resource, comes primarily from Indonesia, Malaysia, India, and the Philippines, though many more additional sources of good quality rubber are becoming available. It is a material that is capable of rapid deformation and recovery, and it is insoluble in a range of solvents, though it will swell when immersed in organic solvents at elevated temperatures. Its many attributes include abrasion resistance, good hysteretic properties, high tear strength, high tensile strength, and high green strength. However, it may also display poor fatigue resistance. It is difficult to process in factories, and it can show poor tire performance in areas such as traction or wet skid compared to selected synthetic elastomers. Given the importance of this material, this section discusses

1. The biosynthesis and chemical composition of NR
2. Industry classification, descriptions, and specifications
3. Typical applications of NR

A. CHEMISTRY OF NATURAL RUBBER

NR is a polymer of isoprene (methylbuta-1,3-diene). It is a polyterpene synthesized in vivo via the enzymatic polymerization of isopentenyl pyrophosphate. Isopentenyl pyrophosphate undergoes repeated condensation to yield *cis*-polyisoprene via the enzyme rubber transferase. Though bound to the rubber particle, this enzyme is also found in the latex serum. Structurally, *cis*-polyisoprene is a highly stereoregular polymer with an –OH group at the alpha terminal and three to four *trans* units at the omega end of the molecule (Figure 1.1). The molecular-weight distribution of *Hevea brasiliensis* rubber shows considerable variation from clone to clone, ranging from 100,000 to over 1,000,000. NR has a broad bimodal molecular-weight distribution. The polydispersity or ratio of weight-average molecular weight to number-average molecular weight, M_w/M_n, can be as high as 9.0 for some varieties of NR [3,4]. This tends to be of considerable significance in that the lower-molecular-weight fraction will facilitate the ease of processing in end product manufacturing, while the higher-molecular-weight fraction contributes to high tensile strength, tear strength, trans-Polyisoprene (repeat units, $n = 100$; $M_w = 7000$)

cis-Polyisoprene (repeat units, $n = 1{,}500-15{,}000$; $M_w = 100{,}000-1{,}000{,}000$)

FIGURE 1.1   *cis-* and *trans-*Isomers of natural rubber.

and abrasion resistance. The biosynthesis or polymerization to yield polyisoprene, illustrated in Figure 1.2, occurs on the surface of the rubber particle(s) [5].

The isopentenyl pyrophosphate starting material is also used in the formation of farnesyl pyrophosphate. Subsequent condensation of *trans-*farnesyl pyrophosphate yields *trans-*polyisoprene or gutta-percha. Gutta-percha is an isomeric polymer in which the double bonds have a *trans* configuration. It is obtained from trees of the genus *Dichopsis* found in Southeast Asia. This polymer is synthesized from isopentenyl pyrophosphate via a pathway similar to that for the biosynthesis of terpenes such as geraniol and farnesol. Gutta-percha is more crystalline in its relaxed state, much harder, and less elastic.

NR is obtained by "tapping" the tree *H. brasiliensis.* Tapping starts when the tree is 5–7 years old and continues until it reaches around 20–25 years of age. A knife is used to make a downward cut from left to right and at about a 20°–30° angle to the horizontal plane, to a depth approximately 1.0 mm from the cambium. Latex then exudes from the cut and can flow from the incision into a collecting cup. Rubber occurs in the trees in the form of particles suspended in a protein-containing serum, the whole constituting latex, which in turn is contained in specific latex vessels in the tree or other plants. Latex constitutes the protoplasm of the latex vessel. Tapping or cutting of the latex vessel creates a hydrostatic pressure gradient along the vessel, with consequent flow of latex through the cut. In this way, a portion of the contents of the interconnected latex vessel system can be drained from the tree. Eventually, the flow ceases, turgor is reestablished in the vessel, and the rubber content of the latex is restored to its initial level in about 48 h.

The tapped latex consists of 30%–35% rubber, 60% aqueous serum, and 5%–10% other constituents such as fatty acids, amino acids and proteins, starches, sterols, esters, and salts. Some of the nonrubber substances such as lipids, carotenoid pigments, sterols, triglycerides, glycolipids, and phospholipids can influence the final properties of rubber such as its compounded vulcanization characteristics and classical mechanical properties. Hasma and Subramanian [6] conducted a comprehensive

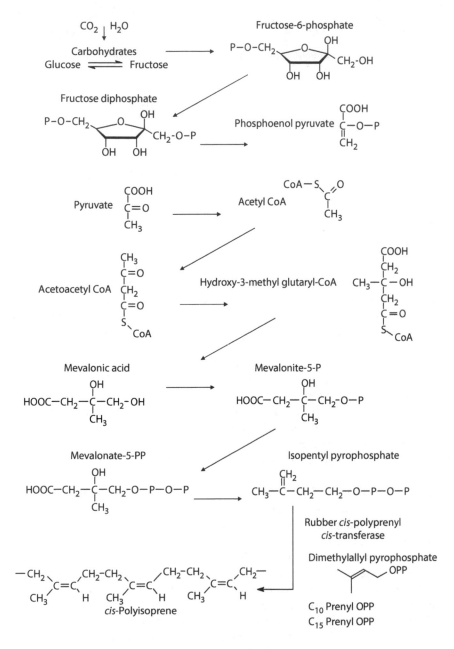

FIGURE 1.2 Simplified schematic of the biosynthesis of natural rubber.

TABLE 1.1
Definition of Natural Rubber Terms

Latex	Fluid in the tree obtained by tapping or cutting the tree at a 20°–30° angle to allow the latex to flow into a collecting cup
Serum	Aqueous component of latex that consists of lower-molecular-weight materials such as terpenes, fatty acids, proteins, and sterols
Whole field latex	Fresh latex collected from trees
Cup-lump	Bacterially coagulated polymer in the collection cup
Lace	Trim from the edge of collecting vessels and cut on tree
Earth scrap	Collecting vessel overflow material collected from the tree base
RSS	Sheets produced from whole field latex
LRP	Large rubber particles
NSR	Nigerian Standard Rubber
SIR	Standard Indonesian Rubber
SLR	Standard Lanka Rubber
SMR	Standard Malaysian Rubber
SRP	Serum rubber particles
SSR	Standard Singapore Rubber
TSR	Technically specified Rubber
TTR	Thai Tested Rubber

study characterizing these materials to which further reference should be made. Lipids can also affect the mechanical stability of the latex while it is in storage, because lipids are a major component of the membrane formed around the rubber particle [7]. NR latex is typically coagulated, washed, and then dried in either the open air or a "smokehouse." The processed material consists of 93% rubber hydrocarbon; 0.5% moisture; 3% acetone-extractable materials such as sterols, esters, and fatty acids; 3% proteins; and 0.5% ash. Raw NR gel can range from 5% to as high as 30%, which in turn can create processing problems in tire or industrial products factories. Nitrogen content is typically in the range of 0.3%–0.6%. For clarity, a number of definitions are given in Table 1.1.

The rubber from a tapped tree is collected in three forms: latex, cup-lump, and lace. It is collected as follows:

1. Latex collected in cups is coagulated with formic acid, crumbed, or sheeted. The sheeted coagulum can be immediately crumbed, aged and then crumbed, or smoke-dried at around 60°C to produce typically ribbed smoked sheet (RSS) rubber.
2. Cup-lump is produced when the latex is left uncollected and allowed to coagulate, due to bacterial action, on the side of the collecting cup. Field coagulum or cup-lump is eventually collected, cut, cleaned, creped, and crumbed. Crumb rubber can be dried at temperatures up to 100°C.
3. Lace is the coagulated residue left around the bark of the tree where the cut has been made for tapping. The formation of lace reseals the latex vessels and stops the flow of rubber latex. It is normally processed with cup-lump.

The processing factories receive NR in one of two forms: field coagula or field latex. Field coagula consist of cup-lump and tree lace (Table 1.1). The lower grades of material are prepared from cup-lump, partially dried small holders of rubber, rubber tree lace, and earth scrap after cleaning. Iron-free water is necessary to minimize rubber oxidation. Field coagula and latex are the base raw materials for the broad range of natural grades described in this review. Fresh *Hevea* latex has a pH of 6.5–7.0 and a density of 0.98 [3,4]. The traditional preservative is ammonia, which in concentrated solution is added in small quantities to the latex collected from the cup. Tetramethylthiuram disulfide and zinc oxide are also used as preservatives because of their greater effectiveness as bactericides. Most latex concentrates are produced to meet the International Organization for Standardization (ISO) 2004 [8]. This standard defines the minimum content for total solids, dry rubber content, nonrubber solids, and alkalinity (as ammonia, NH_3).

B. Production of Natural Rubber

Total global rubber consumption in 2014 was approximately 27.3 million metric tons of which 11.8 million tons (43%) was NR and the remaining was synthetic rubber [9]. The world production of NR was up by nearly 14% from the same period in 2010 as economies improve their performance from the low points in 2008 and 2009. The major regional consumers of NR have undergone a major shift from the period 2000 to 2004 with China being the largest followed by the European Union, North America, India, and Japan. It is also anticipated that western European and Japanese consumption will increase due to economic growth in both areas, with sustained economic expansion in the United States; the net impact will be further growth in consumption toward 14.0 million tons/year. Further, NR consumption will then increase slowly, this being dependent on global economic conditions (Figure 1.3). Globally, NR consumption is split—tires consuming around 75%, automotive mechanical goods at 5%, nonautomotive mechanical goods at 5%–10%, and miscellaneous applications such as medical and health-related products consuming the remaining 5%–10% [10].

There are around 25 million acres planted with rubber trees, and production employs nearly 3 million workers, with the majority coming from smallholdings in order: Thailand, Indonesia, Malaysia, India, Vietnam, China, and West Africa. Many times, the dominance of smallholdings has raised issues regarding quality and consistency, which will be discussed later. Smallholdings produce mainly cup-lump, which is used in block rubber. Sheet rubber is generally regarded to be of higher quality, typically displaying higher tensile and tear strength.

In 1964, the ISO published a set of draft technical specifications that defined contamination, wrapping, and bale weights and dimensions, with the objectives of improving rubber quality, uniformity, and consistency and developing additional uses for contaminated material [11,12].

The three sources leading to crumb rubber (i.e., unsmoked sheet rubber, aged sheet rubber, and field cup-lump) typically provide different grades of TSRs. For example, one grade of TSR (L) is produced from coagulated field latex, TSR 5

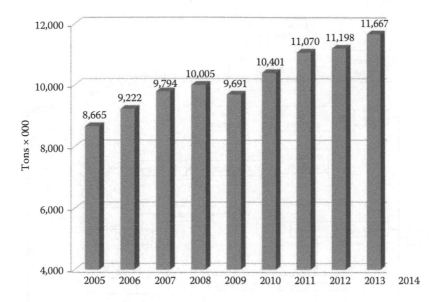

FIGURE 1.3 Global natural rubber productions (millions of tons).

is produced from unsmoked sheets, and lower grades such as TSR 10 and 20 are produced from field coagulum. A simplified schematic of the production process is presented in Figure 1.4.

C. NATURAL RUBBER PRODUCTS AND GRADES

NR is available in the following six basic forms:

1. Sheets
2. Crepes
3. Sheet rubber, technically specified
4. Block rubber, technically specified
5. Preserved latex concentrates
6. Specialty rubbers that have been mechanically or chemically modified

Among these six types, the first four are in a dry form and represent over 90% of the total NR produced in the world. In the commercial market, these four types of dry NR are available in over 40 grades, consisting of RSS; air-dried sheets (ADS); crepes, which include latex-based and field coagulum-derived estate brown crepes and remilled crepes; and TSR in block form. Among the four major types, crepes are now of minor significance in the world market, accounting for less than 75,000 tons/year. Field coagulum grade block rubbers have essentially replaced brown crepes except in India. Only Sri Lanka and India continue to produce latex crepes. Figure 1.4 presents a simplified schematic of the process followed in the production of NR.

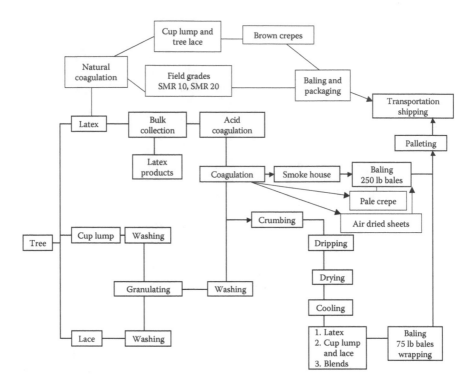

FIGURE 1.4   Simplified schematic of the natural rubber production process.

1.   Sheet Rubber

NR in sheet form is the oldest and most popular type. Being the simplest and easiest to produce on a small scale, smallholders' rubber in most countries is processed and marketed as sheet rubber. From the end user's perspective, two types of sheet rubbers are produced for the commercial market: RSS and ADS. Of the two, RSS is the most popular due to its mechanical properties and high tensile strength.

RSS rubbers are made from intentionally coagulated whole field latex. They are classified by visual evaluation. To establish acceptable grades for commercial purposes, the *International Rubber Quality and Packing Conference* prepared a description for grading, and the details are given in *The Green Book* [13]. Whole field latex used to produce RSS is first diluted to 15% solids and then coagulated for around 16 h with dilute formic acid. The coagulated material is then milled, the water is removed, and the material is sheeted with a rough surface to facilitate drying. Sheets are then suspended on poles for drying in a smokehouse for 2–4 days. Only deliberately coagulated rubber latex processed into rubber sheets, properly dried, and smoked can be used in making RSS. A number of prohibitions are also applicable to the RSS grades. Wet, bleached, undercured, and original rubber and rubber that is not completely visually dry at the time of the buyer's inspection are not acceptable (except slightly undercured rubber as specified for RSS 5). Skim rubber made

TABLE 1.2

Grade Classification of Ribbed Smoked Sheet Rubber

RSS	Rubber Mold	Wrapping Mold	Opaque Spots	Oversmoked Spots	Oxidized Spots	Burned Sheets	Comments
1X	No	No	No	No	No	No	Dry, clean, no blemishes
1	Very slight	Very slight	No	No	No	No	Dry, clean, no blemishes
2	Slight	Slight	No	No	No	No	No sand or foreign matter
3	Slight	Slight	Slight	No	No	No	No sand or foreign matter
4	Slight	Slight	Slight	Slight	No	No	No sand or foreign matter
5	Slight	Slight	Slight	Slight	N/A	No	N/A

of skim latex cannot be used in whole or in part in patches as required under packing specifications defined in *The Green Book*. Prior to grading RSS, the sheets are separated and inspected, and any blemishes are removed by manually cutting and removing defective material. Table 1.2 provides a summary of the criteria followed by inspectors in grading RSS. The darker the rubber, the lower the grade. The premium grade is RSS 1, and the lower quality grade is typically RSS 4. ADS are prepared under conditions very similar to those for smoked sheets but are dried in a shed without smoke or additives, with the exception of sodium bisulfate. Such rubber therefore lacks the antioxidation protection afforded by drying the rubber in a smokehouse. This material can be substituted for RSS 1 or RSS 2 grades in various applications.

2.　Crepe Rubber

Crepe is a crinkled lace rubber obtained when coagulated latex is selected from clones that have a low carotene content. Sodium bisulfite is also added to maintain color and prevent darkening. After straining, the latex is passed several times through heavy rolls called creepers, and the resultant material is air-dried at ambient temperature. There are different types of crepe rubber depending upon the type of starting materials from which they are produced. Sri Lanka is the largest producer of pale crepes and the sole producer of thick pale crepe.

The specifications for the different types of crepe rubbers for which grade descriptions are given in *The Green Book* are as follows:

1. *Pale latex crepes*: Pale crepe is used for light-colored products and therefore commands a premium price. Trees or clones from which the grade is obtained typically have low yellow pigment levels (carotenes) and greater resistance to oxidation and discoloration. There are eight grades

in this category. All these grades must be produced from the fresh coagula of natural liquid latex under conditions where all processes are quality controlled. The rubber is milled to produce both thin and thick crepes. Pale crepes are used in pharmaceutical appliances such as stoppers and adhesives (Table 1.3).

2. *Estate brown crepes*: There are six grades in this category. All six grades are made from cup-lump and other higher grade rubber scrap (field coagulum) generated on the rubber estates. Tree bark scrap, if used, must be precleaned to separate the rubber from the bark. Powerwash mills are to be used in milling these grades into both thick and thin brown crepes (Table 1.4).
3. *Thin brown crepes (remills)*: There are four grades in this class or category. These grades are manufactured on powerwash mills from wet slab unsmoked sheets at the estates or smallholdings. Tree bark scrap, if used, must be precleaned to separate the rubber from the bark. Inclusion of earth scrap and smoked scrap is not permissible in these grades (Table 1.5).
4. *Thick blanket crepes (ambers)*: The three grades in this category are also produced on powerwash mills from wet slab unsmoked sheets, lump, and other high-grade scrap (Table 1.5).

TABLE 1.3
White and Pale Crepes

				Discoloration				
Class	Grade	Color	Uniformity	Spots, Streaks, Bark	Odor	Dust, Sand	Oil Stains	Oxidation
1X	Thin white crepe	White	Uniform	No	No	No	No	No
1X	Thick pale crepe	Light	Uniform	No	No	No	No	No
1X	Thin pale crepe	Light	Uniform	No	No	No	No	No
1	Thin white crepe	White	Slight shade	No	No	No	No	No
1	Thick pale crepe	Light	Slight shade	No	No	No	No	No
1	Thin pale crepe	Light	Slight shade	No	No	No	No	No
2	Thick pale crepe	Slightly darker	Slight shade	Slight, <10% of bales	No	No	No	No
2	Thin pale crepe	Slightly darker	Slight shade	Slight, <10% of bales	No	No	No	No
3	Thick pale crepe	Yellowish	Variation	OK if <20% of bales	No	No	No	No
3	Thin pale crepe	Yellowish	Variation	OK if <20% of bales	No	No	No	No

TABLE 1.4
Estate Brown Crepes

Class	Grade	Color	Uniformity	Discoloration				
				Spots, Streaks	Odor	Dust Bark	Oil Stains	Oxidation
1X	Thick brown crepe	Light brown	Uniform	No	No	No	No	No
1X	Thin brown crepe	Light brown	Uniform	No	No	No	No	No
2X	Thick brown crepe	Medium brown	Uniform	No	No	No	No	No
2X	Thin brown crepe	Medium brown	Uniform	No	No	No	No	No
3X	Thick brown crepe	Dark brown	Variation	No	No	Bark	No	No
3X	Thin brown crepe	Dark brown	Variation	No	No	Bark	No	No

TABLE 1.5
Compo, Thin Brown, Thick Blanket, Flat Bark, Pure Smoked Blanket Crepe

Type	Grade	Color	Discoloration				
			Spots, Streaks	Odor	Dust, Sand, Bark	Oil Stains	Oxidation
Compo crepes	1	Light brown	Yes	No	No	No	No
	2	Brown	Yes	No	No	No	No
	3	Dark brown	Yes	No	No	No	No
Thin, brown crepes	1	Light brown	Slight	No	No	No	No
	2	Medium brown	Yes	No	No	No	No
	3	Medium brown	Yes	No	No	No	No
	4	Dark brown	Yes	No	Bark	No	No
Thick blanket crepes (ambers)	2	Light brown	Slight	No	No	No	No
	3	Medium brown	Slight	No	No	No	No
	4	Dark brown	Slight	No	No	No	No
Flat bark crepes	Standard	Very-dark brown	No	No	Fine bark	No	No
	Hard	Black	No	No	Fine bark	No	No
Pure smoked blanket crepe	Pure smoked	Not specified	No	Smoked odor	No	No	No

5. *Flat bark crepes*: The two grades of rubber in this category are produced on powerwash mills out of all types of scrap NR in uncompounded form, including earth scrap (Table 1.5).
6. *Pure smoked blanket crepe*: This grade is made by milling on powerwash mills smoked rubber derived from RSS (including block sheets) or RSS cuttings. No other type of rubber can be used. Rubber of this type must be dry, clean, firm, and tough and must also retain an easily detectable smoked sheet odor. Sludge, oil spots, heat spots, sand, dirty packing, and foreign matter are not permissible. Color variation from brown to very-dark brown is permissible (Table 1.5).

3. Technical Classification of Visually Inspected Rubbers

The Malaysian Rubber Producers' Research Association has published a technical information sheet describing sheet rubbers that have been further tested and classified with respect to cure characteristics [14]. The cure or vulcanization classes are distinguished by a color coding (i.e., blue for fast cure, yellow for medium cure, and red for slow cure) (Table 1.6) when the rubber is compounded using the American Society for Testing and Materials (ASTM) No. 1A formulation [15]. This color coding is limited to RSS 1 and ADS. Upon cure classification, the rubbers are further tested, and at 0.49 MPa, the strain on the sample is measured after 1 min. This classification scheme has not received wide acceptance, which is clearly unfortunate, for a more quantitative classification scheme is required for visually inspected grades of NR. For example, rubber meeting a specific visually determined grade or classification might display poor mechanical properties when compounded with carbon black and vulcanizing agents owing to a broad or lower-molecular-weight distribution. This may in turn create factory processing difficulties and product performance deficiencies.

4. TSR

The ISO first published a technical specification (ISO 2000) for NR in 1964 [11]. Based on these specifications, Malaysia introduced a National Standard Malaysian Rubber (SMR) scheme in 1965, and since then, all the NR-producing countries have started production and marketing of TSRs based on the ISO 2000 scheme. TSRs are shipped in "blocks," which are generally 33.3 kg bales in the international market and 25.0 kg in India. All the block rubbers are also guaranteed to conform to certain technical specifications, as defined by the national schemes or by ISO 2000 (Table 1.7).

TABLE 1.6
Technical Certification of Sheet Rubber

	Class Limits, % Strain		
	Blue	Yellow	Red
Production classification	55–73	73–85	85–103
Consumer acceptance	55–79	61–91	79–103

TABLE 1.7
TSRs Defined in ISO 2000

Property	Grade					
	TSR CV	TSR L	TSR S	TSR 10	TSR 20	TSR 50
Dirt content, max, wt%	0.05	0.05	0.05	0.1	0.2	0.5
Ash content, max, wt%	0.6	0.6	0.5	0.75	1	1.5
Nitrogen content, max, wt%	0.6	0.6	0.5	0.6	0.6	0.6
Volatile matter, max, wt%	0.8	0.8	0.8	0.8	0.8	0.8
Initial Wallace plasticity P_0	30	30	30	30	30	
Plasticity retention index (min)	60	60	60	50	40	30
Color, max, Lovibond units		6				
Mooney viscosity	60+5					

The nomenclature describing technically specified rubbers consists of a three- or four-letter country code followed by a numeral indicating the maximum permissible dirt content for that grade expressed as hundredths of 1%. In Malaysia, the TSR is designated as SMR. In Indonesia, the designation given is Standard Indonesian Rubber (SIR). In Thailand, the TSRs are called Standard Thai Rubber (STR, sometimes denoted as TTR). In India, the TSRs are designated as Indian standard NR. Grading is based on the dirt content measured as a weight percent. Dirt is considered to be the residue remaining when the rubber is dissolved in a solvent, washed through a 45 μm sieve, and dried.

TSR accounts for approximately 60% of the NR produced worldwide. The advantages claimed for the TSRs over the conventional sheet and crepe grades of rubbers are the following:

1. They are available in a limited number of well-defined grades, intended to ensure a uniform, defined quality.
2. Data on the content of foreign and volatile matter can be provided, again to ensure better uniformity.
3. They are shipped as compact, polyethylene-wrapped bales of standard weight.
4. They can be prepared to prevent degradation of the rubber during storage, handling, and transportation.
5. They have a standard bale size to enable ease of transport through mechanized handling and containerization.

ISO has specified six grades of TSR. The detailed characteristics of the different grades of TSR are discussed next.

TSR CV: TSR CV, the CV designating "constant viscosity," is produced from field latex and is stabilized to a specified Mooney viscosity. The storage hardening of this grade of rubber must be within 8 hardness units. It is shipped in a 1.2 ton pallet, which facilitates handling, transportation, and storage space utilization. Each pallet consists of 36 bales of 33.3 kg net weight, and each bale is wrapped in a polyethylene bag that is dispersible and compatible with rubber when mixed in an internal mixer at temperatures exceeding 110°C, which are, of course, typical in any rubber-mixing facility. TSR CV rubber is generally softer than conventional technically specified grades. Coupled with its CV feature, it can provide a cost advantage by eliminating premastication. When used in open mills, the rubber forms a coherent band almost instantaneously, thus potentially improving milling throughput. Additional claimed benefits of TSR CV include

1. Reduction of mixing times, giving higher throughput
2. Reduction of scraps and rejects due to better material uniformity
3. Better resistance to chipping and chunking for off-the-road (OTR) tires
4. Better green strength

TSR CV rubber is available in different viscosities, with 50 and 60 being the more common. This material can be used for high-quality products such as mechanical mountings for engines and machinery, railway buffers, bridge bearings, vehicle suspension systems and general automotive components, large-truck tire treads, conveyor belt covers, cushion gum for retreading, masking tapes, injection-molded products including rubber–metal bonded components, industrial rolls, inner tubes, and cement.

TSR L: TSR L is a light-colored rubber produced from high-quality latex; it has low ash and dirt content and is packed and presented in the same way as TSR CV. The advantage of TSR L is its light color together with its cleanliness and better heat-aging resistance. Technologically, TSR L shows high tensile strength, modulus, and ultimate elongation at break for both black and nonblack mix.

This material can be used for light-colored and transparent products such as surgical or pressure-sensitive tape, textiles, rubber bands, hot-water bottles, surgical and pharmaceutical products, large industrial rollers for the paper printing industry, sportswear, bicycle tubes, chewing gum, cable sheaths, gaskets, and adhesive solutions and tapes.

TSR 5: TSR 5 is produced from fresh coagulum, RSS, or ADS. It is packed and shipped to the same specifications as TSR CV and TSR L. TSR 5 is typically used for general-purpose (GP) friction and extruded products; small components in passenger vehicles such as mountings, sealing rings, cushion gum, and brake seals; bridge bearings; ebonite battery plates; separators; adhesives; and certain components in tires.

TSR 10: TSR 10 is produced from clean and fresh field coagulum or from unsmoked sheets. It is packed and shipped in the same way as TSR CV, TSR L, and TSR 5. TSR 10 has good technological properties similar to those of RSS 2 and RSS 3, but has an advantage over RSS because of its

1. Lower viscosity
2. Easier mixing characteristics (more rapid breakdown)
3. Technical specifications and packaging in 33.3 kg bales

It can be used for tires, inner tubes, cushion gum stocks, joint rings by injection molding, raincoats, microcellular sheets, upholstery and packing, conveyor belts, and footwear.

TSR 20: This is a large-volume grade of technically specified NR. It is produced mostly from field coagulum, lower grades of RSS, and unsmoked sheets. It is packed and shipped to the same specifications as TSR CV, TSR L, TSR 5, and TSR 10. TSR 20 has good processing characteristics and physical properties. Its low viscosity and easier mixing characteristics (compared with the RSS grades) can reduce the mastication and mixing period considerably. It is used mostly for tires, cushion gum stock, bicycle tires, raincoats, microcellular sheet for upholstery and packing, conveyor belts, footwear, and other general products.

TSR 50: This is the lowest grade of TSR and is produced from old, dry field coagulum or partly degraded rubber. It is packed and shipped in the same way as other grades of TSR. It should be noted that these specifications will continue to be improved as production methods improve. For example, in 1991, the Rubber Research Institute of Malaysia revised the dirt levels of SMR CV60, CV50, and L from 0.05 to 0.025, that of SMR 10 from 0.10 to 0.08, and that of SMR 20 to 0.016.

In addition, Malaysia has produced grades of rubber outside the specific scope of ISO 2000. SMR GP is a standard GP rubber made from a 60:40 mixture of latex-grade sheet rubber and field coagulum. It is viscosity stabilized at 65 Mooney units using hydroxylamine neutral sulfate (HNS). It is similar to SMR 10 in specification.

To illustrate the distribution and consumption of these various grades, shipments of SMR from Malaysia are typically SMR 20, 60%; SMR 10, 27%; SMR CV and SMR L, 5%; SMR GP, 7%; and SMR 5, 1.0%.

D.   Viscosity and Viscosity Stabilization of Natural Rubber

The properties of NR that are most important regarding its use in the manufacture of tires or other products include viscosity, fatty acid bloom, and compliance with the technical specifications. Of these three parameters, viscosity is probably the most important. This property relates to the molecular weight, molecular-weight distribution, and amounts of other materials present in the polymer such as low-molecular-weight resins, fatty acids, and other natural products. It affects the initial mixing of the rubber with other compounding ingredients and subsequent processing of the compounded materials to form the final manufactured product.

NR viscosity is a function of two major factors: viscosity of the rubber produced by the specific clone and the viscosity stabilization method. A range of methods are available to characterize the viscosity of NR. The most popular is Mooney viscosity (V_r), which is obtained by measuring the torque that is required to rotate a disk embedded in rubber or a compounded sample. This procedure is defined in ASTM D 1646, "Standard Test Methods for Rubber—Viscosity, Stress Relaxation, and Prevulcanization Characteristics (Mooney Viscometer)" [16]. The viscosity will typically range from 45 to over 100. The information obtained from a Mooney viscometer can include the following:

1. Prevulcanization properties or scorch resistance for the compounded polymer, a test that is conducted at temperatures ranging from 120°C to 135°C (Figure 1.5).

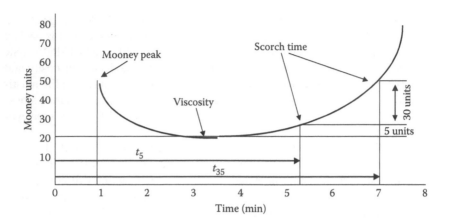

FIGURE 1.5 Mooney scorch typically conducted at 121°C and 135°C.

2. Mooney peak, which is the initial peak viscosity at the start of the test and a function of the green strength and can be a measure of compound factory shelf life.

3. Viscosity (V_r), typically measured at 100°C, provides a measure of the ease with which the material can be processed (Figure 1.6). It depends on molecular weight and molecular-weight distribution, molecular structure such as stereochemistry and polymer chain branching, and nonrubber constituents. Caution is always required when attempting to establish relationships between Mooney viscosity and molecular weight. Mooney viscosity can be expressed as ML 1 + 4 (i.e., Mooney large rotor, with 1 min pause and 4 min test duration).

4. Stress relaxation, which can provide information on gel (T-95), is defined as the response to a cessation of sudden deformation when the rotor of the Mooney viscometer stops. The stress relaxation of rubber is a combination

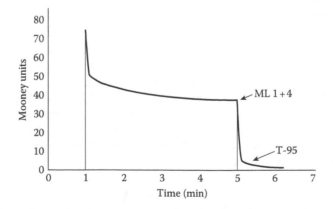

FIGURE 1.6 Mooney viscosity and stress relaxation.

of both elastic and viscous response. A slow rate of relaxation indicates a higher elastic component in the overall response, whereas a rapid rate of relaxation indicates a more highly viscous component. The rate of stress relaxation can correlate with molecular structural characteristics such as molecular-weight distribution, chain branching, and gel content. It can be used to give an indication of *polydispersity* or M_n/M_w. It is determined by measuring the time for a 95% (T-95) decay of the torque at the conclusion of the viscosity test.

5. Delta Mooney, typically run at 100°C, is the final viscosity after 15 min. This provides another measure of the processing characteristics of the rubber. It indicates the ease of processing compounds that are milled before being extruded or calendered (e.g., hot feed extrusion systems).

Much work has been done to establish a relationship between the Mooney viscosity (ML) and molecular weight of NR as well as the molecular-weight distribution. Bonfils et al. [17] measured the molecular weight and molecular-weight distribution of a number of samples of rubber from a variety of clones of *H. brasiliensis* and noted the following trend:

Sample	P_0	ML 1 + 4	M_w (kg/mol)
1	32	57	746
2	41	78	739
3	54	92	799
4	62	104	834

where P_0 is initial Wallace plasticity, ML 1 + 4 is Mooney viscosity after 4 min, and M_w is molecular weight.

Though clearly not linear, there is an empirical relationship between Mooney viscosity and molecular weight. Nair [18,19] explored this, established a relationship between intrinsic viscosity and Mooney viscosity, and determined a correlation coefficient of 0.87. This correlation can be improved by mastication of the test samples, which improves the homogeneity. Mastication or milling also narrows the molecular-weight distribution, which is an important factor in this respect [20].

The cure characteristics of NR are highly variable due to such factors as maturation of the specific trees from which the material was extracted, method of coagulation, pH of the coagulant, preservatives used, dry rubber content, and viscosity stabilization agent.

A standardized formulation has been developed to enable a comparative assessment of different NRs; it is known as the American Chemical Society No. 1. The formulation consists of NR (100 phr), stearic acid (0.5 phr), zinc oxide (6.0 phr), sulfur (3.5 phr), and 2-mercaptobenzothiazole (0.5 phr). This formulation is very sensitive to the presence of contaminants or other materials such as fatty acids, amines, and amino acids, which may influence the vulcanization rate.

NR is susceptible to oxidation. This can affect both the processing qualities of the rubber and the mechanical properties of the final compounded rubber. Natural antioxidants will offer protection from the degradation of NR, which can be measured by the change in the material's plasticity. The Wallace plasticity test reports two measures:

1. Plasticity, P_0, is the initial Wallace plasticity and a measure of the compression of a sample after a load has been applied for a defined time.
2. The plasticity retention index (PRI) measures recovery after a sample has been compressed, heated, and subsequently cooled. PRI% is defined as $(P_{30}/P_0) \times 100$, where P_0 is the initial plasticity and P_{30} is the plasticity after aging for 30 min typically at 140°C. During processing in, for example, a tire factory, NRs with low PRI values tend to break down more rapidly than those with high values.

Various equations have been proposed that provide an empirical relationship between Mooney viscosity V_r and Wallace plasticity P_0. These equations depict a linear relationship between these two parameters and are therefore typically of the form

$$V_r = XP_0 + \text{constant } C \tag{1.1}$$

The numerical coefficient X and constant C are functions of the clone and grade of rubber but normally fall between 1.15 and 1.50 for coefficient X and between 4.0 and 12.5 for C [21].

Other materials can be added to assist in improving the processability of NR. These include peptizers such as 2,2'-dibenzamidodiphenyl disulfide, which when added at levels of around 0.25 phr can significantly improve the productivity of the mixers, allow lower mixing temperatures, improve mixing uniformity, and reduce mixing energy. Synthetic polyisoprene when added at levels of up to 25% of the total polyisoprene content will also reduce the viscosity of the compound with little loss in other mechanical properties. It also allows for better control of component tack, which is important in subsequent product assembly steps such as those in tire building.

NR tends to harden during processing and storage at the plantation processing factory, and also during shipping and prior to use in a rubber products manufacturing facility. This hardening phenomenon is manifested as an increase in viscosity, which is due to oxidation of the polymer chain and cleavage to form the functional groups, ketones $-C(CH_3)=O$ and aldehydes $-C-CH=O$. The aldehyde group can readily react with the $-NH_2$ groups in proteins to form a gel and thereby increase polymer viscosity. This occurs primarily during the latex drying process, which can last for 5–7 days at around 60°C. Materials may be added to NR to suppress this increase in viscosity, and this has been the basis for the development of CV rubbers. Hydroxylamine neutral sulfate ($NH_2OH \cdot H_2SO_4$), denoted as HNS, or propionic hydrazide (PHZ) can be added to NR latex at levels of 0.08–0.30 wt% and 0.20–0.40 wt%, respectively, to prevent gel formation. An accelerated storage-hardening test can measure the hardening of CV rubber that will occur during normal storage. When HNS is added before coagulation, treated rubbers will show a P_0 change of 8 units or less (CV). However, they will tend to display a darker color due to the HNS addition.

$$\begin{array}{c} O \\ \parallel \\ H_2N - NH - C - Et \end{array}$$
Propionic hydrazide (PHZ)

Both HNS and PHZ block the reaction of the aldehyde groups with $-NH_2$ by reacting with the $-C(CH_3)=O$ group to form

$$R'-C(CH_3)=N-NH-CO-R$$

and

$$R'-C(CH_3)-CH=N-CO-R$$

In compounded rubber, the term "bound rubber" has frequently been used to describe this cross-linking condition in both NR and polymers such as polybutadiene. Bound rubber can be found in all synthetic unsaturated elastomers and is due to a variety of factors such as covalent bonding, hydrogen bonding, and strong van der Waals forces. It can be readily measured by solvent extraction to remove polymer and leave a swollen insoluble gel. The formation of bound rubber can result from the use of high-structure carbon blacks, the use of silane coupling agents, or the application of fast to ultrafast accelerators such as zinc diethyldithiocarbamate found in vulcanization systems with low cure temperatures.

A number of production techniques can have an impact on the final viscosity of the rubber. The field methods are documented as follows:

1. *Latex dilution*: The effect is small, with 1:1 dilutions required to have any measurable effect.
2. *Ammonia*: An increase in the ammonia level added initially for preservation from 0.01% to 0.50% can result in a Mooney viscosity increase of up to 10 Mooney units.
3. *Coagulation method*: Coagulation methods can range from natural or bacterial coagulation to the addition of formic acid or heating. Mooney viscosity will range from 65 to 85, with higher Mooney viscosity values being obtained through the use of natural coagulation techniques.
4. *Maturation*: Storage of latex prior to drying and sheeting can cause an increase in Mooney viscosity due to an increase in gel content. This rise in gel content can be due to an increase in pH due to partial hydrolysis of protein and amino acids and subsequent cross-linking or to an increase in bacterial action.
5. *Drying temperature*: Above 60°C, there is a slight increase in Mooney viscosity.

Another factor that can affect viscosity is baling temperature. The age of the tapped rubber tree, yield stimulants, and seasonal effects may also play some role. If baled hot, the rubber can take a considerable time to cool. When hot, the polymer gel content or other cross-linking phenomenon may increase.

Because of the stereoregular structure of the polymer, NR crystallizes when strained or when stored at low temperatures. This phenomenon is reversible and is very different from storage hardening. The rate of crystallization is temperature dependent and is most rapid between $-20°C$ and $-30°C$. The rate of crystallization varies by grade, with pale crepe rubbers tending to show the greatest degree of crystallization. The rapid crystallization of NR is also due to nonrubber constituents present in the rubber. Fatty acids, particularly stearic acid, can act as a nucleating agent in strain-induced crystallization [22,23]. This can influence the end product performance, for example, in tires where strain-crystallized rubber can display reduced fatigue resistance but improved green strength, tensile strength, and abrasion resistance compared to elastomers that do not experience this phenomenon.

E. SPECIAL-PURPOSE NATURAL RUBBERS

A considerable amount of work has been directed toward enhancing the properties of NR through chemical modification. A number of polymers emerged from this work:

Liquid low-molecular-weight rubber: Produced by depolymerization of NR, liquid low-molecular-weight rubber can be used as a reactive plasticizer, processing aid, and base polymer. Molecular weights range between 40,000 and 50,000. This rubber is liquid at room temperature but is also available on a silica carrier [24]. Depolymerized NR finds application in flexible molds for ceramics, binders for grinding wheels, and sealants. It is susceptible to oxidation and therefore requires appropriate compounding techniques for adequate aging resistance. Liquid NR can be produced by a combination of mechanical milling, heat, and the addition of chemical peptizer. Reference may be made to the work of Claramma et al. [25] for a discussion on the effect of liquid low-molecular-weight NR on compounded classical mechanical properties.

Methyl methacrylate grafting: Three grades of rubber with different levels of grafted methyl methacrylate are available (Heveaplus MG 30, 40, and 49). These are prepared by polymerizing 30, 40, and 49 parts of methyl methacrylate, respectively, in the latex before coagulation. They have found application primarily in adhesives due to the effectiveness of the polar methacrylate group and nonpolar isoprene bonding dissimilar surfaces. Such polymers tend to have very high hardness (international rubber hardness degrees [IRHD]), with values up to 96 and have thus had no application in pneumatic tires [7,8]. When blended with regular grades of NR such as RSS 2, vulcanizates with high stiffness are attained but display Mooney viscosities ranging from 60 to 80 at typical factory compound processing temperatures.

Oil-extended NR: Oil-extended NR (OENR) treads are very effective in improving ice grip and snow traction of tires and have been reported to be used for service in northern Europe. OENR is produced by one of several methods: (1) cocoagulation of latex with an oil emulsion prior to coagulation or with the dried field coagulum, (2) Banbury mixing of the oil and rubber, and (3) soaking of the rubber in oil pans followed by milling to facilitate further incorporation and sheeting. Both aromatic (A) and naphthenic (N) oils are used at loadings typically around 65 phr. When compounded, filler loadings can be higher than those typically found in non-oil-extended rubber. The ratio of rubber to oil and oil type are denoted by a code that would read, for example, OENR 75/25N for a 75% rubber, 25% naphthenic oil material.

Deproteinized NR: This is produced by treating NR latex with an enzyme that breaks down naturally occurring protein and other nonrubber material into water-soluble residues. The residues are then washed out of the rubber to leave a polymer with low sensitivity to water. Typically, NR contains around 0.4% nitrogen as protein; deproteinized rubber contains typically 0.07%. Deproteinized NR has found application in medical gloves to protect workers from allergic reactions and in automotive applications, seals, and bushings. The polymer displays low creep, exhibits strain relaxation, and enables greater control of product uniformity and consistency [26].

Epoxidized NR: Compared with NR, epoxidized NR shows better oil resistance and damping and low gas permeability. However, its tear strength is low, which has prevented its use in pneumatic tires. Two grades are available, ENR 25 and ENR 50, that is, 25 mol% epoxidized and 50 mol% epoxidized. Epoxy groups are randomly distributed along the polymer chain. Calcium stearate is required as a stabilizer. These polymers offer a number of advantages such as improved oil resistance (ENR 50 is comparable to polychloroprene), low gas permeability equivalent to that of butyl rubber, and compatibility with PVC. When compounded with silica, epoxidized NR has reinforcement properties equivalent to those of carbon black but without the use of silane coupling agents [27].

Epoxidized natural rubber

Thermoplastic NR: Thermoplastic NR materials consist of blends of NR and polypropylene. No application in tires or other major elastomer-based products has been developed, though that remains one area that offers considerable potential for the future [27].

Superior processing rubber: This consists of a mixture of two types of NR, one cross-linked and the other not. It is prepared by blending vulcanized latex with diluted field latex in levels according to the grade being prepared (SP 20, SP 40, SP 50 with 20%, 40%, and 50% cross-linked phase, respectively). Two grades are also available (PA 57 and PA 80) with a processing aid added to further facilitate factory handling. These two grades contain 80% cross-linked rubber. These two-phase polymer systems display high stiffness with good flow and process qualities.

Guayule: Guayule is a shrub that grows in the southern region of the United States and northern Mexico. A typical 5–10-year-old plant will grow to about 30 in. in height and have a dry weight of approximately 20% resinous rubber.

Rubber of reasonable quality can be obtained after extraction. Though work of any significance has not been conducted in this area for many years, given the advances in genetic engineering and related fields in biotechnology, it is an area that merits further exploration. New clones could be developed that might have improved output and supplement current supplies of NR extracted from *H. brasiliensis* [28].

Ebonite: Ebonite is a rubber vulcanized with very high levels of sulfur. True ebonite has a Young's modulus of 500 MPa and Shore D hardness of typically 75. The term "pseudoebonite" has been used to describe rubber with a Shore A hardness, or IRHD, of 98 or Shore D hardness of 60. Ebonite has a sulfur content of 25–50 phr, and resins may also be used to obtain the required hardness or meet any compounding constraints of concern to the technologist. The principal use of ebonite materials is in battery boxes, linings, piping valves and pumps, and coverings for rollers, where chemical and corrosion resistance is required [29].

Synthetic polyisoprene: Global production of NR is expected to be in the order of 12–13 million tons in the year 2015 and the total capacity growing toward 15–16 million tons a few years after that. Demand, however, is projected to grow at a faster rate, and a potential shortage of supply could develop. Periodic shortages and temporary price peaks have been met by the use of synthetic polymers such as polyisoprene, styrene butadiene rubber (SBR), and PBD. The isoprene unit not only exists in NR but is also the building block for terpenes, camphor, and other natural products, and such materials have been added to synthetic rubber compounds to help meet a set of final compound properties. Isoprene for chemical synthesis is typically recovered from the C_5 streams obtained in the thermal cracking of naphtha. Three organometallic initiators have attained commercial significance in *cis*-polyisoprene production. These are *n*-BuLi, $TiCl_4R_3Al$ anisole and CS_2, and $TiCl_4$ $(HAIN-i-C_3H7)_6$, where HAIN is a poly(*N*-alkylimino alane). The lithium catalyst will produce a polymer with a microstructure that is 92% *cis*-1,4-isoprene and 1% *trans*-1,4-isoprene and 7% vinyl-3,4. Titanium-based catalysts will produce a polymer typically 96% *cis*-1,4-isoprene and 4% vinyl or isopropenyl, though there may be trace amounts of *trans*-1,4-isoprene. With appropriate levels of modifiers, the level of *trans*-1,4-isoprene can be increased. Such polymers have a much higher glass transition temperature (T_g) and therefore tend to find application in tire tread compounds where traction is a required performance parameter. When the required conversion is complete, a terminator (short stop) to deactivate the catalyst and a stabilizer are added. The number-average molecular weight (M_n) of polyisoprene is 350,000–400,000, and its Mooney viscosity (ML 1 + 4) ranges from 55 to 95, depending on the commercial grade. The glass transition temperature is around –70°C. Polyisoprene is more uniform than NR and thus lends itself better to applications requiring good mixing efficiency, high-speed extrusions, mix consistency, and product component uniformity as in tires. It is used in applications requiring high tensile strength, resilience, tear strength, and abrasion resistance [30,31].

F. QUALITY

A number of factors can be considered under the broad category of "quality" and include consistency or uniformity, supply, packaging, and minimum contamination. The following discussion will highlight some general qualities that rubber product manufacturers should expect from the raw NR producers.

1. Consistency and Uniformity

Within a grade, end users of NR require uniformity, little spread in properties such as PRI, and little or no need to warm the rubber prior to mixing. In tire and industrial goods manufacturing, NR uniformity is required for final compound consistency, which in turn yields consistent processing characteristics. Lack of consistency will result in variation in mixing specifications, extrudate uniformity, tack, and product component properties.

The only physical measures that are used to quantify the processing characteristics of NR are original Wallace rapid plasticity (P_0) and the PRI. P_0 tested via the Wallace Plastimeter is used as a rapid means of measuring plasticity. The level of P_0 has been determined to represent approximately half the level of Mooney viscosity, that is, a P_0 of 30 would suggest a Mooney viscosity of about 60. Mooney viscosity and P_0 alter with storage hardening (as a result of the cross-linking of random functional groups such as aldehyde groups in the polyisoprene chain), increasing with time (storage between processing at source and use at tire plant delayed by ocean transit) and unfortunately at an inconsistent rate and level. In an effort to provide consistency and stability, HNS is added to grades such as TSR 10 CV and TSR 20 CV that have been stabilized. It is also possible to stabilize other grades to a Mooney of 65 ± 10 units using HNS if necessary.

2. Packaging

Bales must be wrapped properly to prevent moisture penetration and mold growth, to maintain the quality levels of the rubber at time of purchase, and to avoid contamination. Shipping in metal containers avoids wood contamination and is recommended.

3. Contamination

Considerable work has been done to lower the dirt level in both technically specified and visually inspected rubbers. The last revision to the SMR scheme [12] introduced the following revisions:

1. Two CV grades, SMR 10 CV and SMR 20 CV, were defined whose Mooney viscosities (ML 1 + 4) are 50 ± 5 and 60 ± 5, respectively.
2. Dirt level specifications were reduced from 0.10 to 0.08 for SMR 10 CV and from 0.20 to 0.16 for SMR 20 CV.
3. CV grades of SMR 5 were specified, again with viscosities of 50 and 60 (SMR 50 CV, SMR 60 CV). Dirt levels of 0.03% are now typical.

In the future, emphasis must be placed on reducing contamination. This is in recognition that the "dirt" level has improved significantly over the last few years for all TSR grades. Contaminants include foreign material originating from the field in the form of bark, wood, twigs, leaves, and leaf stems. These have the potential to cause final product failure, because large foreign particles do not disperse during compounding and can provide sites for crack initiation. Although the level of dirt may be measured by the residue weight and as such can be included in technical specifications, contamination by foreign light matter such as wood chips and plastic material are not specifiable at this time. In the washing and cleaning of NR at the processor's factory, the sedimentation process separates heavy material from the floating light rubber crumbs, which reduces dirt content, but it does not separate the light, floating contaminants satisfactorily. The NR industry has focused on reduction of dirt content by the use of sedimentation within the process, but contamination with foreign matter is generally caused by material that floats and therefore is not controlled in the traditional process.

4. Fatty Acids

Excessive levels of fatty acids such as palmitic acid, oleic acid, and stearic acid can bloom to the surface of compounded rubber components prepared for tire building or other engineered products. Tire plants may have component tack difficulties when, for example, a TSR 20 with fatty acid levels of 0.25 wt% is changed to a TSR 20 grade with a fatty acid level of 0.9–1.0 wt%. This is due to bloom. Such bloom can later cause component separations. High levels of fatty acids can also affect vulcanization kinetics. Table 1.8 presents total fatty acid levels for a variety of NRs and shows that they can vary from 0.3% to 0.8%. Synthetic polyisoprene can be used to control bloom, tack, and viscosity. Addition of a polymer such as Natsyn 2200™ (The Goodyear Tire & Rubber Company) of up to 50% of the final product can be used where there are concerns regarding excessive fatty acid levels. Tack-inducing resins such as ExxonMobil Escorez 1102™ may also be used to correct adverse effects of bloom.

Fatty acid levels are to a large degree a function of the amount of washing the raw materials undergo prior to shipping. Malaysian rubbers are produced to clearly defined dirt levels and thus require little washing. In consequence, fatty acid levels can be relatively high. However, materials from other regional sources such as Indonesia initially contain much higher dirt acid levels, require more washing, and as a result have a greater amount of fatty acids removed before baling and shipping.

TABLE 1.8

Examples of Fatty Acid Levels in TSR 20 Natural Rubber

	Weight Percent Fatty Acid			
Acid	Indonesia (SIR 20)	Thailand (TTR 20)	Thailand (RSS 2)	Malaysia (SMR 20)
Linoleic	0.05	0.30	0.40	0.20
Palmitic	0.08	0.10	0.10	0.20
Oleic	0.05	0.10	0.10	0.15
Stearic	0.08	0.15	0.15	0.20
Other	0.05	0.10	0.05	0.20
Total	0.31	0.75	0.80	0.90

G. USE OF NATURAL RUBBER IN TIRES

The amount of NR used in specific tires varies according to the design and size of the tire. In terms of tonnage, consumption of NR in tires is divided as follows: automobile tire production uses around 45%, truck tires use approximately 35%, and the remaining 20% is used for farm, earthmover, or OTR vehicles and aircraft [10,32]. Smithers Scientific Services publishes an analysis of the materials used in various tire lines. From this information, an average NR content may be estimated for each class of tire as illustrated in Figure 1.7. For example, automobile and radial light truck tires, which range in size from 12 in. at the lower end up to 16 in. at the upper end, contain from 10% to 15% by weight NR. Larger tires for commercial applications will have NR contents within a range of 32–40 wt% [19,20]. The higher NR levels found in commercial tires are required to meet the following performance needs:

1. Lower operating temperatures.
2. Reduced rolling resistance.
3. Improved OTR tire rating, ton miles per hour. This is a measure of the number of tons a tire configuration on a vehicle is capable of hauling at the average vehicle speed for the operation shift. It is a function of tire operating temperature and durability.
4. Component-to-component adhesion for durability and tire retreadability.
5. Tear strength, particularly for tires operating in off road conditions.
6. Tread wear.

At temperatures below 100°C, NR can be difficult to break down on mills or internal mixers and subsequently process. However, when first placed in a hothouse or broken down on warm-up mills, NR-based compounds can be processed quite easily. Peptizers enable a lowering of compound mixing temperatures, permit shorter dwell time in internal mixers, and thus save energy. When NR is compounded with a highly reinforcing carbon black such as N121 or N134, strong interactions can occur such as hydrogen bonding, covalent bond formation, and van der Waals forces. When immersed in a solvent such as toluene, free rubber can be extracted, leaving a swollen rubber-filler gel (i.e., bound rubber). Uncured compound that has been stored for long periods will show an increase in bound rubber content with consequent loss in ease of factory processing [33]. Bound rubber content is not a constant property but will evolve until a fixed value is attained. The change in bound rubber content can be readily estimated from Mooney viscosity and Mooney peak data. This provides the factory rubber technologist with a simple tool to determine factory compound shelf life and times between compound mixing and subsequent extrusion, calendering, or other processing step. In the absence of refrigeration, truck tire tread compounds containing 100 phr NR, 50 phr carbon black, and 3–5 phr of process oil will have a shelf life of 3–5 days before extrusion due to the increase in bound rubber.

Radialization has led to significant increases in the use of NR, and this will continue as the use of radial tires increases in farm equipment, large earthmovers, and aircraft tires, and the size of the bias truck tire market decreases. A comparative study was undertaken to obtain an overall assessment of NR use by tire component

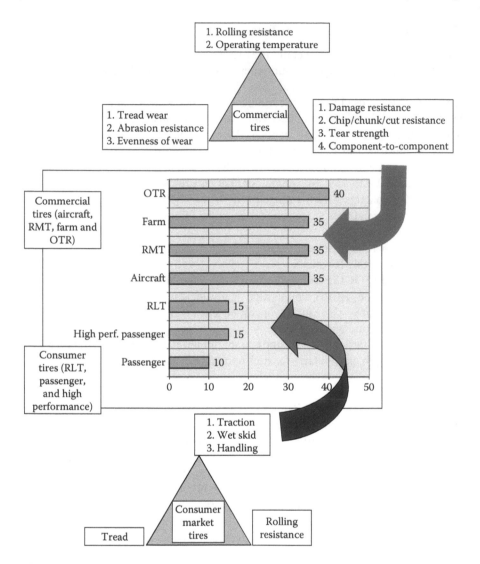

FIGURE 1.7 Natural rubber content by tire line (% of total tire weight) and relationship to tire performance triangles. RLT, radial light truck; RMT, radial medium truck; OTR, off the road. (From Barker, C.S.L., *Rubber Dev.*, 49(3/4), 40, 1996. With permission.)

for both radial medium truck (RMT) and automobile tires. The Malaysian Natural Rubber Producers Association and Smithers Scientific Services have both reported on the use of NR in the various components of a tire. Typical levels of NR in tread compound, base compound, sidewall, and wire coat compounds of three major classes of tires are presented in Table 1.9 [34–36].

The bulk of NR is compounded with other elastomers to produce blends and thereby obtain the desired mechanical properties. The NR content in tread compounds can range from 10 phr, when it has been added to improve processing qualities, to 100 phr, as when it is used in commercial radial truck tires for good

TABLE 1.9
Natural Rubber Levels (phr) in Selected Types of Tires and Tire Components

Component	Automobile Tires	Radial Medium Truck Tires	Bias Truck Tires
Tread	45	90	50
Base	70	100	—
Sidewall	45	55	40
Wire coat (breaker coat)	100	100	70

Sources: Steichen, R., Impact of future tire trends on natural rubber, Indonesia Rubber Research Institute, Bogar, Indonesia, 2000; Smithers Scientific Services, Tire analysis database, Akron, OH, 2000; Barker, C.S.L., *Rubber Dev.*, 49(3/4), 40, 1996.

hysteresis and tear strength. Other polymers typically blended with NR are polybutadiene (BR) for resistance to abrasion and fatigue, styrene butadiene rubber (SBR) for traction and stiffness, butyl rubber (IIR) and halobutyl rubber (CIIR, BIIR) for enhanced tire traction performance, and synthetic polyisoprene (IR) for processing qualities. Tire sidewalls are typically 50:50 blends of NR and polybutadiene for resistance to fatigue, cut growth, and abrasion.

Internal components of tires such as wedges, wire skim or wire coat compounds, fabric skim compounds, and gum strips typically contain 100 phr NR for component-to-component adhesion, tear strength, and hysteretic qualities.

III.   OTHER NATURALLY OCCURRING MATERIALS

Naturally occurring materials fall into two fundamental classes: organic or biotechnology products and inorganic materials. Table 1.10 provides a simple overview of the range of materials of interest that are either already available or are under

TABLE 1.10
Examples of Naturally Occurring Materials for Use in Rubber Compounds

Material	Compounding Ingredient Type	Potential Replacement for Synthetic Material
Guayule	Replacement for NR	Synthetic elastomers
Rice husks	Filler	Silica
Starch	Filler	Silica
Bamboo fillers	Filler	Clay, silica
Pine tar	Tackifying resin	Synthetic resins
Rosin	Tackifying resin	Synthetic resins
Coumarone Indene	Resin	Synthetic resins
Waxes	Processing aid	Synthetic oils
Fatty acids	Vulcanization system	N/A
Cotton	Filler, reinforcement	Carbon black, polyester, nylon

investigation. Though many of these will be discussed elsewhere in this text, it is appropriate to list them and note their potential application. Though all require various degrees of processing prior to use in rubber compounds, they still represent noncarbon sources of raw materials available to the rubber technologist.

Inorganic mineral fillers already find extensive use in rubber compounds.

1. Talc is used in products such as carpet backing and can be effective when blended with reinforcing fillers such as silica or carbon black. Particle sizes can range from 0.5 to 10.0 µm. It has also been used in tire inner liners as a low-cost filler though it can create processing challenges in the tire production factory due to the increase in compound tack.
2. Clays such as kaolin and bentonite can also be used in combination with silica or carbon black. Particle sizes tend to range from 0.5 to 5.0 µm. High-surface-area chemically modified clays will improve the tensile strength, abrasion resistance, and tear strength of the rubber product.
3. Calcium carbonate can be used as a filler even though its reinforcing properties are negligible. Surface modification by use of coupling agents can enhance the properties of compounds containing calcium carbonate. However, it is most effective when blended with carbon black or silica.

Biotechnology fillers offer considerable potential and have attracted attention in recent scientific literature. At this point, they fall into three primary categories: silica ash derived from rice husk waste, starch, and bamboo fibers. Burning of rice husks leaves a waste consisting of SiO_2 (95%), CaO, MgO, Fe_2O, K_2O, and Na_2O. Rice husks that have been milled, filtered, and then treated with sodium hydroxide, hydrochloric acid, and water produce a hydrated silica that when compounded can produce a material with mechanical properties similar to those of silica and carbon black. White rice husk silica contains around 95% SiO_2, whereas black rice husk silica is approximately 55% silica and 45% carbon. Residual carbon cannot be completely eliminated because it is trapped within the amorphous silica structure or is completely coated with silica, so it is impossible to remove it by thermal processes. The reinforcement properties of black rice husk ash are comparable to those of calcium carbonate and not as effective as those of carbon black or silica. White rice husk ash when added up to 20 phr in a NR-based compound did show good compound properties that were nearly equivalent to those found for silica-loaded compounds [37–39].

Starch has considerable potential when blended with carbon black or silica to improve the hysteretic properties of compounded rubber. This has implications for improvement in, for example, tire rolling resistance. Of the range of materials, biofillers hold the most promise for future increases in consumption. However, to achieve the most from such systems, either a resorcinol/hexamethylenetetramine system or silane coupling agent is required.

Bamboo fiber-filled NR has been investigated [40,41]. With the use of a silane coupling agent, workers were reported to obtain good tensile strength, tear strength, and hardness due to bonding between the polymer matrix and fiber. Filler loadings up to 50 phr are feasible. Work of this nature merits further investigation.

Rice bran oil has been evaluated as a substitute for process oils, as a coactivator, and as an antioxidant for NR. Raw rice bran oil contains fatty acids, phosphatides, and wax. This material was evaluated in NR compounds containing a conventional cure system and was found to be an effective substitute for more expensive antioxidants and processing aids and as a coactivator in place of stearic acid, with no toxicity concerns [42].

Other naturally occurring materials that have found application in rubber-based products include silicates, calcium carbonate, rayon, and cotton. Also of considerable importance are waxes and fatty acids, which are discussed elsewhere in this volume.

It is anticipated that there will be a growing emphasis on the use of naturally occurring materials, particularly in tires where vehicle manufacturers desire their products to have a defined level of recycled or potentially renewable resource content. Future government regulations may also require that automotive products and parts contain such materials.

IV. SUMMARY

This chapter has reviewed the classification and major uses of NR. Of the range of naturally occurring materials used in advanced engineered products, NR is among the most extensively used.

Four key factors will determine its use in the future:

1. *Availability*: Given the growth of the global economies and the automotive industry specifically, additional sources of materials will be required to meet the shortages anticipated by the year 2007.
2. *Technical specifications*: Specifications will be needed for visually inspected rubbers such as RSS grades to meet the end users' need for consistency and uniformity in their factories.
3. *Quality*: End product specifications and performance requirements will continue to be refined, thereby necessitating continuing improvement in consistency, absence of foreign materials or other contaminants, and purity.
4. *Chemical modification*: To improve the mechanical properties of current materials and enable their use in novel compounds, new synthetic derivatives of polymers will be required to compete with new functionalized synthetic elastomers.

Research institutes throughout the world are working on these issues, which will ensure the use of NR products long into the future.

REFERENCES

1. Barlow F. *Rubber Compounding*, 2nd edn. New York: Marcel Dekker, 1994.
2. Roberts AD. *Natural Rubber Chemistry and Technology*. New York: Oxford University Press, 1988.
3. Cyr DS. Rubber, natural. In: Kroschwitz JI, ed., *Encyclopedia of Polymer Science and Engineering*, Vol. 14, 2nd edn. New York: Wiley, 1988, pp. 687–716.

4. Baker CSL, Fulton WS. Rubber, natural. In: Kroschwitz JI, Howe-Grant M, eds., *Kirk-Othmer Encyclopedia of Chemical Technology*, Vol. 21, 4th edn. New York: Wiley, 1997, pp. 562–591.
5. Mahler HR, Cordes EH. *Basic Biological Chemistry*. New York: Harper & Row, 1969.
6. Hasma H, Subramanian A. Composition of lipids in latex of *Hevea brasiliensis* clone RRIM 501. *J Nat Rubber Res* 1986;1:30–40.
7. Hasma H. Lipids associated with rubber particles and their possible role in mechanical stability of latex concentrates. *J Nat Rubber Res* 1991;6:105–114.
8. International Standards Organization. ISO 2004. Specifications for natural rubber latex concentrates, 1988.
9. International Rubber Study Group. Worldwide rubber database, 2014. http://www.rubberstudy.com.
10. Barbin W, Rodgers MB. Science of rubber compounding. In: Mark JE, Erman B, Eirich FR, eds., *Science and Technology of Rubber*. New York: Wiley, 1994, Chapter 9.
11. International Standards Organization. ISO 2000. Rubber grades, 1964.
12. Rubber Research Institute of Malaysia. Revisions to Standard Malaysian Rubber Scheme. *SMR Bull. 11*. Kuala Lumpur, Malaysia, 1991.
13. The International Standards of Quality and Packaging for Natural Rubber Grades (The Green Book). *The International Rubber Quality and Packaging Conference*, Office of the Secretariat, January 1979. Washington, DC: Rubber Manufacturers Association, 1979.
14. The Malaysian Rubber Producers Research Association. Technically classified (TC) rubber—Visually graded natural rubber of classified cure characteristics. Natural Rubber Tech Info Sheet D100, 1982.
15. American Society for Testing and Materials. ASTM D 3184. Rubber—Evaluation of NR (Natural Rubber). Annual Book of ASTM Standards, Philadelphia, PA, Vol. 09.01, 1999.
16. American Society for Testing and Materials. ASTM D 1646. Standard test methods for Mooney viscosity, stress relaxation, and prevulcanization characteristics (Mooney viscometer). Annual Book of ASTM Standards, Philadelphia, PA, Vol. 09.01, 1999.
17. Bonfils F, Char C, Garnier Y, Sanago A, Sainte Beuve J. Inherent molar mass distribution of clones and properties of crumb rubber. *J Rubber Res* 2001;3:164–168.
18. Nair S. Dependence of bulk viscosities (Mooney and Wallace) on molecular parameters of natural rubber. *J Rubber Res Inst Malaya* 1970;23:76–83.
19. Nair S. Characterization of natural rubber for greater consistency. *Rubber World* July 1998;9:21.
20. Subramaniam A. Molecular weight and other properties of natural rubber: A study of clonal variations. *International Rubber Conference*, Kuala Lumpur, Malaysia, 1975.
21. Subramaniam A. Viscosity of natural rubber. *Planters Bull* 1984;180:104–112.
22. Kawahara S, Kakubo T, Sakdapipanich JT, Isono Y, Tanaka Y. Characterization of fatty acids linked to natural rubber—Role of linked fatty acids on crystallization of the rubber. *Polymer* 2000;41:7483–7488.
23. Sharples A. *Polymer Crystallization*. London, U.K.: Edward Arnold, 1966.
24. *Element's Tech Bull*. DPR liquid natural rubber, isolene, kalene, kalar. Belleville, NJ, 2003.
25. Claramma NM, Nair NR, Mathew NM. Production of liquid natural rubber by thermal depolymerization. *Indian J Nat Rubber Res* 1991;4:1–7.
26. Aziz A, Kadir SA. Advances and developments in NR. *Rubber World* November 2000;223:44–50.
27. Chai LP, Sang STM. Epoxidized natural rubber in tubeless tyre inner liners. *International Rubber Conference*, Kuala Lumpur, Malaysia, 1985.

28. Cyr DR. Rubber, natural. In: Grayson M, ed., *Kirk-Othmer Encyclopedia of Chemical Technology*, Vol. 20, 3rd edn. New York: Wiley, 1982, pp. 468–491.
29. Pyne JR. Rubber, hard. In: Kroschwitz JI, ed., *Encyclopedia of Polymer Science and Engineering*, Vol. 14, 2nd edn. New York: Wiley, 1988, p. 670–686.
30. Hsieh HL, Wagner PH, Wilder CR. Rubber, synthetic. In: McKetta JJ, Cunningham WA, eds., *Encyclopedia of Chemical Processing and Design*, Vol. 48. New York: Marcel Dekker, 1994, pp. 388–393.
31. Schoenberg E, Marsh HA, Walters SJ, Saltman WM. Polyisoprene. *Rubber Chem Technol* 1979;52:526–604.
32. Rodgers MB. Rubber tires. In: Buschow KHJ, Cann RW, Fleming MC, Ilschner B, Kramer EJ, Mahajan S, eds., *Encyclopedia of Materials: Science and Technology*. London, U.K.: Elsevier Science, 2001, pp. 8237–8242.
33. Leblanc JL, Hardy P. Evolution of bound rubber during the storage of uncured compounds. *Kautsch Gummi Kunstst* 1991;44:1119–1124.
34. Rodgers B, Tracey D, Waddell W. Natural Rubber. Presented at the 165th Technical Meeting of the American Chemical Society Rubber Division CS Rubber Division, Grand Rapids, MI, 2004.
35. Smithers Scientific Services. Tire analysis database, Akron, OH, 2000. http://www.smithersrapratds.com.
36. Barker CSL. Natural rubber. *Rubber Dev* 1996;49(3/4):40–44.
37. Mowdood S, Locatelli JL, De Putdt Y, Serra A. Future performance needs for tire fillers. *Proceedings of Functional Tire Fillers, Intertech Consulting Conference & Studies*, Fort Lauderdale, FL, January 29–31, 2001.
38. Sae-Oui P, Rakee C, Thanmathron P. Use of rice husk ash in natural rubber vulcanizates: In comparison with other commercial fillers. *J Appl Polym Sci* 2002;83:2485–2493.
39. Da Costa HM, Visconte LLY, Numes RCR, Furtado CRG. Mechanical and dynamic mechanical properties of rice husk ash-filled natural rubber compounds. *J Appl Polym Sci* 2002;83:2331–2346.
40. Ismail H, Edyham MR, Wirosentono B. Dynamic and swelling behaviour of bamboo filled natural rubber composites: effect of bonding agents. *Iranian Polym J* 2001;10:377–383.
41. Ismail H, Schuhelmy S, Edyham MR. The effect of silane couple agent on curing characteristics and mechanical properties of bamboo-filled rubber composites. *Eur Polym J* 2002;38:39–47.
42. Kuriakose AP, Rajendran G. Rice bran oil as a novel compounding ingredient in sulphur vulcanization of natural rubber. *Eur Polym J* 1995;31:596–602.

2 General-Purpose Elastomers

Howard Colvin

CONTENTS

I.   Introduction 34
II.   Structure–Property Relationships for General-Purpose Elastomers
        Used in Tire Applications..34
        A.   Laboratory Testing Methods 34
        B.   Glass Transition Temperature..35
        C.   Molecular Weight and Molecular Weight Distribution 37
        D.   Sequence Distribution in Solution SBR...38
III.   Emulsion Polymerization and Emulsion Polymers...39
        A.   Polymerization...39
        B.   Functional Emulsion Polymers..42
        C.   Oil-Extended Emulsion Polymers 42
        D.   Emulsion–Filler Masterbatches...43
        E.   Commercial Emulsion Polymers and Process..43
IV.   Anionic Polymerization and Anionic Polymers...46
        A.   Initiation 46
        B.   Propagation..50
        C.   Termination 53
        D.   Chain Transfer 58
        E.   Other Types of Functionalized Anionic Polymers 58
        F.   Commercial Anionic Polymers and Processes.......................................59
V.   Comparison of SBRs in Tire Compounds 60
VI.   Ziegler–Natta Polymerization and Ziegler–Natta Polymers..........................66
        A.   Mechanism of Butadiene Polymerization 67
        B.   *cis*-Polybutadiene...70
             1.   Titanium Catalysts 70
             2.   Cobalt Catalysts 70
             3.   Nickel Catalysts 70
             4.   Neodymium Catalysts...71
        C.   Syndiotactic Polybutadiene 73
        D.   Commercial Ziegler–Natta Polymers and Processes 75
        E.   New Processes: Gas-Phase Polymerization..75
VII.   Summary 76
References...76

I. INTRODUCTION

General-purpose elastomers played a critical role in the history of the last half of the twentieth century. In 1942, the Rubber Reserve program developed both the basic technology and manufacturing capability to make emulsion styrene-butadiene rubber (SBR) just a few years after World War II had interrupted natural rubber supplies. Historians have noted that the scientific contribution to that effort is comparable to the nuclear research program at Los Alamos that occurred at the same time [1]. After the petroleum shortages of the 1970s, fuel economy became a primary driving force in the automotive industry, and the tire industry was challenged to develop new products that would improve gas mileage. New elastomers based on solution SBR technology proved to be part of the answer.

Today, the tire industry is challenged to meet new environmental standards while maintaining or improving the vehicle handling, ride, and durability that have already been achieved. To meet this challenge, the rubber technologist must have a thorough understanding of how general-purpose elastomers (i.e., polybutadiene, styrene/ butadiene, and styrene/butadiene/isoprene) affect compound processability, tire rolling resistance, tire traction, tire treadwear, and overall cost of tire components. Use of these elastomers outside of the tire industry requires the same type of understanding of fundamental polymer characteristics and how they affect the final application. This review will describe the basic structure–property relationships between general-purpose elastomers and end-use properties, with a focus on the tire industry. The processes used to make the general-purpose elastomers will be described with an emphasis on how the polymerization variables (mechanism, catalyst, process) affect the macrostructure and microstructure of the polymer. It is polymer microstructure and macrostructure that determine whether a polymer is suitable for a particular application, not the type of process or catalyst used to produce the polymer.

Some important terms used in this chapter are defined in Table 2.1.

II. STRUCTURE–PROPERTY RELATIONSHIPS FOR GENERAL-PURPOSE ELASTOMERS USED IN TIRE APPLICATIONS

A. LABORATORY TESTING METHODS

Prediction of tire properties based on laboratory properties has met with various degrees of success, depending on which property was being predicted. There is a good correlation between the rolling resistance of tires and the tread compound tangent delta at 60°C and 40 Hz [2]. There is a reasonable correlation between tire wet traction and tangent delta of the tread compound at 0°C and 40 Hz [2]. Tire wear is more difficult to predict, with one researcher observing, "Despite more than 50 years of effort to devise laboratory abraders that give a good prediction of the wear resistance in real-world situations, no abrasion device currently exists that does an acceptable job" [3]. Typically, DIN abrasion or some type of blade abrader is used as a general indicator, however. Rubber processability has been defined in a number of ways [4] but is usually determined by what type of equipment will be used to process the rubber. Mooney stress relaxation time to 80% decay (MSR t-80) is a rapid,

TABLE 2.1

Definitions

Polymer microstructure: Monomers incorporated into the polymer and the stereochemistry of enchainment (i.e., *cis*, *trans*, vinyl).

Polymer macrostructure: Polymer molecular weight and molecular weight distribution, molecular geometry (linear, branched, comb), and the order in which monomers are incorporated (block, tapered block, or random).

Number-average molecular weight (M_n): Summation of the number of polymer chains (N) with a given molecular weight (m) times the molecular weight of each chain divided by the total number of polymer chains: $\Sigma m_i N_i / \Sigma N_i$.

Weight-average molecular weight (M_w): Summation of the number of polymer chains (N) with a given molecular weight (m) times the square of the molecular weight of each polymer chain divided by the total number of polymer chains times the molecular weight of each chain: $\sum m_i^2 N_i / \sum m_i N_i$.

Molecular weight distribution M_w/M_n: Ratio of weight average to number average molecular weight

Glass transition temperature (T_g): Temperature at which local molecular motion in a polymer chain virtually ceases. General-purpose elastomers behave like a glass below this temperature.

Weight-average T_g: Average T_g of a compound:

$$\sum \left[\left(\frac{\text{wt. polymer } X_n}{\text{total polymer wt.}} \right) (T_g \text{ polymer } X_n) \right]$$

effective processability test that works well with both emulsion [5] and solution SBR [6]. Other more sophisticated instruments such as the rubber processability analyzer or capillary rheometer are now becoming more common.

B. GLASS TRANSITION TEMPERATURE

The most important elastomer variable in determining overall tire performance is the glass transition temperature, T_g. Aggarwal et al. [2] showed that the tangent delta at 60°C of filled rubber vulcanizates made from "conventional rubbers" correlated with tire rolling resistance and then determined that the tangent delta values were approximately a linear function of the compound's T_g value. This was true whether the polymers were made by a solution process or an emulsion process. They did not compare solution and emulsion polymers at the same glass transition temperature.

Oberster et al. [7] showed that traction and wear properties were not dependent on the way the polymer was manufactured but were functions of the overall glass transition temperature of the compound, as shown in Figures 2.1 and 2.2. In actual tire tests, results are more complicated. The weight-average T_g of the tread compound is still a major variable, but it is not as dominant as in laboratory tests. A comprehensive study of tire wear under a variety of environmental and road conditions showed that tire wear improves linearly as the ratio of BR to SBR is increased in BR–SBR tread compounds (lower weight-average T_g). The wear behavior was more complex in BR–NR blends with low carbon black levels and was shown to be a function of ambient test temperature [3].

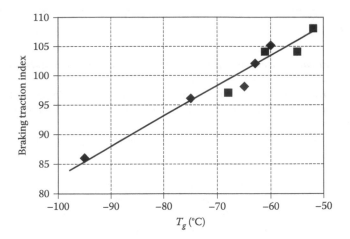

FIGURE 2.1 Effect of T_g on traction of (◆) solution polymers and (■) emulsion polymers. (From Oberster, A. et al., *Angew. Makromol. Chem.*, 29/30, 291, 1973. With permission.)

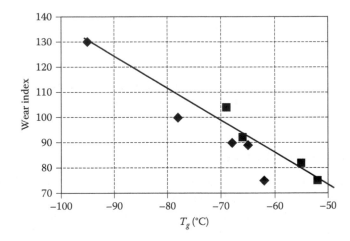

FIGURE 2.2 Effect of T_g on wear of (◆) solution polymers and (■) emulsion polymers. (From Oberster, A. et al., *Angew. Makromol. Chem.*, 29/30, 291, 1973. With permission.)

Nordsiek [8] expanded the concept of using the glass transition temperature to using the entire damping curve to predict tire performance. He divided the damping curve into regions that influenced various tire properties (Figure 2.3). The damping curves for an emulsion SBR, a high-vinyl polybutadiene, and a medium-vinyl SBR at the same T_g were compared and shown to be different at temperatures of 20°C–100°C. This led to the proposal of an "integral rubber" that would have a compilation of damping curves from a number of polymers and would incorporate damping behavior that would lead to the "ideal" elastomer for tread compounds. It was implied that this elastomer consisted of segmented blocks of different elastomers with different glass transition temperatures. An "integral rubber" was prepared

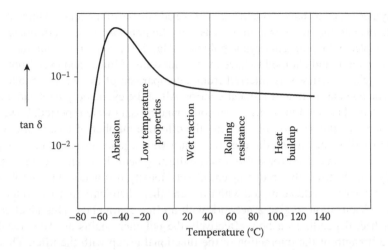

FIGURE 2.3 Damping curve of ESBR 1500 tread compound. (From Nordsiek, K., *Kautsch Gummi Kunstst.*, 38, 178, 1985. With permission.)

and compared to natural rubber and SBR 1500 controls in a laboratory compounding study. The "integral rubber" had a hot rebound within one point of the natural rubber control and was three points higher than the SBR 1500 control. Abrasion resistance was better than that of the natural rubber control but slightly worse than that of the SBR 1500. The 0°C rebound was lower than that of either control.

C. MOLECULAR WEIGHT AND MOLECULAR WEIGHT DISTRIBUTION

The molecular weight aspect of polymer macrostructure affects the rolling resistance (via hysteresis) and processability of the tread compound. As the molecular weight is increased, the total number of free chain ends in a rubber sample is reduced, and energy loss of the cured compound is reduced. This leads to improved rolling resistance, but at the expense of processability. Caution should be used in extrapolating lab data on high-molecular-weight rubbers to factory-mixed stocks, because filler dispersion is not as efficient with large-scale equipment. Thus, low hysteresis in lab compounds may not translate into low hysteresis in commercial tire compounds. There is an optimum balance between molecular weight and processability that is defined by the type of mixing equipment used. Increasing the molecular weight distribution at equivalent molecular weight by branching produces more free chain ends and more hysteresis but at moderate levels can improve other properties. Saito [9] showed that in silicon-branched solution SBR, the effect on hysteresis could be minimized and ultimate tensile strength could be improved because of better carbon black dispersion. In emulsion polymers, the branching is uncontrolled and the polymers have poorer hysteresis than the corresponding solution polymer [10]. From a practical standpoint, some branching in tire polymers is necessary to prevent cold flow and ensure that the elastomer bales will retain their dimensions on storage.

Polymer scientists have worked hard to take advantage of the relationship between free chain ends and hysteresis. In one case, an attempt was made to eliminate chain ends completely by preparing cyclic polymers. Hall [11] polymerized butadiene with a cyclic initiator and claimed to have made a mixture of linear and cyclic polybutadiene. Cyclic structure was inferred from a comparison of the viscous modulus of the cyclic polymer to that of a linear control. All of the cyclic polymers had a lower viscous modulus than the controls. No compounding data were reported, however.

A more popular method of reducing the effective number of free chain ends is to functionalize the end of the polymer chain with a polar group. Functional end groups can enhance the probability of cross-linking near the chain end and interact directly with the filler, thus reducing end effects. Ideally, difunctional low-molecular-weight polymers would be mixed with filler and then chemically react with the filler during vulcanization to give a network with no free chain ends. This ideal can be approached, depending on how effectively the polymer chains are functionalized and the strength of the interaction of the functional group with the filler. This will be discussed further in the section on anionic polymerization and anionic polymers (Section IV).

D. Sequence Distribution in Solution SBR

Day and Futamura [12] compared different 35% styrene solution SBRs at equivalent molecular weights and found that hysteresis is a linear function of the block styrene content. The effect of the polystyrene block length on hysteresis is shown in Figure 2.4.

Sakakibara et al. [13] made block polymers of polybutadiene and SBR with anionic polymerization and compared them to an SBR with the same overall microstructure. They found that the block polymers had broader glass transition temperatures that resulted in better wet skid resistance and lower rolling resistance than the corresponding random SBRs. They also found that blocky styrene in the SBR block was detrimental to overall performance.

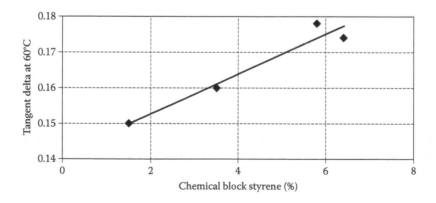

FIGURE 2.4 Effect of block styrene on hysteresis in SBR. (From Day, G. and Futamura, S., *Kautsch Gummi Kunstst.*, 40, 39, 1987. With permission.)

III. EMULSION POLYMERIZATION AND EMULSION POLYMERS

The copolymerization of styrene and butadiene is accomplished by dispersing the monomers in water in the presence of a surfactant, an initiator, and a chain transfer agent. The process offers limited control over polymer microstructure, and the polymers are branched. Emulsion SBR, however, has played and continues to play an important role in tire compounds.

A. POLYMERIZATION

The best way to consider the overall emulsion process is to examine the original recipe used to produce government rubber-styrene (GR-S) emulsion SBR rubber at the beginning of World War II [14] (Table 2.2).

It is important that the polymerization be done in the absence of oxygen. Oxygen is removed from the water by bubbling nitrogen through it prior to the polymerization, and the polymerization is conducted under a nitrogen atmosphere. When the ingredients are mixed, the monomers are partitioned between the water, micelles, and monomer droplets. The water solubility of styrene and butadiene is very low, so there is little of either in the water phase. Micelles are aggregates of surfactant (fatty acid soap) with the polar carboxylic group on the outside oriented toward the polar water and the nonpolar hydrocarbon tail oriented toward the inside of the micelle. The nonpolar styrene and butadiene are "soluble" inside the nonpolar environment of the micelle. Still, only a small portion of the monomer is located in micelles. There are approximately 10^{17}–10^{18} micelles per milliliter of emulsion [15]. Most of the monomer is contained in monomer droplets, which are in lower concentration (10^{10}–10^{11} monomer droplets per milliliter emulsion) and much larger than the micelles [15]. When the mixture is heated to 50°C, the potassium persulfate decomposes into radicals in the aqueous phase. Because the surface area of the micelles is much greater than that of monomer droplets, the radicals are more likely to inoculate the micelles to begin the polymerization. A representation of this is shown in Figure 2.5.

TABLE 2.2
GR-S Recipe for Emulsion SBR[a]

Component	Parts by Weight
Styrene	25
Butadiene	75
Water (deoxygenated)	180
Fatty acid soap	5
Dodecyl mercaptan	0.5
Potassium persulfate	0.3

Source: Odian, G., *Principles of Polymerization*, 2nd edn., Wiley, New York, 1981, pp. 319–337.
[a] Polymerization conducted at 50°C.

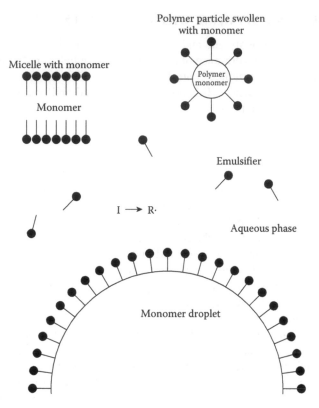

FIGURE 2.5 Species present during emulsion polymerization. (Odian, G.: *Principles of Polymerization*, 2nd edn. 319–337. 1981. Copyright Wiley-VCH Verlag GmbH & Co. KGaA. Reproduced with permission.)

As the polymerization proceeds, monomer migrates from the monomer droplets to the micelles until the monomer droplets are gone. Chain transfer to the mercaptan controls polymer molecular weight. Conversion is stopped at approximately 70% by addition of a radical trap such as the salt of a dithiocarbamate or hydroquinone. The latex is stabilized and then coagulated to give crumb rubber.

A major improvement in this process was the development of the redox initiation system shortly after World War II [16] (Table 2.3). With this recipe, the polymerization could be conducted at 5°C by changing the initiator system from potassium persulfate to cumene hydroperoxide. The iron(II) salt lowers the activation energy for the decomposition of the cumene hydroperoxide and is oxidized to iron(III) during the process. The dextrose is present to reduce the iron(III) back to iron(II) so more peroxide can be decomposed.

The importance of the lower polymerization temperature is shown in Figure 2.6. As the polymerization temperature is decreased, the ultimate tensile strength of cured rubber increases dramatically [17]. This is because there is less low-molecular-weight material and less branching at the lower polymerization temperature [18].

TABLE 2.3
"Custom" Recipe for Emulsion SBR

Component	Parts by Weight
Styrene	28
Butadiene	72
Water	180
Potassium soap of rosin acid	4.7
Mixed tertiary mercaptans	0.24
Cumene hydroperoxide	0.1
Dextrose	1.0
Iron(II) sulfate heptahydrate	0.14
Potassium pyrophosphate	0.177
Potassium chloride	0.5
Potassium hydroxide	0.1

Source: Blackley, D.: Diene based synthetic rubbers, in: Lovell, P. and El-Aasser, M., eds., *Emulsion Polymerization and Emulsion Polymers*. 535. 1997. Wiley-VCH Verlag GmbH & Co. KGaA. Reproduced with permission.

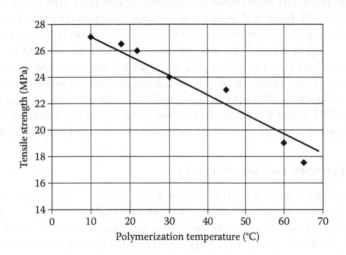

FIGURE 2.6 Effect of polymerization temperature on mechanical properties of ESBR. (Blackley, D.: Diene based synthetic rubbers, in: Lovell, P. and El-Aasser, M., eds., *Emulsion Polymerization and Emulsion Polymers*. 532–533. 1997. Copyright Wiley-VCH Verlag GmbH & Co. KGaA. Reproduced with permission.)

There is little control over butadiene polymer microstructure in the emulsion process. It remains fairly constant at 12%–18% *cis*, 72%–65% *trans*, and 16%–17% vinyl as the polymerization temperature is increased from 5°C to 50°C. Butadiene microstructure does not vary significantly as the styrene content is changed [19]. The glass transition temperature of emulsion SBR is controlled by the amount of styrene in the polymer.

B. FUNCTIONAL EMULSION POLYMERS

It is easy to incorporate a functional monomer into the backbone of an emulsion polymer as long as there is some water solubility. Emulsion butadiene or SBRs containing acrylate, amine, cyano, and hydroxyl groups have been made. Although some recent work has been done in exploring the interaction of functional emulsion rubbers with fillers, more work could be done. Emulsion SBR containing 3%–5% acrylonitrile displays better abrasion resistance than the corresponding unfunctionalized rubber in carbon black compounds [20]. Emulsion SBRs containing one to four parts of copolymerized amines were compounded into silica-containing stocks and showed good processability, improved tensile strength, lower hysteresis, and better abrasion resistance than a corresponding emulsion SBR control [21]. A detailed comparison of vinyl pyridine, cyano, and hydroxyl functionalized emulsion SBR with carbon black and silica and in blends with NR or high-*cis* BR demonstrated that the functional group of choice for maximum tire performance is a 3% hydroxyl functionalized polymer [22]. Comparison of this polymer with a corresponding solution SBR at a comparable T_g in a conventional SBR/BR silica tread formulation showed that the functionalized emulsion polymer is predicted to have better wear, wet performance, and rolling resistance than the solution polymer [22].

Recently, there have been efforts to functionalize emulsion SBR by use of reactive chain transfer [23]. Instead of using a conventional mercaptan to control the molecular weight during the polymerization, a functional mercaptan is used. This puts the functional group at the end of every polymer chain that was formed by initiation with the functionalized mercaptan radical. Emulsion polymers end functionalized with trialkoxysilane, amine, carboxylic acid, ester, and alcohol groups were prepared using this technique. Laboratory indicators of fuel economy showed these polymers to be 10%–25% better than control emulsion polymers in silica formulations. Wet grip and abrasion resistance predictors were also shown to be better than the control.

C. OIL-EXTENDED EMULSION POLYMERS

A substantial percentage of the rubber used in tire compounds is oil-extended emulsion SBR, which is prepared by adding an emulsion of oil to SBR latex prior to coagulation. Oil extension allows higher-molecular-weight elastomers to be used without processing problems, and incorporating the oil into the latex is much easier than putting it in the compound at the mixer. The oils used in compounding rubber are classified as paraffinic, naphthenic, and aromatic depending on the aromatic content of the oil. The different types of oils affect rubber compounds differently, and they cannot be directly substituted for each other without compounding changes. The more paraffinic the oil is, the lower its T_g, which will lead to different compound properties than a higher T_g naphthenic or aromatic oil. Direct comparison of SBR 1712 (37.5 phr aromatic oil) with SBR 1778 (37.5 phr of naphthenic oil) in a sulfur-vulcanized stock showed that the 1778 stock had a six-point higher room temperature rebound and a higher 300% modulus but poorer wet traction [24]. Schneider et al. suggested using a higher surface area black and adding small amounts of a higher T_g SBR to match the 1712 performance. Since the late 1980s, the aromatic oil used in SBR 1712 has come under fire for containing polycyclic aromatics that may be a factor in causing cancer. In 2005, the European Union legislated strict limitations of the amount of these materials allowed in extender oils [25]. This has forced the use of "clean oils" that have been refined so as to remove most of the polycyclic aromatics for tires made or sold in Europe.

D. EMULSION–FILLER MASTERBATCHES

Carbon black and carbon black–oil masterbatches of emulsion SBR have been used commercially for a long time. They are prepared by blending a dispersion of carbon black and oil with latex followed by coagulation. Masterbatching offers the advantages of improved black dispersion and shorter mix times. A major problem with masterbatching is that it limits compound flexibility to compounds that contain the type of black that is in the masterbatch. There can also be unexpected effects on the vulcanization rate [26]. Surprisingly, it has only been recently that a corresponding silica masterbatch has been introduced commercially [27], although there have been a number of patents on the subject [28–32]. In the commercial process, a dispersion of silica is hydrophobated with a water-soluble silane and blended with the latex prior to coagulation. The product reduces mix time, but also eliminates the ethanol produced from the silane using conventional silica mixing.

E. COMMERCIAL EMULSION POLYMERS AND PROCESS

The International Institute of Synthetic Rubber Producers (IISRP) classifies commercial emulsion polymers as shown in Table 2.4. Specifics (soap type, Mooney viscosity, coagulation, and supplier) for different grades of polymers are provided in the detailed section of the IISRP *Synthetic Rubber Manual* [33].

A schematic representation of a commercial continuous emulsion SBR process is shown in Figures 2.7 and 2.8. Most of the ingredients are mixed and cooled and then combined with a solution of initiator immediately before they enter the first reactor.

TABLE 2.4
Numbering System for Commercial Emulsion Polymers

Series No.	Description
1000	Hot nonpigmented rubbers
1500	Cold nonpigmented rubbers
1600	Cold black masterbatch with 14 or less parts of oil per 100 parts SBR
1700	Cold oil masterbatch
1800	Cold oil-black masterbatch with more than 14 parts of oil per 100 parts SBR
1900	Emulsion resin rubber masterbatches

Source: *The Synthetic Rubber Manual*, 18th edn., International Institute Synthetic Rubber Producers, Houston, TX, 2012. With permission.

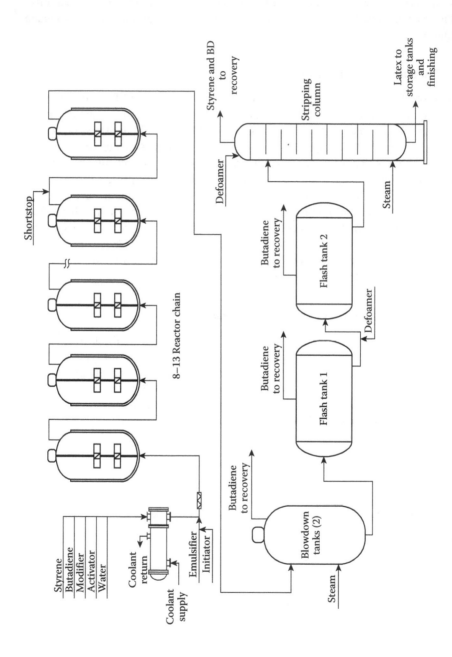

FIGURE 2.7 Emulsion polymer process—polymerization. (Courtesy of G. Rogerson, Goodyear Tire & Rubber Co., Akron, OH. With permission.)

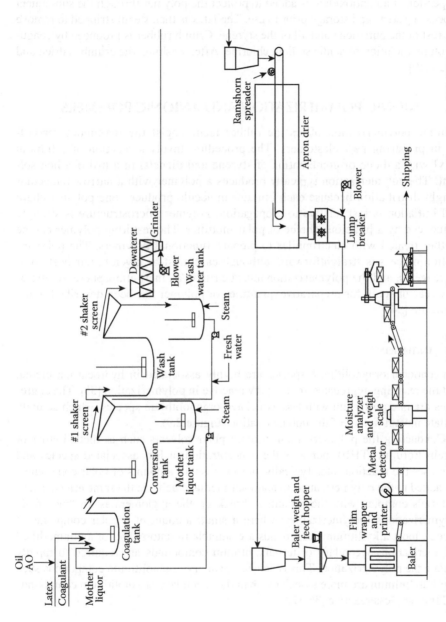

FIGURE 2.8 Emulsion polymer process—finishing. (Courtesy of G. Rogerson, Goodyear Tire & Rubber Co., Akron, OH. With permission.)

The number of reactors is chosen to control the residence time to reach 60%–65% conversion in 10–12 h. The polymerization is shortstopped, and the latex is pumped to a blowdown tank and flash tanks to remove most of the residual butadiene. A dispersion of an antioxidant is added to protect the polymer through the subsequent processing steps and storage prior to use. The latex is then steam stripped to remove the rest of the butadiene and all of the styrene. Crumb rubber is produced by coagulation in a solution of acidic sodium chloride. After washing, the crumb is dried and baled [19].

IV. ANIONIC POLYMERIZATION AND ANIONIC POLYMERS

Anionic polymerization offers the rubber technologist the maximum versatility in preparing new elastomers. The procedure involves reaction of a lithium alkyl with a diene or combination of styrene and diene(s) in a hydrocarbon solvent. The polymerization typically produces a polymer with a narrow molecular weight distribution because each initiator molecule produces one polymer chain and initiation is fast relative to propagation. Polymer microstructure is strongly influenced by a judicious choice of polar modifier. The resulting polymer can be further treated with electrophiles to prepare functional polymers. The polymerization process is straightforward, although care must be given to purification of all reagents, and the polymerization must be run in an inert atmosphere. A laboratory reactor setup for preparative quantities of polymer has been described in the literature [34].

A. INITIATION

Conventional organolithium species are highly associated in hydrocarbon media, and the resulting aggregates are not very reactive in polymerization [35]. The aggregates are in equilibrium with less associated organolithium species, which actually initiate most if not all of the polymerization (Figure 2.9).

Conducting the polymerization in more polar solvents such as diethyl ether or tetrahydrofuran (THF) increases the concentration of less associated species and increases the reaction rate. Typically, however, small amounts of polar compounds are added to the polymerization in nonpolar media to achieve the same effect. These materials complex with the lithium to break up the agglomerates. In "modified" polymerizations (polymerizations where a small amount of a polar compound is added), most alkyllithium compounds are suitable initiators, but for an unmodified polymerization, secondary or tertiary lithium compounds are required to rapidly initiate the polymerization. This is because primary organolithium compounds such as n-butyllithium are more associated than the secondary organolithium compounds and thus are less reactive [36,37].

$$(\text{R-Li})_n \;\rightleftharpoons\; n(\text{R-Li}) \xrightarrow{\text{Monomer}} \text{R-M-Li}$$

FIGURE 2.9 Aggregation of organolithium species.

Functional organolithium reagents are used to make functional polymers [38]. This technique is generally better than functionalizing a living polymer by reaction with an electrophile, because there are fewer side reactions with initiation. The reactivity of the lithium portion of the initiator requires that the functional group be protected in most cases, but the available functionality is surprisingly diverse. The key issues with functional initiators are storage stability and solubility in solvents suitable for polymerization. Lithiated acetals [39] and lithiated trialkylsilyl ethers [40] are used to form hydroxyl-terminated polymers after deprotection. Amine-terminated polymers have proven to be more useful for the preparation of tire elastomers. The synthetic routes diagrammed in Figure 2.10 can prepare these initiators.

The reaction of imine 1 with n-butyllithium produced initiator 2. SBR was prepared with this initiator, but the number-average molecular weight was much higher than predicted, which indicates that the alkyllithium reaction with the imine produced less than 100% of 2 or that the initiator is not completely efficient for initiation. The compounded SBR did exhibit improved hysteresis compared to a butyllithium-initiated control [41,42]. The reaction of secondary amines with butyllithium seems like an easy way to prepare n-lithium amides, but most of them are insoluble in non-polar media. Cheng [43] prepared a series of simple secondary lithium amides, but in all cases, they were insoluble in hexane. The heterogeneous initiators were used to polymerize dienes, but the polymerizations did not go to completion and the resulting polymers most likely had a very broad molecular weight distribution. Lawson et al. [44,45] showed that preparation of lithium amides in the presence of two equivalents of THF gave soluble initiators that could be used to make a medium-vinyl SBR at high conversion. The resulting polymer was coupled with tin tetrachloride and showed a 40% reduction in hysteresis as measured by tan δ at 50°C compared to a

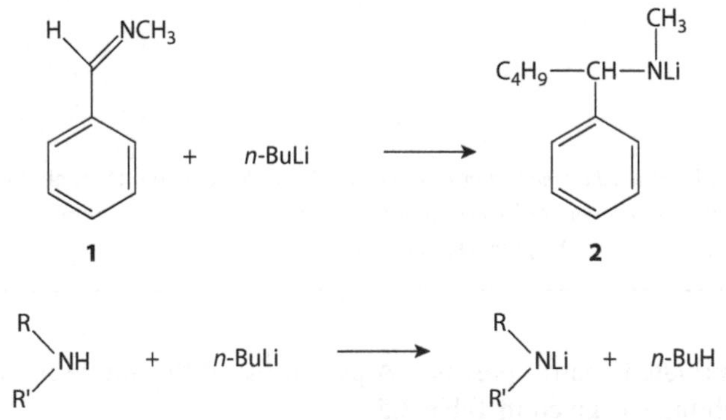

FIGURE 2.10   Synthesis of lithium amide initiators. (From Antkowiak, T. et al., Elastomers and products having reduced hysteresis, U.S. Patent 5,153,159, October 6, 1992; Antkowiak, T. et al., Methods for preparing functionalized polymer and elastomeric compounds having reduced hysteresis, U.S. Patent 5,354,822, October 11, 1994; Cheng, T., Anionic polymerization. VII. Polymerization and copolymerization with lithium-nitrogen-bonded initiator, in: McGrath, J., ed., *Anionic Polymerization, Kinetics, Mechanisms and Synthesis*, ACS Symposium Series No. 166, American Chemical Society, Washington, DC, 1981, p. 513.)

TABLE 2.5
Solubility and Effectiveness of Lithium Amide Initiators

| | Solubility | | | | |
Structure	Hexanes	Cyclohexane	2 equiv THF per Lithium	Initiation Ability M_w/M_n	Hysteresis Reduction Effect
BuLi	+	+	+	<1.1	−
![structure]	+	+	+	No poly-merization	−
![structure]	−	−	+	1.8	−
![structure]	+	+	+	1.1	−
![structure]	−	−	+	<1.1	+
![structure]	−	+	+	1.1	+
![structure 4]	+	+	+	1.5	+

Sources: Lawson, D. et al., *Polymer Preprints*, 37(2), 728, 1996; Lawson, D. et al., in: Quirk, R., ed., *Applications of Anionic Polymerization Research*, ACS Symposium Series No. 696, American Chemical Society, Washington, DC, 1998, pp. 77–87.

butyllithium-initiated control polymer. A partial list of the amide initiators studied and their solubilities is given in Table 2.5.

Interestingly, although almost all of the amide initiators effectively initiated polymerization, not all of the resulting polymers showed reduced hysteresis on compounding.

N-Lithiohexamethyleneimine 3 and *N*-lithio-1,3,3-trimethyl-6-azabicy-clo[3.2.1] octane 4 were studied further. They were both shown to be stable for "several days." Initiator 4 produced polymers with a broader molecular weight distribution than initiator 3 [46]. One difficulty in working with these initiators is that the amine

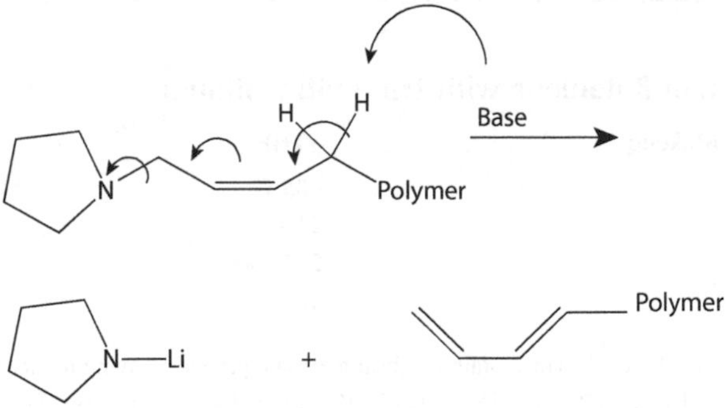

FIGURE 2.11 Head group loss in functional polymers.

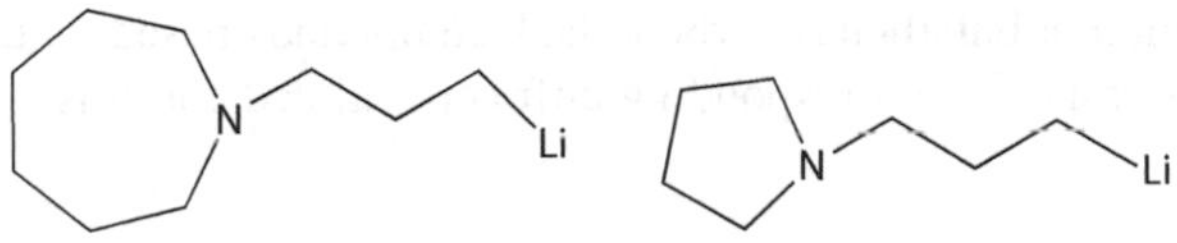

FIGURE 2.12 Functional initiators to avoid head group loss.

group is lost during polymerization by the mechanism shown in Figure 2.11. This reaction becomes more significant in the presence of excess initiator and at temperatures above 80°C.

Initiators **5** and **6** (Figure 2.12) do not lose the head group because the additional carbon atom between the nitrogen and lithium prevents elimination [47].

The difficulty with the lack of solubility of simple lithium amides can be overcome by in situ formation of the initiator. Immediately after charging a reactor with solvent, monomer, randomizer (THF or potassium amylate), and butyllithium, a secondary amine is added to the mixture. The amide is made in situ, and high-molecular-weight polymers are formed that have lower hysteresis than the corresponding polymers made with butyllithium. Approximately 85%–90% of the chains have amine head groups when this procedure is used [48].

Tin-containing initiators are also important compounds used to prepare high-performance tire rubbers. Addition of lithium metal to tributyltin chloride in an ether solvent produces a solution of the desired initiator that is filtered to remove lithium chloride [49] (Figure 2.13). The initiator is stable at room temperature and can be stored for approximately 8 weeks before a loss in activity is observed. Polymer with a lower vinyl content and narrower molecular weight distribution is obtained if the initiator is made in dimethyl ether rather than THF. This is illustrated in Table 2.6 for

$$Bu_3\text{-Sn-Cl} \quad + \quad Li \quad \longrightarrow \quad Bu_3\text{-Sn-Li} \quad + \quad LiCl$$

FIGURE 2.13 Synthesis of tributyltin lithium.

TABLE 2.6
Polymerization of Butadiene with Tributyltin Lithium

Solvent–Initiator Makeup	THF	Dimethyl Ether
T_g (onset)	–85°C	–93°C
Vinyl content	21%	11%
M_n	223,000	206,000
M_w/M_n	1.25	1.11

Source: Hergenrother, W. et al., Tin containing elastomers and products having reduced hysteresis properties, U.S. Patent 5,268,439, December 7, 1993, see also Hergenrother, W. and Bethea, T. (to Bridgestone Corp.), In the synthesis of tributyltin lithium, U.S. Patent 5,877,336, March 2, 1999.

the polymerization of butadiene. Carbon black compounds based on these polymers have lower hysteresis than corresponding unfunctionalized controls.

B. PROPAGATION

Propagation takes place at typical reaction temperatures (20°C–75°C) in inert solvents such as hexane or benzene without chain transfer or termination. At high temperature, however, the growing polymer chain can eliminate lithium hydride, which stops the polymerization and broadens the molecular weight distribution. The mechanism is shown in Figure 2.14.

Elimination of lithium hydride is a first-order process that yields a polymer terminated with a diene. Addition of living polymer doubles the molecular weight of the chain and provides an active site that can react with additional butadiene to form a branched polymer [50].

The ratio of monomer to initiator has a major influence on the *cis/trans* ratio in the homopolymerization of both butadiene and isoprene in unmodified polymerizations, as shown in Table 2.7 [51,52]. The higher the ratio of monomer to initiator, the higher the *cis/trans* ratio produced with both butadiene and isoprene.

Two kinetic factors affect the diene microstructure. The first involves the relative rates of propagation versus isomerization of the initially formed allyl anion. Monomer is inserted initially to form the allyl anion in the *anti* form. If propagation

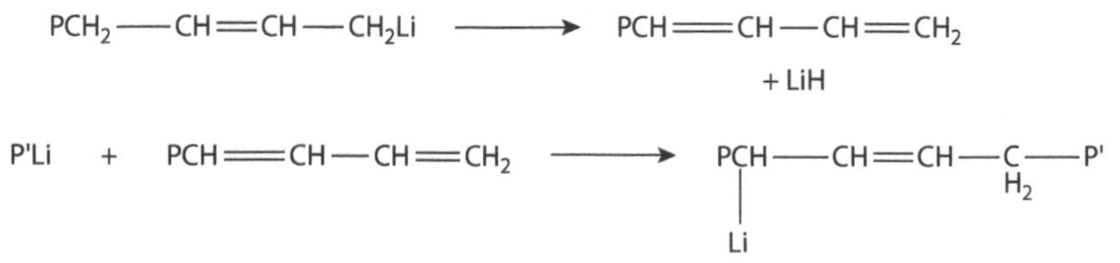

$$PCH_2 - CH = CH - CH_2Li \longrightarrow PCH = CH - CH = CH_2$$
$$+ LiH$$

$$P'Li \;+\; PCH = CH - CH = CH_2 \longrightarrow PCH - CH = CH - \underset{H_2}{C} - P'$$
$$|$$
$$Li$$

FIGURE 2.14 Mechanism for branching in lithium polymerization. (From Hsieh, H. and Quirk, R., *Anionic Polymerization: Principles and Practical Applications*, Marcel Dekker, New York, 1996, pp. 173–180.)

TABLE 2.7
Effect of Monomer/Initiator Ratio on Microstructure

Polymerization Conditions				Microstructure			
Monomer	Solvent	Initiator	Monomer/Initiator	Cis	Trans	1,2	3,4
Butadiene	Hexane	Li	5×10^4	0.68	0.28	0.04	N/A
			17	0.30	0.62	0.08	N/A
Isoprene	Cyclohexane	Li	$>5 \times 10^4$	0.94	0.01	—	0.05
			15	0.76	0.19	—	0.05

Sources: Morton, M., *Anionic Polymerization: Principles and Practice*, Academic Press, New York, 1983; McGrath, J., ed., *Anionic Polymerization: Kinetics, Mechanisms and Synthesis*, ACS Symposium Series No. 166, American Chemical Society, Washington, DC, 1981, Chapter 5.

is rapid, the microstructure of the penultimate unit will be *cis*. If, however, the allyllithium has sufficient time to isomerize to the thermodynamically more stable *syn* form, then the penultimate unit will be *trans*. Thus, at a high monomer/initiator ratio that favors rapid propagation, the microstructure is primarily *cis*. As the monomer is depleted and the monomer/initiator ratio decreases, more *trans* microstructure will be formed (Figure 2.15) [53,54]. The second factor is the relative rate of addition of

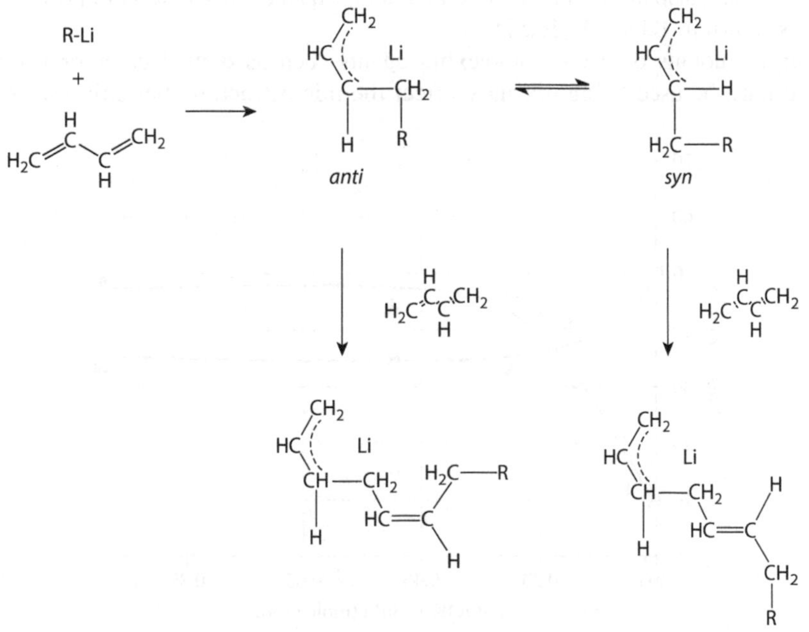

FIGURE 2.15 Microstructure formation during lithium polymerization. (Adapted from Gerbert, W. et al., *Die Makromol. Chem.*, 144, 97, 1971; Wosfold, D. and Bywater, S., *Macromolecules*, 11, 582, 1978.)

TABLE 2.8

Effect of Polar Modifiers on Polybutadiene Microstructure during Lithium Polymerization

Modifier	Modifier/Li	% 1,2-Addition at		
		30°C	50°C	70°C
Triethylamine	270	37	33	25
Diethyl ether	12	22	16	14
	96	36	26	23
Tetrahydrofuran	5	44	25	20
	85	73	49	46
Tetramethylethylenediamine	1.14	76	61	46
1,2-Dipiperidinoethane	1	99	68	31
	10	99	95	84

Source: Halasa A, Lohr D, Hall J. *J Polym Sci Polym Chem Ed.*, 19(6), 1357, 1981; Antkowiak T, Oberster A, Halasa A, Tate D. *J Polym Sci A-1*, 10, 1319, 1972.

monomer to the *syn* or *anti* isomer. Butadiene will add approximately twice as fast to the *anti* form as to the *syn* form. With isoprene the factor is eight times as fast [55].

In addition to increasing the rate of polymerization, polar solvents or polymerization modifiers also affect the vinyl content and sequence distribution in polybutadiene, as shown in Table 2.8 [56,57].

Large amounts of weak complexing agents such as diethyl ether or triethylamine must be used to significantly affect the microstructure, but only low levels

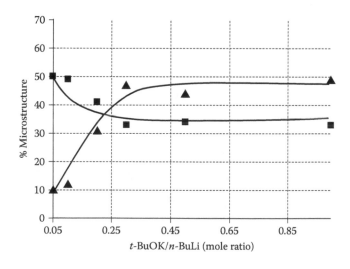

FIGURE 2.16 Effect of potassium butoxide/lithium ratio on polybutadiene microstructure. (■) Percent *trans*, (▲) percent vinyl. (From Hsieh, H. and Wofford, C., *J. Polym. Sci. Part A-1*, 7, 449, 1969. With permission.)

of strongly chelating modifiers such as tetramethylethylenediamine (TMEDA) or 1,2-dipiperidinoethane are needed to increase the vinyl content dramatically. The effect of polymerization temperature and its interaction with modifier is also illustrated by the data. Vinyl content is increased as the temperature is reduced for all polymerizations, but the effect is more pronounced at low modifier/lithium ratios.

In the copolymerization of styrene and butadiene, the sequence distribution is strongly affected by the addition of polar modifiers or salts. In hydrocarbon solvents without polar materials, most of the butadiene will polymerize first, followed by the styrene. This process is used to prepare "tapered" block polymers where there is a butadiene block, a mixed butadiene–styrene block, and a styrene block [58].

Addition of polar compounds will randomize the styrene and increase the rate of polymerization. Choice of modifier is critical to get the proper degree of randomization and control the vinyl content. Modifiers such as potassium *tert*-butyl alkoxide (*t*-BuOK) are used to randomize the styrene without significantly increasing the vinyl content. At a ratio of *t*-BuOK/*n*-BuLi of 0.1, there is only a small increase in vinyl content (Figure 2.16) over an unmodified control, but this is sufficient to randomize styrene in an SBR [59].

For higher-vinyl SBR, a more powerful randomizer such as TMEDA is used that produces high-vinyl polymers at relatively low modifier/lithium ratios [60]. Very-high-vinyl SBR and polybutadiene can be prepared with a modifier consisting of a mixture of TMEDA and an alkali metal salt of an alcohol [61].

C. TERMINATION

Termination is easily accomplished by reaction of the living polymer with an electrophile. In early anionic polymerization studies, the electrophile was a proton donor and termination resulted in a hydrocarbon polymer. Reaction with other electrophiles such as carbon dioxide (carboxylic acid), sultones (sulfonates), ethylene oxide (alcohol), or imines (amines) produce functional polymers, but unless conditions are carefully controlled, the functional polymer is contaminated with other materials [62]. Virtually every electrophile known has been tested as a terminating agent for lithium polymerizations. In one patent alone, the following were claimed for terminating a living *trans*-polybutadiene polymerization— isocyanates, isothiocyanates, isocyanuric acid derivatives, urea compounds, amide compounds, imides, *N*-alkyl-substituted oxazolidinones, pyridyl-substituted ketones, lactams, diesters, xanthogens, dithio acids, phosphoryl chlorides, silanes, alkoxysilanes, and carbonates [63]. Amine- and tin-containing electrophiles provide the greatest interaction with carbon black. Epoxy compounds and alkoxysilanes are most beneficial for silica-filled compounds. The early work focused on termination with amine-containing functional groups such as EAB [4,4′-bis-(diethylamino)benzophenone] [64–66]. Black compounds made with these polymers showed higher rebound, lower heat buildup, higher compound Mooney viscosity, and more bound rubber than the corresponding control rubber. Another study by Kawanaka et al. [67] suggested that the mechanism of the rubber–filler interaction was through an iminium salt formed from the reaction product of the amide and living polymer chain end (Figure 2.17). The authors inferred this

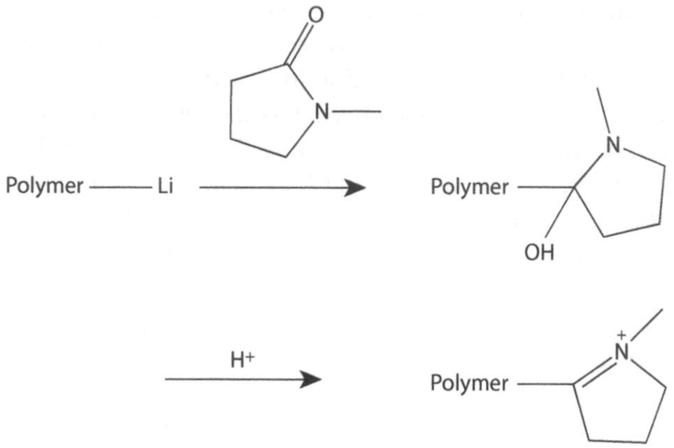

FIGURE 2.17 Termination of lithium polymerization with a cyclic amide. (From Kawanaka, N. et al., Analysis of chain end modified rubber structure, in: *Meeting of ACS Rubber Division*, Detroit, MI, 1989, Paper 118. With permission.)

because rubber functionalization with amides that could not easily form iminium salts did not interact well with carbon black.

Termination with tin-containing compounds provides more flexibility than with amine compounds. $R_x SnCl_y$ (where $x + y = 4$) can be chosen to give different levels of branching and thus assist in macrostructure control. Phillips pioneered the coupling of solution polymers with tin halides to make radial polymers in the 1960s, but the Japanese Synthetic Rubber (JSR) Company was the first to use the nature of the carbon–tin bond for tire compounds. Tsutsumi et al. [68] outlined the synthesis of tin-coupled solution SBR, the mechanism of how it improves hysteresis, structure–property relationships to maximize the effect of tin, and pitfalls to avoid in compounding.

Tsutsumi first demonstrated that coupling solution SBR with tin tetrachloride provided a superior polymer compared to other coupling agents (Table 2.9). The SBR polymerization was terminated with tin tetrachloride such that 50% of the chain ends were coupled. The only major difference in performance among the coupling agents was the low hysteresis exhibited by the tin-coupled polymer. He then compared a series of tin-coupled polymers with a polymer containing trialkyltin groups along the backbone. Only tin located at the end of the polymer chain (or branch point) was effective in reducing hysteresis (Figure 2.18).

In another study, the same group showed that putting the tin group on a butadienyl chain end was more effective in reducing compound hysteresis than putting it on a styryl chain end. Finally, they postulated that the mechanism of interaction with carbon black is by formation of a bond between the polymer chain and the quinone groups on the carbon black. This was based on a model study of the reaction of tributyltin-capped low-molecular-weight polybutadiene with a series of compounds containing functionality found on carbon black. Only the quinones reacted to any extent. The ease of cleavage of the tin–carbon bond is the reason this chemistry can take place, but it also puts some restrictions on how tin-containing polymers can be

TABLE 2.9
Coupling of Solution SBR[a,b]

Coupling Agent	ML 1+4 (100°C)	Compounded ML 1+4 (100°C)	Tensile Strength (MPa)	Elongation at Break	tan δ at 50°C	tan δ at 0°C
None	54	93	22.3	400	0.121	0.235
Divinylbenzene	51	70	22.5	400	0.125	0.241
Diethyladipate	47	74	21.6	410	0.126	0.237
Silicon tetrachloride	57	89	23.5	400	0.126	0.240
Tin tetrachloride	57	76	25.0	400	0.096	0.239

Source: Tsutsumi, F. et al., *Rubber Chem. Technol.*, 63, 8, 1990.

[a] *Formulation (phr):* polymer 100, HAF black 50, zinc oxide 3, stearic acid 2, antioxidant 1.8, accelerator 1.8, sulfur 1.5.

[b] *SBR:* 24% bound styrene, 40% vinyl.

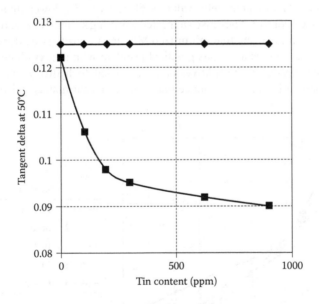

FIGURE 2.18 Effect of tin content position on dynamic properties of tin-coupled SBR. (◆) Polymer modified on backbone. (■) Polymer modified at chain end. (From Tsutsumi, F. et al., *Rubber Chem. Technol.*, 63, 8, 1990. With permission.)

isolated and compounded. Acid will cleave the bond and materials such as stearic acid should be avoided until late in the compound mix cycle. Mooney viscosity of coupled polymers will drop if the tin–carbon bond is broken, but if the polymer is capped with a trialkyltin halide, there will be little change in Mooney viscosity.

The importance of complete functionalization is illustrated by a study by Quiteria et al. [69]. They examined the effect of the polymer end group and the effect of unfunctionalized polymer (via incomplete coupling) on the dynamic properties of tin-coupled polymer in a simple black formulation. They synthesized a 25% styrene SBR with 32% vinyl via adiabatic polymerization and reacted the living polymer with a small amount of monomer (butadiene, styrene, isoprene, or α-methylstyrene) to ensure a specific end group. Tin tetrachloride was added to couple 40% of the polymer. The residual polymer chains were terminated with tributyltin chloride. The loss tangent as a function of temperature for these polymers is shown in Figure 2.19. The most important feature of the graph is the effect of unfunctionalized polymer on hysteresis (run 5). Compound made with polymer from run 5 (uncoupled polymer terminated with a proton) had 15%–20% higher tangent delta values at 80°C than runs 1–3 (uncoupled polymer terminated with tributyltin chloride).

For silica compounds, different functional groups are required for polymer–filler interaction. Alkoxysilanes such as 3-triethoxysilylpropyl chloride, chlorodimethylsilane, and bis-(3-triethoxysilylpropyl) tetrasulfide react with a living isoprene–butadiene chain to give a polymer that is claimed to interact well with silica [70]. Gorce and Labauze [71] showed that 3-glycidoxylpropyl-trimethoxysilane reacted primarily at the silicon instead of at the epoxy group when used to terminate SBR polymerization. The tangent delta value at 60°C was 28% lower than that of the unfunctionalized control. They also suggested a mixing system for reacting the polymer with the coupling agent that minimized Mooney viscosity rise after steam stripping and storage. This is a serious practical problem with alkoxysilane-terminated polymers. Hydrolysis of the alkoxysilane group led to hydroxysilyl end groups that condensed to increase the molecular weight and ultimately gel the polymer.

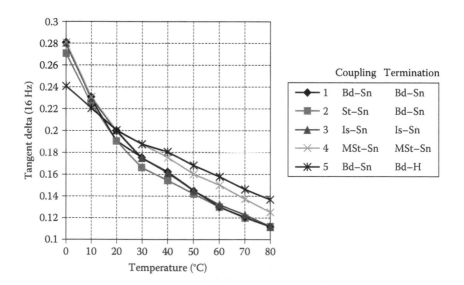

FIGURE 2.19 Loss tangent versus temperature for different tin–carbon bonds. Bd, butadiene; St, styrene; Is, isoprene; MSt, α-methylstyrene; H, hydrogen; Sn, tin. (From Quiteria, V. et al., Tin coupled SBRs. Relationship between coupling type and properties, in: *Meeting of ACS Rubber Division*, Cleveland, OH, 1995, Paper 78. With permission.)

TABLE 2.10

Terminating Agents for Interaction with Silica

Modifier	Probable Structure after Functionalization

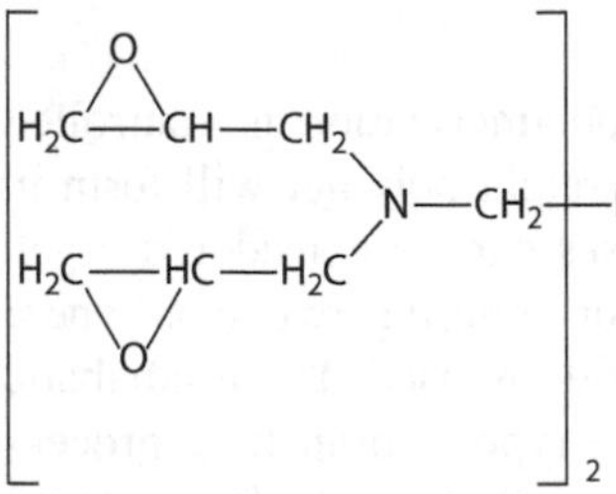

Commercial compound with diglycidiyl amino groups

$SBR-CH_2-CH(OH)-CH_2-N<$

$H_2C-CH(O)-CH_2-N<$

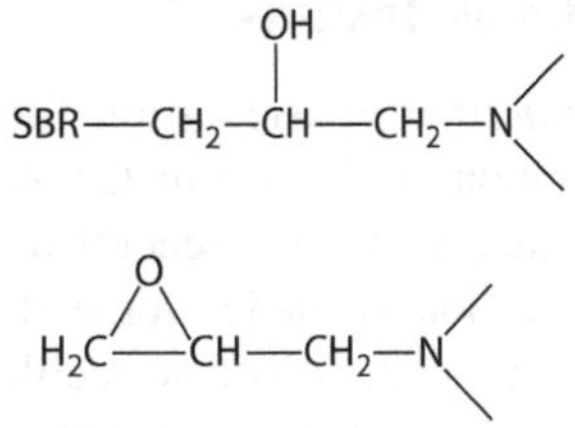

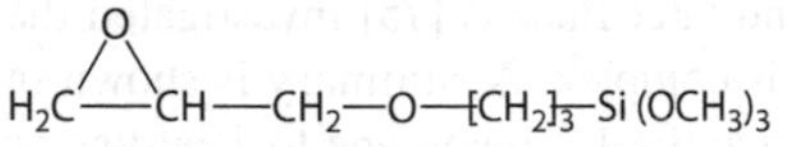

Glycidoxypropyltrimethoxysilane

$H_2C-CH(O)-CH_2-O-[CH_2]_3-Si(OCH_3)_3$

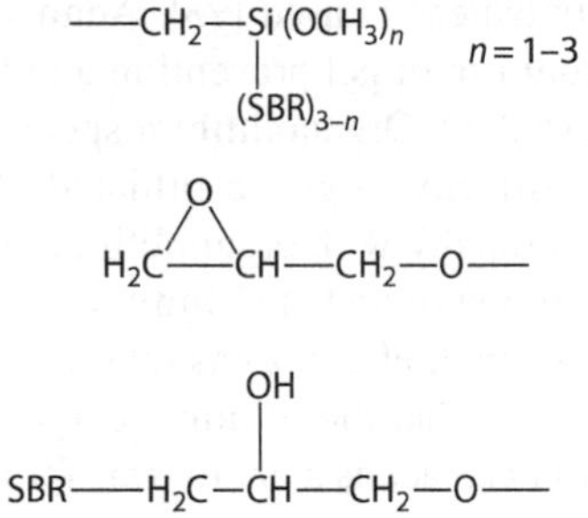

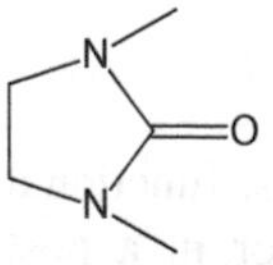

$-CH_2-Si(OCH_3)_n$; $(SBR)_{3-n}$; $n = 1-3$

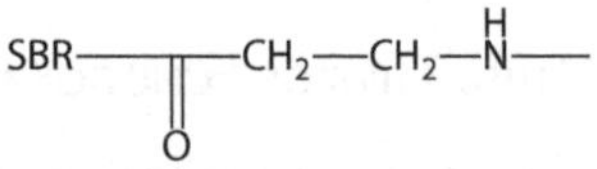

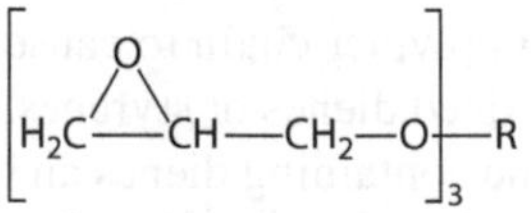

$H_2C-CH(O)-CH_2-O-$

$SBR-H_2C-CH(OH)-CH_2-O-$

Dimethylimidazolidinone

$SBR-C(=O)-CH_2-CH_2-N(H)-$

Commercial compound with triglycidoxy group

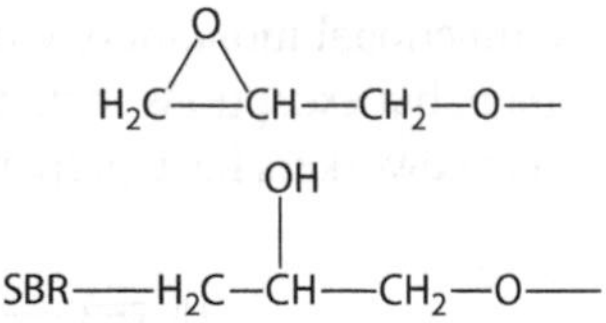

$H_2C-CH(O)-CH_2-O-$

$SBR-H_2C-CH(OH)-CH_2-O-$

Source: Saito, A. et al., Improvement of rolling resistance of silica tire compounds by modified S-SBR, in: *Meeting of ACS Rubber Division*, Savannah, GA, 2002, Paper 39. With permission.

Saito et al. [72] compared a number of different types of functional groups (Table 2.10) in a 35% styrene, 38% vinyl SBR to determine which ones interacted most strongly with silica and improved compound performance. In addition to the structures shown in Table 2.10, they also studied SBRs terminated with tin tetrachloride and silicon tetrachloride. Compounds made from the polymers containing the diglycidylamine group, glycidoxypropyltrimethoxysilane, and dimethylimidazolidinone had very low hysteresis and better abrasion resistance than the control polymer.

The viscosity of the glycidoxypropyltrimethoxysilane- and dimethylimidazolidi-none-modified polymers rose on storage, however, and would not be suitable for commercial production.

D. CHAIN TRANSFER

An important consideration for continuous anionic polymerization is controlled termination or chain transfer. A very-high-molecular-weight polymer will form in unagitated areas of a reactor and for practical purposes can be considered a gel. The situation is made worse if the reactor is used for coupling reactions where divinylbenzene, silicon tetrachloride, or tin tetrachloride is used. In an adiabatic process, the high temperature can lead to a gel via the type of branching process shown in Figure 2.14. To prevent reactor fouling, a small amount of a material that can act as a chain transfer agent or a slow "poison" must be added. Typically, 1,2-butadiene is used [73]. Adams et al. [74] and later Puskas [75] investigated the mechanism of gel prevention and found that it is complex. A summary is shown in Figure 2.20. Organolithium species can isomerize the 1,2-butadiene to 1-butyne or react directly to give a lithiated allene that can be further lithiated. The 1-butyne reacts rapidly with organolithium compounds to give a lithium acetylide. The reaction of poly(butadienyl)lithium with the 1,2-butadiene is slow at 50°C, as evidenced by the small effect it has on conversion or molecular weight of the polymer. It was suggested that the lithium acetylide or the allenyllithium could reinitiate polymerization and act as a chain transfer agent.

E. OTHER TYPES OF FUNCTIONALIZED ANIONIC POLYMERS

In addition to functionalization using initiators and terminating agents, functional anionic polymers can be prepared using functionalized monomers or in a post polymerization reaction step.

Most functional monomers will react with a propagating polymer chain to cause termination, but exceptions include tertiary amine functionalized dienes or styrenes. Halasa and coworkers have prepared a series of tertiary amine containing dienes and

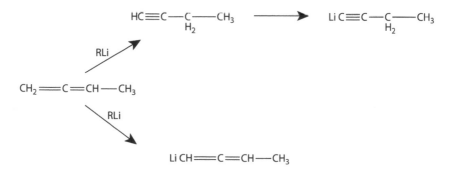

FIGURE 2.20 Reaction of alkyllithium with 1,2-butadiene. (Adapted from Puskas, J., *Makromol, Chem.*, 194, 187, 1993.)

styrenes that have been copolymerized effectively into SBR [76]. Incorporation of these types of monomers along the backbone of an SBR reduced the tangent delta value of silica compounds at 60°C thus showing better polymer–filler interaction. Wong [77] and coworkers showed that vinyl benzyl pyrrolidine could be incorporated into low-*cis* polybutadiene and was effective in improving rolling resistance in tire tests. At 0.5% incorporation of the pyrrolidine monomer, substitution of the low-*cis* functional polymer for high-*cis* BR improved rolling resistance but was detrimental to treadwear in both miscible and immiscible blends of BR/SBR.

Solution SBR has been functionalized in a post polymerization process with functional mercaptans. The mercapto group will react with pendent vinyl groups on the backbone to provide the functional polymer. This reaction can be done in a solution or during bulk mixing. Examples include functionalization of a medium vinyl solution SBR dissolved in toluene with either thioglycolic acid or ethyl mercapto acetate. An azo or peroxy initiator is required for the reaction and reasonable conversions can be obtained. The ester functionality is easier to incorporate than the carboxylate functionality [78]. Derivatives of the naturally occurring mercapto amino acid L-cysteine were added to solution SBR in both solution and in the bulk phase in the presence of a radical initiator [79]. Peroxide initiators led to gelling of the SBR, but very little change in molecular weight was seen if azo initiators were used. About 30% of the mercapto amino acid derivatives could be added to the backbone under optimized reaction conditions.

F. Commercial Anionic Polymers and Processes

IISRP does not make a distinction between commercial polymers produced anionically and polymers produced with a Ziegler–Natta catalyst. A classification of both types of polymers is provided in Table 2.11 [33]. Thus a non-oil-extended polybutadiene would carry a 1200–1249 designation, regardless of whether it was made anionically or with a Ziegler–Natta catalyst.

TABLE 2.11

Numbering System for Commercial Ziegler–Natta and Anionic Polymers

Polymer Form	Butadiene and Copolymers Series No.	Isoprene and Copolymers Series No.	Styrene, Isoprene, Butadiene Terpolymers Series No.
Dry polymer	1200–1249	2200–2249	2250–2599
Oil extended	1250–1299	2250–2299	
Black masterbatch	1300–1349	2300–2349	
Oil-black masterbatch	1350–1399	2350–2399	
Latex	1400–1449	2400–2499	
Miscellaneous	1450–1499	2450–2499	

Source: *The Synthetic Rubber Manual*, 18th edn., International Institute Synthetic Rubber Producers, Houston, TX 2012. With permission.

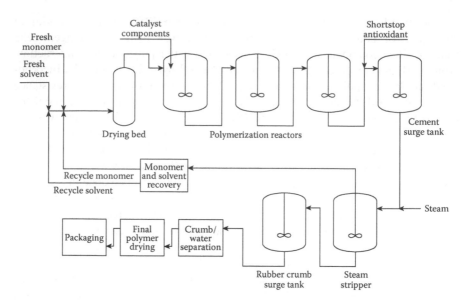

FIGURE 2.21 Continuous lithium polymerization process. (Courtesy of S. Christian, The Goodyear Tire & Rubber Co., Akron, OH. With permission.)

There are several commercial processes for producing anionic polymers. Although some polymers are made by batch polymerization, most are made with continuous polymerization. One method to make continuous random low-vinyl SBR controls the butadiene addition in such a way as to ensure random incorporation of styrene [80]. This process does not require a modifier and gives a very-low-vinyl (approximately 10%) polymer. Another continuous process uses a styrene–butadiene premix and randomizes the styrene in the polymer with a modifier such as an ether, alkoxide, or amine. An example of this' type of process is shown in Figure 2.21. A fresh monomer is diluted with a hydrocarbon solvent, mixed with recycle monomer, and passed through a drying bed to remove moisture, stabilizers, and impurities. The organolithium initiator and modifiers are added in the first reactor. The number of reactors is chosen such that the polymerization is run to high conversion. When the polymerization is complete, the lithium is neutralized, antioxidant is added, and the solvent and unreacted monomer are removed by steam stripping. The resulting crumb rubber is skimmed from the water, dried, and baled.

V. COMPARISON OF SBRs IN TIRE COMPOUNDS

Moore and Day [10] compared emulsion SBRs to solution SBRs in an oil-extended SBR/BR formulation with approximately the same weight-average T_g. Compounds with the solution polymers had lower hysteresis, which was demonstrated in both laboratory and tire testing. This was expected because of the microstructural and macrostructural differences in the polymers. The solution polymers cured faster than the emulsion polymers because the soap residues in the emulsion polymers retard the cure. The solution SBR compounds had higher modulus, but the compounds made with emulsion SBR had higher tensile strength.

Anionic polymerization is flexible enough that the effect of microstructure can be separated from the effect of T_g. Saito [9] compared compounds made from two series of solution SBRs, one with a vinyl content of 30% and varying styrene and another with 15% styrene and varying vinyl. Compounding results for traction, wear, and rolling resistance are shown in Figures 2.22 through 2.24 in a carbon black all SBR stock.

Traction and wear vary similarly as T_g is increased, regardless of the styrene or vinyl content. The lower-styrene, higher-vinyl polymers give superior rolling resistance at higher T_g values, as shown by the plot of rebound versus glass transition temperature.

Colvin and coworkers looked at a series of SBRs at three different T_gs, where the styrene and vinyl content differed at each T_g in a conventional BR/SBR highly-loaded silica tread formulation [81]. They found that at any T_g, the higher-styrene polymer provided significantly lower Mooney viscosity compounds and better laboratory indicators for wet traction than the corresponding high-vinyl polymer. The high-styrene polymers also were tougher than the high-vinyl polymers at all T_gs. The high-vinyl compounds showed better rolling resistance than the high-styrene polymers. These subtleties can be used by compounders to develop improved tread compounds depending on which performance/processing aspect is being targeted.

Day and Futamura [12] also looked at the effect of microstructure and macrostructure of solution polymers on compounded properties. They demonstrated the importance of polymer molecular weight on compound hysteresis in solution SBR

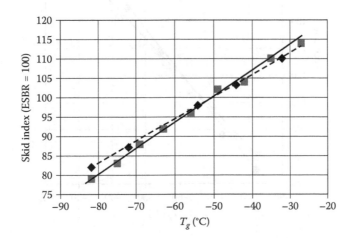

FIGURE 2.22 Effect of T_g on wet skid resistance in solution SBR. (◆) Change in vinyl content (15% styrene), (■) change in styrene content (30% vinyl). (From Saito, A., *Int. Polym. Sci. Technol.*, 26(6), T19, 1999. With permission of the copyright holders Rapra Technology Ltd.)

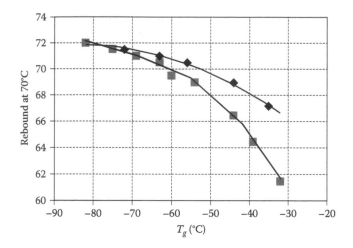

FIGURE 2.23   Effect of T_g on hot rebound in solution SBR. (◆) Change in vinyl content (15% styrene), (■) change in styrene content (30% vinyl). (From Saito, A., *Int. Polym. Sci. Technol.*, 26(6), T19, 1999. With permission of the copyright holders Rapra Technology Ltd.)

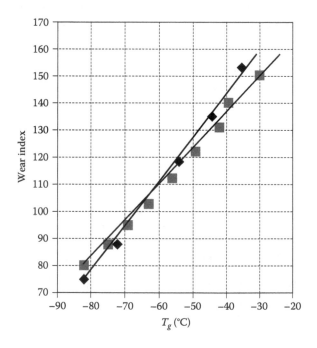

FIGURE 2.24   Effect of T_g on wear resistance in solution SBR (lower is better). (◆) Change in vinyl content (15% styrene), (■) change in styrene content (30% vinyl). (From Saito, A., *Int. Polym. Sci. Technol.*, 26(6), T19, 1999. With permission of the copyright holders Rapra Technology Ltd.)

TABLE 2.12
Effect of Styrene Level on Compounded Properties

Properties	Polymer				
	A	B	C	D	E
Raw polymer properties					
Styrene%	0	18	25	35	35
Vinyl%	12.0	9.8	9.0	7.8	7.8
T_g (°C)	−96	−75	−65	−53	−45
MWD	1.75	2.05	1.85	1.8	1.8
ML 1+4 (100°C)	55	100	110	148	148
M_n	133,000	200,000	200,000	225,000	225,000
Compounded properties					
Rheometer at 150°C					
TS_2	8.0	9.5	10.5	10.3	9.0
T_{50}	15.8	16.0	17.0	17.6	16.7
T_{90}	21.3	22.5	25.0	24.8	25.0
300% modulus (MPa)	6.99	9.30	8.27	10.68	11.02
Elongation (%)	515	515	550	535	510
Die C (kN/m)	36.9	41.1	44.8	42.9	42.9
tan δ at 20°C	0.196	0.207	0.224	0.238	0.213
tan δ at 66°C	0.175	0.161	0.177	0.158	0.140

Source: Day, G. and Futamura, S., *Kautsch Gummi Kunstst.*, 40, 39, 1987. With permission.

and showed that in some cases it could override the effect of polymer T_g (Table 2.12). A compound made with an SBR of 200,000 molecular weight and T_g of −65°C (polymer C) had the same loss tangent at 66°C as a compound made with a 133,000 molecular weight SBR and a T_g of −96°C (polymer A) [12].

Day and Futamura also compared two solution SBRs and a high-vinyl polybutadiene with the same glass transition temperature and similar macrostructures. Cure time was longer, and tensile strength, elongation, and tear strength were poorer as the vinyl content of the polymers increased (Table 2.13 and Figure 2.25).

Finally, all rubber technologists should appreciate that solution SBRs with similar T_g values and styrene and vinyl contents do not necessarily process the same or show exactly the same compounded properties. This is due to macrostructural differences and was illustrated by Kerns and Henning [6] in their study of the effect of synthesis parameters on polymer structure. Some common initiator systems, such as the alkali metal alkoxide-butyllithium system, can behave as a "superbase" and abstract a proton from the backbone of the polymer. This creates a site for branching and results in higher hysteresis in compounds. They developed a method for measuring the ability of different initiators to induce branching and then studied these initiators in a series of high-vinyl SBRs. Branching was characterized by MSR, gel permeation chromatography, and crossover frequency of elastic and storage shear modulus. The relative

TABLE 2.13

Effect of Styrene and Vinyl at Constant T_g

	Polymer		
Properties	F	G	H
Raw polymer properties			
Styrene%	0	17.6	28.3
Vinyl%	54	31	15
T_g (°C)	−52	−51	−51
ML 1+4 (100°C)	97	102	97
M_n	195,000	182,000	165,000
Compounded properties			
Rheometer at 150°C			
TS$_2$	9.5	10	10
T_{50}	20.5	18	17.0
T_{90}	32	27.7	23.3
300% modulus (MPa)	8.96	8.96	8.96
Elongation (%)	465	545	555
Die C tear (kN/m)	40.1	43.6	45.7
tan δ at 20°C	0.226	0.241	0.247
tan δ at 66°C	0.159	0.165	0.171

Source: Day, G. and Futamura, S., *Kautsch Gummi Kunstst.*, 40, 39, 1987. With permission.

order of branching was the same regardless of which technique was used. The data are shown in Table 2.14 and Figures 2.26 and 2.27.

The polymers in runs 1–3 are virtually identical in most ways that are included in a typical specification sheet but would be expected to process differently as evidenced by the Mooney force decay (T_{80}, time to 80% relaxation). The higher value indicates higher branching. Thus, the *n*-butyllithium/sodium *t*-amylate initiator produces a polymer that is more branched than the *n*-butyllithium/sodium dodecylbenzenesulfonate initiator, which in turn is more branched than the polymer produced by the dibutylmagnesium/sodium *t*-amylate initiator. The same effect is seen in the root-mean-square (RMS) radius versus molar mass in that the polymer with the smaller radius at a given molecular weight is more branched than a polymer with a larger radius at that same molecular weight.

Runs 4 and 5 (Figure 2.27) illustrate the microstructural differences between two low-vinyl SBRs. In run 4, the styrene is randomized by use of sodium *t*-amylate/ *n*-butyllithium, whereas in run 5, randomization takes place by controlling the monomer concentration by "distributing" the butadiene in such a way that blocky styrene does not occur. No modifier is used in run 5. The T_{80} for the run with no modifier is only one-third that of the modified run, although the rest of the typical raw polymer properties are virtually identical. The difference in branching is also seen in the RMS radius versus molar mass curve.

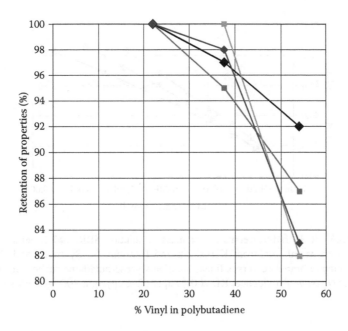

FIGURE 2.25 Effect of vinyl content on solution polymers. (◆) Tangent delta, (■) die C tear strength, (◆) elongation, (■) tensile strength. (Reprinted from Day, G. and Futamura, S., *Kautsch Gummi Kunstst.*, 40, 39, 1987. With permission.)

TABLE 2.14
Characterization and Stress Relaxation of SBR Prepared by Selected Initiators

Sample	Type	M_w	M_n	M_w/M_n	ML 4	T_g (°C)	Styrene (%)	Vinyl (%)	T_{80}
1	n-BuLi/SMT[a]	5.20E+05	2.42E+05	2.15	58	−14	26	53	0.033
2	n-BuLi/SDBS[b]	3.29E+05	1.92E+05	1.71	54	−13	27	51	0.016
3	Bu₂Mg/SMT[c]	2.89E+05	1.82E+05	1.59	57	−12	26	58	0.013
4	n-BuLi/SMT	3.78E+05	1.74E+05	2.17	70	−70	18	15	0.020
5	Dist. feed	3.15E+05	1.98E+05	1.59	68	−72	18	10	0.007

Source: Kerns, M. and Henning, S., Synthesis and rheological characterization of branched versus linear solution styrene-butadiene rubber, in: *Meeting of ACS Rubber Division*, Providence, RI, 2001, Paper 52. With permission.

[a] n-Butyllithium/sodium t-amylate.
[b] n-Butyllithium/sodium dodecylbenzenesulfonate.
[c] Dibutylmagnesium/sodium t-amylate.

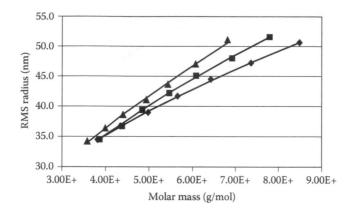

FIGURE 2.26 RMS radius versus molar mass of solution SBRs (◆) 1, (■) 2, and (▲) 3. (Reprinted with permission from Kerns, M. and Henning, S., Synthesis and rheological characterization of branched versus linear solution styrene-butadiene rubber, in: *Meeting of ACS Rubber Division*, Providence, RI, 2001, Paper 52. Copyright 2001 American Chemical Society.)

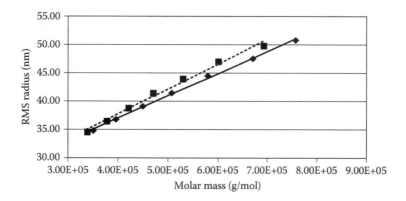

FIGURE 2.27 RMS radius versus molar mass of solution SBRs (◆) 4 and (■) 5. (Reprinted with permission from Kerns, M. and Henning, S., Synthesis and rheological characterization of branched versus linear solution styrene-butadiene rubber, in: *Meeting of ACS Rubber Division*, Providence, RI, 2001, Paper 52. Copyright 2001 American Chemical Society.)

VI.   ZIEGLER–NATTA POLYMERIZATION AND ZIEGLER–NATTA POLYMERS

Ziegler–Natta catalysts are prepared from transition metal compounds (halides, alcoholates, acetylacetonates, or long-chain carboxylic acid salts) and an aluminum alkyl. Ziegler–Natta polymerizations are usually developed for a specific polymer, and it is difficult to modify a specific catalyst to make others. This is in contrast to anionic polymerization, where temperature, modifier, and feed rate can be controlled to make a predictable variety of polymers. Small changes in a Ziegler–Natta catalyst system can have major unpredictable effects on polymer macrostructure and microstructure. Although there are general principles, much catalyst development work is still empirical. The two most important general-purpose elastomers prepared by Ziegler–Natta polymerization are high-*cis* polybutadiene and high-*cis* polyisoprene.

A. MECHANISM OF BUTADIENE POLYMERIZATION

Reaction of a transition metal salt with alkylaluminum in the presence of butadiene can lead to a *syn-* or *anti-π*-allyl complex as shown in Figure 2.28. The thermodynamically less stable *anti* form is the primary reaction product in many systems [82–84] where a preformed π-allyl complex is reacted with a diene substrate.

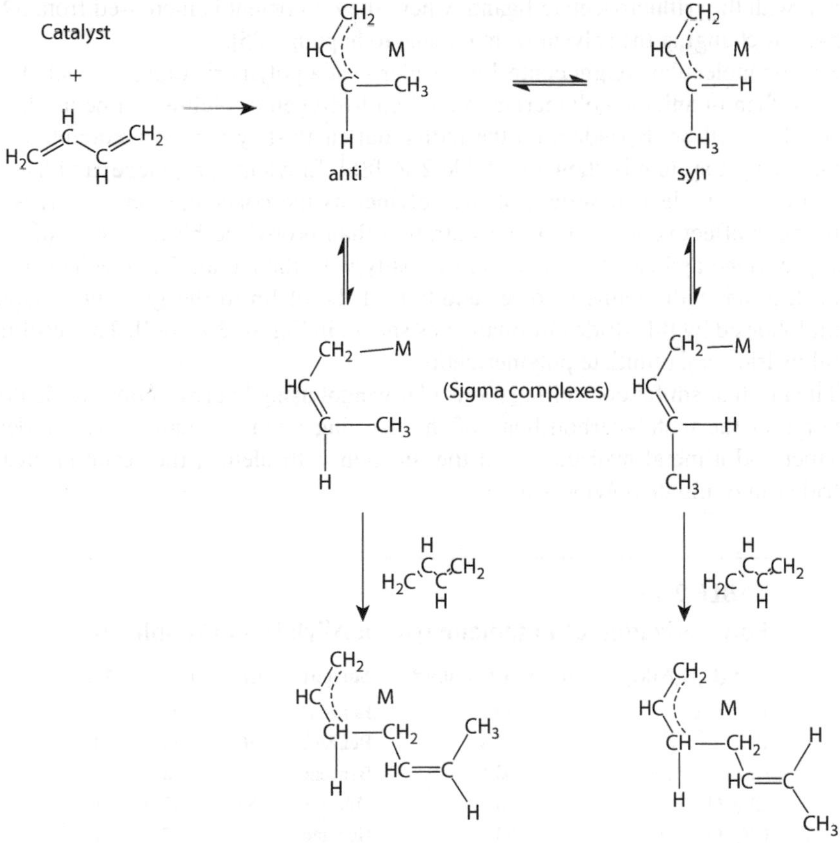

FIGURE 2.28 Mechanism of polymerization of butadiene with a Ziegler–Natta catalyst. (Adapted from Porri, L. and Giarrusso, A., Conjugated diene polymerization, in: Eastmond, G., Ledwith, A., Russo, S., Sigwalt, P., eds., *Comprehensive Polymer Science*, Vol. 4, Part II, Pergamon Press, Oxford, U.K., 1989, p. 93.)

The π-allyl complex is in equilibrium with both the *syn* complex and σ complex, where the metal is directly bonded to one of the carbon atoms. The σ complex will coordinate with another butadiene molecule, and then a new carbon–carbon bond will be formed to the carbon bonded to the metal migrating to the terminus of the newly coordinated butadiene. The microstructure of the polymer is set in this reaction. If the original complex is *anti*, a *cis* double bond will be formed. If the original complex is *syn*, a *trans* double bond will be formed. If the reaction of the *anti* form with monomer is faster than the equilibrium between *syn* and *anti*, a predominantly *cis* polymer will be formed. Ligands (anions from the original transition metal salt, reaction products from alkylaluminum, or added ligands) and solvents play a major role in determining the rate and microstructure of the polymer by complexing with the metal and affecting the various equilibria. This is illustrated in Table 2.15, where a preformed π-allylnickel complex is used to polymerize butadiene. As the anion in the complex varies from iodide to chloride, the rate of polymerization decreases, whereas the *cis* content increases from virtually none to 92%. The solvent effect is shown with the trifluoroacetate ligand, where the *cis* content is increased from 59% to 94% by changing the solvent from toluene to heptane [85].

Precise molecular weight control in Ziegler–Natta polymerizations is much more difficult than in anionic polymerization. Often hydrogen or olefins can be used, but their effect will be dependent on the individual catalyst system. An example for a cobalt catalyst system is shown in Table 2.16 [86]. Ethylene, propylene, and hydrogen lower the molecular weight of the polymer as the concentration is increased. Ethylene is effective at a lower concentration than propylene because it is smaller than propylene and can coordinate more easily with the metal. The mechanism of chain transfer with olefins involves addition of the olefin to the growing polymer chain followed by β-hydride elimination as shown in Figure 2.29 [87]. The resulting metal hydride can reinitiate polymerization.

The mechanism for controlling molecular weight using hydrogen involves hydrogenation of the metal–carbon bond of the growing polymer chain to give a dead polymer and a metal hydride. As in the reaction with olefins, the resulting metal hydride can reinitiate polymerization.

TABLE 2.15
Polymerization of Butadiene with π-Allylnickel Complexes

X in $(C_3H_5NiX)_2$	Turnover Number[a]	Solvent	Cis	Trans	1,2
I	30	Benzene	—	95	5
Br	2.4	Benzene	46	53	3
Cl	0.1	Benzene	92	6	2
CF_3CO_2	50	Toluene	59	40	1
CF_3CO_2	30	Heptane	94	5	1

Source: Adapted from Taube, R. et al., *Makromol. Chem. Macromol. Symp.*, 66, 245, 1993. With permission.

[a] A measure of polymerization rate (mol butadiene/mol Ni)/h.

TABLE 2.16

Effect of Chain Transfer Agents on Intrinsic Viscosity of cis-PBD from the $AlEt_2Cl-CoCl_2-2Py$ Catalyst in Benzene

Transfer Agent	Intrinsic Viscosity (dL/g)
None	5.57
Ethylene (2.1 mol/100 mol butadiene)	2.72
Ethylene (5.1 mol/100 mol butadiene)	1.78
Propylene (72.2 mol/100 mol butadiene)	3.43
Hydrogen (0.5 MPa)	2.95
Hydrogen (2 MPa)	1.37

Source: Porri, L. and Giarrusso, A., Conjugated diene polymerization, in: Eastmond, G., Ledwith, A., Russo, S., Sigwalt, P., eds., *Comprehensive Polymer Science*, Vol. 4, Part II, Pergamon Press, Oxford, U.K., p. 68, 1989.

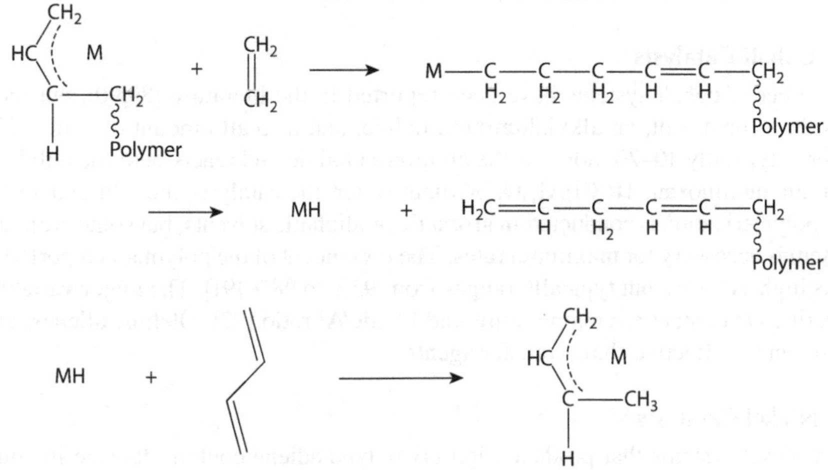

FIGURE 2.29 Chain transfer in a Ziegler–Natta polymerization using an olefin. (From Porri, L. and Giarrusso, A., Conjugated diene polymerization, in: Eastmond, G., Ledwith, A., Russo, S., Sigwalt, P., eds., *Comprehensive Polymer Science*, Vol. 4, Part II, Pergamon Press, Oxford, U.K., p. 68, 1989. With permission.)

It is important to distinguish between a catalyst poison and a chain transfer agent. A catalyst poison may lower the molecular weight but seriously affect conversion. The ideal chain transfer agent will not affect rate or conversion. In addition, the chain transfer agent should not affect the polymer structure, something that is difficult to avoid in Ziegler–Natta polymerizations. Ziegler–Natta diene polymers vary in molecular weight distribution from 2 to over 10. The broader molecular weight distributions are probably caused by multiple active catalytic species that are present in different concentrations and have different propagation or chain transfer rates.

When the distributions for these species are superimposed in the final polymer, the overall distribution is broad. One of the challenges in Ziegler–Natta research is to eliminate the catalytic species that give very-high- and very-low-molecular-weight products, because these adversely affect processing and mechanical properties, respectively.

B. *CIS*-POLYBUTADIENE

1. Titanium Catalysts

The first Ziegler–Natta polybutadiene catalysts were similar to those used for polyolefins. Titanium tetrachloride/trialkylaluminum produced a 65%–70% *cis* polymer at an Al/Ti ratio of about 1 and a mixture of *cis* polymers and *trans*-polybutadiene at higher ratios. Titanium tetraiodide is required to make a high-*cis* polymer (92%–94%) [88]. A gel-free polymer can be made in batch polymerization, but gel slowly builds up in the reactor in continuous polymerization. An aromatic solvent is required to get commercially viable rates [89], which makes it difficult to isolate the polymer in an environmentally responsible manner.

2. Cobalt Catalysts

A number of cobalt systems have been reported in the literature [89,90], but most contain a cobalt salt, an alkylaluminum halide, and a small amount of water. The water is typically 10–20 mol% of the aluminum halide and reacts with the halide to form an aluminoxane $[R(Cl)Al]_2O$. Without water, the catalysts show little activity. The polymerization is conducted in aromatic or aliphatic solvents, but some aromatic content is necessary for maximum rates. The *cis* content of the polymer is reported to be as high as 99.5% but typically ranges from 95% to 98% [91]. The major variables affecting *cis* content are temperature and halide/Al ratio [92]. Olefins, allenes, and hydrogen are effective chain transfer agents.

3. Nickel Catalysts

Most nickel systems that produce high-*cis* polybutadiene contain fluorine in some form. The "Bridgestone" process uses a nickel salt (typically a carboxylate), boron trifluoride etherate, and an alkylaluminum [93] and produces a 96%–98% *cis* polymer. In the patent examples, the polymerization is conducted in aromatic solvents. The mechanism for the polymerization was elucidated much later and is shown in Figure 2.30 [85]. The active catalyst is formed from a nickel(0) compound deposited on aluminum fluoride. The aluminum fluoride is made by the reaction of the alkylaluminum with the boron trifluoride. Butadiene reacts with the nickel(0) compound to form the active π-allyl cationic nickel complex that is coordinated to a polymeric fluoroaluminate anion. The presence of aluminum fluoride is significant because this strong Lewis acid can lead to branching.

The second major type of nickel catalyst is the nickel carboxylate/trialkylaluminum/hydrogen fluoride system [94] developed by Goodyear. This catalyst also produces about 97% *cis* polymer, and rates are high in either aliphatic or aromatic solvents. An HF/trialkylaluminum ratio of 2.5:3.0 gives good yields of

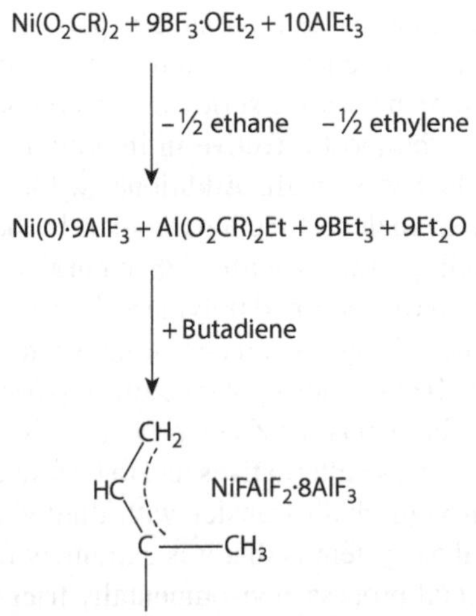

$$Ni(O_2CR)_2 + 9BF_3 \cdot OEt_2 + 10AlEt_3$$

$$\downarrow \quad -\tfrac{1}{2} \text{ ethane} \quad -\tfrac{1}{2} \text{ ethylene}$$

$$Ni(0) \cdot 9AlF_3 + Al(O_2CR)_2Et + 9BEt_3 + 9Et_2O$$

$$\downarrow \quad + \text{Butadiene}$$

FIGURE 2.30 Polymerization of butadiene with the "Bridgestone" nickel catalyst. (From Taube, R. et al., *Makromol. Chem. Macromol. Symp.*, 66, 245, 1993. With permission.)

polymer. As the polymerization temperature increases, the rate increases and the molecular weight decreases. Temperature has little effect on *cis* content. Olefins [95,96] can be used as chain transfer agents. The molecular weight distribution can also be influenced by addition of *p*-styrenated diphenylamine (SDPA) as a catalyst component [97]. Castner [98] showed that *p*-SDPA alters the distribution of alkyl-aluminum fluorides formed by the reaction of the aluminum alkyl with hydrogen fluoride. This catalyst alters the molecular weight distribution so that the resulting polymer has lower cold flow and better mill processability than polybutadiene made with an olefin chain transfer agent. Thermal field flow fractionation chromatography revealed that a high-molecular-weight component that was present in the olefin-modified polymerization was absent when SDPA was used as a modifier. The SDPA polymer was more branched than the control.

4. Neodymium Catalysts

Neodymium catalyst components are typically a neodymium carboxylate, a trialkyl-aluminum, and a chloride source such as an alkylaluminum chloride, although many variations are possible. These catalysts make a relatively linear high-*cis* polybutadiene (98.2%), which has been claimed to give better fatigue to failure than polybutadienes made from cobalt- or titanium-based catalysts [99]. The high linearity has led to processing problems in the tire plants and this had caused suppliers to develop "better processing" grades. These newer grades involve some modification of the polymer macrostructure and are claimed to process as well as the typical cobalt polybutadiene [100].

The *cis* content of neodymium polybutadiene varies with molecular weight and decreases as molecular weight decreases [101]. The polymer molecular weight increases with reaction time, which is in contrast with cobalt or nickel catalysts, whose molecular weight levels out at 40%–50% conversion [102]. This "pseudo-living" character of the polymerization makes it possible to partially functionalize the polymer when the polymerization is complete. Hattori et al. [103] functionalized neodymium high-*cis* polybutadiene with a series of tin compounds. They found that diphenyltin dichloride was the most effective in improving abrasion resistance and modulus, although the effect was small. Additional work by Tanaka et al. of JSR improved the technology to include functionalization by heterocumulenes such as isocyanates. In compounding studies with both natural rubber and SBR, carbon black compounds made with the modified polymers showed 20%–30% improvement in both laboratory wear and rolling resistance testing when compared to compounds containing the unfunctionalized control polybutadiene [104].

The typical chain transfer agents for Ziegler–Natta polymerization are not effective with neodymium catalysts, so alternatives had to be found. The molecular weight is typically controlled through chain transfer with dialkylaluminum hydride. One key advantage to this catalyst system is that it is extremely fast in aliphatic solvents, which makes the commercial process environmentally friendly.

Clearly, with so many different catalyst systems, it is not surprising that the high-*cis* polybutadienes do not all behave the same. Kumar et al. [105] studied a series of commercial high-*cis* polybutadienes and compared their raw polymer properties and performance in different types of tire compounds. A list of the polymers and raw polymer properties is shown in Table 2.17.

Polymers made with cobalt, neodymium, nickel, and titanium were evaluated at approximately the same Mooney viscosity. The *cis* content of the polymers ranged from 96.3% to 97.6%, with the exception of the polymer made with a titanium catalyst, which was 91.6% *cis*. The branching index G is a ratio of measured intrinsic viscosity to theoretical intrinsic viscosity determined from a GPC curve. Polymers with

TABLE 2.17
Comparison of Raw Polymer Properties of Polybutadienes

Polymer	*Cis* Content	Vinyl Content	Branching Index, G	M_w ($\times 10^{-3}$)	M_n ($\times 10^{-3}$)	M_w/M_n	ML 1+4 at 100°C	Mooney Relaxation
Co-BR-1	97.3	1.3	0.75	321	125	2.57	46	9.0
Co-BR-2	97.2	1.4	0.71	303	108	2.81	44	14.0
Co-BR-3	97.2	1.3	0.75	318	131	2.43	45	7.5
Co-BR-4	97.4	1.1	0.88	309	113	2.73	42	11.0
Co-BR-5	97.3	0.9	0.70	338	156	2.17	47	4.5
Co-BR-6	97.0	1.5	0.90	458	125	3.66	48	11.0
Nd-BR-7	97.5	0.8	0.96	412	99	4.16	46	7.5
Nd-BR-8	96.3	0.4	0.92	353	186	2.10	50	5.0
Nd-BR-9	97.6	0.8	0.98	381	103	3.70	42	8.0
Ni-BR-10	96.6	1.2	0.71	368	86	4.28	42	10.0
Ni-BR-11	96.3	2.0	0.80	347	87	3.99	44	9.0
Ti-BR-12	91.6	3.9	0.91	337	126	2.67	44	5.0

Source: Kumar, N. et al., *Int. J. Polym. Mater.*, 34, 91, 1996. With permission.

a branching index close to 1 are more linear (less branched) than those having a lower value. By this measure, the neodymium catalysts produce the most linear polymers of those tested. There is quite a difference in the Mooney relaxation data among the polybutadienes produced with cobalt catalysts, although this difference was not observed in the processing studies the authors conducted. A 60:40 BR/NR truck tread formulation made with neodymium BR was more difficult to process than the same formulation made with the other polymers. In the same formulation, the cobalt-catalyzed polybutadiene showed slightly higher modulus at 300% elongation and higher tensile strength than the other polybutadienes. Heat buildup was slightly higher for polymers made with nickel and titanium, but this is not surprising in light of the relatively low M_n of these polymers. Although some generalizations can be made on the basis of the metal used to prepare the polymer, it is best to look at the macrostructural and microstructural features of a polymer to predict compound properties.

C. SYNDIOTACTIC POLYBUTADIENE

Syndiotactic polybutadiene is a unique material in that it functions as both a plastic and an elastomer. The melting point ranges from 70°C up to approximately 200°C and is controlled by the stereoregularity of the polymerization. There are three commercial grades of material, and their properties are shown in Table 2.18 [106].

The polymer can be made by solution, suspension, or emulsion polymerization. The catalyst for the solution process is made from a cobalt salt, an organic phosphine, a trialkylaluminum, and a small amount of water [107]. The polymerization is conducted at 5°C–10°C in a halogenated hydrocarbon such as methylene chloride. Polymer yields are as high as 95%. Crystallinity is controlled by the type of phosphine used and ranges from 25% to 35% [108]. Triaryl phosphines substituted in the meta position give polymer with higher crystallinity. Molecular weight is controlled by the use of reactive organic (allyl, benzyl, or tertiary) halides or alkylaluminum halides [109]. Syndiotactic polybutadiene with a melting point higher than 126°C is not made on a commercial scale using solution polymerization, probably because of difficulty with precipitation due to higher crystallinity of the polymer.

TABLE 2.18
Characteristics of Commercial Syndiotactic Polybutadiene

Product	Specific Gravity	1,2 Content (%)	Melting Point (°C)
JSR RB 810	0.90	90	71
JSR RB 820	0.91	92	90
JSR RB 830	0.91	93	105
JSR RB 840	0.91	94	126

Source: JSR product data sheet, syndiotactic polybutadiene, Japan Synthetic Rubber Co. With permission.

A slightly different catalyst is used in aqueous polymerization. Reducing a cobalt salt with a trialkylaluminum in the presence of butadiene forms the allylcobalt catalyst component. For suspension polymerization, this component is mixed with butadiene and dispersed in water along with a reagent to control the melting point of the polymer. A number of materials can be used to control the melting point, but *N,N*-dibutylformamide is one of the most efficient [110]. Addition of carbon disulfide to the suspension initiates polymerization [111,112]. For emulsion polymerization, soap is added to the solution of reduced cobalt, and it is emulsified using ultrasound or a high-shear mixer [113,114]. More butadiene followed by carbon disulfide is added to start the polymerization. In both the suspension and emulsion processes, the aqueous dispersion of allylcobalt is remarkably stable prior to addition of the carbon disulfide. The technique is so versatile that it can be used to make syndiotactic polybutadiene within previously formed latex [113]. The dispersion of allylcobalt is absorbed into an SBR, NR, or PBD latex, and butadiene/carbon disulfide is added to start the polymerization. The syndiotactic polybutadiene forms a "needlelike" structure inside the diene latex particle.

Syndiotactic polybutadiene can be used in adhesives, films, gloves, golf balls, and a multitude of other items. A number of uses in tires have been claimed, including treads [115], innerliners [116], adhesives for splices [117], and sidewall decorations [118]. An example of how inclusion of a high-melting syndiotactic PBD into a foamed passenger tread formulation can improve performance is shown in Table 2.19 [115]. The syndiotactic polybutadiene was blended with the carbon black, sulfur, and accelerators and then pulverized. This composite was compounded into the rubber

TABLE 2.19
Syndiotactic Polybutadiene in Foamed Tread Compound

Formulation[a]	Control	Example 1	Example 2	Example 3	Example 4	Example 5
Natural rubber	40	40	40	40	40	40
High-*cis* BR	60	60	60	60	60	60
N-220 black	60	40	50	45	55	60
Foaming agent	5.2	4.6	5.0	4.8	5.2	5.5
Syndiotactic PBD (mp 170°C)	0	20	20	20	20	20
Properties						
Storage modulus (MPa) at −20°C	9.0	5.0	9.0	7.0	15	22
Braking index	100	95	123	125	115	91
Wear index	100	90	101	100	105	108

Source: Teratani, H. and Aoyama, M. (To Bridgestone Corp.), Pneumatic tire with tread of matrix foamed rubber containing resin, U.S. Patent 5,571,350, November 5, 1996.

[a] *Common ingredients*: stearic acid 1.5 phr, zinc oxide 3.0 phr, antioxidant 1.0 phr, accelerator DM2 0.2 phr, vulcanization accelerator CZ3 0.5 phr, sulfur 1.0 phr.

formulation. In tire tests, the composite dramatically improved braking performance without affecting treadwear. Storage modulus of the formulation was a critical factor in performance. Although there is extensive patent literature on the preparation and use of the higher melting syndiotactic polybutadiene (mp > 110°C), there is no commercial source for the material.

D. COMMERCIAL ZIEGLER–NATTA POLYMERS AND PROCESSES

As mentioned earlier, the IISRP does not have a separate category for stereoregular polymers made by Ziegler–Natta polymerization. All of these polymers are classified in the categories shown in Table 2.11.

The process equipment and process steps used for anionic polymerization can also be used to make Ziegler–Natta polymers (see Figure 2.21). As with anionic polymerization, it is important that the solvent and monomer be free from impurities that will interfere with the polymerization. The importance of each impurity is dependent on the specific catalyst system used. Catalyst preparation and addition can be more complicated with Ziegler–Natta systems. Catalysts are either preformed (the metal salt, alkylaluminum, and other additives are mixed prior to being added to the reactor) or made in situ (individual catalyst components are added to the reactor and react in the presence of monomer to form the active catalyst). The way the catalyst is formed can have an important effect on the polymerization. Gel formation is a problem with some Ziegler–Natta systems, so enough reactors are in place that one can be taken off line for cleaning if necessary. Drying and baling the polymer are similar to procedures used for anionic polymers.

E. NEW PROCESSES: GAS-PHASE POLYMERIZATION

A number of companies have experimented with a gas-phase polymerization process to make polybutadiene. This process is used commercially to make polyethylene and polypropylene. Catalyst is put onto a support that is then fluidized with a combination of inert gas and butadiene. Polymerization takes place on the surface of the supported catalyst. There is no solvent, so emissions are minimized and there is no energy cost for solvent recovery. The problem with the process is that the growing particles agglomerate and quickly foul the reactor. This can be avoided by heavily partitioning the particles with something that will be present in the final compound such as carbon black or silica [119]. The difficulty with this approach is that very high levels of black or silica are required to avoid fouling. Gas-phase reactors make large quantities of material at a time, and it would be difficult to make a variety of products with different carbon blacks. This significantly limits the flexibility of the compounder in the choice of reinforcing agents. This process has been used with nickel catalysts [120], cobalt catalysts [121], rare earth (neodymium) catalysts [122], and lithium catalysts [123]. A pilot plant using this type of process to make EPDM was built by Dow (Union Carbide), but has been shut down. The outlook for commercial application for this technology to polybutadiene is bleak.

VII. SUMMARY

General-purpose elastomers have played a critical role in the world economy and will continue to do so for the foreseeable future. Understanding the relationships between polymer microstructure and/or macrostructure and ultimate properties has spurred efforts to control polymer architecture through polymer synthesis. The tools that polymer scientists have provided for tailoring polymers have been an important part of the long-lasting fuel-efficient tires that we ride on today and in the design of rubber articles in general. Future advances in technology will allow for more control over polymer processes and decrease the cost of production. Compounding with general-purpose elastomers will focus on cost reduction with efforts to reduce the number of mixing cycles per batch of final compound. More research into silica compounds and other alternative fillers is expected to push the performance window of the general-purpose elastomers even further.

REFERENCES

1. Love S, Giffels D. *Wheels of Fortune.* Akron, OH: University of Akron Press, 1999, p. 110.
2. Aggarwal S, Hargis I, Livigni R, Fabris H, Marker L. Structure and properties of tire rubbers prepared by anionic polymerization. In: Lai J, Mark JE, eds., *Advances in Elastomers and Rubber Elasticity.* New York: Plenum Press, 1986, pp. 17–36.
3. Veith A. A review of important factors affecting treadwear. *Rubber Chem Technol* 1992;65:601–658.
4. White J, Soos I. Development of elastomer rheological processability quality control instruments. *Rubber Chem Technol* 1993;66:435–444.
5. Male F. Mooney stress testing for SBR processability, *Rubber World* November 1996;215(2):39.
6. Kerns M, Henning S. Synthesis and rheological characterization of branched versus linear solution styrene-butadiene rubber. In: *Meeting of ACS Rubber Division*, Providence, RI, 2001, Paper 52.
7. Oberster A, Bouton T, Valaitis J. Balancing wear and traction with lithium catalyzed polymers. *Angew Makromol Chem* 1973;29/30:291–305.
8. Nordsiek K. The "integral rubber" concept—An approach to an ideal tire tread rubber. *Kautsch Gummi Kunstst* 1985;38:178–185.
9. Saito A. Solution polymerized styrene-butadiene rubber. *Int Polym Sci Technol* 1999;26(6):T19–T28.
10. Moore D, Day G. Comparison of emulsion vs. solution SBR on tire performance. In: *Meeting of ACS Rubber Division*, Washington, DC, 1985, Paper 49.
11. Hall J. (To Bridgestone Corp.) Synthesis of macrocyclic polymers with group IIA and IIB metal cyclic organometallic initiators. U.S. Patent 5,677,399, October 14, 1997.
12. Day G, Futamura S. A comparison of styrene and vinyl butadiene in tire tread polymers. *Kautsch Gummi Kunstst* 1987;40:39–43.
13. Sakakibara M, Tsutsumi F, Hattori I, Hongu Y. Structure and properties of diene rubbers. In: *Meeting of ACS Rubber Division*, Detroit, MI, October 1991, Paper 31.
14. Blackley D. Diene based synthetic rubbers. In: Lovell P, El-Aasser M, eds., *Emulsion Polymerization and Emulsion Polymers.* New York: Wiley, 1997, p. 534.
15. Odian G. *Principles of Polymerization*, 2nd edn. New York: Wiley, 1981, pp. 319–337.
16. Blackley D. Diene based synthetic rubbers. In: Lovell P, El-Aasser M, eds., *Emulsion Polymerization and Emulsion Polymers.* New York: Wiley, 1997, p. 535.

17. Howland L, Messer W, Neklutin V, Chambers V. The effect of low polymerization temperatures on some properties of GR-S vulcanizates. *Rubber Age* 1949;64:459–464.
18. Blackley D. Diene based synthetic rubbers. In: Lovell P, El-Aasser M, eds., *Emulsion Polymerization and Emulsion Polymers*. New York: Wiley, 1997, pp. 532–533.
19. Blackley D. Diene based synthetic rubbers. In: Lovell P, El-Aasser M, eds., *Emulsion Polymerization and Emulsion Polymers*. New York: Wiley, 1997, pp. 531–545.
20. Senyek M, Colvin H. (To Goodyear Tire & Rubber Co.) Polymers derived from a conjugated diolefin, a vinyl-substituted aromatic compound, and olefinically unsaturated nitrile. U.S. Patent 5,310,815, May 10, 1994.
21. Takagishi Y, Nakamura M. (To Nippon Zeon Co.) Diene rubber composition. U.S. Patent 6,114,432, September 5, 2000.
22. Thielen G. Chemically modified emulsion polymers in tire treads. In: *Meeting of the ACS Rubber Division*, Cleveland, OH, October 16–18, 2007, Paper 58.
23. Inoue S, Nishioka K. (To Sumitomo Rubber Industries) Diene polymer and production method thereof. European Patent Application 2511106A1, February 28, 2012.
24. Schneider W, Huybrechts F, Nordsiek K. Process oils in oil extended SBR. *Kautsch Gummi Kunstst* 1991;44:528–536.
25. Directive 2005/69/EC of the European Parliament and of the Council, November 16, 2005.
26. Blackley D. Diene based synthetic rubbers. In: Lovell P, El-Aasser M, eds., *Emulsion Polymerization and Emulsion Polymers*. New York: Wiley, 1997, pp. 544–545.
27. Colvin H, Hardiman C. Answering the challenge of silica mixing. *Rubber World* 2014;249(6):24–30.
28. Lightsey J, Kneiling D, Long J. (To DSM Copolymer Inc.) Process for producing improved silica-reinforced masterbatch of polymers prepared in latex form. U.S. Patent 5,763,388, June 9, 1998.
29. Raines C, Starmer P. (To Zeon Chemicals USA) Free flowing particles of an emulsion polymer having SiO_2 incorporated therein. U.S. Patent 5,166,227, November 24, 1992.
30. Burke O. Elastomer-silica pigment masterbatches and production processes relating thereto. U.S. Patent 3,689,451, September 5, 1972.
31. Burke O. Silica pigments and preparation thereof. U.S. Patent 3,855,394, December 17, 1974.
32. Wallen P, Bowman G, Colvin H, Hardiman C, Reyna J. (To Industrias Negromex and Cooper Tire & Rubber Company) Process for making silane, hydrophobated silica, silica masterbatch and rubber products. U.S. Patent 8,357,7333 B2, January 22, 2013.
33. *The Synthetic Rubber Manual*, 18th edn. International Institute of Synthetic Rubber Producers, Houston, TX, 2012.
34. McGrath J, Wilkes G, Ward R, Broske A, Lee B, Yilgor I, Bradley D, Hoover J, Long T. *New Elastomer Synthesis for High Performance Applications*. Park Ridge, NJ: Noyes Data Corp., 1988, Chapter 5.
35. Arest-Yakubovich A. The kinetics of lithium-initiated anionic polymerization in nonpolar solvents. *J Polym Sci Polym Chem Ed* 1997;35(16):3613–3615.
36. Quirk R, Monroy V. Anionic initiators. In: Kraschwitz J, ed., *Kirk-Othmer Encyclopedia of Chemical Technology*. New York: Wiley, 1995, pp. 461–476.
37. Wakefield B. *The Chemistry of Organolithium Compounds*. Oxford, U.K.: Pergamon Press, 1974.
38. Quirk R, Jang S. Recent advances in anionic synthesis of functionalized elastomers using functionalized alkyllithium initiators. *Rubber Chem Technol* 1996;69:444–461.
39. Schulz D, Halasa A, Oberster A. Anionic polymerisation initiators containing protected functional groups and functionally terminated diene polymers. *J Polym Sci Polym Chem Ed* 1974;12(1):153–166.

40. Shepherd N, Stewart M. (To Secretary of State for Defense for the U.K.) Polymerisation of olefinic-containing monomers employing anionic initiators. U.S. Patent 5,362,699, November 8, 1994.
41. Antkowiak T, Lawson D, Koch R, Stayer M. (To Bridgestone/Firestone, Inc.) Elastomers and products having reduced hysteresis. U.S. Patent 5,153,159, October 6, 1992.
42. Antkowiak T, Lawson D, Koch R, Stayer M. (To Bridgestone/Firestone, Inc.) Methods for preparing functionalized polymer and elastomeric compounds having reduced hysteresis. U.S. Patent 5,354,822, October 11, 1994.
43. Cheng T. Anionic polymerization. VII. Polymerization and copolymerization with lithium-nitrogen-bonded initiator. In: McGrath J, ed., *Anionic Polymerization, Kinetics, Mechanisms and Synthesis*. ACS Symposium Series No. 166. Washington, DC: American Chemical Society, 1981, p. 513.
44. Lawson D, Brumbaugh D, Stayer M, Schreffler J, Antkowiak T, Saffles D, Morita K, Ozawa Y, Nakayama S. Anionic polymerization of dienes using homogeneous lithium amide initiators. *Polymer Preprints* 1996;37(2):728–729.
45. Lawson D, Brumbaugh D, Stayer M, Schreffler J, Antkowiak T, Saffles D, Morita K, Ozawa Y, Nakayama S. Anionic polymerization of dienes using homogeneous lithium amide initiators, and determination of polymer-bound amines. In: Quirk R, ed., *Applications of Anionic Polymerization Research*. ACS Symposium Series No. 696. Washington, DC: American Chemical Society, 1998, pp. 77–87.
46. Lawson D, Morita K, Ozawa Y, Stayer M, Fujio R. (To Bridgestone Corp.) Soluble anionic polymerization initiators and preparation thereof. U.S. Patent 5,329,005, July 12, 1994.
47. Lawson D, Antkowiak T, Hall J, Stayer M, Schreffler J. (To Bridgestone Corp.) Alkyllithium compounds containing cyclic amines and their use in polymerization. U.S. Patent 5,496,940, March 5, 1996.
48. Morita K, Nakayama A, Ozawa Y, Fujio R. (To Bridgestone Corp.) Process for preparing a polymer using lithium initiator prepared by in situ preparation. U.S. Patent 5,625,017, April 29, 1997.
49. Hergenrother W, Bethea T, Doshak J. (To Bridgestone/Firestone.) Tin containing elastomers and products having reduced hysteresis properties. U.S. Patent 5,268,439, December 7, 1993. See also Hergenrother W, Bethea T. (To Bridgestone Corp.) In the synthesis of tributyltin lithium. U.S. Patent 5,877,336, March 2, 1999.
50. Hsieh H, Quirk R. *Anionic Polymerization: Principles and Practical Applications*. New York: Marcel Dekker, 1996, pp. 173–180.
51. Morton M. *Anionic Polymerization: Principles and Practice*. New York: Academic Press, 1983.
52. McGrath J, ed. *Anionic Polymerization: Kinetics, Mechanisms and Synthesis*. ACS Symposium Series No. 166. Washington, DC: American Chemical Society, 1981, Chapter 5.
53. Gerbert W, Hinz J, Sinn H. Umlagerungen bei der durch lithiumbutyl initiierten Polyreaktion der Diene Isopren und Butadien. *Die Makromol Chem* 1971;144:97–115.
54. Wosfold D, Bywater S. Lithium alkyl initiated polymerization of isoprene. Effect of cis/trans isomerization of organolithium compounds on polymer microstructure. *Macromolecules* 1978;11:582–586.
55. Glaze W, Hanicak J, Moore M, Chaudhuri J. 3-Neopentylallyllithium 1. The 1,4-addition of tert-butyllithium and 1,3-butadiene. *J Organomet Chem* 1972;44:39–48.
56. Halasa A, Lohr D, Hall J. Anionic polymerisation to high vinyl polybutadiene. *J Polym Sci Polym Chem Ed* 1981;19(6):1357–1360.
57. Antkowiak T, Oberster A, Halasa A, Tate D. Temperature and concentration effects on polar-modified alkyllithium polymerizations and copolymerizations. *J Polym Sci A-1* 1972;10:1319–1334.

58. Hsieh H, Glaze W. Kinetics of alkyllithium initiated polymerizations. *Rubber Chem Technol* 1970;43:22–73.
59. Hsieh H, Wofford C. Alkyllithium and alkali metal *tert*-butoxide as polymerization initiator. *J Polym Sci Part A-1* 1969;7:449–460.
60. Langer A. (To Esso Research and Engineering.) Polymerization catalyst and uses thereof. U.S. Patent 3,451,988, June 24, 1969.
61. Halasa A, Hsu W. (To Goodyear Tire & Rubber Co.) Synthesis of high vinyl rubber. U.S. Patent 6,140,434 October 31, 2000.
62. Quirk R, Yin J, Guo S, Hu X, Summers G, Kim J, Zhu L, Schock L. Anionic synthesis of chain-end functionalized polymers. *Makromol Chem Macromol Symp* 1990;32:47–59.
63. Hattori I, Shimada N, Oshima N, Sakakibara M, Mouri H, Fujimaki T, Hamada T. (To Japan Synthetic Rubber and the Bridgestone Corp.) Rubber compositions from modified *trans*-polybutadiene and rubber for tires. U.S. Patent 5,017,636, May 21, 1991.
64. Akita S, Namizuka T. (To Nippon Zeon Co.) Rubber composition. U.S. Patent 4,550,142, October 29, 1985.
65. Noguchi K, Yoshioka A, Komuro K, Ueda A. Structure and properties of newly developed chemically modified high vinyl polybutadiene and solution polymerized styrene-butadiene rubbers. In: *Meeting of ACS Rubber Division*, New York, April 8–11, 1986, Paper 36.
66. Nagata N, Kobatake T, Watanabe H, Ueda A, Yoshioka A. Effect of chemical modification of solution-polymerized rubber on dynamic mechanical properties in carbon-black-filled vulcanizates. *Rubber Chem Technol* 1987;60:837–855.
67. Kawanaka N, Yosioka A, Nagata N, Watanabe H. Analysis of chain end modified rubber structure. In: *Meeting of ACS Rubber Division*, Detroit, MI, 1989, Paper 118.
68. Tsutsumi F, Sakakibara M, Oshima N. Structure and dynamic properties of solution SBR coupled with tin compounds. *Rubber Chem Technol* 1990;63:8–22.
69. Quiteria V, Sierra C, Fatou J, Galan C, Fraga L. Tin coupled SBRs. Relationship between coupling type and properties. In: *Meeting of ACS Rubber Division*, Cleveland, OH, 1995, Paper 78.
70. Hsu W, Halasa A. (To Goodyear Tire & Rubber Co.) Rubbers having improved interaction with silica. U.S. Patent 5,652,310, July 29, 1997.
71. Gorce J, Labauze G. (To Compagnie Generale des Establissements Michelin-Michelin & Cie.) Functional diene polymers, their method of preparation and their use in silica-filled elastomeric compositions which can be used for tires. U.S. Patent 5,665,812, September 9, 1997.
72. Saito A, Yamada H, Matsuda T, Kubo N, Ishimura N. Improvement of rolling resistance of silica tire compounds by modified S-SBR. In: *Meeting of ACS Rubber Division*, Savannah, GA, 2002, Paper 39.
73. Firestone Tire and Rubber Co. Method of polymerizing butodiene and monomer mixtures containing same. U.K. Patent 1,142,101, 1968.
74. Adams H, Bebb R, Eberly K, Johnson B, Kay E. Stereopolymerization of butadiene and styrene in the presence of acetylenes and ketones. *Kautsch Gummi Kunstst* 1965;18:709–716.
75. Puskas J. Investigation of the mechanism of chain termination and transfer by 1,2-butadiene in the butyllithium-initiated polymerization of 1,3-butadiene in non-polar solvents. *Makromol Chem* 1993;194:187–195.
76. Halasa A, Hsu W, Zhou J, Jasiunas C, Yon C. (To the Goodyear Tire & Rubber Co.) Functionalized monomers for synthesis of rubbery polymers U.S. Patent 6,933,358, August 23, 2005.
77. Alwardt C, Brace L, Francik W, Ramanathan A, Rodewald S, Wong T. In chain functionalized polybutadiene. In: *Meeting of the ACS Rubber Division*, Louisville, KY, March 24–26, 2014.

78. Romani F, Passaglia E, Aglietto M, Ruggeri G. Functionalization of SBR copolymer by free radical addition of thiols. *Macromol Chem Phys* 1999;200:524–530.
79. Passaglia E, Donati F. Functionalization of a styrene/butadiene random copolymer by radical addition of L-cysteine derivatives. *Polymer* 2007;48:35–42.
80. Keckler N. (To The Firestone Tire & Rubber Co.) Process for the production of copolymers free of block polystyrene. U.S. Patent 3,558,575, January 26, 1971.
81. Colvin H, Douglas J, Singh H, Snider M, Steinhauser N, Hardy D, Gross T. Influence of SSBR microstructure on silica mixing and compound performance. In: *Meeting of ACS Rubber Division*, Louisville, KY, March 24–26, 2014.
82. Tolman C. Chemistry of tetrakis(triethyl phosphite)nickel hydride, HNi [P(OEt)$_3$]$_4$$^+$. II. Reaction with 1,3-butadiene. Catalytic formation of hexadienes. *J Am Chem Soc* 1970;92(23):6777–6784.
83. Kormer V, Lobach M, Klepikova V, Babitskii B. Stereochemical control of 1,3-diene polymerization. *J Polym Sci Polym Lett Ed* 1976;14:317–322.
84. Kormer V, Lobach M. NMR studies of polymerization of 1,3-dienes with bis (π-crotylnickel iodide). *Macromolecules* 1977;10:572–579.
85. Taube R, Schmidt U, Gehrke J, Bohme P, Langlotz J, Wache S. New mechanistic aspects and structure activity relationships in the allyl nickel complex catalysed butadiene polymerization. *Makromol Chem Macromol Symp* 1993;66:245–260.
86. Porri L, Giarrusso A. Conjugated diene polymerization. In: Eastmond G, Ledwith A, Russo S, Sigwalt P, eds., *Comprehensive Polymer Science*, Vol. 4, Part II. Oxford, U.K.: Pergamon Press, 1989, p. 68.
87. Porri L, Giarrusso A. Conjugated diene polymerization. In: Eastmond G, Ledwith A, Russo S, Sigwalt P, eds., *Comprehensive Polymer Science*, Vol. 4, Part II. Oxford, U.K.: Pergamon Press, 1989, p. 88.
88. Phillips Petroleum Co. Polymerization process and product. Br Patent 848,065, 1960.
89. Cooper W. Polydienes by coordination catalysts. In: Saltman W, ed., *The Stereo Rubbers*. New York: Wiley, 1977, pp. 21–78.
90. Porri L, Giarrusso A. Conjugated diene polymerization. In: Eastmond G, Ledwith A, Russo S, Sigwalt P, eds., *Comprehensive Polymer Science*, Vol. 4, Part II. Oxford, U.K.: Pergamon Press, 1989, pp. 53–108.
91. Horne S. Diene polymerization: Some facts and unsolved problems. In: Quirk R, ed., *Transition Metal Catalyzed Polymerization, Alkenes and Dienes*, Vol. 4, Part B. New York: Harwood Academic, 1983, p. 527.
92. Porri L, Giarrusso A. Conjugated diene polymerization. In: Eastmond G, Ledwith A, Russo S, Sigwalt P, eds., *Comprehensive Polymer Science*, Vol. 4, Part II. Oxford, U.K.: Pergamon Press, 1989, p. 66.
93. Ueda K, Onishi A, Yoshimoto T, Maeda K, Yokohama T, Hosono J, Yokohama K. (To Bridgestone Tire Co.) Production of *cis*-1,4 polybutadiene with a Ni-BF$_3$ etherate-AlR$_3$ catalyst. U.S. Patent 3,170,904, February 23, 1965.
94. Throckmorton M, Farson F. An HF-nickel-R$_3$Al catalyst system for producing high *cis*-1,4-polybutadiene. *Rubber Chem Technol* 45:268–277.
95. Donbar K, Saltman W, Throckmorton M. (To Goodyear Tire & Rubber Co.) Controlling the molecular weight of polybutadiene. U.S. Patent 5,698,643, December 16, 1997.
96. Castner K. (To Goodyear Tire & Rubber Co.) Molecular weight regulation of *cis*-1,4-polybutadiene. U.S. Patent 4,383,097, May 10, 1983.
97. Castner K. (To Goodyear Tire & Rubber Co.) Technique for reducing the molecular weight and improving the processability of *cis*-1,4-polybutadiene. U.S. Patent 5,451,646, September 19, 1995.
98. Castner K. Improved processing of *cis*-1,4-polybutadiene. In: *Meeting of ACS Rubber Division*, Chicago, IL, 1999, Paper 3.

99. Lauretti E, Miani B, Mistrali F. Improving fatigue resistance with neodymium polybutadiene. *Rubber World* May 1994;210:34–37.
100. Saeid K, Kloppenburg H. Processing high end tire rubber made easier: Just 30 phr changes everything. *Rubber World* 2014;249(5):23–26.
101. Porri L, Giarrusso A. Conjugated diene polymerization. In: Eastmond G, Ledwith A, Russo S, Sigwalt P, eds., *Comprehensive Polymer Science*, Vol. 4, Part II. Oxford, U.K.: Pergamon Press, 1989, p. 69.
102. Hsieh H, Yeh H. Polymerization of butadiene and isoprene with lanthanide catalysts; characterization and properties of homopolymers and copolymers. *Rubber Chem Technol* 1985;58:117–145.
103. Hattori I, Tsutsumi F, Sakakibara M, Makino K. Modification of neodymium high *cis*-1,4-polybutadiene with tin compounds. *J Elastomers Plast* 1991;23:135–151.
104. Tanaka R, Kouichirou T, Sone T, Tadaki T. (To JSR Corporation) Method for producing modified conjugated diene polymer, modified conjugated diene polymer and rubber composition. U.S. Patent 8,404,785, March 26, 2013.
105. Kumar N, Chandra A, Mukhopadhyay R. A correlation between micro and macrostructure of high *cis*-polybutadiene and its performance in tyre compound. *Int J Polym Mater* 1996;34:91–103.
106. JSR product data sheet, syndiotactic polybutadiene. Japan Synthetic Rubber Co.
107. Makino K, Ishikawa T, Komatsu K. (To Japan Synthetic Rubber Co.) Process for the production of 1,2-polybutadiene with regulated molecular weight. U.S. Patent 4,176,219, November 27, 1979.
108. Makino K, Komatsu K, Takeuchi Y, Endo M. (To Japan Synthetic Rubber Co.) Process for the preparation of 1,2-polybutadiene. U.S. Patent 4,182,813, January 8, 1980.
109. Makino K, Miyabayashi T, Ohshima N, Takeuchi Y. (To Japan Synthetic Rubber Co.) Process for the preparation of 1,2-polybutadiene. U.S. Patent 4,255,543, March 10, 1981.
110. Burroway G. (To Goodyear Tire & Rubber Co.) Syndiotactic 1,2-polybutadiene synthesis in aqueous medium utilizing *N,N*-dibutylformamide as a modifier. U.S. Patent 5,405,816, April 11, 1995.
111. Henderson J, Donbar K, Barbour J, Bell A. (To Goodyear Tire & Rubber Co.) Microencapsulated aqueous polymerization catalyst. U.S. Patent 4,506,031, March 19, 1985.
112. Henderson J, Donbar K, Barbour J, Bell A. (To Goodyear Tire & Rubber Co.) Microencapsulated aqueous polymerization catalyst. U.S. Patent 4,429,085, January 31, 1984.
113. Ono H, Ito N, Kassi K, Sakurai N, Okuya E. (To Japan Synthetic Rubber Co.) Polymer particles and process for producing the same. U.S. Patent 4,742,137, May 3, 1988.
114. Burroway G. (To Goodyear Tire & Rubber Co.) Syndiotactic 1,2-polybutadiene latex. U.S. Patent 4,902,741, February 20, 1990.
115. Teratani H, Aoyama M. (To Bridgestone Corp.) Pneumatic tire with tread of matrix foamed rubber containing resin. U.S. Patent 5,571,350, November 5, 1996.
116. Sandstrom P, Maly N, Marinko M. (To Goodyear Tire & Rubber Co.) Tire compounds containing syndiotactic-1,2-polybutadiene. U.S. Patent 4,790,365, December 13, 1988.
117. Tuttle J, Vannan F, Head W. (To Goodyear Tire & Rubber Co.) Methods of securing splices in curable rubber articles. U.S. Patent 5,824,383, October 20, 1998.
118. Gartland R, Finelli A, Bell A. (To Goodyear Tire & Rubber Co.) Tire having decorative appliqué on sidewall and method for preparing same. U.S. Patent 5,058,647, October 22, 1991.
119. Rhee S, Baker E, Edwards D, Lee K, Moorhouse J, Scarola L, Karol F. (To Union Carbide Chemicals and Plastics Co.) Process for producing sticky polymers. U.S. Patent 4,994,534, February 19, 1991.

120. Calderon N, Muse J, Colvin H, Castner K. (To Goodyear Tire & Rubber Co.) Vapor phase synthesis of rubbery polymers. U.S. Patent 5,859,156, January 12, 1999.
121. Windisch H, Sylvester G. (To Bayer AG.) Supported cobalt catalyst, production thereof and use thereof for the polymerization of unsaturated compounds. U.S. Patent 6,093,674, July 25, 2000.
122. Windisch H, Sylvester G, Taube R, Maiwald S. (To Bayer AG.) Compounds of the rare earths and their use as polymerization catalysts for unsaturated compounds. U.S. Patent 6,284,697 B1, September 4, 2001.
123. Brady M, Cann K, Dovedytis D. (To Union Carbide Chemicals and Plastics Technology Corp.) Gas phase anionic polymerization of dienes and vinyl-substituted aromatic compounds. U.S. Patent 5,728,782, March 17, 1998.

3 Special-Purpose Elastomers

Sudhin Datta and Syamal S. Tallury

CONTENTS

I.   Introduction to Specialty Elastomers..83
II.   Need for Specialty Synthetic Elastomers 85
III. Vulcanization of Specialty Elastomers..89
IV. Synthesis and Manufacturing Process..89
V.   Solvent-Resistant Specialty Elastomers..92
     A. NBR Derivatives..94
     B. Polychloroprene...95
     C. Acrylic Elastomers 97
     D. Chlorosulfonated Polyethylene..98
     E. Ethylene–Acrylic Elastomers 99
VI. Temperature-Resistant Elastomers 100
     A. Ethylene–Propylene Copolymers and
         Ethylene–Propylene–Diene Terpolymers.....................................101
     B. Silicone Rubber 102
     C. Fluorocarbon Elastomers...104
     D. Phosphazenes...105
     E. Polyethers 105
     F. Ring-Opened Polymers 106
VII. Olefin Block Copolymers...106
VII. Summary 108
References..108

I. INTRODUCTION TO SPECIALTY ELASTOMERS

The worldwide production capacity for all synthetic elastomers was around 15,500 kilotons [1] per year (kt/yr) in 2012 and is projected to grow at 3.4% in the next 3 years. Historically, the split between natural and synthetic elastomers has ranged from 40% to 50% natural rubber (NR), though the level of NR being used has been decreasing due to the emergence of new solution-polymerized elastomers and special-purpose polymers. Specialty elastomers make up a small but important portion of the arena of synthetic elastomers and consist of elastomers that have an annual estimated consumption of about 1000 kt/yr or less. These elastomers occupy critical application areas principally because they have one or more unique attributes. This distinction in the attributes of the specialty elastomers compared to general-purpose rubbers (GPRs) and the structural features responsible for this distinction are illustrated in Table 3.1. Morton [2] and Dick [3] compiled excellent reviews of special-purpose elastomers. The purpose of this chapter is therefore to provide a brief overview of this important class of polymers and, as appropriate, provide references for further reading.

It is important to realize that specialty elastomers have a variety of components and structures that have been selected to provide key performance attributes. These compositions have been recognized by the ISO 1629 nomenclature procedures (Table 3.2) and have been commonly used to distinguish them. For example, the word "rubber" is inserted after the name of the monomers from which the rubber was prepared and is applicable to elastomers that contain high levels of unsaturation. Polybutadiene is thus denoted as butadiene rubber (BR) and styrene–butadiene copolymer as styrene–butadiene rubber (SBR). Elastomers that have substituted carboxylic acid groups (–COOH) on the polymer chain begin with an X. For example, carboxylated chloroprene rubber would be denoted as XCR. Rubbers containing a

---

TABLE 3.1
Specialty Elastomers: Estimated Volumes in 2001 and Key Attributes

Elastomer[a]	Key Attribute	Structural Feature	Estimated Volume (kt/yr)
EPM, EPDM	Environmental resistance	Saturated backbone	1200
Plastomer	Environmental resistance; compatibility with iPP	Saturated backbone; higher α-olefin copolymer	1000
CR	Resistance to hydrocarbon solvents	Chlorination of isoprene rubber	310
CSM	Strong resistance to solvents	Ionic interactions and chlorinated backbone	280
EAM	Resistance to hydrocarbon solvents; heat resistance	Ethylene backbone with functionality	75
HNBR	Environmental resistance; oxidation resistance	Saturated backbone; nitrile groups by copolymerization	45
ACM	Oxidation resistance	Saturated backbone	8
Silicone (Q)	Wide temperature service	Stable silicon–oxygen bonds, low T_g for polymers	42
Polyether (O)	Excellent low-temperature properties	Ether linkage	11
Fluoroelastomers	Chemical resistance	Chemically unreactive carbon–fluorine bonds	8

Source: De, S.K. and White, J.R., eds., *Rubber Technologist's Handbook*, Vol. 1, iSmithers Rapra Publishing, Shawbury, U.K., 2001. With permission.

[a] EPM, ethylene–propylene copolymer; EPDM, ethylene–propylene–diene terpolymer; CR, chloroprene rubber; CSM, chlorosulfonated polyethylene; EAM, ethylene-vinyl acetate copolymer; HNBR, hydrogenated acrylonitrile–butadiene rubber; ACM, acrylic rubber; Q, polysiloxane rubber; O, oxygenated rubber.

TABLE 3.2

Generic Nomenclature of Synthetic Elastomers

M	Saturated polymethylene chain
N	Nitrogen on the polymer chain
O	Carbon and oxygen in the polymer chain
R	Unsaturated carbon chain
Q	Silicon and oxygen in the polymer chain
T	Carbon, oxygen, and sulfur in the polymer chain (polysulfide elastomers)
U	Carbon, nitrogen, and oxygen in polymer

Source: De, S.K. and White, J.R., eds., *Rubber Technologist's Handbook*, Vol. 1, iSmithers Rapra Publishing, Shawbury, U.K., 2001. With permission.

halogen begin with the element employed. Bromobutyl rubber is therefore denoted as bromo-isobutene-isoprene (bromobutyl) rubber (BIIR). Chlorobutyl rubber is described as chloro-isobutene-isoprene (chlorobutyl) rubber (CIIR) [4]. Table 3.3 presents the nomenclature for over 20 commercial elastomers [5].

II. NEED FOR SPECIALTY SYNTHETIC ELASTOMERS

A wide variety of specialty synthetic elastomers have been developed to overcome some of the performance deficiencies of NR and the larger-volume GPRs such as SBR and BR. Some of these deficiencies are

1. Poor resistance to light, oxygen, and ozone weathering
2. Relatively poor heat resistance
3. Poor resistance to organic fluids

A more recent innovation is the replacement of larger-volume GPRs by specialty synthetic elastomers to exactly match the performance of the elastomer to the intended use. This is mostly due to the ease of manipulation of the structure of the specialty elastomers in low-volume applications, because corresponding changes for GPR and NR on a small scale are economically unattractive.

Specialty elastomers are valued for particular properties or combinations of properties typically unavailable in the large-volume elastomers. This premium on property enhancement affords the polymer designer considerable latitude in the selection of monomer(s) and polymer chain architecture as well as the process to be used for the synthesis. Thus, specialty elastomers typically consist of more than one monomer in an effort to deliver a combination of properties not available from a single monomer. A notable example of this trend is the development of ethylene–propylene copolymer (EPM) elastomers, which contain both ethylene and propylene as the principal monomers. In these copolymers, the absence of propylene would lead to crystalline polyethylene, whereas the absence of ethylene would lead to thermally unstable polypropylene.

TABLE 3.3

International Abbreviations for Elastomers

ACM	Copolymer of ethyl acrylate or other acrylate plus a low level of other unsaturated monomer for vulcanization
AU	Polyester polyurethane
BR	Butadiene rubber
BIIR	Bromo-isobutene–isoprene (bromobutyl) rubber
CFM	Polychlorotrifluoroethylene
CIIR	Chloro-isobutene–isoprene (chlorobutyl) rubber
CM	Chloropolyethylene
CR	Chloroprene rubber
CSM	Chlorosulfonyl polyethylene
EAM	Ethylene-vinyl acetate copolymer
EPDM	Terpolymer of ethylene, propylene, and a diene with unsaturation to facilitate vulcanization
EPM	Ethylene–propylene copolymer
EU	Polyether polyurethane
HNBR	Hydrogenated acrylonitrile–butadiene rubber
IIR	Isobutene–isoprene rubber (butyl rubber)
IM	Isobutene
IR	Polyisoprene (synthetic)
MQ	Silicone rubber with methyl groups on the polymer chain (e.g., dimethyl polysiloxane)
NBR	Acrylonitrile–butadiene rubber
E-SBR	Emulsion styrene–butadiene rubber
S-SBR	Solution styrene–butadiene rubber
XNBR	Carboxylated acrylonitrile–butadiene rubber
XSBR	Carboxylated styrene–butadiene rubber
YAU	Thermoplastic polyester polyurethane
YEU	Thermoplastic polyether polyurethane
YSBR	Block copolymer of styrene and butadiene
YSIR	Block copolymer of styrene and isoprene

Source: De, S.K. and White, J.R., eds., *Rubber Technologist's Handbook*, Vol. 1, iSmithers Rapra Publishing, Shawbury, U.K., 2001. With permission.

There are six distinct and important tools employed in matching the structure of the synthetic specialty elastomer to its intended use:

1. The composition of the elastomer
2. The microstructure and orientation of the monomers
3. The use of a combination of monomers
4. Segregation of the different monomers into portions of a single chain (block copolymers)

5. The architecture of the elastomer as defined by the distribution of composition and molecular weight
6. The use of long-chain branching to improve the fabrication and processing of elastomers

The use of the composition of the synthetic specialty elastomer as a tool in matching the properties of the polymer is shown in Figure 3.1. Two of the most important properties of elastomers are the ability to withstand weathering due to oxygen, ozone, and light (as shown by the continuous-use temperature) and the ability to withstand organic fluids. Figure 3.1 shows graphically the temperature resistance of a variety of synthetic elastomers. Elastomers with a saturated backbone are more resistant to weathering than those with unsaturated backbones. Thus, EPM is significantly better than BR or NR. The ultimate weathering is for elastomers where the C–C backbone is replaced with the Si–O–Si backbone of the Q elastomers. Figure 3.1 also shows the ordering of these elastomers with respect to resistance to organic solvents. Elastomers with strong polarity are more resistant to polar organic fluids and oils than those that are composed entirely of hydrocarbons. Thus nitrile (acrylonitrile–butadiene rubber [NBR]), acrylic rubber (ACM), and fluorinated rubber (fluorocarbon elastomer [FKM]) are more solvent resistant than SBR, ethylene–propylene–diene terpolymer (EPDM), or butyl rubber (IIR). Chlorinated polymers, such as polychloroprene (CR), which are intermediate in polarity, are intermediate between these extremes.

The microstructure and the orientation of the monomer units in the synthetic elastomer can be altered by the choice of reaction conditions and catalysts. NR exists as a single, naturally derived isomer. It is challenging to manipulate the properties of NR by changing the stereochemistry. *cis*-1,4-Isoprene rubber (IR) is the synthetic analog of NR, whereas its isomeric form *trans*-1,4 IR is a tough,

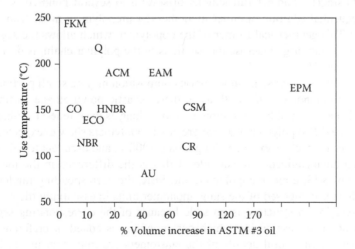

FIGURE 3.1 Graphical correlation of solvent and temperature resistance of elastomers. For abbreviations see text.

semicrystalline polymer. IR polymers containing intermediate amounts of these isomers display intermediate properties. A similar example is the difference between isotactic and atactic polypropylene. These isomers differ in the stereochemical orientation of the methyl group on the propylene monomer. The former is a thermoplastic, whereas the latter can be used as an elastomer. In general, the ability of a polymer to exhibit elastomeric properties depends on the flexibility of rotation around its backbone. Rigid stereochemical isomers often crystallize in ordered crystals, which prevent these polymers from exhibiting elastomeric properties.

A number of synthetic specialty elastomers contain two or more monomers. EPM, which contains ethylene and propylene, and SBR, which contains styrene and butadiene, are examples. The mixture of monomers is used to lower either the glass transition temperature (T_g) (for SBR) or the crystallinity (for EPM). In addition, the presence of the two monomers allows a tailoring of the properties of the synthetic elastomer that cannot be attempted with homopolymers such as NR. Normally, the two monomers are randomly mixed in the copolymer. The resulting copolymer, which contains a mixture of the two monomers, has a compositional microstructure determined by the reactivity ratio. The compositional microstructure is the number of monomer units in runs of one, two, three, and more. The reactivity ratio (which could be defined as the relative tendency of the monomers to cross-react) for a pair of monomers A and B is indicated by r_{AB} and is the product $(R_{AA}R_{BA}/R_{AB}R_{BB})$, where R_{IJ} is the kinetic rate constant for the insertion of the J monomer into a growing polymerization chain terminated by the I monomer. The reactivity ratio is largely determined by the structure of the polymerization catalyst and the monomers as well as by the physical conditions of the polymerization such as temperature and concentrations. Changes in the reactivity ratio rarely lead to run lengths of any monomer greater than 7–10 units in length for an equimolar copolymer. A more direct way to control the polymerization sequence or the run length of a particular monomer leading to single-monomer run lengths of several to several hundred is the practice of sequential addition of either monomers or prepolymers made from a single monomer. This geometrical isomer of the copolymer, which allows the segregation of monomers into large macroscopic sections of the polymer chain, is designated a block polymer [6].

These sections of the same monomeric composition may be small portions of the chain several monomers in length as in thermoplastic polyurethane elastomers or sections that are several hundred monomer units long as in styrene-block-butadiene-block-styrene (SBS) polymer. These segregated copolymers show unexpected elastomeric properties such as extreme elongation (~1000%) and excellent tensile strength even when unvulcanized. An example of this is the difference in the mechanical properties of SBS, a block copolymer, and SBR, the corresponding random copolymer, which is composed of the same monomer units in essentially the same ratio. This segregation in structure is possible because of the corresponding segregated polymerization process during which one monomer is added in preference to the other. These architectural details of the elastomers are important in fulfilling the needs of particular applications and cannot be achieved for larger-volume rubbers, for which neither the selection of the monomer nor the polymerization condition can be altered.

The use of copolymers in specialty elastomers often leads to additional structural features such as intermolecular compositional differences by inducing differences in the comonomer distribution within the copolymer. These compositional differences typically lead to improvements in the mechanical properties of the vulcanized copolymers beyond those that are accessible for the average composition in the absence of compositional differences.

Copolymerization of two or more monomers can lead to several geometric isomers of the same polymer. The practical specialty polymers are variations of three limiting cases:

1. *Alternating polymerization.* The two types of monomer units alternate in the polymer chain until the monomer in the minor concentration is exhausted.
2. *Statistical polymerization.* The two monomers enter into a polymer chain in a statistically random manner, with their average concentration in the chain corresponding to their feed ratio.
3. *Block polymerization.* Complete polymerization of one monomer occurs prior to polymerization of the second monomer.

III. VULCANIZATION OF SPECIALTY ELASTOMERS

Elastomers are vulcanized after compounding with fillers, plasticizers, curatives, accelerators, and minor amounts of antioxidants. The vulcanization reaction is conducted by heating the formed or extruded rubber part to a temperature such that the cross-linking reaction is initiated. Most synthetic elastomers have specific curing sites, formed by the introduction of a special monomer, where the cross-linking occurs. These special monomers include isoprene for isobutylene–isoprene rubber (IIR), halogenation of IIR to form CIIR or BIIR, diene for EPDM, and alkenoic acids for ethylene acrylate rubber (ethylene-vinyl acetate copolymer [EAM]). Because the number of cross-linking sites is limited compared to GPR, the conditions and chemistry for cross-linking are typically more severe for synthetic elastomers. The reaction conditions are maintained until the cross-link density is significant before the temperature is lowered. Press cure, transfer molding, steam cure, hot air cure, and injection molding are acceptable methods of curing.

IV. SYNTHESIS AND MANUFACTURING PROCESS

All of the commonly known polymerization processes are used for the synthesis of specialty elastomers. These include both the chain growth process and condensation polymerization. The chain growth process can be propagated by (1) free radicals, (2) ionic (cationic or anionic) polymerization, and/or (3) coordination or metal complex polymerization.

In free radical polymerization, heat, light, or electrochemical reactions produce free radical generators (initiators) from molecules added to the monomer for this purpose. The best chemical initiators are formed by the decomposition of peroxides, hydroperoxides, or azo compounds. The most frequently used electrochemical initiators are redox systems, where the reaction of reducing and oxidizing agents forms free radicals. After initiation of the polymerization, the propagation step occurs, during which the polymer chain grows through stepwise addition of monomer molecules until the chain is terminated. The growing polymer chain can react with other molecules (chain transfer agents) that are present to start new chains. The polymer radical becomes deactivated, and a chain transfer radical forms that, in turn, starts a new monomer radical and thus a new polymer chain. Alcohols, alkyl halides, mercaptans, and xanthogen disulfide are examples of practical chain modifiers.

Anions or cations, as opposed to free radicals, are used as chain initiators in ionic polymerization. In ionic polymerization there is a transfer of charge from the initiator ion to the monomer. Therefore, positively or negatively charged initiator ions result in cationic or anionic polymerization, respectively. Brønsted or Lewis acids can initiate a cationic polymerization. For anionic polymerizations, initiators such as alkyllithium compounds are used. The activation energies for the initiation and polymerization reactions are often small, and ionic polymerizations are often conducted at low temperatures. The extent of formation of the dissociated initiator, and thus its efficiency, is dictated by the dielectric constant of the solvent. Chain transfer or termination is the principal feature of ionic polymerization, which limits the molecular weight of the elastomer. However, at low polymerization temperatures for cationic polymerization or low-temperature ambient conditions for anionic polymerization, these chain transfer or termination processes do not occur. Thus, very narrow molecular-weight distributions can be obtained.

A particular case of coordination polymerization is commonly known as anionic polymerization. This is used mostly for the polymerization of conjugated dienes and styrene. These polymerizations are initiated by a carbon-centered organic anion of an alkali metal. Polymerization proceeds by the stepwise addition of the diene to the anion, which always leaves the growing terminus as an anion coordinated to the alkali metal cation. Homopolymers and copolymers of styrene and 1,3-dienes are made by this procedure. A recent innovation is living anionic polymerization. Under carefully selected polymerization conditions where the initiation of polymerization is instantaneous and all termination pathways are suppressed, the polymerization leads to intermolecular uniform polymers similar in both composition and molecular weight ($Mw/Mn = 1$). However, modifying the composition of the monomers during polymerization can change the intramolecular composition. These techniques lead to the formation of styrene–butadiene block polymers.

In coordination polymerization reactions, the initiator exists as a metal-centered complex attached to the polymer chain and the polymerization progresses by insertion of new monomer molecules into this complex. The most common initiators for these insertions are the Ziegler–Natta catalysts. These are reaction products of aluminum alkyl halides with transition metal salts of group IV–VIII elements. Because a monomer unit enters the polymer chain in a sterically controlled coordination, sterically and geometrically regular polymers are obtained with coordination catalysts. Thus in the polymerization of butadiene, pure *cis*-1,4, *trans*-1,4, or *trans*-1,2 microstructures can be made.

In contrast to chain growth polymerization, in which each chain grows from a single insertion point where fresh monomer is introduced and then incorporated into the chain, in step growth polymerization individual monomer molecules are incorporated into small prepolymer units that are dormant for a while and then assemble into high-molecular-weight polymers. Thus, insertion in step growth processes may occur at hundreds of points for any single polymer chain. Several intermediate-molecular-weight polymers often exist simultaneously, and the average molecular weight increases continuously during polymerization. Polysulfide rubber (TM), polyurethane rubber (AU), and polysiloxane rubber (Q) are examples of polymers obtained by condensation polymerization. In all of these polymerizations, multifunctional compounds react to form elastomers. If bifunctional compounds are used exclusively, linear molecules result. The presence of at least one trifunctional component per macromolecule yields branched or cross-linked elastomers.

There are several processes for the production of specialty elastomers:

1. Emulsion
2. Bulk
3. Solution
4. Suspension

Emulsion polymerization is primarily used in free radical chain polymerization. At least four components are required here—water-insoluble monomers, water, emulsifiers, and water-soluble initiators—in addition to chain modifiers and polymerization stoppers. The emulsifier causes the monomer to be emulsified into small droplets, which aggregate into micelles. Initiator radicals react with the monomer in the micelles. The micelles grow larger by absorbing new monomer from the surroundings to form latex. Emulsion polymerization leads to a very high conversion of monomer because the termination reactions are avoided. In comparison with homogeneous polymerization processes, in emulsion polymerization, (1) water is efficient at dissipating the heat of polymerization; (2) viscosity is independent of the molecular weight of the polymer, leading to elastomers with very high molecular weight; and (3) polymerization can be conducted in simple reaction vessels. Although originally developed for the use of peroxide initiators (reaction temperature > 100°C) to form "hot polymers," the use of redox initiators at low polymerization temperatures (near room temperature) has led to more uniform and thus more valuable "cold polymers."

In bulk polymerization, the polymerization takes place in the pure liquid monomer as a reaction medium. The heat of polymerization is dissipated through external or evaporative cooling. The first synthetic BR elastomers were made by bulk polymerization.

In solution polymerization, the monomer is dissolved in an organic solvent. Heat of polymerization is removed by refrigerating the solvent. After initiation of the polymerization reaction, the viscosity of the reaction medium rises with the degree of polymerization, and this limits the extent of chain growth.

Suspension polymerization is carried out using a liquid monomer as the solvent. However, in contrast to bulk polymerization, the resulting polymer is insoluble in the monomer and precipitates as a suspension in the reaction medium. In suspension polymerization, the viscosity of the reaction medium increases only slightly, and thus very high-molecular-weight elastomers can be made. For instance, suspension polymerization processes are of importance for the production of ultrahigh-molecular-weight

EPDM for which ethylene, the diene monomer, and polymerization aids are dissolved in the liquefied propylene and then polymerized.

The recent increase in the amounts of synthetic elastomers made in solution reflects some inherent advantages. In general, compared to other synthetic processes, solution polymerization leads to

1. Greater control of both intermolecular and intramolecular composition
2. Attenuation of competing intermolecular reactions, for example, branching
3. Easier removal of catalyst residues
4. Easier diffusion of monomers for elastomers with high melting points (T_m) such as those that are semicrystalline, high glass transition temperatures (T_g), or very high bulk viscosities
5. Easier removal of the heat of polymerization

Although directed polymerization is the synthesis method of choice for a majority of specialty elastomers, a significant number of these elastomers are made by the postpolymerization chemical modification of existing polymers. The value of these polymers is sufficient to justify the additional cost of a second chemical transformation step in their synthesis. The chlorination of polyethylene and the bromination of IIR are notable examples. These chemical modification reactions are conducted almost exclusively in solution. Solution processes are favored over the cheaper slurry or emulsion polymerization processes because they afford better control over the extent and speed of the chemical modification reactions when all of the reactant is in solution.

A recent development in coordination polymerization [7] utilizes a dual catalyst system (metal-centered complexes), one capable of stereoregular growth and a chain shuttling agent thought to facilitate the transfer of the growing polymer chains from one catalyst to the other. This mechanism is commonly known as "chain shuttling polymerization." The catalysts are typically hafnium- or zirconium-based complexes, whereas the chain shuttling agents are metal alkyl compounds such as $ZnEt_2$.

V. SOLVENT-RESISTANT SPECIALTY ELASTOMERS

Solvents encompass a broad class of apolar organic fluids. This includes paraffins, aromatics, cycloalkanes, olefins, chlorinated hydrocarbons, and fatty acids and their esters but not small polar molecules such as methanol. Vulcanizates of hydrocarbon elastomers easily swell in these solvents and lose most of their tensile strength, elongation, and resistance to set or abrasion. Thus, specialty elastomers that resist hydrocarbon solvents are an important commercial development. The class of solvent-resistant elastomers incorporates large amounts of strongly polarizing groups such as esters, nitriles, and halogens to raise the solubility parameter of the elastomer such that it is no longer miscible with the solvents.

The polarity of acrylonitrile makes NBR resistant to common hydrocarbon solvents (Figure 3.2). Increasing the acrylonitrile concentration of NBR improves its resistance to swelling in oil (Figure 3.3). However, the trade-off is a deterioration

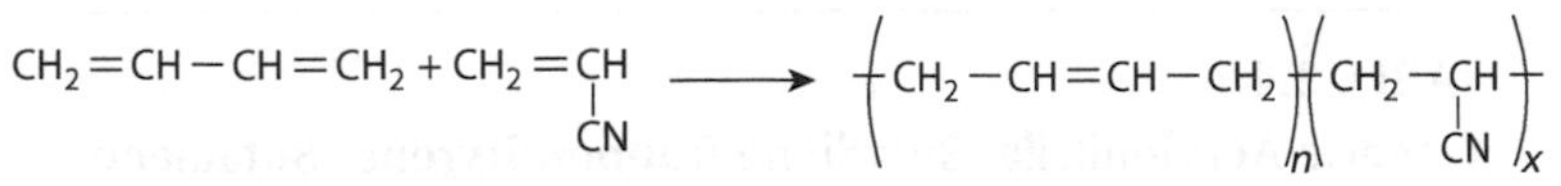

$$CH_2=CH-CH=CH_2 + CH_2=\underset{\underset{CN}{|}}{CH} \longrightarrow \left(CH_2-CH=CH-CH_2\right)_n\left(CH_2-\underset{\underset{CN}{|}}{CH}\right)_x$$

FIGURE 3.2 Acrylonitrile–butadiene rubber or nitrile rubber (acrylonitrile–butadiene copolymer).

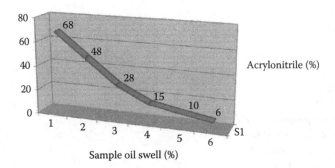

FIGURE 3.3 Acrylonitrile–butadiene rubber acrylonitrile content and oil swell (ASTM #3).

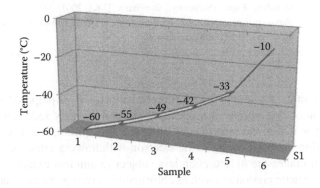

FIGURE 3.4 Acrylonitrile–butadiene rubber acrylonitrile content and brittle point.

in crack resistance, elasticity, and low-temperature properties. Figure 3.4 illustrates the effect of acrylonitrile on the brittle temperature of the elastomer [8]. However, NBR is an unsaturated elastomer and is easily degraded by weathering. Most of these shortcomings can be alleviated by hydrogenating the double bonds to form hydrogenated NBR (HNBR).

Table 3.4 displays a generic compound formulation for an innertube of a high-pressure hydraulic hose [6]. Nitrile polymers are typically compounded with low-structure, large particle size carbon blacks such as N660 or N762. In addition, a plasticizer is added along with stearic acid and zinc oxide for the cure system.

Acrylic elastomers (ACMs) are copolymers of various alkyl acrylates differing in the nature of their alkyl group. They contain the strongly polar ester group.

TABLE 3.4

Generic Acrylonitrile–Butadiene Rubber/Styrene–Butadiene Rubber Hydraulic Hose Tube

Formula (phr)	
NBR (40% acrylonitrile)	80
E-SBR (1500)	20
N660	110
Stearic acid	1
Zinc oxide	5
Dioctyl phthalate	16
Sulfur	2
TBBS	1
TMTD	0.1
Typical property targets	
Tensile strength (MPa)	17.00
Ultimate elongation (%)	170.00
Volume swell (%)	25
(ASTM #3 oil, aged 3 days at 70°C)	

Sources: De, S.K. and White, J.R., eds., *Rubber Technologist's Handbook*, Vol. 1, iSmithers Rapra Publishing, Shawbury, U.K., 2001; Stevens, M.P., *Polymer Chemistry: An Introduction*, Oxford University Press, New York, 1999. With permission.

Acrylate elastomers are saturated. Chloroprene rubber (CR) is prepared by emulsion polymerization of chloroprene. It resists hydrocarbon solvents and shows improved flame resistance due to the incorporated chlorine. In addition, it is also more resistant to oxidation than NR owing to the electron-withdrawing effect of the chlorine groups, which makes the unsaturation less subject to ambient oxidation. Saturated chlorinated synthetic elastomers such as chlorinated polyethylene are made by reacting polyethylene with chlorine. Chlorosulfonated polyethylene (CSM) is made similarly by reacting polyethylene with both chlorine and sulfur dioxide.

A. NBR DERIVATIVES

The most important specialty solvent-resistant elastomer is HNBR made by the almost complete catalytic hydrogenation of the unsaturation derived from butadiene moieties. Nippon Zeon makes HNBR almost exclusively. The small amount of residual unsaturation is used for vulcanization. This results in a product with much improved resistance to oxidation and weathering but with little or no sacrifice of solvent resistance or elastomer performance. As in NBR, HNBR is also available with different ACN contents. The HNBR has a higher glass transition temperature though it has significantly enhanced solvent resistance. Applications for HNBR take advantage of the excellent resistance to organic fluids. However, the increased

TABLE 3.5

Blending of Other Polymers with Acrylonitrile–Butadiene Rubber

Polymer Blend	Effect	Comment
SBR	Reduces cost	Noncompatible due to differences in solubility parameters
BR	Improves low-temperature properties	Improves flex resistance
EPDM	Improves ozone resistance	70/30 NBR/EPDM blends optimum
CR	Improves flex properties	Best at low NBR AN content
Resins	Increase hardness	Improve abrasion

high-temperature performance of the HNBR makes it far superior to NBR in critical applications.

Chemically modified NBR containing pendant carboxylic acids is made by copolymerizing methacrylic or acrylic acid with butadiene and acrylonitrile (carboxylated acrylonitrile–butadiene rubber [XNBR]). These polymers, which contain 2–6 wt% acid monomer, can be vulcanized with polyvalent metal ions, which results in a large improvement in abrasion resistance. This acid functionality leads to excellent polymer–filler interaction in compounds. XNBR polymers require care when mixing, and temperatures typically should not exceed 135°C. The compounds tend to be hard, display good abrasion resistance, and have high tensile and tear strength compared to conventional nitrile compounds.

Nitrile polymers are often blended with other elastomers to (1) improve ozone resistance, (2) lower temperature properties, (3) improve aging resistance, and (4) improve abrasion resistance. Table 3.5 is a simple illustration on how blends with other elastomers will impact nitrile compound properties.

B. POLYCHLOROPRENE

Polychloroprene (CR) was the first commercially developed synthetic elastomer. The worldwide capacity of CR is around 370 kt as of 2012 [1]. DuPont Performance Elastomers and DENKA have ~27% of the capacity each with facilities in the United States and Japan, respectively [1]. There are five other producers of CR. Commercial CR polymerization is typically conducted in aqueous emulsion (Figure 3.5). Historically, it was produced first by the dimerization of acetylene, though in more recent years the chloroprene monomer has been produced from the halogenation of butadiene. Molecular weight is controlled with chain transfer agents. Mercaptan chain transfer agents are used, but other materials such as xanthogen disulfides have also been employed. The latter provide reactive end groups. Molecular weight is also controlled by copolymerization with sulfur. This copolymer contains polysulfide units that are then cleaved chemically to lower the molecular weight. The predominant microstructure of CR is the head-to-tail *trans*-1,4-CR, although other structural units are also present. The high concentration of this repeat unit is responsible for the crystallinity of CR and for the ability of the material to crystallize under strain.

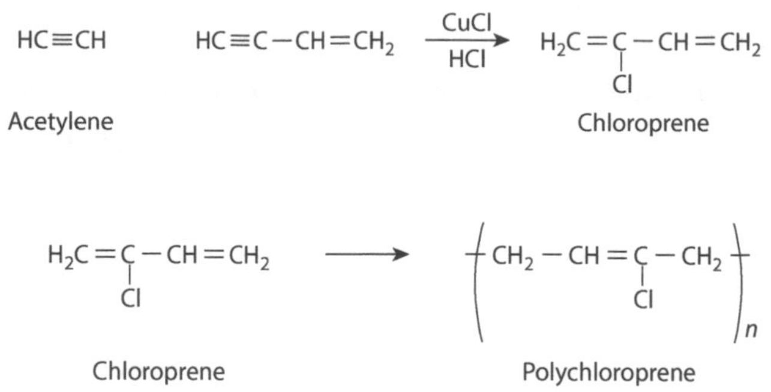

$$HC\equiv CH \qquad HC\equiv C-CH=CH_2 \xrightarrow[HCl]{CuCl} H_2C=\underset{\underset{Cl}{|}}{C}-CH=CH_2$$

Acetylene   Chloroprene

$$H_2C=\underset{\underset{Cl}{|}}{C}-CH=CH_2 \qquad\longrightarrow\qquad \left(CH_2-CH=\underset{\underset{Cl}{|}}{C}-CH_2\right)_n$$

Chloroprene   Polychloroprene

FIGURE 3.5 Chloroprene production from dimerization of acetylene and polymerization. (From Barbin WW, Rodgers MB. The science of rubber compounding. In *Science and Technology of Rubber*, Mark JE, Erman B, Eiri FR, Eds., Academic Press, Orlando, FL, 1994. With permission.)

Other insertion pathways such as 3,4 insertion in minor amounts lead to easily vulcanized sites on the polymer. The presence of these structural irregularities does lead to lower crystallinity and tensile strength. The total amount of insertion modes other than *trans*-1,4 increases with polymerization temperature from 5% at −40°C to about 30% at 100°C. CR made with high *trans*-1,4 content is tough and is used for adhesives. CR made at higher temperature, with more chain irregularities, tends to crystallize much more slowly and is more suitable for elastomer uses (Figure 3.5).

Polychloroprene polymers are often differentiated by the use of an appended letter. These designations do not reflect the composition of the polymer but are derived from the manufacturing and process conditions. Ultimately, the process conditions affect the microstructure—that is, branching, molecular-weight distribution, and isomer distribution—and thus the compounding and ultimate physical properties of the polymer. The G™ family of DuPont polymers, containing residual thiuram disulfide, can be cured with metallic oxides. The W™, T™, and xanthate-modified families require the addition of an organic accelerator, often in combination with a cure retarder, for practical cures.

Polychloroprene rubber has been successful because it combines environmental resistance and toughness, especially in dynamic applications involving heat buildup and resistance to flex cracking. The high strength of CR vulcanizates is a result of the tendency of CR to crystallize under stress. The rate of crystallization and the melting point of CR both increase with increasing *trans*-1,4 microstructure. The resistance of CR to aging is higher than that of other diene elastomers because of the presence of an electronegative chlorine atom on the repeat unit double bond. Polyester plasticizers are more effective than hydrocarbon oils in reducing the T_g of vulcanizates. However, the plasticizers are also effective in promoting crystallization. For many applications, where cost and processability are the objective, naphthenic and aromatic oils are used. CR has a well-balanced combination of properties including processability, strength, flex and tear resistance, flame resistance, and adhesion, together with sufficient heat, weather, and ozone resistance for most applications. Typical applications involve power transmission and timing belts, automotive boots

TABLE 3.6
Generic Hydraulic Hose Cover Compound

Formula (phr)	
Polychloroprene	100
N762	100
Aromatic oil	20
MgO	4
Antiozonant	2
Wax	3
DPG	1
TMTD	1
Sulfur	0.75
ZnO	5
Typical target properties	
Tensile strength (MPa)	14
Ultimate elongation (%)	300

Source: Dunn, J.R. and Vara, R.G., *Rubber Chem. Technol.*, 56, 557, 1983. With permission.

(rubber piece around the constant-velocity joint in the front wheels of cars), air springs, and truck engine mounts. A large amount of CR is used in adhesives. CR tends to have high uncured strength due to strain-induced polymer crystallization. CR latex and solvent adhesives have been used in foil laminating adhesives, facing adhesives, and construction mastics. CR latex can be used in a variety of special applications to make binders, coatings, dipped goods, elasticizers, and foam where either dry or solvent-based processes would be impractical. The water-based systems are now preferred because they have high solids content and minimal solvent emissions but are unaffected by polymer rheology and have extremely small particle size.

Over the years, almost every vinyl and diene monomer has been tested with CR in free radical polymerization. Copolymers usually contain only a limited amount of random comonomer. Methacrylic acid, for example, promotes adhesion to substrates and increases cohesive strength. A large number of graft polymers of CR have been described, but the only ones of commercial significance are those made with acrylates and methacrylate. These are particularly useful for adhesion to elasticized polyvinyl chloride. The graft polymers may be made by either solution or emulsion polymerization. For a good review of the compounding of polychloroprene elastomers, further reference should be made to Colbert [9] and Dunn and Vara [10]. Table 3.6 displays a simplified hydraulic hose cover compound formulation that might serve as a starting point in a new compound development study.

C. ACRYLIC ELASTOMERS

ACMs are a limited volume specialty elastomer with a worldwide capacity of about 8 kt. This volume is included in the worldwide capacity for the more common EAM.

ACM manufacture is distributed among the United States, Western Europe, and Japan. Its production is intimately associated with automotive production. The major producer for ACM is Zeon Chemical. Cray Valley produces a small amount of low-molecular-weight ACM polymer (under trade name Ricon™) for adhesive applications.

ACMs are produced by free radical polymerization using principally aqueous suspension and emulsion polymerization. ACM elastomers consist of a majority (~97–99 wt%) of ethyl acrylate, butyl acrylate, and 2-methoxy ethyl acrylate monomers. In addition, a small amount of cure site monomers is added to facilitate vulcanization. The acrylic esters constituting the majority of the polymer chain determine the physical, chemical, and mechanical properties of the polymer and its vulcanizates. Cure site monomers have an acrylate double bond for polymerization into the ACM and a reactive group for the vulcanization process. Two of the most important classes of cure site monomers are reactive chlorine-containing monomers and epoxy/carboxyl-containing monomers.

ACM has both a saturated backbone, which is responsible for the heat and oxidation resistance, and ester side groups, which contribute to the marked polarity. The cure sites present in the ACM also affect the expected properties. Reactive chlorine cure sites generally give excellent heat aging as shown by retention of elongation. The alkali metal carboxylate–sulfur cure system has been widely used for this type of cure monomer since its introduction in the early 1960s. Epoxy/carboxyl cure monomers give good compression set and good resistance to hydrolysis. New vulcanization systems based on quaternary ammonium salts have been introduced to vulcanize epoxy/carboxyl cure sites. These have been found to be effective also in chlorine-containing ACM.

Its resistance to aging at high temperatures and insensitivity to organic fluids makes ACM useful in automotive underhood parts. These include lip and shaft seals, O-rings, oil pan and cover valve gaskets, and hoses.

D. CHLOROSULFONATED POLYETHYLENE

The worldwide production capacity of CSM (Figure 3.6) elastomers is about 38 kt [1]. The major producer of this elastomer is TOSOH with plants in Japan [1]. There are a few other producers based in the United States and China.

CSM is not made from component monomers but from the chemical modification of polyethylene. The chlorination and chlorosulfonation of polyethylene are carried out simultaneously or sequentially using carbon tetrachloride as the reaction solvent. The elastomeric character of CSM arises from the inherent flexibility of the polyethylene chain due to reduced crystallinity. Substitution of chlorine in the polymer chain provides sufficient molecular irregularity to reduce crystallinity. The sulfonyl chloride groups provide cross-linking sites for nonperoxide vulcanization.

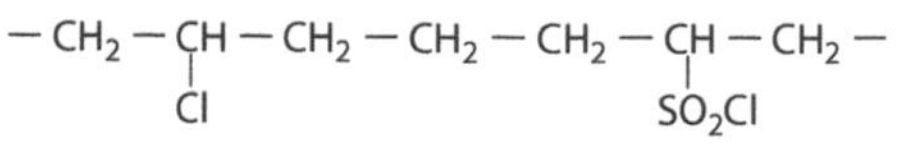

$$- CH_2 - \underset{\underset{Cl}{|}}{CH} - CH_2 - CH_2 - CH_2 - \underset{\underset{SO_2Cl}{|}}{CH} - CH_2 -$$

FIGURE 3.6 Simplified chemical structure of chlorosulfonated polyethylene.

TABLE 3.7

Model Formula for a Chlorosulfonated Polyethylene–Based Compound Development Study

Material	Loading (phr)	Purpose
CSM	100	Polymer
MgO	4	Acid acceptor
Pentaerythritol	3	Activator for acid acceptor
Clay	80	Filler (carbon black preferred)
Oil	25	Plasticizer
Waxes	3	Processing aid
Sulfur	1	Vulcanizing agent
TMTD, TMTS	2	Accelerator

Source: ExxonMobil Chemical, 2014, http://www.exxonmobilchemical.com/ Chem-English/brands/vistalon-ethylene-propylene-diene-epdm-rubber. aspx?ln=productsservices. With permission.

CSM properties can vary from elastomeric to plastic depending on the amount of chlorine and sulfonyl chloride substitution.

The presence of chlorine and sulfonyl chloride in the chain requires weak bases in CSM compounds to react with acidic by-products. Acceptable bases include magnesia, litharge, organically bound lead oxide, calcium hydroxide, synthetic hydrotalcite, and epoxy resins. The presence of sulfonyl chloride groups contributes to adhesion and mechanical reinforcement by reaction with fillers. In addition, the presence of chlorine imparts significant flame and light resistance to CSM compounds. However, all of these properties depend on the extent and distribution of the chlorination and chlorosulfonation. CSM polymers can be easily functionalized by the replacement of the halogen either on the backbone or on the sulfonyl group. Acid, ester, and amide derivatives of the sulfonyl chloride groups have been prepared.

The sulfonyl chloride group is the cure site for CSM and determines the rate and state of cure. Carbon black fillers give the best reinforcement as well as significant resistance to photochemical degradation. Ester plasticizers provide the best combination of low-temperature flexibility, heat resistance, and mechanical properties.

CSM elastomers are used in specialty end-use applications such as coating, adhesives, roofing membranes, pond and reservoir liners, electrical wiring, insulation, automotive and industrial hoses, tubing and belts, and molded goods. A typical starting formulation for compound development is compiled in Table 3.7 for reference purposes.

E. ETHYLENE–ACRYLIC ELASTOMERS

The worldwide capacity for acrylate rubbers is about 75 kt. E.I. DuPont is the sole producer of ethylene–acrylic (EAM) polymers under the trade name Vamac™. EAM polymers are made by the free radical polymerization of ethylene, methyl acrylate,

Ethyl acrylate (EA)

$$CH_2 = CH - \overset{\overset{\displaystyle O}{\|}}{C} - O - CH_2 - CH_3$$

Butyl acrylate (BA)

$$CH_2 = CH - \overset{\overset{\displaystyle O}{\|}}{C} - O - CH_2 - CH_2 - CH_2 - CH_3$$

Methoxyethyl acrylate (MEA) $CH_2 = CH - \overset{\overset{\displaystyle O}{\|}}{C} - O - CH_2 - CH_2 - O - CH_3$

FIGURE 3.7   Monomers for acrylic-based elastomers.

and an alkenoic acid. A small amount (1–5 mol%) of an alkenoic acid is incorporated to provide sites for cross-linking with diamines. Copolymers of ethylene and methyl acrylate have recently been commercialized.

Ethylene–ACM is an amorphous polymer due to the random placement of the acrylate comonomer along the ethylene backbone. The polymer is saturated, making it highly resistant to aging and weathering even in the absence of antioxidants. In addition, the methyl acrylate-to-ethylene ratio determines both the low-temperature properties and the resistance to organic fluids. EAM elastomers have aging, heat, and fluid resistance in addition to acceptable elastomeric properties. In particular, they show largely temperature-stable vibrational damping properties and the ability to form flame-resistant elastomeric compounds with combustion products that have low toxicity and corrosiveness. Figure 3.7 illustrates some of the monomers that can potentially be used to produce such polymers.

VI.   TEMPERATURE-RESISTANT ELASTOMERS

Temperature-resistant elastomers can be classified into two groups: (1) those that are suitable for automotive underhood applications where only transient temperatures are greater than 100°C and (2) those that are suitable for much higher temperature use such as for gasket sealant for engines. Polyolefin elastomers with saturated backbones are the material of choice for the first kind of application and have gradually supplanted other unsaturated elastomers in spite of the difficulty in achieving physical targets with these elastomers. In the truly temperature-resistant elastomers of the second kind, the hydrocarbon structure has been replaced with more refractory building blocks. The most common of these are silicone rubbers (Q). These are based on chains of Si–O–Si rather than C–C units and owe their temperature resistance to their unique structure. Silicone rubbers are almost exclusively polydimethylsiloxanes. The other heat-resistant elastomer, FKM, is derived from perfluorinated versions of common polyolefins such as polypropylene and polyethylene. Its thermal resistance arises from the nonreactivity of the carbon–fluorine bond. Elastomers with a C–C skeleton that has a saturated backbone chain but retains a small amount of pendant unsaturation (for vulcanization) resist environmental aging better than diene elastomers. Degradation by oxygen, ozone, and light is slow on the saturated backbone and causes only infrequent chain cleavage.

A. ETHYLENE–PROPYLENE COPOLYMERS AND ETHYLENE–PROPYLENE–DIENE TERPOLYMERS

The worldwide production of EPMs and EPDMs is approximately 1150 kt [1]. In the United States the producers are ExxonMobil Chemical (38%), Dow Chemical (30%), Lanxess (11%), and Lion Copolymer (21%). European producers are Lanxess (56%), ExxonMobil Chemical (23%), and Versalis (21%). In Asia, the producers are Mitsui Petrochemical (70%) and Sumitomo Chemical (30%). In addition, Lanxess produces about 4% of worldwide capacity in Latin America.

Even though EPM and EPDM elastomers have been available for more than 30 years, the technology for these products, both their production and their application, is still under development. The most widely used process is a solution polymerization in which the polymer is produced in a hydrocarbon solvent. EPM and EPDM rubbers are produced in continuous processes. Slurry polymerization is conducted in liquid propylene. Special reactor designs with multiple feeding locations to achieve special molecular structures have been developed. Gas-phase polymerization of EPDM is possible as an extension of the ubiquitous gas-phase processes for polyethylene and polypropylene.

As manufactured today, EPM and EPDM are polymers based on the early work of Natta and coworkers. Generically, an EPM is 60 mol% ethylene and 40 mol% propylene. Analogous EPDM polymer contains in addition 1.5 mol% nonconjugated diene such as ethylidene norbornene or dicyclopentadiene (Figure 3.8). The comonomers are statistically distributed along the molecular chain. EPM is a saturated synthetic elastomer because it does not contain any unsaturation. It is inherently resistant to degradation by heat, light, oxygen, and, in particular, ozone. EPDM, which contains pendant unsaturation, is only slightly less stable to aging than EPM. The properties of EPM copolymers are dependent on the relative content of ethylene units in the copolymer chain and the variation in the comonomer composition of different chains. EPM and EPDM polymers with greater than 60 mol% ethylene are increasingly crystalline and tough.

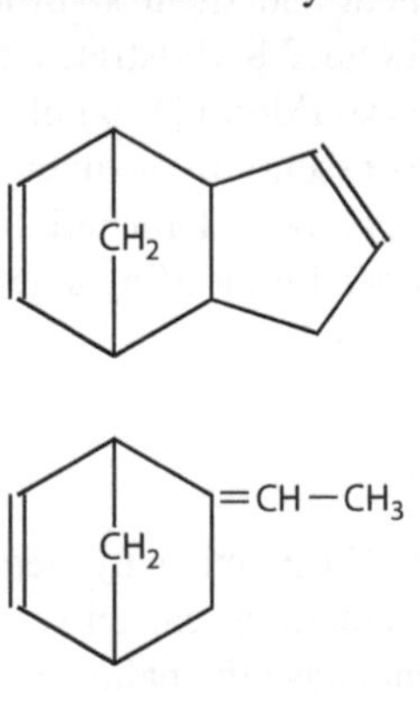

Dicyclopentadiene (DCPD)

Ethylidene norbornene (ENB)

$H_2C = CH - CH_2 - CH = CH - CH_3$ 1,4-hexadiene (1,4-HD)

FIGURE 3.8 Nonconjugated diene comonomers used in ethylene-propylene-diene terpolymers.

EPM can be vulcanized with peroxides. The small amount of a third diene monomer in EPDM permits conventional vulcanization with sulfur and other vulcanization systems such as resins at the pendant sites of unsaturation. EPM and EPDM elastomers have to be compounded with reinforcing fillers, for example, carbon black, if enhanced mechanical properties are required. Paraffinic oils are widely used as plasticizers. In EPM and EPDM compounds, mechanical properties depend on the composition of the elastomers and the types and amounts of fillers. In general, the elastic properties are better than those of many other synthetic rubbers but not as good as those of NR or SBR. The resistance of EPM and EPDM to heat and aging is much better than that of SBR and NR. EPM and EPDM vulcanizates have excellent resistance to inorganic or highly polar fluids such as dilute acids, alkaline, and alcohol. However, the resistance to aliphatic, aromatic, or chlorinated hydrocarbons is very poor. The electrical insulating and dielectric properties of the pure EPM and EPDM are very good, but in compounds they are also strongly dependent on the choice of compounding ingredients.

Among the synthetic elastomers, EPM and EPDM have the fastest growing markets [1], owing to their excellent ozone resistance in comparison to the diene elastomers. This growth still comes from the replacement of these commodity rubbers by virtue of their better ozone and thermal resistance. Another facet of the growth is that EPDM rubber can be extended with fillers and plasticizers to an extremely high level in comparison with the other elastomers and still maintain excellent processability and properties in end-use articles. The main uses of EPM or EPDM are in (1) automotive applications such as profiles, hoses, and seals (41% of worldwide consumption); (2) building and construction as profiles, roofing membranes, and seals (21%); (3) cable insulation and jacketing (6%); and (4) molded appliance parts. An important application for EPDM is in blends with GPRs. Considerable amounts of EPM and EPDM are also used in blends with thermoplastics (28%), and substantial amounts of EPM are used as additives to lubrication oils (12%) because of its excellent heat and shear stability under the operating conditions of automobile engines. Table 3.8 illustrates the range of EPDM polymers available commercially from ExxonMobil [11]. Polymers vary in termonomer content; Mooney viscosity, which is essentially a function of molecular weight; and ethylene content. Two model compound formulations are displayed in Table 3.9, which can be used as a reference for building a new compound formulation.

B. SILICONE RUBBER

The majority of silicone rubber (Q) elastomers (>60%) are made by General Electric. Laur and Dow Corning have about 10% of the remaining capacity, while smaller manufacturers (Wacker, Bayer, and Rhone-Poulenc) have the balance.

Q rubbers are made commercially either by the multistep hydrolysis of dimethyldichlorosilane or by the ring-opening polymerization of the cyclic oligomer octamethylcyclosiloxane. In the hydrolysis procedure, the chlorine atoms are hydrolyzed and replaced with oxygen atoms bonding a pair of silicon atoms. Water is frequently replaced with methanol, which leads to the formation of methyl chloride

TABLE 3.8
ExxonMobil Chemical Vistalon™ Ethylene–Propylene–Diene Terpolymer Grades

Vistalon Grade	Wt%
407–878	
Copolymer of Mooney range	18–51
Ethylene range	44–78
1703–3708	
Terpolymer ENB range	0.9–3.8 (1–4)
Mooney range	25–52
Ethylene range	55–70
2504–7800	
Terpolymer ENB range	4.2–6.0 (4–6)
Mooney range	20–82
Ethylene range	55–79
6505–9500	
Terpolymer ENB range	8–11
Mooney range	51–83
Ethylene range	53–69

Source: ExxonMobil Chemical, 2014, http://www.exxonmobilchemical.com/Chem-English/brands/vistalon-ethylene-propylene-diene-epdm-rubber.aspx?ln=productsservices.

TABLE 3.9
Model Formulas for Initiation of New Compound Development Work

Roofing Cover or Sheeting		Radiator Hose	
EPDM	100	EPDM	100
N330 carbon black	120	N660 carbon black	100
Clay or talc	30	Calcium carbonate	30
Paraffinic oil	90	Paraffinic oil	100
Zinc Oxide	5	Zinc oxide	3
Stearic acid	2	Stearic acid	1
MBTS	2.50	DTDM	2
TMTD	0.50	ZDBDC	2
TETD	0.50	TMTD	2
Sulfur	0.80	Sulfur	0.75

Source: http://www.exxonmobilchemical.com/Chem-English/productsservices/vistalon-automotive-formulations.aspx. With permission.

rather than the more corrosive hydrochloric acid. In the ring-opening polymerization, strong acid or strong base catalysts are used to produce high-molecular-weight
Q. The ring-opening polymerization process can be conducted in an aqueous emulsion procedure using dodecylbenzenesulfonic acid as the catalyst.

Most Q polymers have the repeat unit empirical formula $((CH_3)_2SiO)_n$ and are
referred to as polydimethylsiloxanes. The elastomer consists of alternating silicon
and oxygen atoms with two methyl groups on each silicon. A significant departure
from most other elastomers is the absence of carbon from the backbone. Three reaction types are predominantly employed for the formation of vulcanized Q: peroxide-
induced free radical vulcanization, hydrosilylation addition cure, and condensation
cure. Silicones have also been cross-linked using radiation to produce free radicals
or to induce photoinitiated reactions.

Q elastomers do not crystallize even under strain and have very poor physical
properties. Unfilled silicone rubber has a tensile strength of only 0.345 MPa. Q vulcanizates are reinforced with ~25% finely divided fumed silica. This reinforcing
filler increases tensile strength, tear properties, and abrasion resistance. The Si–O–
Si bonds in Q have a much lower energy of activation for rotation than C–C or C–O
bonds. This makes Q elastomers flexible and rubbery even at very low temperatures,
and this elastomeric property is little affected by temperature changes. This feature
combined with their refractory characteristics makes these elastomers useful over a
wide temperature range.

Q elastomers are used for electrical insulation, medical devices, seals, surface-
treated fillers, elastic textile coatings, and foams. Liquid injection molding is used
for electrical connectors, O-ring seals, valves, electrical components, health-care
products, and sporting equipment such as goggles and scuba diving masks.

C. FLUOROCARBON ELASTOMERS

The principal producers of fluorocarbon (FKM) polymers are 3M, Ausimont, and
E.I. DuPont Dow in the United States. In the Far East, production is maintained by
Asahi Glass and Daikin. The annual worldwide FKM usage is about 8 kt. About
40% of this is in the United States, 30% in Europe, and 20% in Japan.

High-pressure, free radical, and aqueous emulsion polymerizations are typically
used to prepare these elastomers. The initiators are organic or inorganic peroxides,
for example, ammonium persulfate. The emulsifying agent is usually a fluorinated
acid soap. FKM elastomers are the perfluoro derivatives of the common polyolefins. The monomers are the perfluorinated derivatives of ethylene and propylene.
Copolymers of different olefins are available; they are noncrystalline polymers that
are elastomeric when cross-linked. The vulcanized FKM polymers are dimension-
ally stable and chemically inert in hostile environments and in a variety of organic
fluids such as oils and solvents. This chemical resistance spans a wide temperature
range. In addition, vulcanized FKM polymers show extraordinary self-lubricating
properties due to their low surface energy. FKMs can be vulcanized with one of
three distinct procedures, with diamine, bisphenol–onium, and peroxide curing
agents. The bisphenol–onium cure system is the most widely used. FKMs are resistant to heat, chemicals, and solvents.

The major use of FKM polymers is in the automotive industry for such items as engine, gasket, and fuel system components (hoses and O-rings). This application is fueled by increased demands from higher use temperatures, alcohol-containing fuels, and aggressive lubricants. Other major segments include petroleum, petrochemical and industrial pollution control, and industrial hydraulic and pneumatic applications.

D. PHOSPHAZENES

The worldwide capacity for phosphazenes (FZ) is less than 0.1 kt. Albemarle (Ethyl Corporation) is the sole supplier. FZ polymers are made in a two-step process. First, the trimer hexachlorocyclotriphosphazene is polymerized in bulk to polydichlorophosphazene, a chloropolymer. The chloropolymer is then dissolved and reprecipitated to remove unreacted starting material. Nucleophilic substitution with alkyl or aryloxide substitution of the halide in solution provides the elastomer.

Polyphosphazenes have a backbone of alternating nitrogen and phosphorus atoms with two substituents on each phosphorus atom. The backbone is isoelectronic with that of silicones; these polymer backbones share the characteristics of thermal stability and high flexibility.

Two elastomers with unique property profiles have been commercialized. One has fluoroalkoxy substituents that provide resistance to many fluids, especially to hydrocarbons. FZ elastomer is a translucent pale brown gum with a glass transition temperature of $-68°C$ to $-72°C$. The gum can be cross-linked by using peroxides such as dicumyl peroxide and α,α'-bis(t-butylperoxy)diisopropylbenzene. FZ elastomers have excellent resistance to hydrocarbons and inorganic acids, as is expected for a fluorinated elastomer. They are strongly affected by polar solvents but are more resistant to amines than most other fluorinated elastomers. This material also has a broad use temperature range and useful dynamic properties. The other elastomer (aryloxyphosphazene) has phenoxy and p-ethylphenoxy substituents. It has flame-retardant properties without containing halogens. It may be cured using either peroxides or sulfur.

Varying the polymer can produce coatings, fluids, elastomers, and thermoplastic materials. These variations include changes in the molecular weight and the substituents on the phosphorus. These materials have been suggested for use in biomedical devices, including implants and drug carriers. However, initial applications have been largely in military and aerospace areas.

E. POLYETHERS

Nippon Zeon, with plants in the United States and Japan, has a near monopoly on polyether rubbers. These include epichlorohydrin (ECH) homopolymer rubber (CO), ECH–ethylene oxide copolymer (ECO), ECH–ethylene oxide terpolymers, propylene oxide homopolymer rubber (PO), and propylene oxide–allyl glycidyl ether copolymer (GPO). There are no production facilities in Western Europe for these elastomers. Worldwide production is about 11 kt/yr, with Nippon Zeon having about 90% of the market share and Daiho the balance.

Polymerization is conducted either in solution or in slurry at 40°C–130°C in toluene, benzene, heptane, or diethyl ether. Trialkylaluminum–water and trialkylaluminum–water–acetylacetone catalysts are used. Chain propagation is by a cationic charge transfer mechanism. The polyethers include a group of minor elastomers made by ring-opening polymerization of epoxides and include CO, ECO, ETR, PO, and GPO. CO and ECO are linear and amorphous. Because it is asymmetrical, the ECH monomer can polymerize in the head-to-head, tail-to-tail, or head-to-tail fashion. The commercial polymer is 97%–99% head to tail and atactic. The commercial products are essentially amorphous.

Polyethers are remarkable because of an exceptional combination of properties. The ECH homopolymer has low gas permeability and is better than IIR. It is resistant to ozone and has low hysteresis. The polymer is flame retardant owing to its high chlorine content. Its resilience is poor at room temperature but improves upon heating. ECO is less flame retardant owing to its lower chlorine content. It has some impermeability to gases but has better low-temperature flexibility. It also exhibits good heat resistance. Polyethers are resistant to apolar organic fluids such as oils and aliphatic/aromatic solvents.

The polyethers are important in automotive applications such as fuel, air, and vacuum hoses; vibration mounts; and adhesives. Other uses include drive and conveyor belts, hoses, tubing, and diaphragms; pump parts including inner coatings, seals, and gaskets; printing rolls and blankets; fabric coatings for protective clothing; pond liners; and membranes in roofing material. In the automotive areas, they are used as constant-velocity boots, dust and fuel hose covers, mounting isolators, and. hose and wire covers.

F. RING-OPENED POLYMERS

Ring-opening metathesis polymerization of cyclopentene is carried out in either chlorinated or aromatic solvents to give *trans*-polypentenamer. As of 2014, this polymer is not commercially available. The same polymerization technique is used to convert cyclooctene to *trans*-cyclooctenamer (TOR). Degussa-Hüls is the sole producer of TOR. Their capacity is estimated at 13.5 kt.

VII. OLEFIN BLOCK COPOLYMERS

INFUSE™ olefin block copolymers (OBCs) are new classes of elastomers commercially introduced by Dow Chemical in the United States and Europe using polyethylene chain architecture. Polypropylene-based elastomers on the same lines were announced recently under the trade name INTUNE™. The elasticity in OBCs arises due to the rubbery amorphous domains tethered between the crystalline regions. INFUSE™ products are relatively low-density elastomer resins in the range 0.86–0.89 g/cc with a wide range of melt indices [12].

Although these elastomers currently occupy small capacity, there are a lot of potential applications that could utilize these products. The key advantages of OBCs over established polymers such as EVA are low odor and color, better processability, and lower cost. Hygiene and adhesives markets are the major areas that have immediate applications for OBCs. Although, injection molded and extruded parts are also envisioned for the OBC class of materials. With the advent of the polypropylene OBCs (INTUNE), a wider portfolio of markets and applications such as nonwovens and films are possible with improvements in medical and hygiene sectors.

OBCs can be categorized under thermoplastic elastomers. Similar to styrene block copolymers (SBCs), these materials can be melt-processed with ease. But in contrast to SBCs that contain glass-forming domains of styrenic blocks, OBCs utilize domains capable of crystallizing, which act as physical cross-links below their melting point. Also, SBCs are synthesized using living anionic polymerization and have a narrow molecular-weight distribution ($Mw/Mn \sim 1.5$), while the chain shuttling catalyst process results in broader molecular-weight distribution ($Mw/Mn \sim 2.0$). But due to the absence of unsaturation in OBCs they are significantly amenable to processing compared to SBCs.

In addition to the processing controls such as molecular weight and molecular-weight distribution, the choice of the type of comonomer, and hard segment content significantly influence the mechanical properties of the OBCs. Crystal morphology (schematically distinguished in Figure 3.9) is also thought aid in the mechanical performance of these elastomers. The morphology of OBCs is thought to be lamellar [13] unlike the ethylene octene random copolymers (containing fringed micelles) [14]. The enhanced elastic recovery while preserving the softness and low modulus characteristics can be attributed to such a change in the crystalline morphology [15].

OBCs combine the advantages of polyolefins such as low cost, ease of processing and resistance to permeants, and the mechanical properties of the thermoplastic elastomers. Such a set of properties enables the use of OBCs as a specialty elastomer with great potential to tune the properties for individual application such as adhesive formulation but also makes it suitable for compatibilization of dissimilar commodity polymers and additives.

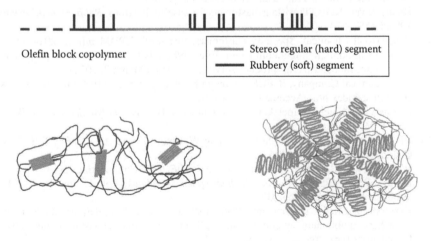

FIGURE 3.9 Chain structure in olefin block copolymers and schematics of crystal morphologies.

VII. SUMMARY

Specialty elastomers have structures and compositions designed to have specific attributes. The principal characteristic of these polymers, in comparison to GPRs, is that they have either resistance to solvents or resistance to elevated temperatures or, preferably, both. In comparison to the high-volume rubbers, the quantities of these specialty polymers are small. However, as the mission profiles for products using special-purpose elastomers become more demanding, the need for such materials can be expected to increase.

REFERENCES

1. International Institute of Synthetic Rubber Producers, Inc. *Worldwide Rubber Statistics.* Houston, TX: International Institute of Synthetic Rubber Producers, 2012.
2. Morton M. *Rubber Technology*, 3rd edn. New York: Van Nostrand Reinhold, 1987.
3. Dick J. *Rubber Technology: Compounding and Testing for Performances.* Berlin, Germany: Hanser, 2001.
4. International Standards Organization. Rubbers and latices—Nomenclature. ISO 1629. New York: ANSI, 1995.
5. International Institute of Synthetic Rubber Producers, Inc. *The Synthetic Rubber Manual*, 14th edn. Houston, TX: International Institute of Synthetic Rubber Producers, 1999.
6. Stevens MP. *Polymer Chemistry: An Introduction.* New York: Oxford University Press, 1999.
7. Arriola DJ, Carnahan EM, Hustad, PD, Kuhlman RL, Wenzel TT. Catalytic production of olefin block copolymers via chain shuttling polymerization. *Science* 2006;312:714–719.
8. Barbin WW, Rodgers MB. The science of rubber compounding. In: Mark JE, Erman B, Eiri FR, eds., *Science and Technology of Rubber*. Orlando, FL: Academic Press, 1994.
9. Colbert GP. Solvent resistant elastomers—Neoprene, hypalon, and chlorinated polyethylene. In: Barawal KC, Stevens HL, eds., *Basic Elastomer Technology*. Washington, DC: Rubber Division, American Chemical Society, 2001.
10. Dunn JR, Vara RG. Oil resistant elastomers for hose applications. *Rubber Chem Technol* 1983;56:557–574.
11. ExxonMobil Chemical. Specialty elastomers: Vistalon™ EPDM rubber, 2014. http://www.exxonmobilchemical.com/Chem-English/brands/vistalon-ethylene-propylene-diene-epdm-rubber.aspx?ln=productsservices. Accessed March 7, 2014.
12. Dow Chemical Company, INFUSE™ olefin block copolymers, 2014. http://www.dow.com/infuse/index.htm. Accessed February 10, 2014.
13. Bensason S, Minick J, Moet A, Chum S, Hiltner A, Baer E. *J Polym Sci, Part B: Polym Phys* 1996;34:1301–1315.
14. Hiltner A, Wang H, Khariwala D, Cheung W, Chum S, Baer E. *ANTEC, SPE* 2006, Charlotte, NC, pp. 1000–1004.
15. Wang H, Chum S, Hiltner A, Baer E. *ANTEC, SPE* 2007, Cincinnati, OH.
16. De SK, White JR, eds. *Rubber Technologist's Handbook*, Vol. 1. Shawbury, U.K.: iSmithers Rapra Publishing 2001.
17. ExxonMobil Chemical Company, Vistalon™ automotive formulations, 2013. http://www.exxonmobilchemical.com/Chem-English/productsservices/vistalon-automotive-formulations.aspx. Accessed March 7, 2014.

4 Butyl Rubbers

Walter H. Waddell and Andy H. Tsou

CONTENTS

I. Introduction 109
II. Synthesis and Manufacture... 110
    A. Butyl Rubber 110
    B. Halobutyl Rubbers... 111
    C. Star-Branched Butyl Rubber 113
    D. Brominated Isobutylene-*co-para*-Methylstyrene.................................... 113
III. Structure 114
    A. Polyisobutylene 114
    B. Butyl Rubber 114
    C. Halogenated Butyl Rubber 114
    D. Star-Branched Butyl Rubber 115
    E. Brominated Isobutylene-*co-para*-Methylstyrene.................................... 116
IV. Physical Properties 116
    A. Permeability 116
    B. Dynamic Damping... 118
V. Chemical Properties 118
    A. Solubility 118
    B. Stability 119
    C. Vulcanization 119
VI. Applications.. 120
    A. Tire Innerliner 121
    B. Tire Black Sidewall 125
    C. Tire White Sidewall and Cover Strip 126
    D. Tire Treads 126
    E. Tire Curing Bladders and Envelopes 128
    F. Automotive Hoses 129
    G. Dynamic Parts... 130
    H. Pharmaceuticals 130
References... 134

I. INTRODUCTION

Isobutylene-based elastomers include butyl rubber, halogenated butyl rubbers, star-branched versions of these polymers, and the terpolymer brominated isobutylene-*co-para*-methylstyrene (BIMSM). A number of recent reviews on the manufacture,

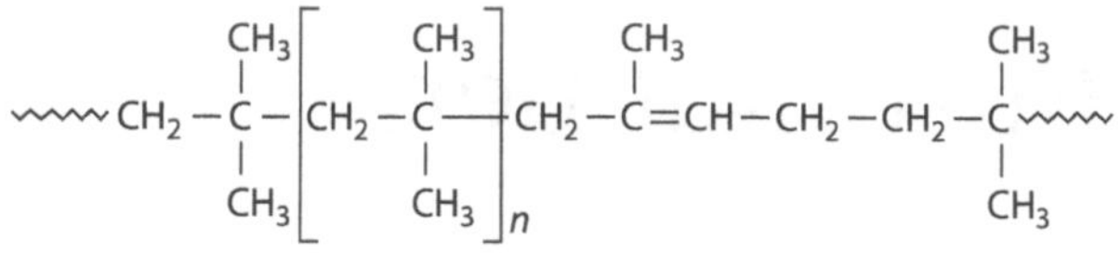

$$\text{\textbackslash}\text{CH}_2-\underset{\underset{CH_3}{|}}{\overset{\overset{CH_3}{|}}{C}}-\left[\text{CH}_2-\underset{\underset{CH_3}{|}}{\overset{\overset{CH_3}{|}}{C}}\right]_n \text{CH}_2-\underset{}{\overset{\overset{CH_3}{|}}{C}}=\text{CH}-\text{CH}_2-\text{CH}_2-\underset{\underset{CH_3}{|}}{\overset{\overset{CH_3}{|}}{C}}\text{\textbackslash}$$

FIGURE 4.1   Butyl rubber: poly(isobutylene-*co*-isoprene).

physical and chemical properties, and applications of isobutylene-based elastomers are available [1–7].

Butyl rubber (IIR) is the copolymer of isobutylene and a small amount of isoprene (see Figure 4.1). Patented in 1937 and first commercialized in 1943, the primary attributes of butyl rubber are excellent impermeability for use as an air barrier and good flex fatigue properties. These properties result from low levels of unsaturation in between the long polyisobutylene chain segments. Tire innertubes were the first major use of butyl rubber, and this continues to be a significant market today.

The development of halogenated butyl rubbers started in the 1950s. These polymers greatly extended the usefulness of butyl rubbers by having faster curing rates and increased polarity. This enabled covulcanization with general-purpose elastomers such as natural rubber (NR), butadiene rubber (BR), and styrene butadiene rubber (SBR) that are used in tire compounds. The enhanced cure properties do not affect the desirable impermeability and fatigue properties, thus permitting the development of more durable tubeless tires in which the air barrier is an innerliner compound chemically bonded to the carcass ply. Today, tire innerliners are the largest application for halobutyl rubber. Both chlorobutyl (CIIR) and bromobutyl (BIIR) rubbers are used commercially.

In addition to tire applications, isobutylene-based elastomers' good impermeability; resistance to ultraviolet light degradation, oxidation, and ozone; viscoelastic (dampening) characteristics; and thermal stability make butyl rubbers the polymers of choice for pharmaceutical stoppers, construction sealants, hoses, vibration isolation, and mechanical goods.

II. SYNTHESIS AND MANUFACTURE

A. Butyl Rubber

Kresge et al. [1] reviewed the synthesis and manufacture of isobutylene-based elastomers, which are summarized here. Butyl rubber (IIR) is prepared from high purity isobutylene (2-methylpropene, >99.5 wt%) and isoprene (2-methyl-1,3-butadiene, >98 wt%). The mechanism of polymerization consists of complex cationic reactions [8–10]. The catalyst system is a Lewis acid coinitiator and an initiator. Typical Lewis acid coinitiators include aluminum trichloride, alkylaluminum dichloride, boron trifluoride, tin tetrachloride, and titanium tetrachloride. Initiators are Brønsted acids such as water, hydrochloric acid, organic acids, or alkyl halides.

The isobutylene monomer reacts with the Lewis acid catalyst to produce a positively charged carbocation called a carbenium ion in the initiation step. Monomer units continue to be added in the propagation step until chain transfer or termination reactions occur. Temperature, solvent polarity, and the presence of counter ions affect the propagation of this exothermic reaction.

In the chain transfer step that terminates the propagation of a macromolecule, the carbenium ion of the polymer chain reacts with the isobutylene or isoprene monomers or with other species such as solvents or counter ions to halt the growth of this macromolecule and form a new propagating polymer chain. Lowering the polymerization temperature retards this chain transfer and leads to higher molecular weight butyl polymers. Isoprene is copolymerized mainly (>90%) by *trans*-1,4-addition. 1,2-Addition or branched 1,4-addition products are also observed. Termination also results from the irreversible destruction of the propagating carbenium ion either by the collapse of the ion pair, by hydrogen abstraction from the comonomer, by formation of stable allylic carbenium ions, or by reaction with nucleophilic species such as alcohols or amines. Termination is imposed after polymerization to control the molecular weight of the butyl rubber and to provide inactive polymer for further halogenation.

In the most widely used manufacturing process, a slurry of fine particles of butyl rubber dispersed in methyl chloride is formed in the reactor after Lewis acid initiation. The reaction is highly exothermic, and a high molecular weight can be achieved by controlling the polymerization temperature, typically between −90°C and −100°C. The most commonly used polymerization process uses methyl chloride as the reaction diluent and boiling liquid ethylene to remove the heat of reaction and maintain the low temperature needed. The final molecular weight of the butyl rubber is determined primarily by controlling the initiation and chain transfer reaction rates. Water and oxygenated organic compounds that can terminate the propagation step are minimized by purifying the feed systems.

The methyl chloride and unreacted monomers are flashed and stripped overhead by the addition of steam and hot water. They are then dried and purified in preparation for recycle to the reactor. Slurry aid (zinc or calcium stearate) and antioxidant are introduced to the hot water-polymer slurry to stabilize the polymer and prevent agglomeration. The polymer is then screened from the hot water slurry and dried in a series of extrusion dewatering and drying steps. Fluid bed conveyors and/or airvey systems are used to cool the hot polymer crumb to an acceptable packaging temperature. The resultant dried polymer is in the form of small crumbs, which are subsequently weighed and compressed into 75 lb bales before being wrapped in EVA film and packaged. Figure 4.2 is a schematic of the butyl rubber manufacturing process.

B. HALOBUTYL RUBBERS

Chlorobutyl (CIIR) and bromobutyl (BIIR) rubbers are commercially the most important derivatives of butyl rubber. The polymerization process for halobutyl rubber starts with exactly the same processes as for butyl rubber. A subsequent halogenation step is added. Reactor effluent polymer, in-process rubber crumb, or butyl product bales must be dissolved in a suitable solvent (e.g., hexane or pentane), and all unreacted monomer removed in preparation for halogenation. Bromine liquid or chlorine vapor is added to the butyl solution in highly agitated reaction vessels. These ionic halogenation reactions are fast. One mole of hydrobromic or hydrochloric acid

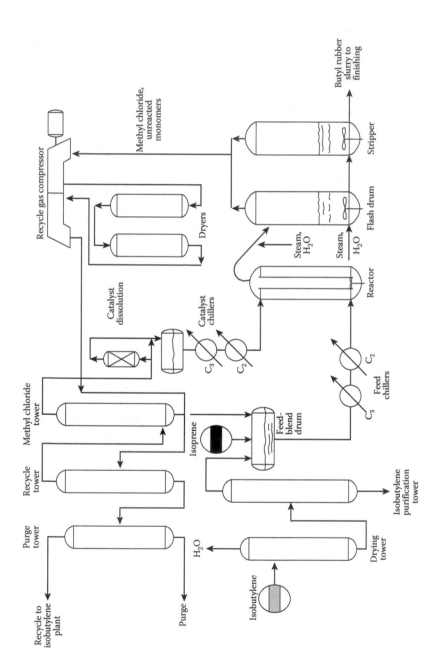

FIGURE 4.2 Commercial butyl rubber slurry polymerization process. (From Kresge, E.N. et al., Isobutylene polymers, in: Kroschwitz, J.I., ed., *Encyclopedia of Polymer Science and Engineering*, Vol. 8, 2nd edn., Wiley, New York, 1987, p. 423. With permission.)

is released for every mole of halogen that reacts; therefore, the reaction solution must be neutralized with caustic such as sodium hydroxide. The solvent is then flashed and stripped by steam or hot water, with calcium stearate added to prevent polymer agglomeration. The resultant polymer–water slurry is screened, dried, cooled, and packaged in a process similar to that of regular (unhalogenated) butyl rubber.

C. STAR-BRANCHED BUTYL RUBBER

Star-branched butyl rubbers (SBBs) have a bimodal molecular weight distribution [11] (e.g., see Figure 4.3). High molecular weight branched components and low molecular weight linear components are both present.

Star-branched butyl rubber is prepared by conventional cationic copolymerization of isobutylene and isoprene at low temperature in the presence of a polymeric branching agent. The high molecular weight branched molecules are formed during the polymerization via a graft mechanism. Useful SBBs comprise 10%–20% high molecular weight components [12]. A star molecule contains 20–40 butyl branches.

Star-branched butyl rubbers have viscoelastic properties that result in measurably improved processability. Improvements include dispersion of the polymer during mixing, higher mixing rates, higher extrusion rates, lower die swell, reduced shrinkage, and improved surface quality. The balance between green strength and stress relaxation properties at ambient processing temperatures is also improved [13]. Thus, operations such as shaping the innerliner compound during tire building are easier.

D. BROMINATED ISOBUTYLENE-*CO-PARA*-METHYLSTYRENE

As is the case with isoprene to form butyl rubber, *para*-methylstyrene is copolymerized with isobutylene in a cationic polymerization using a Lewis acid at low

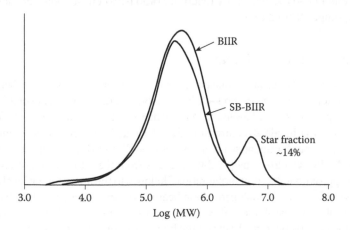

FIGURE 4.3 Molecular weight distribution of bromobutyl and star-branched bromobutyl rubbers.

temperature. Because of the similar reactivities, the resultant copolymer has a random incorporation of comonomer and has the composition of the feed monomer ratio. A reactive benzyl bromide functionality, $C_6H_5CH_2Br$, is introduced by the selective free radical bromination of the methyl group of the pendant methylstyryl group in the copolymer. This new functionalized copolymer preserves polyisobutylene properties such as excellent impermeability and vibration damping while increasing the resistance to oxidative, ozone, and heat aging.

III. STRUCTURE

A. POLYISOBUTYLENE

Isobutylene polymerizes in a head-to-tail sequence, producing a rubber that has no asymmetrical carbon atoms. The *geminal*-dimethyl group has two methyl groups bonded to the same carbon atom $[-C(CH_3)_2)-]$ on alternative chain atoms along the polyisobutylene backbone, producing a steric crowding effect. Distorting the hydrogen atoms of the methylene carbon $(-CH_2-)$ from the normal tetrahedral $109.5°$ to $124°$ and the dihedral angle of the carbon–carbon single bond backbone by about $25°$ relieves some strain [14–16]. Polyisobutylene has a glass transition temperature (T_g) of about $-70°C$ [17]. It is an amorphous elastomer in the unstrained state but crystallizes upon stretching at room temperature. The molecular weight distribution is the most probable, M_w/M_n of 2.

B. BUTYL RUBBER

In butyl rubber, the isoprene is enchained predominantly (90%–95%) by 1,4-addition in a head-to-tail arrangement [18–21]. Depending on the grade, the unsaturation in butyl rubber due to isoprene incorporation is between 0.5% and 3 mol%. T_g is approximately $-60°C$. A random distribution of unsaturation is achieved because of the low isoprene content and the near-unity reactivity ratio between isoprene and isobutylene [9]. M_w/M_n ranges from 3 to 5.

C. HALOGENATED BUTYL RUBBER

The *geminal*-dimethyl groups adjacent to the unsaturation in butyl rubber prevent halogen addition across the carbon–carbon double bond. Rather, halogenation at the isoprene site proceeds by a halonium ion mechanism, leading to the formation of an exomethylene alkyl halide structure in both chlorinated and brominated rubbers (see Figure 4.4). This predominant structure is about 90% based on ^{13}C nuclear magnetic resonance (NMR) spectroscopy [22,23]. It results from the introduction of bromine or chlorine at approximately a unit molar ratio of halogen to the unsaturation level to afford a product with 1.5–2 mol% halogen. Upon heating, the *exo*-allylic halide rearranges to give an equilibrium distribution of exo and endo structures [24–26] (see Figure 4.5). Halogenation has no apparent effects on the butyl backbone structure or on the T_g value. However, cross-linked halobutyl rubbers do not crystallize upon extension, probably because of backbone irregularities introduced by the halogenation process.

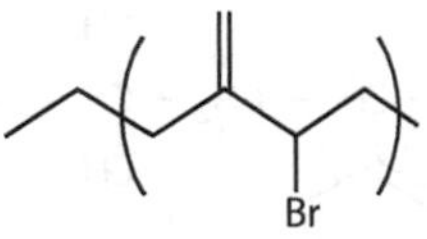

FIGURE 4.4 Most abundant isomer of bromobutyl rubber. (Cl in place of Br for chlorobutyl rubber.)

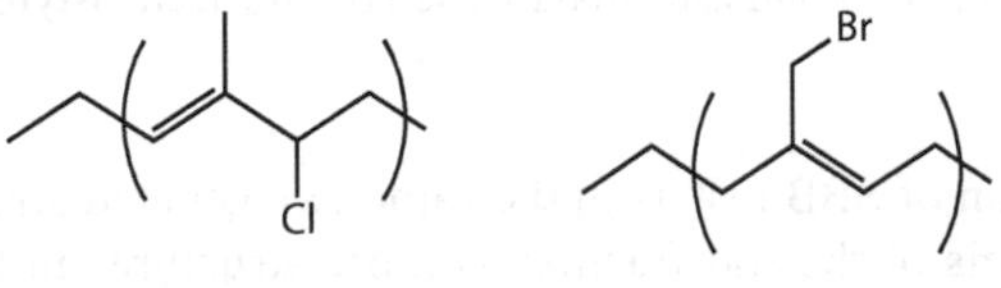

FIGURE 4.5 Minor isomers of chlorobutyl rubber or bromobutyl rubber.

D. STAR-BRANCHED BUTYL RUBBER

Introduction of a styrene butadiene styrene (SBS) block copolymer during the polymerization of butyl rubber leads to a star-branched rubber. SBB is a reactor blend of linear polymers and star polymers (generally 10%–20% by weight [12]); the star molecules were synthesized during polymerization by the cationic grafting of propagating linear butyl chains onto the branching agent (see Figure 4.6). A broad molecular weight distribution is achieved with $M_w/M_n > 8$.

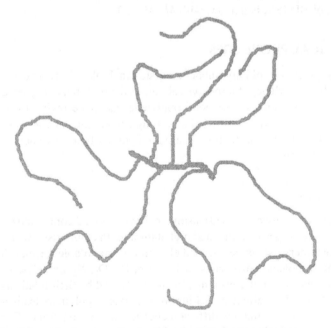

FIGURE 4.6 Schematic drawing of a star-branched butyl rubber chain.

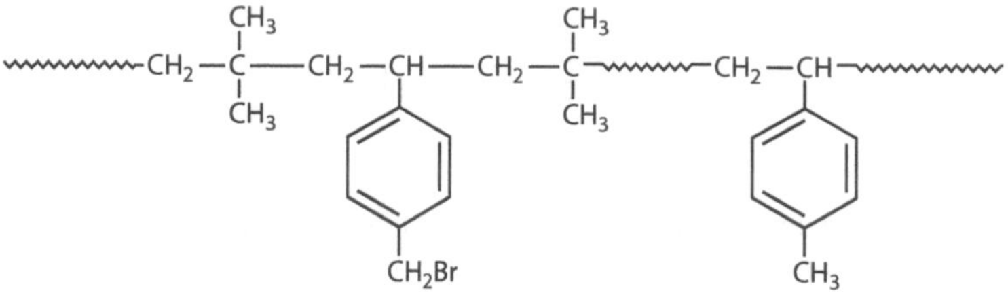

FIGURE 4.7 Structure of brominated isobutylene-*co-para*-methylstyrene (BIMSM).

The halogenation of SBB results in the same halogenated structures in the linear butyl chain arms of the star fraction as those structures in halogenated butyl rubber.

E. BROMINATED ISOBUTYLENE-*CO-PARA*-METHYLSTYRENE

Copolymerization of isobutylene with *para*-methylstyrene produces a saturated copolymer backbone with randomly distributed pendant *para*-methylstyrene substituted aromatic rings. During radical bromination after polymerization, some of the substituted *para*-methylstyrene groups are converted to reactive bromomethyl groups for vulcanization and functionalization [27]. These saturated terpolymers contain isobutylene, 1–8 mol% *para*-methylstyrene, and 0.5–2.5 mol% brominated *para*-methylstyrene (see Figure 4.7). Their T_g values increase with increasing *para*-methylstyrene content and are around –58°C. The molecular weight distribution of BIMSM is narrow, with $M_w/M_n < 3$.

IV. PHYSICAL PROPERTIES

The physical properties of butyl rubber are listed in Table 4.1 [1]. The physical properties of polyisobutylene, chlorobutyl rubber, and bromobutyl rubber are similar. The rotational restriction of the polyisobutylene backbone owing to the presence of the *geminal*-dimethyl groups results in a high interchain interaction and unique William–Landel–Ferry constants compared to hydrocarbon elastomers of similar T_g such as natural rubber.

A. PERMEABILITY

Primary uses of isobutylene-based elastomers in vulcanized compounds rely on their properties of low air permeability and high damping. In comparison with many other common elastomers, isobutylene-based elastomers are notable for their low permeability to small-molecule diffusants such as He, H_2, O_2, N_2, and CO_2 as a result of their efficient intermolecular packing [28], as evidenced by their relatively high density (0.917 g/cm³). This efficient packing in isobutylene polymers leads to their low fractional free volumes and low diffusion coefficients for penetrants. The diffusivities of gases in butyl rubber and natural rubber are given in Table 4.2 [29].

TABLE 4.1
Physical Properties of Butyl Rubber

Property	Value	Composition
Density, g/cm^3	0.917	B
	1.130	CBV
Coefficient of volume expansion, $(1/V)\,(V/T)$, K	560×10^{11}	BV
	460×10^{11}	CBV
Glass transition temperature, °C	−75 to −67	B
	1.95	B
Heat capacity, C_p, kJ/(kg·K)[a]	1.85	BV
	0.130	BV
Thermal conductivity, W/(m·K)	0.230	CBV
Refractive index, n_p	1.5081	B

Source: Data from Kresge, E.N. et al., Isobutylene polymers, in: Kroschwitz, J.I., ed., *Encyclopedia of Polymer Science and Engineering*, Vol. 8, 2nd edn., Wiley, New York, 1987, p. 423.
Note: B, butyl rubber; BV, vulcanized butyl rubber; CBV, vulcanized butyl rubber with 50 phr black.
[a] To convert J to cal, divide by 4.184.

TABLE 4.2
Diffusivity for Gases in Butyl Rubber and Natural Rubbers at 25°C

	Diffusivity (cm^2/s) $\times 10^6$	
Gas	Butyl Rubber	Natural Rubber
He	5.93	21.6
H$_2$	1.52	10.2
O$_2$	0.081	1.58
N$_2$	0.045	1.10
CO$_2$	0.058	1.10

Source: Data from Kresge, E.N. et al., Isobutylene polymers, in: Kroschwitz, J.I., ed., *Encyclopedia of Polymer Science and Engineering*, Vol. 8, 2nd edn., Wiley, New York, 1987, p. 423.

As shown in Figure 4.8, diffusion coefficients of nitrogen in both various diene rubbers and butyl rubber increase with increasing differences between the measurement temperature and the corresponding rubber's glass transition temperature. However, although the rate of increase in diffusion coefficient with $T–T_g$ is about the same for diene rubbers and butyl rubber, the absolute values of the diffusion coefficient in butyl rubber are significantly less than those of diene rubbers. Isobutylene copolymers contain only small amounts of comonomers, and their temperature-dependent permeability values follow the same curve as for butyl rubber (see Figure 4.8). BIMSM has the highest T_g value among isobutylene copolymers and has the lowest permeability at a given temperature.

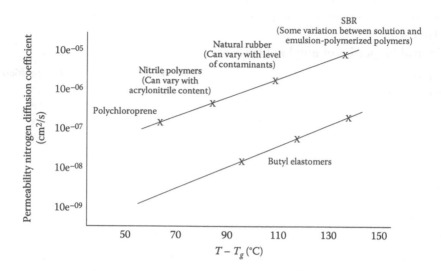

FIGURE 4.8 Diffusion coefficients of nitrogen in diene rubbers and in butyl rubber as a function of $T-T_g$. T, measurement of temperature; T_g, glass transition temperature. (After Boyd, R.H. and Krishna Pant, P.V., *Macromolecules*, 24, 6325, 1991; 25, 494, 1992; 26, 679, 1993.)

B. DYNAMIC DAMPING

Polyisobutylene and isobutylene copolymers are high damping at 25°C, with loss tangents covering more than eight decades of frequencies even though their T_g values are less than −60°C [30,31]. This broad dispersion in polyisobutylene's dynamic mechanical loss modulus is unique among flexible-chain polymers and is related to its broad glass–rubber transition [32]. The broadness of the glass–rubber transition, as defined by the steepness index, for polyisobutylene is 0.65, which is much smaller than that of most polymers. In addition, polyisobutylene has the most symmetrical and compact monomer structure among amorphous polymers, which minimizes the intermolecular interactions and contributes to its unique viscoelastic properties [33,34]. As a result, a separation in time scale between the segmental motion and the Rouse modes is broader in glass–rubber transition, leading to the appearance of the sub-Rouse mode [32,35]. Considering the differences in temperature dependences of these motions, the glass transitions of polyisobutylene and its copolymers are thermorheologically complex, and they do not follow time–temperature superposition. Polyisobutylene and its copolymers have high entanglement molecular weights [36] and correspondingly low plateau moduli, which contribute to their high tack or self-adhesion in the uncross-linked state.

V. CHEMICAL PROPERTIES

A. SOLUBILITY

Polyisobutylene and its copolymers, including butyl, halobutyl, and BIMSM, are readily soluble in nonpolar solvents; cyclohexane is an excellent solvent, benzene is a moderate solvent, and dioxane and pyridine are non-solvents [1].

B. STABILITY

Polyisobutylene and butyl rubber have the chemical resistance expected of saturated hydrocarbons. The in-chain unsaturations of butyl rubbers can be slowly attacked by atmospheric ozone, leading to degradation, and therefore require protection by antioxidants. Oxidative attack results in a loss of molecular weight rather than embrittlement.

Chlorobutyl rubbers are thermally more stable than bromobutyl rubbers. Upon thermal exposure up to 150°C, no noticeable decomposition takes place in chlorobutyl rubber except for some allylic chlorine rearrangement, whereas the elimination of HBr occurs in bromobutyl rubber concurrently with isomerization to produce conjugated dienes that subsequently degrade [25,26]. BIMSM has no unsaturation and is the most thermally stable isobutylene copolymer. In addition, the strong reactivity of the benzylic bromine functionality in BIMSM with nucleophiles allows the functionalization and grafting of BIMSM in addition to its uses for vulcanization [11,12].

C. VULCANIZATION

In butyl rubber, the hydrogen atoms positioned α to the carbon–carbon double bond permit vulcanization into a cross-linked network with sulfur and organic accelerators [37]. The low degree of unsaturation requires the use of ultra-accelerators such as thiuram or thiocarbamates. Phenolic resins, bias-zidoformates [38], and quinone derivatives can also be employed. Vulcanization introduces a chemical cross-link approximately every 250 carbon atoms along the polymer chain, producing a covalent network. Sulfur cross-links have limited stability at elevated temperature and can rearrange to form new cross-links. This rearrangement results in a permanent set and creep for vulcanizates exposed to high temperature for long periods of time. Resin cure systems provide carbon–carbon cross-links and heat-stable vulcanizates; alkyl phenol-formaldehyde derivatives are usually employed. Typical vulcanization systems are shown in Table 4.3 [1].

The presence of allylic halogens in halobutyl elastomers allows cross-linking by metal oxides and enhances the rate of sulfur vulcanization over that of butyl rubber. Halobutyl elastomers can be cross-linked by the same curatives as are used for butyl rubber and by zinc oxide, bismaleimides, diamines, peroxides, and dithiols. The allylic halogen allows more cross-linking than is possible in elastomers with only allylic hydrogens. Halogen is a good leaving group in nucleophilic substitution reactions. When zinc oxide is used to cross-link halobutyl rubber, carbon–carbon bonds are formed through dehydrohalogenation to form a zinc halide catalyst [25]. A very stable cross-link system is obtained for the retention of properties and low compression set. Typical vulcanization systems are also shown in Table 4.3 [1].

Brominated isobutylene-*co-para*-methylstyrene cross-linking involves the formation of carbon–carbon bonds, generally through alkylation chemistry or the formation of zinc salts such as zinc stearate [39,40]. Sulfur vulcanization is achieved by using thiazoles, thiurams, and dithiocarbamates. Diamines, phenolic

TABLE 4.3

Some Typical Vulcanization Systems for Butyl and Halobutyl Rubbers[a]

	Butyl Rubber			Halobutyl Rubber			
	Sulfur/ Accelerator	Resin	Quinone	Sulfur/ Accelerator	Resin	RT Cure	Amine
Ingredient							
Zinc oxide	5	5	5	5	3	5	—
Lead oxide	—	—	2	—	—	—	—
Stearic acid	2	1	—	—	—	—	—
Sulfur	2	—	—	0.5	—	—	—
MBTS[b]	0.5	—	—	1.5	—	—	—
TMTD[c]	1.0	—	—	0.25	—	—	—
Magnesium oxide	—	—	—	0.5	—	—	3
Hexamethylene diamine carbamate	—	—	—	—	—	—	1
SP-1045 resin	—	—	—	—	5	—	—
SP-1055 resin	—	12	—	—	—	—	—
Benzoquinone dioxime	—	—	2	—	—	—	—
Tin chloride	—	—	—	—	—	2	—
Zinc chloride	—	—	—	—	—	2	—
Conditions							
T, °C	155	180	180	160	160	25	160
t, min	20	80	80	20	15	—	15

[a] Concentrations are in parts per 100 parts of rubber.
[b] Benzothiazyl disulfide.
[c] Tetramethylthiuram disulfide.

resins, and thiosulfates [41] are also used to cross-link BIMSM elastomers. The stability of these bonds combined with the chemically saturated backbone of BIMSM yields excellent resistance to heat and oxidative aging and to ozone attack. Table 4.4 is a summary [5].

VI. APPLICATIONS

Isobutylene-based elastomers are used commercially in a number of rubber components and products. Rogers and Waddell [5] reviewed their use in tires and in automotive parts. Commercial tire applications include use in the innerliner, nonstaining black sidewall, white sidewall, white sidewall coverstrip, and tread compounds.

TABLE 4.4

Vulcanization Systems for Brominated Isobutylene-*co-para*-Methyl-Styrene Rubber[a]

Ingredient	Metal Oxide	Sulfur/ Accelerator	Ultra-Accelerator	Resin	Amine
Zinc oxide	2	1	1	1	1
Zinc stearate	3	—	—	—	—
Stearic acid	—	2	2	2	2
Sulfur	—	1	—	1.5	—
MBTS[b]	—	2	—	1.5	—
ZDEDC[c]	—	—	2	1	—
Triethylene glycol	—	—	2	1	—
SP-1045 resin	—	—	—	5	—
DPPD[d]	—	—	—	—	0.5
Conditions					
T, °C	160	160	160	160	160
t, min	25	20	10	20	10

Source: Rogers, J.E. and Waddell, W.H., *Rubber World*, 219(5), 24, 1999.

[a] Concentrations are in parts per 100 parts of rubber.

[b] Benzothiazyl disulfide.

[c] Zinc diethyldithiocarbamate.

[d] Diphenyl-*para*-phenylenediamine.

A. TIRE INNERLINER

The innerliner is a thin layer of rubber laminated to the inside of a tubeless tire to ensure retention of air (see Figure 4.9). It is generally formulated with halobutyl rubber to provide good air and moisture impermeability, flex-fatigue resistance, and durability [42]. The integrity of the tire is improved by using halobutyl rubber in the innerliner because it minimizes the development of intercarcass pressure, which could lead to belt edge separation, adhesion failures, and the rusting of steel tire cords [43].

Innerliners for passenger tires can be formulated with a blend of chlorobutyl rubber and natural rubber (e.g., see Table 4.5 [44]) or bromobutyl rubber (see Table 4.6 [5]). Many factors favor the use of bromobutyl rubber over chlorobutyl rubber [45]. These include (1) superior adhesion to carcass compounds, (2) better balance of properties, (3) increasing use of speed-rated tires with lower profiles having higher ratios of surface area to air volume, (4) requirement for lighter tires to reduce rolling resistance for fuel efficiency, (5) use of high-pressure space-saver spare tires requiring a more impermeable liner, (6) better flex-cracking resistance after aging, and (7) cheaper material costs. A chlorobutyl rubber–natural rubber innerliner would have to be thicker than a 100 phr chlorobutyl rubber liner to obtain the same air

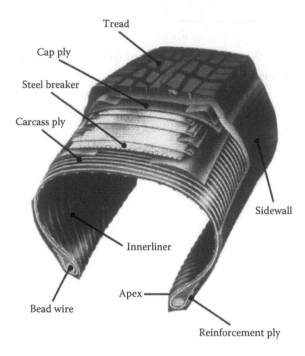

FIGURE 4.9   Cross section of a tubeless radial tire.

TABLE 4.5
Chlorobutyl Rubber/Natural Rubber Innerliner Formulation (phr)

Chlorobutyl rubber	90
Natural rubber	10
GPF carbon black, N660	70
Stearic acid	2
Zinc oxide	3
Lubricant	11
Tackifier	10
Activator	1.3
Sulfur	0.5

Source: Hopkins, W. et al., Bromobutyl and chlorobutyl. A comparison of their chemistry, properties and uses, in: *International Rubber Conference Proceedings*, Kyoto, Japan, October 15–18, 1985, p. 205, Paper 16A10. With permission.

impermeability (see Table 4.7). The permeability increases essentially linearly with increasing natural rubber content [43].

Star-branched bromobutyl rubber (BrSBB) was developed for use in tire innerliner compounds to improve the processability of bromobutyl rubber [11,13]. BIMSM has been evaluated in off-the-road tires (see Table 4.8 [46]) because heat buildup and flex

TABLE 4.6
Bromobutyl Rubber Innerliner Formulation (phr)

Bromobutyl rubber	100
N660 carbon black	60
Naphthenic processing oil, Flexon 876	15
Stearic acid	1
Zinc oxide	3
MBTS accelerator	1.5
Sulfur	0.5

Source: Rogers, J.E. and Waddell, W.H., *Rubber World*, 219(5), 24, 1999.

TABLE 4.7
Effect of Blending Halobutyl Rubber with Natural Rubber

	Halobutyl Content							
	100 phr		80 phr		60 phr		40 phr	
	BIIR	CIIR	BIIR	CIIR	BIIR	CIIR	BIIR	CIIR
Unaged								
300% Modulus, MPa	4.2	3.7	5.7	5.1	7.1	5.7	8.9	4.3
Tensile, MPa	9.3	9.9	10.9	10.7	12.8	10.3	14.7	9.7
Elongation at break, %	740	770	620	620	560	560	490	580
Air aged 168 h at 100°C								
300% Modulus, MPa	6.8	5.5	7.6	7.9	8.4	7.7	6.7	3.6
Tensile, MPa	10.0	10.9	9.8	11.0	9.3	9.2	8.8	5.8
Elongation at break, %	550	640	420	465	320	365	370	475
Permeability to air, 50 psi at 65°C ($Q \times 10^{-8}$)	2.9	2.9	5.4	5.7	9.2	7.5	13.8	13.2
Adhesion at 100°C								
To self, kN·m	16.8	4.4	14.7	4.7	15.2	9.1	15.4	5.2
To NR, kN·m	7.5	1.3	6.2	6.2	14.7	1.9	20.8	2.9
Flex fatigue, air-aged 168 h at 120°C, Cam No. 24 (kilocycles to failure)	61.8	72.7	23.6	3.9	0.3	0.1	0.0	0.0

Source: Data from Hopkins, W. et al., Bromobutyl and chlorobutyl. A comparison of their chemistry, properties and uses, in: *International Rubber Conference Proceedings*, Kyoto, Japan, October 15–18, 1985, p. 205, Paper 16A10.

Recipe: Halobutyl/NR, 100 phr; N660 black, 60; paraffinic oil, 7; pentalyn A, 4; stearic acid, 1; zinc oxide, 3; MBTS, 1.25; sulfur, 0.5.

TABLE 4.8
Brominated Isobutylene-*co-para*-Methylstyrene Innerliner Formulation (phr)

BIMSM (Exxpro™ MDX 89-4)	100
N660 carbon black	60
Naphthenic processing oil, Flexon 641	8
Tackifying resin, Escorez	2
Phenolic resin	2
Resin, Struktol 40 MS	7
Stearic acid	2
Zinc oxide	3
MBTS accelerator	1.5
Sulfur	0.5

Source: Data from Costemalle, B. et al., *J. Elastomers Plast.*, 27, 39, 1995.

TABLE 4.9
Comparison among 100 phr Innerliners

Property	CIIR 1066	BIIR 2222	BIMSM
Mooney viscosity, ML 1 + 4 at 100°C	46	44	56
Mooney scorch			
T5 at 135°C, min	13	16	22
T90 at 160°C, min	15	12	12
Hardness, Shore A	40	42	40
100% Modulus, MPa	1.0	1.0	1.0
Tensile strength, MPa	9.2	10	9
Elongation at break, %	715	745	950
Strain energy			
(tensile strength × elongation)			
Initial	6578	7450	8550
After 3 days at 125°C	3791	4878	7986
After 4 days at 100°C	4034	4075	7769
After 7 days at 180°C	0	0	2682
Monsanto flex, kilocycles			
Initial	360	85	660
After 3 days at 125°C	53	23	260
After 4 weeks at 100°C	25	11	200

Source: Data from Jones, G.E., in: *ITEC'98 Select*, July 1999, p. 13.

TABLE 4.10
Butyl Rubber Tire Innertube Formulation (phr)

Butyl rubber	100
N660 carbon black	70
Paraffinic process oil	25
Zinc oxide	5
Stearic acid	1
MBT accelerator	0.5
TMTDS accelerator	1
Sulfur	2

Source: Data from Jones, G.E. et al., Butyl rubber, in: Dick, J.S., ed., *Rubber Technology, Compounding and Testing for Performance*, Hanser, Munich, Germany, 2001, p. 173.

---

characteristics are improved compared to halobutyl rubbers (see Table 4.9 [47]). A butyl rubber innertube formulation is shown in Table 4.10 [6].

B. TIRE BLACK SIDEWALL

The black sidewall is the outer surface of the tire that protects the casing against weathering. It is formulated for resistance to weathering, ozone, abrasion and tear, and radial and circumferential cracking and for good fatigue life [42]. Traditionally, blends of natural rubber and butadiene rubber are used, but high concentrations of antidegradants are required to provide weather resistance. However, an in-service surface discoloration occurs upon exposure to ozone when using *para*-phenylenediamine antiozonants as protectants [48].

To achieve a stain-resistant black sidewall over the life of a tire, inherently ozone-resistant, saturated-backbone polymers are used in blends with diene rubbers. BIMSM is used in nonstaining passenger tire black sidewalls [46,49–53]. At least 40 phr of BIMSM rubber is needed to protect the natural rubber from ozone attack in order for it to form a co-continuous inert phase [49]. Black sidewalls with BIMSM blends outperformed sidewalls with EPDM blends [52]. The bromination and the *para*-methylstyrene comonomer levels are important factors for ozone resistance. The BIMSM rubber phase must be highly dispersed to minimize crack growth [51], and a three-step remill type of mixing sequence is generally needed to achieve dispersion and co-continuity. Use of a BIMSM rubber with a low bromination level and high *para*-methylstyrene comonomer content resulted in property improvements [51,53]. Tires having BIMSM elastomers in the black sidewall enhanced the tire appearance. A nonstaining black sidewall formulation is shown in Table 4.11 [53].

TABLE 4.11

BIMSM Elastomer Black Sidewall Compound (phr)

BIMSM (Exxpro™ MDX 96-4)	50
Polybutadiene rubber	41.67
Natural rubber	8.33
N330 carbon black	40
Oil, Flexon 641	12
Tackifying resin, Escorez 1102	5
Resin, Struktol 40 MS	4
Resin, SP-1068	2
Stearic acid	0.5
Sulfur	0.32
Zinc oxide	0.75
Rylex 3011 accelerator	0.6
MBTS accelerator	0.8

Source: Tisler, A.L. et al., New grades of BIMSM for non-stain tire sidewalls, in: *Meeting of ACS Rubber Division*, Cleveland, OH, October 21–24, 1997, Paper 66. With permission.

C. Tire White Sidewall and Cover Strip

Chlorobutyl rubber–EPDM rubber–natural rubber blends are used in tire white sidewall compounds [54] (see Table 4.12) and in white sidewall cover strip compounds [55] (see Table 4.13). The chlorobutyl rubber imparts resistance to ozone aging, flex fatigue, and staining to the compounds.

D. Tire Treads

The tread is the wear-resistant component of a tire that comes in contact with the road. It is designed for abrasion resistance, traction, speed, stability, and casing protection. The tread rubber is compounded for wear, traction, low rolling resistance, and durability [42]. For passenger tires, it is normally composed of a blend of SBR and BR elastomers.

Butyl rubbers are used in blends with BR and NR (see Table 4.14) to improve the braking of a winter tire on ice, snow, and/or wet road surfaces; to lower rolling resistance; and to maintain wear resistance [56]. Superior grip and durability are obtained for a CIIR/SBR blend in high-speed tires [57]. Blends of bromobutyl rubber with BR and NR improve lab wear resistance, the coefficient of friction on ice, and tire operating stability on wet road surfaces [58]. Bromobutyl rubber, star-branched bromobutyl rubber, and BIMSM blends with SBR and BR increase tangent delta values at low temperatures (–30°C to +10°C), which is used as a lab

TABLE 4.12
Passenger Tire White Sidewall Recipe (phr)

Chlorobutyl rubber, 1066	55
Natural rubber, SMR 5	25
EPDM rubber, Vistalon 6505	20
Filler, Vantalc 6 H	34
Whitener, Titanox 1000 titanium dioxide	35
Clay, Nucap 200	32
Stearic acid	2
Resin, SP-1077	4
Ultramarine blue	0.2
Zinc oxide	5
Sulfur	0.8
Vultac 5 accelerator	1.3
Altax accelerator	1

Source: Prescott, P.I. and Rice, C.A. (to E.C.C. America), U.S. Patent 4,810,578, March 7, 1989.

TABLE 4.13
Passenger Tire White Sidewall Cover Strip Recipe (phr)

Natural rubber	50
Chlorobutyl rubber	30
Ethylene–propylene diene terpolymer	20
HAF carbon black	25
MT carbon black	75
Magnesium oxide	0.5
Stearic acid	1
Wax	3
Naphthenic oil	12
Zinc oxide	5
Sulfur	0.4
Alkyl phenol disulfide vulcanizing agent	1.34
Benzothiazyl disulfide accelerator	1

Source: Wilson, R.B. (to General Tire), U.S. Patent 3,508,595, April 28, 1970.

predictor of tire traction properties, and decreases tangent delta values at higher temperatures (>30°C), which is used as a lab predictor of rolling resistance [59]. BIMSM/BR/NR winter treads (see Table 4.15 [60,61]) had shorter braking distances on indoor ice, Alpine snow, and wet and dry road surfaces and improved traction on snow and wet asphalt surfaces compared to an SBR/BR/NR reference.

TABLE 4.14
Winter Passenger Tire Tread Recipe (phr)

Natural rubber	50
Polybutadiene rubber	35
Chlorobutyl or bromobutyl rubber	15
Carbon black, N339	80
Aromatic oil	35
Stearic acid	1
Antioxidant (IPPD)	1
Zinc oxide	3
Sulfur	1.5
Vulcanizing agents	1

Source: Takiguchi, E. (to Bridgestone), Gr Brit Patent 2,140,447 A, May 9, 1984; U.S. Patent 4,567,928, February 4, 1986.

TABLE 4.15
BIMSM Winter Tire Tread Compound (phr)

BIMSM, Exxpro™ 3745	20
BR, Buna CB 23	40
NR, SMR 20	40
Silica, Zeosil 1165 MP	60
Silane, X50S	10.2
Silica, Zeosil 1165 MP	15
Processing oil, Mobilsol 30	30
DPG accelerator	2
Stearic acid	1
Antiozonant, Santoflex 6PPD	1.5
Antioxidant, Agerite Resin D	1
Zinc oxide	2
Sulfur	1
TBBS accelerator	1.5

Source: Waddell, W.H. et al., Evaluation of isobutylene-based elastomers in a model winter tire tread, in: *Meeting of ACS Rubber Division*, Cleveland, OH, October 16–19, 2001, Paper 113; *Rubber Chem. Technol.,* 76(2), 348, 2003.

E. TIRE CURING BLADDERS AND ENVELOPES

Butyl rubber curing bladder recipes are given in Table 4.16 [62]. Because sulfur vulcanizates tend to soften during prolonged exposure to high temperatures (300°F–400°F), butyl rubber curing bladders are generally formulated with a heat-resistant resin cure system [2].

TABLE 4.16

Butyl Rubber and Brominated Isobutylene-*co-para*-Methylstyrene Tire Curing Bladder Formulations (phr)

Component		BIMSM
Butyl rubber	100	–
Chloroprene	5	–
BIMSM (Exxpro™ 3035)	–	100
N330 carbon black	50	55
Castor oil	5	5
Methylol phenol	7.5	–
Zinc oxide	5	2
Stearic acid		0.5
Resin, SP-1045		5
MBTS accelerator		1.5
Sulfur		0.75
Magnesium aluminum hydroxycarbonate		0.8

Sources: Costemalle, B. et al., Exxpro™ polymers, in: *Rubbercon'95*, Gothenburg, Sweden, May 9–12, 1995, Paper G11; Larson, L.C. et al., *J. Elastomers Plast.*, 22, 190, 1990.

BIMSM is used to fabricate longer life tire curing bladders (see Table 4.16) [50,63]. The BIMSM bladder formulation also serves as a curing envelope.

F. AUTOMOTIVE HOSES

Hose for automotive applications requires an elastomer that is resistant to the material it is transporting and has low permeability, low compression set, and resistance to increasingly higher under-the-hood temperatures. Applications of isobutylene-based elastomers include air-conditioning hose [64–68], coolant hose [69], fuel line hose [70], and brake line hose [71].

A polymer for an air-conditioning hose requires good barrier properties to minimize refrigerant loss and reduce moisture ingression, good compression set to help ensure coupling integrity, and high-temperature stability. Damping of compressor vibration and noise is also desirable. The hose is typically a composite of rubber layers and reinforcing yarn. Halobutyl rubber is used in hose covers because of its barrier properties and its resistance to moisture ingression. Chlorobutyl rubber as a cover for an air-conditioning hose provides better resistance to moisture ingression than EPDM and is compatible with operating temperatures up to 120°C [64]. The use of a butyl–halobutyl rubber blend as a layer between the nylon and cover eliminates the need for an adhesive (see Table 4.17) [65]. A BIMSM hose composition exhibits good physical property retention [66].

A bromobutyl rubber formulation affords better resistance to alternative fuels such as methanol and an 85:15 methanol–gasoline blend than a nitrile compound

TABLE 4.17
Bromobutyl Compound for Air-Conditioning Hose (phr)

Brominated butyl rubber	100	75
Butyl rubber	—	25
N330 carbon black	30	30
N774 carbon black	30	30
Precipitated silica, HiSil 233	20	20
Zinc oxide	5	5
Stearic acid	1	1
Antioxidant	1	1
Paraffinic oil, Sunpar 2280	2	2
Brominated alkyl phenol formaldehyde resin	10	10
Hardness, JIS K6262	74	75
Tensile strength, kg/cm^2	142	151
Elongation at break, %	250	260
Permanent set, 25% deflection, 72 h at 140°C	52.9	51.1
Adhesion to innermost layer, kg/in.	17.0	16.8

Source: Shiota, A. and Kitani, K. (to Nichirin Co., Ltd), U.S. Patent 5,488,974, February 6, 1996.

(see Table 4.18) [70]. It also provides the most resistance and is impermeable to Delco Supreme II brake fluid (see Tables 4.19 and 4.20) [71].

G. DYNAMIC PARTS

Isobutylene-based polymers are used for various types of automotive mounts because of their ability to damp vibrations from the road or engine, including body mounts and medium-damping engine mounts. Exhaust hanger straps use halobutyl rubber because of its heat resistance (see Table 4.21) [72]. A bromobutyl rubber–natural rubber blend affords a soft, fatigue-resistant compound. Polyisobutylene is also used as an additive to improve durability and fatigue resistance (see Table 4.22) [73].

Natural rubber–BIMSM blends improve heat aging. BIMSM use increases the damping at low temperatures without affecting properties at room and elevated temperatures [74,75].

H. PHARMACEUTICALS

Butyl and halobutyl rubbers are used in the pharmaceutical industry owing to their low permeability; resistance to heat, oxygen, ozone, and ultraviolet light; and inertness to chemicals and biological materials. Bromobutyl rubber can also be cured in the absence of sulfur and zinc compounds, thus providing for a nontoxic vulcanization system (see Table 4.23) [2].

TABLE 4.18
Comparison of Bromobutyl and Nitrile Compounds in Alternative Fuels

	Bromobutyl	Nitrile
Component, phr		
Bromobutyl rubber		100
NBR[a]		100
Stearic acid	1	1
N550 carbon black	70	
N762 carbon black		75
Atomite	30	
Magnesium oxide	0.3	
DOP		5
MBTS accelerator	1	
Zinc oxide	3	5
Sulfur		1.25
TMTD accelerator	0.4	
TMTM accelerator		0.5
Physical properties, cured 10 min at 166°C		
Hardness, Shore A	75	68
100% Modulus, MPa	2.9	3.4
300% Modulus, MPa	9.0	15.4
Tensile, MPa	9.5	19.5
Elongation, %	320	440
Aged in methanol, 168 h at RT, change in		
Hardness, pt	−2	−8
Tensile strength, %	+4	−22
Elongation, %	+5	−31
Volume, %	−2	+11
Aged in M85, 168 h at RT, change in		
Hardness, pt	−26	−16
Tensile strength, %	−21	−37
Elongation, %	−22	−44
Volume, %	+29	+24
Aged in Fuel C, 168 h at RT, change in		
Hardness, pt	−43	−26
Tensile strength, %	−63	−56
Elongation, %	−67	−59
Volume, %	+220	+51
Permeability (weight loss in grams after 14 days)		
Methanol	0.2	1.48
M*%	0.42	4.20

Source: Dunn, J.R., *Elastomerics*, 15, January 1991. With permission.

[a] NBR, Polysar Krynac 3450.

* Percent modulus change after 14 days.

TABLE 4.19
Comparison of Elastomer Resistance to Delco Supreme II Brake Fluid

Polymer	Volume Change	Durometer Change	Permeability Constant K_p, (g·cm)/(cm^2·h)	Loss, g/h
Nitrile rubber	+84	–0		
Chlorinated polyethylene	+10	–11	32.53×10^{-5}	0.110
Neoprene	+9	–8	66.02×10^{-5}	0.200
Silicone	+3	–4	59.14×10^{-5}	0.191
Butyl rubber	+1	–6	4.38×10^{-5}	0.021
EPDM	–12	+4	20.13×10^{-5}	0.063

Source: Trexler, H.E., The development of automotive brake hose elastomer compositions, in: *Meeting of ACS Rubber Division*, Denver, CO, October 23–26, 1984. With permission.

TABLE 4.20
Bromobutyl Compounds for Brake Hose Application

Component, phr		
Bromobutyl rubber	100	100
N330 carbon black	60	
N774 carbon black		80
Oil, Sunpar 2280	15	
Zinc oxide	3	
MgO	0.5	
Resin, SP-1055	4	
Stearic acid	1	
MBTS accelerator	3	
HVA-2		1.5
Di-Cup 40KE vulcanizing agent		1.5
Physical properties		
Hardness, Shore A	55	56
Tensile strength, MPa	11.9	11.9
Elongation, %	720	240
Clash Berg brittleness, °C (ASTM D 1043)	–70	–63
Aged properties at 125°C		
Permeability K_p, cc-mm/(m^2-hour)	2.21	1.16
Volume change, 70 h, Delco Supreme II brake fluid, %	+8	+6
Compression set, 70 h, %	67	20

Source: Trexler, H.E., The development of automotive brake hose elastomer compositions, in: *Meeting of ACS Rubber Division*, Denver, CO, October 23–26, 1984. With permission.

TABLE 4.21
Heat-Resistant Diamine-Cured Bromobutyl Compound

Component, phr		
Bromobutyl rubber	100	100
N550 carbon black	50	50
Stearic acid	1	
Zinc oxide	3	3
Diamine resina[a]	2.5	2.5
Physical properties		
Hardness, Shore A	69	67
100% Modulus, MPa	4.9	4.9
Tensile strength, MPa	11.5	12.3
Elongation, %	300	300
Aged physical properties, aged 168 h at 150°C		
Hardness change, pts	−3	+3
100% Modulus change, %	+8.2	+16.3
Tensile change, %	+19.1	−8.9
Elongation change, %	−33.4	−26.7

Source: Dunn, J.R., *Elastomerics*, 29, February 1989. With permission,
[a] Agerite White, di-2-naphthyl-*p*-phenylenediamine.

TABLE 4.22
Fatigue Resistance of Natural Rubber and Bromobutyl Blend Engine Mounts

NR/BIIR Ratio	Tensile Strength (MPa)	Tear Elongation (%)	Strength (kN/m)	Hardness (Shore A)	Comp. Set (%)	Tan Delta	Fatigue (kilocycles)
100/0	19.6	580	42.4	41	28	0.076	31
80/20	16.8	595	38.9	41	32	0.135	63
70/30	15.4	590	38.9	41	31	0.162	88
60/40	16.0	625	30.6	40	29	0.181	88
50/50	13.9	600	25.5	39	28	0.221	83

Source: Tabar, R.J. and Killgoar, P.C. (to Ford Motor Co), U.S. Patent 4,419,480, December 6, 1983.
Recipe includes (phr): PIB, 20; N765, 25; stearic acid, 2; TMQ, 2; 6-PPD, 1; aromatic oil, 5; zinc oxide, 5; sulfur, 0.6; *N*-oxydiethylene thiocarbamyl-*N*-oxydiethylene sulfenamide, 1.4; N-oxydiethylene 2-benzothiazole sulfenamide, 0.7.

TABLE 4.23

Bromobutyl Rubber Pharmaceutical Closure Recipe (phr)

Bromobutyl rubber	100
Whitetex 2	60
Primol 355 oil	5
Polyethylene AC617A	3
Paraffin wax	2
Vanfre AP2	2
Stearic acid	1
Diak 1 vulcanizing agent	1

Source: Fusco, J.V. and Hous, P., Butyl and halobutyl rubbers, in: Morton, M., ed., *Rubber Technology*, 3rd edn., Van Nostrand Reinhold, New York, 1987, p. 284. With permission.

TABLE 4.24

BIMSM Rubber Pharmaceutical Closure Recipe (phr)

BIMSM, Exxpro™ MDX 89-1	100
Polestar 200R	90
Parapol 2255 plasticizer	5
Polyethylene wax	3
TiO_2	4
MgO	1
Diak 1 vulcanizing agent	0.75

Source: Costemalle, B. et al., Exxpro™ polymers, in: *Rubbercon'95*, Gothenburg, Sweden, May 9–12, 1995, Paper G11. With permission.

Brominated isobutylene-*co-para*-methylstyrene offers potential advantages over halobutyl rubber in health-care applications: lower volatiles and chemical additive levels, lower polymer bromine levels, and a higher clarity product. Because BIMSM is a totally saturated elastomer, it is also more stable to gamma radiation, which is often used as a sterilization treatment and can be cured using a sulfur- and zinc-free system (see Table 4.24) [50].

REFERENCES

1. Kresge EN, Schatz RH, Wang H-C. Isobutylene polymers. In: Kroschwitz JI, ed., *Encyclopedia of Polymer Science and Engineering*, Vol. 8, 2nd edn. New York: Wiley, 1987, p. 423.
2. Fusco JV, Hous P. Butyl and halobutyl rubbers. In: Morton M, ed., *Rubber Technology*, 3rd edn. New York: Van Nostrand Reinhold, 1987, p. 284.
3. Fusco JV, Hous P. Butyl and halobutyl rubbers. In: Ohm RF, ed., *The Vanderbilt Rubber Handbook*, 13th edn. Norwalk, CT: RT Vanderbilt Co., 1990, p. 92.

4. Kresge EN, Wang H-C. Butyl rubber. In: Howe-Grant M, ed., *Kirk-Othmer Encyclopedia of Chemical Technology*, Vol. 8, 4th edn. New York: Wiley, 1993, p. 934.
5. Rogers JE, Waddell WH. *Rubber World* 1999;219(5):24.
6. Jones GE, Tracey DS, Tisler AL. Butyl rubber. In: Dick JS, ed., *Rubber Technology, Compounding and Testing for Performance*. Munich, Germany: Hanser, 2001, p. 173.
7. Webb RN, Shaffer TD, Tsou AH. Commercial isobutylene polymers. In: Kroschwitz JI, ed., *Encyclopedia of Polymer Science and Technology*, Online edition. New York: Wiley, 2003.
8. Plesch PH, Gandini A. *The Chemistry of Polymerization Process*. Monograph 20. London, U.K.: Society of Chemical Industry, 1966.
9. Kennedy JP, Marechal M. *Carbocationic Polymerization*. New York: Wiley, 1982.
10. Matyjaszewski K, ed. *Cationic Polymerizations*. New York: Marcel Dekker, 1996.
11. Wang H-C, Powers KW, Fusco JV. Star branched butyl—A novel butyl rubber for improved processability. I. Concepts, structure and synthesis. In: *Meeting of ACS Rubber Division*, Mexico City, Mexico, May 9–12, 1989, Paper 21.
12. Powers KW, Wang H-C, Handy DC, Fusco JV (to Exxon Chemical). U.S. Patent 5,071,913, February 10, 1991.
13. Duvdevani I, Gursky L, Gardner IL. Star branched butyl—A novel butyl rubber for improved processability. II. Properties and applications. In: *Meeting of ACS Rubber Division*, Mexico City, Mexico, May 9–12, 1989, Paper 22.
14. Boyd RH, Breitling SM. *Macromolecules* 1972;5:1.
15. Suter UW, Saiz E, Flory PJ. *Macromolecules* 1983;16:1317.
16. Cho D, Neuburger NA, Mattice WL. *Macromolecules* 1992;25:322.
17. Wood LA. *Rubber Chem Technol* 1976;49:189.
18. Chu CY, Vukov R. *Macromolecules* 1985;18:1423.
19. Kuntz I, Rose KD. *J Polym Sci Part A* 1989;27:107.
20. Cheng DM, Gardener IJ, Wang H-C, Frederick CB, Dekmezian AH. *Rubber Chem Technol* 1990;63:265.
21. Puskas JE, Wilds C. *Rubber Chem Technol* 1994;67:329.
22. Vukov R. *Rubber Chem Technol* 1984;57:275.
23. Van Tongerloo A, Vukov R. Butyl rubber–halogenation mechanisms. In: *Proceedings of the International Rubber Conference*, Venice, Italy, 1979, p. 70.
24. Chu CC, Vukov R. *Macromolecules* 1985;18:1423.
25. Vukov R. *Rubber Chem Technol* 1984;57:284.
26. Parent JS, Thom DJ, White G, Whitney RA, Hopkins W. *J Polym Sci A: Polym Chem* 2001;39:2019.
27. Powers KW, Wang H-C, Chung T-C, Dias AJ, Olkusz JA (to Exxon Chemical). U.S. Patent 5,162,445, November 10, 1992.
28. Boyd RH, Krishna Pant PV. *Macromolecules* 1991;24:6325, 1992;25:494, 1993;26:679.
29. Stannett V, Crank J, Park GS, eds. *Diffusion in Polymers*. Orlando, FL: Academic Press, 1968, Chapter 2.
30. Fitzgerald ER, Grandine LD Jr, Ferry JD. *J Appl Phys* 1953;24:650.
31. Ferry JD, Grandine LD Jr, Fitzgerald ER. *J Appl Phys* 1953;24:911.
32. Plazek DJ, Chay I-C, Ngai KL, Roland CM. *Macromolecules* 1995;28:6432.
33. Plazek DJ, Ngai KL. *Macromolecules* 1991;24:1222.
34. Rizos AK, Ngai KL, Plazek DJ. *Polymer* 1997;38:6103.
35. Ngai KL, Plazek DJ. *Rubber Chem Technol* 1995;68:376.
36. Ferry JD. *Viscoelastic Properties of Polymers*, 3rd edn. New York: Wiley, 1980.
37. Coran AY. Curing behavior. In: Eirich FR, ed., *Science and Technology of Rubber*. Orlando, FL: Academic Press, 1978, p. 297.
38. Breslow DS, Willis WD, Amberg LO. *Rubber Chem Technol* 1970;43:605.
39. Bielski R, Frechet JMJ, Fusco JV, Powers KW, Wang H-C. *J Polym Sci A* 1993;31:755.

40. Frechet JMJ, Bielski R, Fusco JV, Powers KW, Wang H-C. *Rubber Chem Technol* 1993;66:98.
41. Duvdevani I, Newman NF. *Rubber World* 1997;216(5):28.
42. Bhakuni RS, Mowdood SK, Waddell WH, Rai IS, Knight DL. Tires. In: Kroschwitz JI, ed., *Encyclopedia of Polymer Science and Engineering*, Vol. 16, 2nd edn. New York: Wiley, 1989, p. 834.
43. Hopkins W, Jones RH, Walker J. Bromobutyl and chlorobutyl. A comparison of their chemistry, properties and uses. In: *International Rubber Conference Proceedings*, Kyoto, Japan, October 15–18, 1985, p. 205, Paper 16A10.
44. Morehart CL, Ravagnani FJ (to Bridgestone Corp.). Eur Patent 0,604,834 A1, December 15, 1993.
45. Voigtlander K. *Rubber S Afr* 1995;10(6):10.
46. Costemalle B, Fusco JV, Kruse DF. *J Elastomers Plast* 1995;27:39.
47. Jones GE. Exxpro innerliners for severe service tire applications. In: *ITEC'98 Select*, July 1999, p. 13.
48. Waddell WH. *Rubber Chem Technol Rubber Rev* 1998;71:590.
49. Flowers DD, Fusco JV, Gursky LJ, Young DG. *Rubber World* 1991;204(5):26.
50. Costemalle B, Hous P, McElrath KO. Exxpro™ polymers. In: *Rubbercon'95*, Gothenburg, Sweden, May 9–12, 1995, Paper G11.
51. McElrath KO, Tisler AL. Improved elastomer blend for tire sidewalls. In: *Meeting of ACS Rubber Division*, Anaheim, CA, May 6–9, 1997, Paper 6.
52. Mouri H. Improvement of tire sidewall appearance using highly saturated polymers. In: *Meeting of ACS Rubber Division*, Cleveland, OH, October 21–24, 1997, Paper 65.
53. Tisler AL, McElrath KO, Tracey DS, Tse MF. New grades of BIMS for non-stain tire sidewalls. In: *Meeting of ACS Rubber Division*, Cleveland, OH, October 21–24, 1997, Paper 66.
54. Prescott PI, Rice CA (to E.C.C. America). U.S. Patent 4,810,578, March 7, 1989.
55. Wilson RB (to General Tire). U.S. Patent 3,508,595, April 28, 1970.
56. Takiguchi E (to Bridgestone). Gr Brit Patent 2,140,447 A, May 9, 1984; U.S. Patent 4,567,928, February 4, 1986.
57. Takiguchi E, Kikutsugi T (to Bridgestone). U.S. Patent 4,945,964, August 7, 1990.
58. Yamaguchi H, Nakayama M, Tobori H (to Toyo). Jpn Patent 3-210345, September 13, 1991.
59. Mroczkowski TS (to Pirelli Armstrong). U.S. Patent 5,063,268, November 10, 1992.
60. Waddell WH, Kuhr JH, Poulter RR. Evaluation of isobutylene-based elastomers in a model winter tire tread. In: *Meeting of ACS Rubber Division*, Cleveland, OH, October 16–19, 2001, Paper 113; *Rubber Chem Technol* 2003;76(2):348.
61. Waddell WH, Rouckhout DF, Steurs M. Traction performance of a brominated isobutylene-*co*-para-methylstyrene winter tire tread. In: *International Tire Exhibition Conference*, September 12–14, 2002, Paper 35C.
62. Larson LC, Danilowicz PA, Ruffing CT. *J Elastomers Plast* 1990;22:190.
63. Tracey DS, Gardner IJ. *Rubber World* 1994;211:19.
64. Pilkington MV, Cole RW, Schisler RC (to Goodyear Tire & Rubber Co.). U.S. Patent 4,633,912, January 6, 1987.
65. Shiota A, Kitani K (to Nichirin Co., Ltd). U.S. Patent 5,488,974, February 6, 1996.
66. Costemalle BJ, Keller RC, Kruse DF, Fusco JV, Steurs MA (to Exxon Chem Patents, Inc). U.S. Patent 5,246,778, September 21, 1993.
67. McElrath KO, Measmer MB. Reversion resistant EXXPRO elastomer compounds. *Rubber Plast News*, January 29, 1996.
68. McElrath KO, Measmer MB. Reversion resistant EXXPRO™ elastomer compounds. In: *Meeting of ACS Rubber Division*, Louisville, KY, October 8–11, 1996, Paper 5.
69. Stefano GE, Tally DN (to Gates Rubber Co), U.S. Patent 4,158,033, June 12, 1979.

70. Dunn JR. *Elastomerics* January 1991;15.
71. Trexler HE. The development of automotive brake hose elastomer compositions. In: *Meeting of ACS Rubber Division*, Denver, CO, October 23–26, 1984.
72. Dunn JR. Compounding elastomers for tomorrow's automotive market, part II, *Elastomerics* February 1989;29.
73. Tabar RJ, Killgoar PC (to Ford Motor Co). U.S. Patent 4,419,480, December 6, 1983.
74. McElrath KO, Measmer MB, Yamashita S. Dynamic properties of elastomer blends. In: *Meeting of ACS Rubber Division*, Montreal, Canada, May 4–8, 1996, Paper 44.
75. Measmer MB, McElrath KO. Elastomer blend approach to extend heat life of natural rubber based engine mounts. In: *Meeting of ACS Rubber Division*, Anaheim, CA, May 6–9, 1997, Paper 30.

5 Thermoplastic Elastomers
Fundamentals

Tonson Abraham

CONTENTS

I. Introduction 140
    A. Definition of Thermoplastic Elastomer 141
    B. Classification of Commercially Available Thermoplastic Elastomers 141
II. Segmented Block Copolymer TPEs... 143
    A. Thermoplastic Polyurethanes 146
        1. TPU Morphology and Microstructure.................................... 148
        2. Thermal Characteristics of TPUs 149
        3. Aliphatic TPUs 154
    B. Elastomeric Copolyesters and Copolyamides............................. 155
III. Olefinic Block Copolymers... 157
IV. Styrenic Block Copolymers 160
    A. Effect of Paraffinic Oil on T_{ODT} of SBCs 163
    B. SBCs as Compounded Materials 164
    C. SBC Morphology 165
    D. SEBS Compound Upper Service Temperature Improvement 167
    E. Selected New Developments in SBCs 171
V. Thermoplastic Vulcanizates 173
    A. Definition of Dynamic Vulcanization 174
    B. Development of Dynamic Vulcanizates: Historical Perspective............ 174
    C. Principles of Dynamic Vulcanization... 177
        1. Principle I: Rubber and Plastic Compatibility....................... 177
        2. Principle II: Interphase Structure 185
        3. Principle III: Plastic Phase Crystallinity 185
        4. Principle IV: Rubber Vulcanization 190
        5. Principle V: Morphology Control... 192
        6. Principle VI: Melt Viscosity Control..................................... 195
    D. Rationalization of PP/EPDM TPV Elastic Recovery 197
    E. Thermoplastic Vulcanizate Processability 199
    F. Thermoplastic Vulcanizate Hardness Control............................. 199
VI. Final Comments... 199
Acknowledgments... 200
References.. 200

I. INTRODUCTION

In the fifteenth century, Christopher Columbus witnessed South Americans playing a game centered around a bounceable "solid" mass that was produced from the exudate of a tree they called "weeping wood" [1]. This material was first scientifically described by C.-M. de la Condamine and François Fressneau of France following an expedition to South America in 1736 [2]. The English chemist Joseph Priestley gave the name "rubber" to the material obtained by processing the sap from *Hevea brasiliensis*, a tall hardwood tree (angiosperm) originating in Brazil, when he found that it could be used to rub out pencil marks [2]. A rubber is a "solid" material that can readily be deformed at room temperature and that upon release of the deforming force will rapidly revert to its original dimensions.

Rubber products were plagued by the tendency to soften in the summer and turn sticky when exposed to solvents. This problem associated with natural rubber was overcome by Charles Goodyear in the 1840s by subjecting the rubber to a vulcanization (after Vulcanus, the Roman god of fire) process. Natural rubber was vulcanized by heating it with sulfur and "white lead" (lead monoxide) [2]. In May 1920, the German chemist Hermann Staudinger published a paper that demonstrated that natural rubber was composed of a chain of isoprene units, that is, a polymer (from the Greek *poly*, many, and *mer*, part) of isoprene [3]. In vulcanization, the rubber macromolecules are chemically bonded to one another ("cross-linked" in a thermosetting process) to form a 3D network composing a giant molecule of infinite molecular weight (MW). At present, the word "rubber" is associated with macromolecules that exhibit glass transition below room temperature and have "long-chain," "organic," carbon-based backbones," or "inorganic" backbones typified by polysiloxanes and polyphosphazenes.

"Elastomer" is always used in reference to a cross-linked rubber that is elastic (Greek *elastikos*, beaten out, extensible). An elastomer is highly extensible and reverts rapidly to its original shape after release of the deforming force. Entropic forces best describe rubber elasticity [4]. However, it should be noted that under relatively much smaller deformation, plastic materials and even metals can exhibit elasticity due to enthalpic factors [4]. Gases and liquids also exhibit elastic properties due to reversible volume changes as a result of pressure and/or heat [4]. Nevertheless, the term "elastomer" is always used in reference to rubber elasticity.

A plastic material is one that can be molded (Greek *plastikos*), and a thermoplastic can be molded by the application of heat. A rubber compound (a blend of rubber, process oil, filler, cross-linking chemicals, etc.) is thermoplastic and is "set" after several minutes in a hot mold, with loss of thermoplasticity. A thermoplastic material can be molded in a matter of seconds, and the molded part can be reprocessed. The viscous character of the thermoplastic melt readily allows control of the appearance of the surface of finished goods. In comparison, the effect of "melt elasticity" of a rubber compound on end product surface appearance is not as readily controlled.

The origin of the first thermoplastic material can be traced to Christian Schonbein, a Swiss scientist who broke a beaker containing a mixture of nitric and sulfuric acid and used his wife's cotton apron to clean up the spillage! Unfortunately for his wife, but fortunately for science, he left the washed apron near a fireplace to dry.

The cotton apron soon combusted without leaving any residue! Schonbein realized that the cotton of the apron was converted to "gun cotton," a nitro derivative of the naturally occurring polymer cellulose [1]. This learning may have been instrumental in the preparation of the first plastic by the English chemist and inventor Alexander Parkes in 1862. First called Parkesine, it was later renamed Xylonite. This substance was nitrocellulose softened by vegetable oils and a little camphor. During this time, elephant tusks, which were used to make ivory billiard balls, among other things, became scarce. In 1869, motivated by the need to find a suitable substitute for ivory, John W. Hyatt in the United States recognized the vital plasticizing effect of camphor on nitrocellulose and developed a product that could be molded by heat. He named this product obtained from cellulose "Celluloid" (Greek *oid*, resembling). Though primarily regarded as a substitute for ivory and tortoiseshell, Celluloid, despite its flammability, found substantial early use in carriage and automobile windshields and motion picture film [3].

A. Definition of Thermoplastic Elastomer

A thermoplastic elastomer (TPE) is a nano- or microphase- separated polymeric material that exhibits rubber elasticity over a specified service temperature range, but at elevated temperature can be processed as a thermoplastic (because of the thermoreversible physical cross-links present in the material). It offers the processing advantages of both a highly viscous and shear thinning melt behavior and a short product cycle time in manufacturing due to rapid melt hardening on cooling.

B. Classification of Commercially Available Thermoplastic Elastomers

The TPE products of commerce listed in Table 5.1 are classified in Table 5.2 on the basis of their polymer structure. Representative examples are included for each polymer class. Segmented block copolymers, triblock copolymers, and thermoplastic vulcanizates (TPVs) represent a significant portion of the TPE family.

The fundamental aspects of structure–property relationships in ethylene/octene (EO) segmented block copolymers, thermoplastic polyurethanes (TPUs), styrenic block copolymers (SBCs) (with emphasis on styrene/ethylene-1-butene/styrene [SEBS] copolymers and SEBS compounds), and TPVs produced from polypropylene (PP) and ethylene/propylene/diene monomer (EPDM) rubber were selected for review in this chapter, as representative of the most commercially significant and the closest in performance to thermoset elastomers.

TPVs possess sufficient elastic recovery to challenge thermoset rubber in many applications, and insights into TPE elastic recovery and processability are presented based upon the latest developments in the field. The poor elastic recovery of TPEs at elevated temperature is a key deficiency that has prevented these materials from completely replacing their thermoset counterparts.

TPEs owe their existence as products of commerce to the fabrication economics and environmental advantage they offer over thermoset rubber. TPEs, of course, are designed to flow under the action of heat; hence, their upper service temperature is limited in comparison to thermoset rubber. Thus, a major hurdle to overcome in the

TABLE 5.1

Major Thermoplastic Elastomer Products of Commerce

Product	First Commercialized (Year, Company)
Plasticized poly(vinyl chloride)	1935, B.F. Goodrich
Thermoplastic polyurethane	1943, Dynamit AG
PVC/NBR blends	1947, B.F. Goodrich
Styrenic block copolymers	1965, Shell
Thermoplastic polyolefin elastomers	1972, Uniroyal
Styrenic block copolymers (hydrogenated)	1972, Shell
Copolyester elastomers	1972, DuPont
Thermoplastic vulcanizates (PP/EPDM)	1981, Monsanto
Copolyamide elastomers	1982, Atochem
PP/NBR TPV	1984, Monsanto
Chlorinated polyolefin/ethylene interpolymer rubber	1985, DuPont
UHMW PVC/NBR	1995, Teknor Apex
Ethylene/ethylene–butene/ethylene Triblock copolymer[a]	1999, JSR
Nylon/acrylate rubber TPV[a]	2004, Zeon Chemicals
Styrene/isobutylene/styrene triblock Copolymer[a]	2004, Kaneka
Ethylene/octene segmented block Copolymer[a]	2006, Dow

[a] Selected low-volume products.

TABLE 5.2

Thermoplastic Elastomer Classification

Segmented Block Copolymers	Triblock Copolymers	Thermoplastic Vulcanizates	Polymer Blends
TPU	SBC	PP/EPDM	PVC/NBR
COPE	Hydrogenated SBC	PP/NBR	
COPA	Styrene/isobutylene/styrene	Nylon/acrylate rubber	
Ethylene/octene	Ethylene/ethylene–butene/ethylene		

replacement of thermoset rubber with TPEs is the improvement in elastic recovery, particularly at elevated temperature, especially compression set, because in many applications elastomers are subjected to compression. The scope of this chapter includes those TPEs that in our opinion come reasonably close in properties to thermoset elastomers, as listed in Table 5.1. Not included, for example, are plasto-mers that are ethylene/α-olefin copolymers generally produced using metallocene catalysts [5]. These materials can be rubberlike only at room temperature. They are thermoplastic owing to the thermoreversible cross-links provided by crystallization of the ethylene sequences in the polymer but are deficient in elastomeric character above room temperature or when under excessive strain. TPEs based on melt-blended polyolefins, ethylene/vinyl acetate copolymers, and ethylene/styrene copolymers are also omitted from the list [6,7]. TPEs based on high propylene, propylene/ethylene copolymers, were commercialized in 2002 as Vistamaxx™ by ExxonMobil. Crystallinity due to isotactic PP sequences in Vistamaxx is reduced due to stereo and regio errors in propylene insertion caused by the catalyst and also due to copolymerized ethylene. T_g as low as −30°C can be achieved to allow good low-temperature elastomeric properties, but high-temperature properties are compromised due to low crystallinity and low melting point (~50°C) of the thermoreversible cross-links (crystallites). Of course, the upper use temperature of this TPE can be increased by reducing product ethylene content, at the expense of poorer elastic properties. These statistical propylene/ethylene copolymers with narrow composition and narrow molecular weight distribution (MWD, $M_w/M_n \sim 2.5$) can be considered as segmented block copolymers. A competitive product from Dow, namely, Versify™, which was commercialized in 2004, is also a propylene/ethylene copolymer with a narrow-MWD but with a broad inter polymer-chain composition distribution [8]. Vistamaxx and Versify where the copolymerized ethylene content is 25 wt% or less, depending upon the grade, are generally used in polyolefin compounding, and are not discussed here. Thermoplastic olefins (TPOs) represent a commercially important class of materials, namely, PP that is impact modified with various combinations of polyolefinic elastomers (plastomer, ethylene/propylene rubber, EPDM). These plastic materials are usually reinforced with talc, for use in automotive, appliance, and other applications. However, the TPO materials referred to here are high-rubber-content PP/EPDM/oil melt blend precursors to TPVs. These are included as comparative materials to their more elastomerically performing counterpart TPVs.

Plasticized poly(vinyl chloride) (PVC) is used as a flexible plastic and not an elastomer, but is included in Table 5.1, because it was the first commercially produced TPE. PVC, produced by free radical polymerization, contains crystallizable syndiotactic segments, the crystallization of which is enhanced on mobilization of the polymer chain in the presence of a plasticizer. The differential scanning calorimetry (DSC) trace of an as-polymerized PVC powder exhibited a continuous melting endotherm, beginning at T_g (89°C), to over 200°C [9]. This melting behavior is due to the presence of crystallites of different sizes. Plasticizers lower the T_g of PVC and makes the material elastomeric. However, the very low crystallinity [9–11] of the plasticized material (usually 10%), variation in crystallite melting point due to crystal thickness variation, and the lowering of melting point due to plasticizer ingress into crystallites with increasing temperature limit the use temperature of PVC as an elastomer to about 70°C [12].

II. SEGMENTED BLOCK COPOLYMER TPEs

The segmented block copolymer TPEs included in Tables 5.1 through 5.3 contain sequences of "hard" and "soft" blocks within the same polymer chain. Solubility differences between the polymer segments and association and/or crystallization of the hard blocks produce phase separation in the molten elastomer as it cools. The hard

TABLE 5.3
TPE Property Comparison

Manufacturer: Trade Name	TPE Type	Hardness (Shore A or D)	Compression Set (%, 22 h, ASTM D 395B, Constant Deflection)	T_g (°C, DSC)	T_m (°C, DSC Peak)
Noveon: Estane® 58134	TPU (ester)	45D	62 (70°C)	−47	Multiple m.p. peaks, highest at 218°C
Noveon: Estane® 58137	TPU (ester)	70D	82 (70°C)	−28	227
EMS-Chemie: Grilon® ELX2112	COPA	60D	85 (24 h, 70°C)	30 (dry)	215
DuPont: Hytrel® 7246	COPE	72D	80 (100°C), 5 (100°C ASTM D395A, constant load)	20 (DMA)	218
Hytrel® 5526	COPE	55D	80 (100°C), 8 (100°C, constant load)	−25 (DMA)	203
Hytrel® 4056	COPE	40D	89 (100°C), 12 (100°C, constant load)	−40 (DMA)	150
Zeon: YT355	NBR (30 wt% AN) sulfur-cured thermoset	76A	12 (100°C)	−30	Amorphous

blocks form the thermoreversible cross-links and reinforcement (increasing stiffness) of the elastomeric soft phase. The rate of crystallization or association of the hard blocks will impact product fabrication time. Polymer microstructure and morphology is depicted in Figure 5.1. These TPEs are produced by condensation or addition step-growth polymerization and have low-molecular-weight segments. Although this is desirable, segment MW and MWD cannot be readily controlled. In a 40 Shore D copolyester (COPE) elastomer based upon poly(butylene terephthalate) (PBT) hard blocks and poly(tetramethylene oxide/terephthalate) soft blocks, the hard sequence length varies from 1 to 10 [13]. PBT MW of sequence length 10 is 2,200, whereas high-molecular-weight PBT that is commercially available could easily have an M_n of 50,000! Thus, a sufficient number of hard blocks have to associate to produce a high-enough melting crystal phase to provide a reasonably high elastomer upper service temperature. This necessitates increasing the hard-phase content of the TPE, which results in a hard elastomer ("filler" effect). Note that for a given hard-phase content, the lower the number of hard domains (more hard segments per domain), the greater the entropic penalty imposed on the elastomeric phase and the less favored the phase-separated morphology.

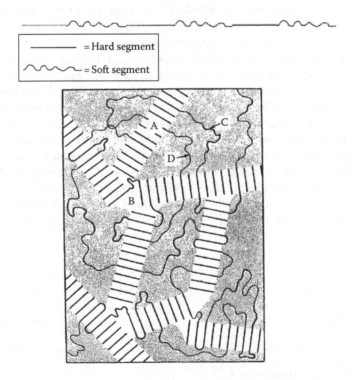

FIGURE 5.1 Polymer microstructure and morphology of segmented block copolymers (thermoplastic polyurethane, copolyester, copolyamide). A, crystalline domain; B, junction area of crystalline lamella; C, polymer hard segment that has not crystallized; D, polymer soft segment.

Increased hard-phase content also causes more hard segments to be rejected into the amorphous elastomeric phase, thus raising the rubber glass transition temperature (T_g) and therefore also the TPE lower service temperature. In the case of an increase in the number of hard domains, the soft-phase T_g is also elevated owing to the increased "cross-link density." These considerations allow the commercial viability of only hard COPEs. This is a major deficiency in this class of TPEs as the softest product available has a hardness of 35 Shore D. Also based on the aforementioned discussion, the more or less continuous hard phase in commercially available COPEs where fibrillar crystalline lamellae (due to short hard segments) are connected at the growth faces by short tie molecules can readily be rationalized. The amorphous phase is also continuous [14].

It is difficult to produce useful soft elastomeric products from segmented block copolymers except in the case of TPUs. The strong association of hard blocks even at low hard block content allows the preparation of soft elastomeric TPUs. TPUs with hardness as low as 55 Shore A are available commercially.

A.   THERMOPLASTIC POLYURETHANES

TPU was the first thermoplastic product that could truly be considered an elastomer [15]. The bulk of commercially available TPUs are produced from hard segments based on 4,4'-diphenylmethane diisocyanate (MDI) and 1,4-butanediol (BDO, a "chain extender"), with either poly(tetramethylene oxide) (PTMO) glycol or poly(1,4-tetramethylene adipate) glycol or poly(ε-caprolactone) glycol as the soft elastomeric segment [15]. TPUs can be produced by a "one-pot" method or a "two-stage" process. In the former, the diisocyanate, chain-extender diol, and soft segment diol are mixed and heated to yield the final product, whereas in the latter, the soft segment diol is first "endcapped" by using an excess of diisocyanate and the chain-extending short-chain diol is subsequently added to form the hard segments and to attach them to the soft segments in an alternating manner to yield a TPU of high MW by addition step-growth polymerization. A representation of a TPU molecule is presented in Figure 5.2. A TPU's M_w can be as high as about 200,000, with M_n about 100,000, although the individual hard and soft segments are of much lower MW. For example, PTMO glycol of M_n 1000 or 2000 is used commercially for

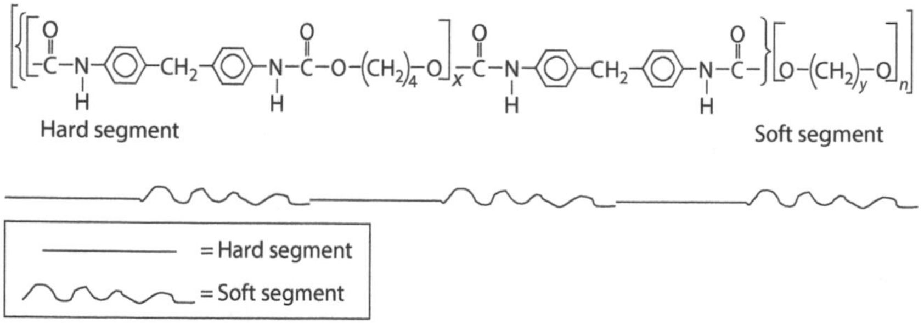

FIGURE 5.2   Polyether-based thermoplastic polyurethane.

TPU production, thereby fixing the soft block length. The longer the soft segment, the lower its hydroxyl end group concentration, which would allow preferential step-growth of the hard segments by reaction of the short-chain diol with the diisocyanate. Hence, the longer the soft segment, the longer the hard segment. Because the number of soft segments will equal the number of hard segments, for a large number of alternating segments,

$$\frac{\text{Weight\% SS}}{M_{\text{nss}}} = \frac{\text{Weight\% HS}}{M_{\text{nhs}}}$$

or

$$\overline{M_{\text{nhs}}} = \text{weight\% HS} \times \frac{\overline{M_{\text{nss}}}}{\text{weight\% SS}}$$

where soft segments and hard segments are abbreviated SS and HS, respectively. For a given soft segment MW, the number-average MW of the hard segment is directly proportional to the hard segment content and inversely proportional to the soft segment content [16]. M_{wss} can be obtained by measurements on the polyol, but obtaining the weight-average MW of the hard segment is difficult. Bonart developed a theoretical method to calculate M_{whs} [17]. The average number of hard segments for a TPU (MDI/BDO hard segments; polyoxypropylene endcapped with poly-oxyethylene soft segments) with a 50 wt% hard phase has been calculated to be 6 [18]. Peebles mathematically modeled the soft and hard segment length distribution in TPUs [19,20]. The representation of a TPV structure in Figure 5.2 is simplistic. There is a distribution of the hard and soft segment lengths in these materials, arising due to difference in reactivity of the chain-extender diol and soft segment diol with the diisocyanate. The second isocyanate group may exhibit a different reaction rate after urethane bond formation at the first group. A greater regularity of structure is expected in the two-stage process since no competition between two different hydroxyl groups exist. The one-pot method yielded fewer single hard segments in the melt polymerized MDI/BDO/PTMO TPV over the two-stage process, allowing better phase separation due to the longer hard segments obtained in the former preparation method [21]. Narrowing of the hard segment length distribution increases TPV modulus, tensile strength, and extension set [22]. Ideally, during and after polymerization, a single-phase melt should be preset during TPV preparation. Degradation and side reactions should also be absent [23].

The infrared studies of Cooper demonstrated that the urethane N–H is hydrogen bonded to the oxygen atoms of the urethane moiety as well as to the oxygen atoms of the polyether or polyester soft segments [24]. This hydrogen bonding and soft segment polarity can retard and lower the ultimate degree of phase separation in TPUs. Poor phase separation is reflected in the increase in T_g of the mostly amorphous soft phase due to the presence of dissolved hard segments. The hard microphase is formed by association of the relatively short hard segments and by their crystallization into fibrillar microcrystals. The poorer phase separation in polyester TPUs compared with polyether TPUs is presumably due to the greater polarity of and stronger hydrogen bonding (with the N–H of the hard segments) in the soft phase of the former compared with the latter [25]. A 1:2:1 (molar polyester polyol–MDI–BDO) TPU (polyester polyol $M_n = 1000$) exhibited a single phase, but the corresponding polyether-based TPU system was phase-separated [26]. The degree of phase mixing is also dependent upon soft segment content. For a polyether-based TPU, complete phase mixing was observed at 80 wt% soft segment content [27,28]. Phase mixing is also dependent upon segment MW, as demonstrated in the case of TPUs containing low-molecular-weight polycaprolactone soft segments [29].

Phase separation in TPUs is driven by the solubility parameter difference between the polymer segments and by association and/or crystallization of the hard segments and is limited by the geometry of the molecule and the hydrogen bonding and polarity effects discussed. In addition, the kinetics of TPU phase separation will also be influenced by the mobility [T_g] of the polymer segments.

1. TPU Morphology and Microstructure

The mechanical behavior (Young's modulus, elastic recovery, elongation, flexural modulus, heat sag, thermomechanical penetration probe behavior) of TPUs suggests a transition from discrete to continuous hard microdomain morphology at hard segment content above about 45 wt% [16,27–32]. The small-angle x-ray studies of Abouzahr et al. [28] and Cooper and coworkers [29] and the small-angle x-ray and neutron scattering analysis of Leung and Koberstein [27] suggested an interlocking hard domain morphology at high hard segment content. Depending upon processing conditions and hard-phase type and content, crystalline TPU systems may exhibit a fringed micellar texture of thickness equal to the hard segment length or clear-cut connectivity of the crystalline hard phase. The hard domain diameter in a TPU produced from a 1:6:5 polycaprolactone ($M_n = 2000$)/MDI/BDO mole ratio was estimated to be 400 Å by transmission electron microscopy (TEM) [32] (hence, "hard nanodomain"), although for the typical TPU materials mentioned in this review, this number is expected to be about 100 Å.

Using small-angle x-ray scattering (SAXS), Leung and Koberstein [27] studied the hard segment nanodomain thickness (which corresponds to the length of the hard segments) in TPUs in which the hard segment content varied from 30 to 80 wt%. The SAXS measurement provided an overall characterization of the domain morphology averaged over crystalline and noncrystalline structures. The hard domain thickness varied from 2 nm (corresponding to a hard segment length containing two MDI residues) to 5.4 nm (hard segment length with four MDI residues) for the 60 wt% hard segment content TPU, after which the thickness did not increase further with increased TPU hard segment content. Since the hard segment length increases with increased TPU hard segment content, chain folding via the flexible BDO segments to accommodate longer hard sequences within the crystal is thought to occur. Other possible explanations for this phenomenon have been discounted. The extended chain crystal structure, irrespective of TPU hard segment length, that has been demonstrated to occur by wide-angle x-ray diffraction (WAXD) may well be characteristic of the TPU samples studied that were annealed to maximize crystallinity, so as to be amenable to analysis by the WAXD method [27].

Spherulitic structure for high hard segment content (>40 wt%) TPUs have been observed in samples crystallized in the laboratory [16,32,33]. In one case, because of the large spherulite diameter (several micrometers) and the absence of a hard-phase T_g, the spherulites may have contained occluded soft phase [16]. Hard-phase T_g is rarely discernible even in high hard-phase-content TPUs. A hard-phase T_g was observed in a melt-quenched TPU with 80 wt% hard segment content [16]. Owing to the tendency of the relatively short TPU hard segments to associate or crystallize or to be miscible in the TPU soft phase, amorphous hard segments may exist only as tie molecules connecting microfibrillar crystalline segments. Low TPU amorphous hard-phase content would preclude T_g detection. Moreover, hard-phase T_g observation would be obscured by other transitions (discussed later). Spherulitic soft segment structure in a high PTMO soft segment content TPU has been observed [16]. Generally, TPU parts that are fabricated by commercial processing equipment exhibit crystallinity but no spherulitic structure [34].

2. Thermal Characteristics of TPUs

Although the structure of TPUs changes constantly during DSC, DSC coupled with SAXS has proven to be a powerful tool in uncovering TPU microstructure and thermal behavior, as in the masterful research work of Koberstein and coworkers [18,27,35,36], who studied polyether TPUs with MDI/BDO hard segments. Molten TPUs from a homogeneous melt state were rapidly quenched to and held at various annealing temperatures for specific time periods. Generally, three distinct endotherms were observed by DSC of the annealed samples. The first endotherm (T_I) is dependent upon the annealing temperature, annealing time, and TPU hard segment content. This endotherm is observed at 20°C–40°C above the annealing temperature, which was varied from 30°C to 170°C, depending upon TPU hard segment content. Higher hard segment content TPUs gave higher T_I values. The exact origin of T_I is still unknown, but it is linked to a short-range order dissociation endotherm in the hard microphase and not in the interphase, because this transition is also observed in pure hard segment materials as suggested by Cooper and coworkers [37,38]. For a soft TPU with a discrete hard phase and a total hard-phase content of 30 wt%, the T_g of the soft phase kept increasing with increased annealing temperature up to 170°C. Annealing above 170°C did not change the soft-phase T_g, indicating that the domain structure is completely disordered above this temperature [18]. The T_g increase of the soft phase was related to increased solubilization of hard segments into the soft phase. Increasing annealing temperature caused the solubilization of hard segments of high MW into the soft phase that already contained lower-molecular-weight hard segments. It has also been suggested that "cross-linking" by soft segment–hard segment hydrogen bonding is another factor that contributes to increased soft-phase T_g in addition to the physical presence of TPU hard segments in the soft phase [39]. By studying the change in TPU heat capacity at its glass transition temperature, it was concluded that below an annealing temperature of 80°C, hard segment solubilization into the soft phase occurs, and above 80°C, which is near the hard segment T_g, soft segments that are trapped in the hard microphase also enter the bulk soft phase in addition to further hard segment dissolution into the soft phase.

The T_{II} endotherm is also dependent upon annealing temperature, and for the soft TPU under discussion, the T_{II} maximum is 175°C. This transition was identified by Koberstein as the microphase separation transition (MST), where the partially ordered "noncrystalline" segments in the hard domain are mixed into the soft TPU phase. The TPU with 30 wt% hard segment content did not exhibit a microcrystalline melting T_{III} endotherm, which is observed for higher hard segment content TPUs. The identification of T_{II} as the MST was further confirmed by simultaneous DSC/SAXS measurements in a TPU with 50 wt% hard segment content [35]. The TPU interdomain spacing increased dramatically beginning at T_{II}. This TPU exhibited a higher T_{III} endotherm corresponding to the melting of a crystalline hard phase within the "noncrystalline-"ordered hard domain.

For the TPU with 50 wt% hard segment content, the T_I endotherm merged with the T_{II} endotherm when annealing took place at 155°C. Annealing above 155°C raised the T_{II} endotherm and decreased its intensity, whereas the intensity of the T_{III} microcrystalline peak melting endotherm increased. T_{III} was the only DSC peak endotherm observed at 210°C when annealing was conducted at 175°C. At annealing temperatures of 175°C–190°C, the T_{III} endotherm diminished in magnitude and the T_{II} endotherm reappeared. These findings are consistent with an expected decrease in crystallinity at low undercoolings where crystallization is controlled by nucleation. Above the MST, TPU crystallization occurs from a homogeneous mixed melt phase ("solution" crystallization). Crystallization occurs within the hard microdomains ("bulk" crystallization) below the MST. For harder TPUs (70 wt% hard segment content), melting endotherms corresponding to different crystal structures have been observed, depending upon annealing conditions.

The thermogravimetric analysis trace of the TPUs of the Hu and Koberstein study [36] demonstrates initial weight loss around 300°C, which is well above the annealing temperatures used to probe the TPU microstructure. A small change in annealing temperature (from 190°C to 195°C) exhibited a dramatic increase in TPU M_n and M_w values (gel permeation chromatography measurements). The increased MW is presumably the result of "trans-urethanation" reactions that result from cleavage of the urethane bond in a polymer segment back to the isocyanate and alcohol and subsequent allophanate formation by addition of the newly formed isocyanate to the urethane N–H bond of another polymer chain, thus creating a branched structure. Crystallization of the branched TPU molecules appears to be hindered in comparison with their linear counterparts. Reduction in the heat of fusion is observed for TPU samples where MW was increased by annealing at high temperature, due to "trans-urethanation" reactions. It should also be reiterated here that the sequence length of the hard segments that are incorporated into the soft phase increases with increased annealing temperature. For more on trans-urethanation reactions and TPU thermal degradation mechanisms, the reader is referred to the work of Macosko and coworkers [40].

According to Koberstein, all three TPU endotherms T_I, T_{II}, and T_{III} are accompanied by the mixing of hard and soft microphases. The Koberstein schematic model for the morphological changes that occur during the DSC scans of TPUs is presented in Figure 5.3. It should be noted that Koberstein's work is grounded on the pioneering TPU research work of Wilkes and Cooper and coworkers, who

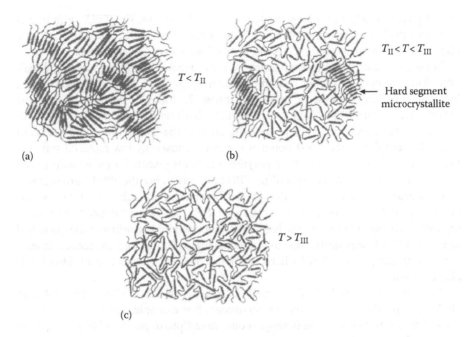

FIGURE 5.3 Schematic model for the morphological changes that occur during DC scans of polyurethane elastomer (a) below the microphase mixing transition temperature, (b) between the microphase mixing temperature and the melting temperature, and (c) above the melting temperature. The microcrystalline hard segment domains are indicated. (From Saiani, A, et al., *Macromolecules*, 34, 9059, 2001. With permission.)

had previously recognized the time- and temperature-dependent morphological and mechanical properties of TPUs [37,41–47]. The increased mutual solubility of TPU hard and soft phases with increasing temperature was recognized, as was the influence of hydrogen bonding and soft-phase T_g on phase mixing and demixing over a broad temperature range. Both phase mixing and demixing have been observed on TPU mechanical deformation, depending upon sample thermal history, including changes in phase continuity [45]. TPU morphology is complex, and a small change in the polymer segment type can result in diverse melting behavior. For example, TPUs produced from MDI/BDO hard segments and poly(hexa-methylene oxide) soft segments exhibited five melting endotherms that were attributed to hard segment sequences containing one to five MDI-derived units [48]. Saiani and coworkers have studied (DSC, SAXS, WAXS (Wide Angle X-Ray Scattering), TEM) morphology development in high hard segment (≥50 wt%) TPUs based on MDI/2-methyl-1,3-propanediol hard segments and poly (propylene oxide) soft segments that are endcapped with ethylene oxide [49–51]. These TPUs have melting endotherms in the 180°C–190°C range, as opposed to TPUs with MDI/BDO hard segments with a 210°C–230°C melting range. The melt memory of the TPUs of this work could then be readily erased by holding at 220°C for 2 min, without degradation or increase in product MW. Quenching molten samples to −130°C resulted in a homogeneous mixed phase for a hard segment content of 65 wt% or less. For TPUs with higher than 65 wt% hard segment content, melt quenching resulted in a nearly pure hard phase and a homogeneous mixed phase with 65 wt% hard-phase content.

Subsequent morphology development varied, depending upon annealing conditions (time, temperature). For example, annealing for 72 h at 120°C resulted in a hard-phase melting endotherm above hard-phase T_g, followed by the phase mixing endotherm. For the 90 wt% hard segment TPU, both hard-phase melting and hard-phase T_g were observed. Complete crystallization of the hard phase could not occur due to limited flexibility of the polymer segments caused by low material soft segment content. Note that amorphous polymer segments (except loops in chain folded crystals) are not present in a crystalline TPU hard phase. For the 100% hard segment TPU, annealing at 120°C precluded crystallization due to lack of mobility of this phase, and only a T_g could be observed by DSC, from the associated hard-phase segments. The morphology of the hard TPUs obtained by annealing of the quenched samples at 120°C was similar due to microphase separation from an identical homogeneous mixed phase (65 wt% hard segment content), irrespective of TPU hard-phase content.

Annealing the quenched TPU samples at lower than 120°C afforded varied morphology, depending upon annealing conditions. For example, under certain conditions, phase separation in the homogeneous mixed phase produced higher melting crystals (due to hard-phase mobility allowed by the interconnecting soft segments) than those from the initially present hard phase. In some cases, a large interphase consisting of hard and soft segments was produced, with undetectable soft-phase T_g (DSC), due to broadening of the signal caused by restriction of soft segment movement nearer to the closely spaced hard segment junctions and due to soft-phase hydrogen bonding with the hard segments that are mixed into the soft phase.

Even at low annealing temperatures, the observed microphase mixing transition implies phase separation in these TPUs. Well-phase-separated TPUs still contain some soft phase mixed into the hard phase and vice versa.

It is now readily understood how TPU morphology is dependent upon processing conditions and what thermally induced phase transitions can occur that would be detrimental to product elastic recovery at elevated temperature.

Based upon the information presented so far, it would appear that TPUs that are designed for improved phase separation (decreased hard- and soft-phase compatibility) should provide improved elastic recovery. However, TPU mechanical properties are adversely affected when the desired polymer segment structure is difficult to achieve due to incompatibility of the TPU building blocks under the industrially desirable, solvent-free polymerization conditions, including incompatibility of the reactants with the polymer produced. This is the case for TPUs with nonpolar soft segments such as polybutadiene diol and polar hard segments based on MDI and toluene diisocyanate (TDI) where microphase and macrophase separation has been demonstrated to occur during polymerization [21,52–54]. Molecular heterogeneity in chemical composition and average hard segment length is expected to be the key factor contributing to the poor mechanical properties of these hydrocarbon soft segment TPUs compared with conventional TPUs, based on, for example, MDI/BDO/PTMO. In fact, these undesirable features have been demonstrated to occur to a lesser extent in TPUs derived from poly(propylene oxide) endcapped with ethylene oxide soft segments and MDI/BDO hard segments [53]. The use of a solvent (50 vol% tetrahydrofuran in dimethylacetamide) allowed the preparation of a polybutadiene diol/MDI/ BDO) TPU with excellent microphase separation as indicated by the low observed T_g (approx. $-108°C$) of the α,ω-hydroxyl-terminated, high-cis polybutadiene soft phase [55]. Compatibility among system components could be improved for polybutadiene diol/MDI/2-ethyl-1,6-hexane diol TPU over that with MDI/BDO as the hard phase. In this case, the polybutadiene diol ($M_n=2100$, 1.9 OH/chain; 65 wt% vinyl, 12.5 wt% cis, 22.5 wt% $trans$; $T_g=-35°C$) is produced with a bifunctional initiator in a polar solvent and terminated with propylene oxide. An 80 Shore A gel-free TPU 7840 produced by the "one-pot" procedure, and TPU 2035 containing gel due to allophanate cross-links from the "two-stage" process, is commercially available from total. Owing to the microheterogeneity in the system as discussed earlier, and due to limited attraction between the nonpolar soft segments, the physical properties of these polybutadiene diol-based TPUs are poorer than those of comparable hardness conventional TPUs [56–58]. Hydrocarbon diols are being promoted for nonelastomeric polyurethane applications, as in the preparation of moisture-resistant, tie-layer adhesives between thermoset rubber and polar plastics, coatings, and electrical potting compounds [59,60]. TPUs with physical properties similar to that of TPU 7840 can be obtained from hydrogenated polybutadiene diols (with at least 40 wt% vinyl content in order to reduce product melt viscosity)/MDI/BDO [61]. These TPUs would have the moisture resistance of TPU 7840 and improved heat resistance.

TPUs produced with 2,6-TDI hard segments with BDO as chain extender and PTMO as the soft phase undergo cleaner phase separation than the corresponding 2,4-TDI-based TPUs [39] as symmetrical diisocyanates associate or crystallize more readily than their unsymmetrical counterparts. The use of 2,6-TDI as the hard-phase isocyanate may provide TPUs with excellent elastic recovery, but difficulty in 2,6-TDI/2,4-TDI isomer separation makes this approach commercially unfeasible. Moreover, the volatility of TDI over MDI makes the latter isocyanate preferable because of toxicity considerations. However, TDI, the first isocyanate developed for the thermoset polyurethane industry, is still used in North America in the manufacture of thermoset polyurethane foam [62–64]. TPUs produced with aromatic diol chain extenders such as hydroquinone bis(2-hydroxyethyl) ether instead of BDO allow improved high-temperature elastic recovery due to a more thermally stable hard-phase morphology [65–67]. However, the melting point of the hard domain exceeds the material thermal and thermooxidative stability in this case and also for highly crystalline, high-melting PU hard domains formed from dibenzyldiisocyanate, with, for example, certain chain extenders such as BDO [68]. These PUs are best suited for polymerization in a mold to yield a molded part directly (cast polyurethanes).

Aliphatic and aromatic diamines can be used as chain extenders to form TPU ureas with high-melting-point hard segments, but these materials melt with some decomposition, and well above the processing temperature of TPUs [15], and hence are not commercially feasible as TPEs with improved elastic recovery.

However, owing to improved elastic recovery after high strain and a higher use temperature due to the urea hard segments, solution-processed aromatic polyurethane ureas are preferable to conventional melt-processed aromatic polyurethanes in fiber applications (clothing, upholstery, and carpet). Spandex is the generic trade name given by the Federal Trade Commission to synthetic elastomeric fibers that contain at least 85% segmented polyurethane. In comparison with natural rubber threads, Spandex fibers are readily dyeable, lightweight materials with excellent abrasion resistance, tensile strength, and tear strength. They have better resistance to oxidation, sunlight, and dry cleaning fluids than natural rubber threads and are also tolerant to bleach containing a low chlorine level. Although cured natural rubber fibers have the advantage of low hysteresis and stretch crystallinity, they are being replaced by Spandex, which can also be cured during the fiber-forming process [69]. The role of polymer segment interactions on the mechanical properties of TPUs has been published [68,70,71].

3. Aliphatic TPUs

Aliphatic TPUs are used in light-stable (nonyellowing) applications and can have mechanical properties comparable to those of aromatic TPUs [72]. These materials are synthesized from hydrogenated MDI diisocyanate/BDO or hexamethylene-diamine diisocyanate/BDO hard segments and polyester soft segments. (Polyether soft phase would reduce TPU UV resistance.) Conventional MDI-based aromatic TPUs yellow on exposure to UV light owing to the formation of quinone imides. The quinone imides are UV absorbers that dissipate UV energy as heat and hence retard further TPU degradation. On UV exposure, the aliphatic TPUs undergo a greater reduction in mechanical properties than their aromatic counterparts but without color change or loss of transparency. Hence, UV-stabilized aliphatic TPUs are used in outdoor applications where the abrasion resistance of TPUs is necessary. For example, some outdoor signs enclosed in transparent acrylic are laminated with aliphatic TPUs. Aircraft canopies are fabricated with high-impact-resistant layered structures produced from polycarbonate (PC) and a "flexibilizing" aliphatic TPU "glue."

As illustrated by the data in Table 5.3, the compression set of TPUs is much poorer than that of thermoset rubber. Under compression at elevated temperature, irreversible deformation in TPUs occurs by continued phase separation and/or reorganization of the hard and soft segments over that established after part manufacture.

Hydrogen bonding in the hard phase and in the interphase (the region where the polymer composition changes from 100% hard segment to 100% soft segment) between the hard and soft domains provides a ready mechanism for chain slip because hydrogen bonds can reorganize readily by the partial formation of "new" hydrogen bonds as the "old" hydrogen bonds are partially broken. Increasing the amount of the hard phase (to provide more secure thermoreversible cross-links at the TPE upper service temperature) increases compression set because the now higher modulus material is subjected to much higher stress under compression compared to the corresponding softer material (under constant deflection). Increased hard-phase volume fraction in TPUs also restricts polymer motion in the soft phase (increased elastomer cross-link density), and there is an increased presence of hard segments in the soft phase. These factors cause an increase in the soft-phase T_g that raises the product's lower use temperature. The hard TPU product, of course, would have an advantage in constant load applications.

TPUs may also contain thermoreversible allophanate branch points resulting from the reaction of the urethane N–H bond with excess diisocyanate. It is not feasible to design allophanate bonds into a TPU, but these fortuitously present cross-links may contribute to improved TPU elastic recovery. Nevertheless, elastic recovery in the various types of TPUs does not approach that of thermoset rubber. In some cases, the elastic recovery of a soft product can be worse than that of a harder product because of product design. For example, it may be necessary to produce a soft TPU with a low rate of crystallization to achieve desirable processing characteristics in film applications. This may be accomplished by the use of a low-molecular-weight soft segment in which the TPU crystallization rate is lowered owing to increased phase mixing. Continued phase separation in the finished product is one factor that would raise set.

Amorphous materials exhibit a gradual decrease in viscosity with increasing temperature beyond T_g, compared with crystalline materials, in which viscosity drops sharply on melting due to the T_m being much greater than the T_g. In crystalline hard-phase TPUs, the viscosity drop on crystal phase melting may not be as precipitous as expected because of association among the hard-phase molecules that are still present just after melting because of incompatibility with the soft phase. Even so, this viscosity drop in a crystalline hard-phase TPU may cause it to lack desirable processing characteristics, and TPUs with a high amorphous hard segment content may be designed for an improved processing window and for transparency. The excellent impact properties, processability, and transparency of Lubrizol's Isoplast™ are credited to the amorphous hard segment that makes up most of this TPU engineering plastic. In the finished product, elastic recovery is controlled by both raw material properties and part design.

B. ELASTOMERIC COPOLYESTERS AND COPOLYAMIDES

Elastomeric COPEs [14] and elastomeric copolyamides (COPAs) [73] are similar in structure to TPUs and suffer similar drawbacks in rubber performance. The hydrogen bonding present in TPUs and COPAs is absent from COPEs. Commercially available COPEs are based upon crystalline polybutylene terephthalate (PBT) hard segments and PTMO soft segments. PBT monofilaments exhibit only a 1% permanent set after 11% extension at room temperature, owing to a reversible α- to β-crystal transition [74,75]. This reversible crystal transition, which would be beneficial in the elastic recovery of COPEs, has been observed in PBT/PTMO COPEs with a high-enough level of the PBT hard phase that the amorphous phase is hard enough (due to the presence of PBT hard segments in the amorphous phase) to bear the level of tensile stress necessary to cause the reversible deformation behavior in the hard phase [76]. Although it is generally thought that segmented block copolymers have a homogeneous amorphous phase consisting of hard and soft blocks, experimental evidence indicates that a biphasic amorphous phase consisting of a PTMO phase and a mixed PBT/PTMO phase exists in this COPE [77]. No amorphous PBT phase could be detected in this case. For COPEs produced with PBT as hard phase and ethyleneoxide/ethylene-stat-butylene/ethylene oxide triblock copolymer as soft phase, a pure PEB phase, a PEO-rich phase, and a PEO/PBT mixed phase were observed for compression-molded samples. Additionally, in high hardness compositions, a PBT amorphous phase was also detected. These COPEs have better elastic recovery than PBT/PTMO-based COPEs, presumably due to softness, and thus higher deformability under low stress, due to the softening of the PEB phase [78]. The lack of hydrogen bonding in COPEs and the reversible crystal transformation possible in the PBT hard phase are responsible for the modest improvement in elastic recovery of these materials over TPUs and COPAs. However, at elevated temperature, the motion of the soft segments cannot be adequately restrained by the crystalline polymer chains, thus causing reorganization in the hard phase that leads to irreversible deformation. COPEs cannot match the elastic recovery of thermoset rubber (Table 5.3). The significantly higher instantaneous elastic recovery at room temperature of poly (trimethylene terephthalate) (PTT) over PBT has been credited to the helical conformation of this polymer in the crystal lattice, and not due to a reversible crystal phase transition as observed for PBT [79,80]. Hence, the use of PPT as the COPE hard phase may improve product elastic recovery. At present, PTT is commercially available as CORTERRA™ (from SW/CO) and finds applications as fibers in carpet and fabrics, due to the excellent elastic recovery and stain resistance of PTT.

In addition to the disadvantage of poor elastic recovery at elevated temperature that is characteristic of most TPEs, the segmented block copolymers suffer the additional disadvantage of the lack of commercially available soft products due to inadequate physical properties as already discussed. In the case of TPUs, the association of the hard segments is strong enough to confer excellent physical properties to soft products (55 Shore A), but difficulty in pelletization of the soft product during manufacture and pellet agglomeration on storage have to be overcome.

Addition of plasticizer to hard segment block copolymers is not a viable option for the production of soft products, because the plasticizer would lower the melting point of the polar hard phase in addition to softening the polar elastomeric phase, which, in any case, cannot hold a high level of added plasticizer. Moreover, continued phase separation after processing can cause the exudation of plasticizer from the molded product. The bulk of the commercially available segmented block copolymer TPEs are plasticizer-free.

Elastic recovery is an important property for elastomer performance. Because of the price and performance requirements in diverse applications, the hydrocarbon oil-resistant segmented block copolymers discussed are successful products of commerce.

The most important end use of the polyurethane-elastomer, polyamide-elastomer, and polyester-elastomer block copolymers has been in thermoset rubber replacement. Their crystalline hard segments make them insoluble in most liquids. Products feature exceptional toughness and resilience, creep and flex fatigue resistance, impact resistance, and low-temperature flexibility. All three types are generally used uncompounded, and the final parts can be metallized or painted. Thus, they are often used as replacements for oil-resistant rubbers such as neoprene because they have better tensile and tear strength at temperatures up to about 100°C. Automotive applications include flexible couplings, seal rings, gears, timing and drive belts, tire chains, and brake hose. Special elastomeric paints have been developed that match the appearance of automotive sheet metal; such parts have been used in car bodies

[14,15,73]. Flexible membranes, tubing, hydraulic hose, and wire and cable jackets are included in the long list of applications for TPUs, COPEs, and COPAs.

III. OLEFINIC BLOCK COPOLYMERS

TPEs with a saturated carbon backbone that can be synthesized directly from low-cost α-olefins would be commercially attractive. The advent of metallocene catalysts allows the random copolymerization of ethylene with propylene or other higher α-olefins, at any desired composition. The crystallinity of polyethylene can then be reduced in ethylene/α-olefin copolymers to yield elastomeric products, where the crystallinity due to ethylene sequences is dependent upon the amount and branch length of the α-olefin, as does copolymer T_g. For EPR, the lowest T_g occurs at $-53°C$, when the material is completely amorphous, which precludes its use as a TPE. EO copolymers will allow even lower T_g while still providing long-enough ethylene sequences (between monomer units) that can crystallize and act as TPE thermoreversible cross-links [81,82]. However, when enough octene has been copolymerized with ethylene to provide good low-temperature elastomeric properties, the ethylene sequences are not long enough to crystallize via the chain folding process. As the octene content increases in EO copolymers, crystallite lamella became shorter, then segmented, and eventually the ethylene sequences do not chain fold, but associate into bundles to form fringed micelles [83,84]. The low crystallinity and presence of only low-melting imperfect crystallites makes EO elastomeric copolymers unsuitable in TPE applications. For example, a 74 Shore A EO copolymer has a DSC melting point of 66°C and a T_g of $-51°C$ [81].

TPEs can be obtained by the reduction in PP crystallinity by control of catalyst isospecificity, to allow increased stereo error frequencies during polymer chain growth. In one case, the material thus obtained contained mixtures of isotactic, atactic, and isotactic/atactic block copolymer PP. The block copolymer prevented macrophase separation in these TPEs [85,86]. TPEs have also been obtained by the synthesis of isotactic PP containing statistically distributed stereo errors only [87]. Isotactic/atactic/isotatic PP triblock copolymers should have improved mechanical properties over the stereo defect PPs [88]. All these TPEs have limited lower use temperature due to the high T_g (~0°C) of PP. Upper service temperature is also compromised due to the low crystalline melting points (or 120°C or less) caused by low isotacticity of the propylene sequences in these polymeric materials. Moreover, the need for a catalyst molecule for each polymer chain in triblock copolymer preparation by living polymerization may be cost prohibitive for product commercialization. Living polymerization has also been used to obtain TPEs by the introduction of regio defects into PP. Block copolymers have been produced by initially generating isotactic PP chains at a low propylene polymerization temperature ($-78°C$), with subsequent temperature increase (up to room temperature) for introduction into the growing polymer chain, of regio defects from 2,1, and subsequent catalyst "chain walking" to give 3,1 insertions ((CH_2)$_3$ linkages). Polymerization at room temperature allowed low T_g (approx. $-60°C$) chain segments due to the presence of increased 3,1 defects in the PP chain. However, the low isospecificity of the catalyst, along with a long reaction time at low temperature ($-78°C$) that is needed to generate the low melting (~137°C) isotactic PP chain sequences, makes this block copolymer route to PP based TPEs commercially unfeasible [89,90].

Very low isospecificity is obtained in the iPP segment of iPP-*b*-EP produced by another living polymerization process [91].

If polymer-chain initiation by the catalyst molecule is rapid relative to propagation, and if chain transfer and termination are absent, or very low in the time scale of the experiment, then the polymerization can be considered "living" for the said period. This "time scale" varies from fractions of a second for Ziegler–Natta to a few seconds for metallocene catalysts, in olefin polymerization. Using catalysts that are "living" for a few minutes at practical (>20°C) polymerization temperatures, iPP segments with high isotacticity have been produced in iPP-*b*-polyethylene, but no information on iPP-*b*-EPR was presented [92,93]. Highly syndiospecific sPP/PE, sPP/EPR, and sPP/EPR/sPP block copolymers have also been produced by living coordination polymerization [94–96], but, once again, the need for one catalyst molecule for every polymer chain makes this approach commercially unattractive. PE/EPR and PP/E-higher α-olefin rubber block copolymers with a high-melting-point polyethylene phase have also been produced by living coordination polymerization [96–98].

Commercialization by Dow, of Infuse™, which is catalytically produced segmented olefinic block copolymers (OBCs), consisting of both hard and soft segments derived from EO copolymers, represents a breakthrough in polyolefin technology. These materials are produced using two catalysts with varying comonomer incorporating capabilities and a chain shuttling agent that transfers the growing polymer chain from one catalytic site to the other. Narrow MW (~2.0) distribution copolymers are obtained. However, the block length and number are polydisperse [99], while within the blocks, the comonomer distribution is statistical and homogeneous. The Baer group has characterized several experimental OBCs provided by Dow [100]. These materials had 2 and 48 wt% copolymerized octene respectively in the hard and soft segments.

Unlike statistical EO copolymers (see previous discussion), the crystallizable ethylene sequence length in the polymer chain remains constant with increased octene level, for OBCs. Thus, the polymer melting point drops only from about 124°C for about a 10 wt% total octene content to about 120°C for about a 40 wt% total octene content OBC. For the corresponding statistical EO copolymers, a melting point change from 120°C to 52°C is observed [100]. Although the T_g of the soft block in the OBC is low ($T_g = -44$°C, dynamic mechanical thermal analysis [DMTA]) per se, restricted movement of these polymer segments in high hard block content OBCs raises T_g, and the soft block T_g of -43°C is observed only for an OBC with about 18 wt% hard block content and a total octene content of about 40 wt%. Soft block T_g increase also occurs due to the presence of mixed-in hard blocks. A spherulitic crystalline structure is observed even for an OBC with low (18 wt%) hard block content. However, WAXD indicated a distorted orthorhombic unit cell, due to the octene chain branches. Owing to low crystallinity, and slightly imperfect crystals (in comparison with HDPE) in OBCs with good low-temperature elastomeric properties, high-temperature elastic recovery is compromised [101]. The hard segments in the OBCs are about 58% crystalline, irrespective of hard block content. Hence, some portion of the hard block is present as amorphous material, in the OBC amorphous phase that is dominated by the soft blocks. The low MW (overall M_w ~100,000, PDI ~ 2) of the hard and soft segments, and insufficient solubility parameter difference between these segments, allows these OBCs to form a homogeneous melt from which the hard polyethylene segments crystallize.

A different set of higher-MW (overall M_n ~ 80,000, PDI ~ 2) OBCs, supplied by Dow, was studied by the Register group [102]. An exemplary OBC of this group contained 48 wt% hard segment (5.4 wt% copolymerized octene, m.p. 116°C) and 60.3 wt% copolymerized octene in the soft block. Flow-aligned specimens were characterized by simultaneous WAXS and SAXS measurements at the three perpendicular cross-sectional views. In this case, a phase-separated polymer melt was observed. Lamellar melt morphology was preserved, with polyethylene crystallization occurring within the confines of the molten hard domains, with the polymer c axis oriented perpendicular to the lamellar normal, so that the crystal spacing does not disrupt the lamellar thickness. Soft domains were also expected to contain a significant amount of polyethylene crystals, which would enlarge the said domains. By comparison of the domain spacing of polydisperse diblock OBCs with narrow-MW diblock copolymers such as polyethylene-b-poly (ethylene-alt-propylene), it was determined that block polydispersity contributed largely to phase mixing. By the study of OBCs with similar multiblock distribution, but with varying differences in segmental compatibility, it was demonstrated that increased hard/soft segment compatibility had a greater effect on increased soft domain size over multiblock architecture.

Based on the aforementioned facts, it is difficult to see how OBCs can compete with TPVs and oil-extended SEBS compounds, in elastomeric applications, as claimed [101]. Certain SEBS products can absorb several times its mass of paraffinic oil, whereas the OBCs can hold only a maximum of 50 wt% of oil. Melt blending of paraffinic oil into polyolefinic compounds improves product processability, softens the product, and reduces cost. It is appropriate to mention here that a series of ethylene/ethylene–butylene/ethylene triblock copolymers (Dynaron™ 6100P, 6200P, 6201B) has been commercially available since about Y2000 from JSR. These materials have crystalline polyethylene end blocks (T_m ~ 100°C) and low T_g (–50°C to –55°C) and noncrystalline rubber midblocks and have been produced by the anionic polymerization of butadiene, followed by polymer hydrogenation. The melting point of the crystalline block is low (compared to HDPE) as a minimum of 10 wt% of butadiene is polymerized in a 1,2 fashion by anionic polymerization, instead of the desirable 1,4 polymerization that would yield strictly polyethylene hard segments on hydrogenation. Much higher MW has been achieved in Dynaron, in comparison with Infuse. Also, the triblock copolymer architecture is much more amenable to phase separation than that of segmented block copolymers. These factors may allow, over OBCs, lower high-temperature compression set for the E/EB/E materials (in spite of the lower melting crystalline polyethylene blocks) and increased oil-holding capacity. Both Infuse and Dynaron can be used as a compatibilizer in the melt blending of PP and high-density polyethylene [103,104].

IV. STYRENIC BLOCK COPOLYMERS

The advent of hydrogenated styrene/butadiene/styrene (SBS), that is, SEBS, tri-block copolymer compounds represented an advance in the elastic performance of TPEs at elevated temperature. SEBS is almost always compounded: processable soft compositions (0–30 Shore A) that are not possible in the case of segmented block copolymers can be formulated. The key features of SEBS will be described before discussion of SEBS compounds. Phase separation in these triblock copolymers is more complete and occurs more readily than in the segmented block copolymers. This is reflected in the T_g of the rubber phase, which is nearly unaffected by the polymer styrene content. The T_g of the styrene phase depends upon its MW. More phase mixing with the rubber can be expected with decreasing styrene MW when the material is heated to the T_g of styrene [105]. Both the polystyrene (PS) end block content and PS MW in SEBS are designed to be lower than that of the rubber midblock. For example, Kraton R G1651 (SEBS) of Kraton Polymers has a plastic block of MW 29,000 (33 wt%) and a rubber block of MW 116,000 (68 wt%) [106]. The rubber block is designed to have a 40 wt% butene content in order to limit polyethylene crystallinity for maximum oil absorption by the rubber phase and to lower T_g (low T_g for improved low-temperature performance) [107]. Simplistically, SEBS has a "spaghetti and meatball" morphology, in which the styrenic nanodomains (200–300 Å) are dispersed in a continuous rubber matrix [108]. The PS nanodomain size reflects the entropic penalty that would be imposed on the rubber in the case of larger plastic domains. The higher-molecular-weight and narrower-MWD of SEBS than those of the segmented block copolymers are factors that favor improved phase separation in the former system in spite of the smaller solubility parameter difference between the phases in SEBS versus the segmented block copolymers [109,110]. Molecular architecture also favors better phase separation in SEBS than in the segmented block copolymers. The PS phase will flow above its T_g (~95°C), and these nanodomains form the thermoreversible cross-links in the SEBS TPE. The styrenic cross-links, however, do not contribute much to the "cross-link" density of the rubber phase that is dominated by the trapped entanglements within it [110,111]. This can readily be inferred by a comparison of the modulus (initial slope of the stress–strain curve and also the plateau modulus) of SEBS with other SBCs such as SBS and styrene/isoprene/styrene (SIS). The modulus in these systems is directly related to the MW between entanglements in the rubber phase [108]. The modulus of SEBS (lowest MW between entanglements and highest entanglement density) is greater than that of SBS, which in turn has a higher modulus than SIS (highest MW between entanglements and lowest entanglement density).

Thus, the function of the styrenic domains is to prevent disentanglement of the rubber segments when these SBCs are subjected to load. For example, Kraton G1651 has a 33.3 wt% PS content and a rubber MW of 116,000 $M_n \approx M_w$. Neglecting the interphase, the total PS phase volume in 100 g of SEBS would be 31.71 cm³ (PS density = 1.05 g/cm³). Assuming spherical 200 Å diameter PS domains, the volume per domain is 4.19×10^{-18} cm³, which translates to 7.57×10^{18} domains in the SEBS sample.

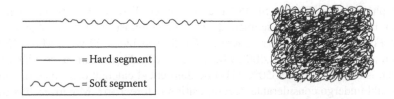

FIGURE 5.4 High-rubber-content styrene/ethylene-1-butene/styrene triblock copolymer microstructure and morphology.

The number of PEB macromolecules is 34.63×10^{19} ($66.7/116,000 = 5.75 \times 10^{-4}$ g·mol $= 5.75 \times 10^{-4} \times 6.023 \times 10^{23}$ macromolecules). Assuming a MW between entanglements for PEB of 1,800, the number of entanglements per chain is 64 (116,000/1,800). If entanglements occur only by the crossing of two different rubber chains, the total number of entanglements in the rubber is 1108×10^{19} ($34.63/2 \times 10^{19} \times 64$), which results in 1464 entanglements in the rubber phase per PS domain. A representation of SEBS polymer microstructure and morphology, reflecting this high entanglement density in the rubber phase, is presented in Figure 5.4. Note that in SBS and SIS, the rubber block has a high 1,4-copolymerized diene content that maximizes phase separation (due to maximized incompatibility between the plastic and rubber phases) for improved elastic properties, but which is also detrimental to product processability. On the other hand, SEBS is produced by the hydrogenation of high "vinyl" (low 1,4-copolymerized diene) SBS for reasons already discussed. Hydrogenation of commercially available SBS would yield a crystalline plastic instead of an elastomeric polymer midblock.

The foregoing discussion is based upon the "spaghetti and meatball" SEBS morphology described earlier. In the case of lower-molecular-weight SEBS, a higher modulus has been observed compared to those of the corresponding higher-molecular-weight counterparts. This has been attributed to the presence of a larger interphase in the former case due to greater phase mixing [112]. If the TPE hard block content is high enough to form a continuous phase, a higher modulus can be expected.

Upon increasing PS content, the discrete plastic phase morphology in SEBS can change to a cocontinuous rubber and plastic phase and further to a discrete rubber phase in a plastic matrix. Also, the shape of the plastic phase can change from spheroidal to cylindrical to platelike with increasing SEBS PS content. These regular shapes can be achieved only under carefully controlled annealing or shearing conditions or by slow (over about 3 days) evaporation of solvent from an SEBS solution.

Compared with a corresponding low-molecular-weight polymer, high-molecular-weight SEBS exhibits superior mechanical properties and can be used to produce lower-cost end products owing to its ability to absorb large amounts of paraffinic oil. However, high-molecular-weight SEBS is not processable, because this polymer alone does not flow well under polyolefin plastic processing conditions [113]. This is due to phase incompatibility that necessitates high-temperature and high-shear (for increased phase-mixing kinetics) conditions to transform biphasic SEBS to a molten single-phase system. That is, SEBS has a high order–disorder transition temperature (T_{ODT}) that is related to the segmental MW and composition of this triblock copolymer. For example, the T_{ODT} of Kraton G1650 (PS block 29 wt%, MW = 13,500; PEB block 71 wt%, MW = 66,400), which is considered a medium-molecular-weight product, is estimated to be 350°C [114] by theoretical calculations, since this material would undergo considerable decomposition before T_{ODT}. Moreover, this transformation would not occur instantaneously at this temperature; it is expected to be retarded due to the highly entangled nature of the rubber phase. Hence, experimental work has been conducted with only low-MW SEBS. In one method, T_{ODT} is the temperature at which there is a precipitous drop in polymer elastic modulus when measured as a function of temperature at a fixed frequency, although this approach may not yield the true T_{ODT}, because some order may still exist in the polymer melt at this temperature. Moreover, this method is inapplicable when morphological changes occur in SEBS, on heating, prior to formation of a single-phase system. For example, a SEBS (Asahi) with a PS volume fraction of 0.084 (50 mol% butene content in rubber block, $M_n = 6.7 \times 10^4$) exhibited a lattice disordering transition temperature (T_{LDT}) at 150°C and a domain dissolution temperature (T_{DDT}) at about 210°C, before T_{ODT} (~232°C). At T_{LDT}, the body-centered cubic structure of the PS spheres at room temperature (solution cast SEBS) converted to disordered spheres with liquid-like short-range order. At T_{DDT}, the PS spheres began to dissolve into the rubber matrix. This process was complete at T_{ODT}. The aforementioned determinations were made by annealing the solution cast SEBS at various temperatures and then quenching the sample to room temperature for study by SAXS and TEM. T_{ODT} was determined by measuring G' and G'' as a function of frequency at various temperatures and then plotting G' (proportional to ω^2) versus G'' (proportional to ω) at progressively increasing temperature. The temperature at which this plot (G' vs. G'') just becomes independent of temperature is T_{ODT} [115]. This is a reliable experimental method to identify T_{ODT} provided the SEBS is thermally and thermooxidatively stable [116].

SEBS Kraton G1657 (13 wt% PS, MW = 7.0×10^4, 30 wt% diblock) exhibited an order–order transition (T_{OO}) from PS cylinders to PS spheres at 152°C, followed by T_{LDT} (170°C) and T_{ODT} (202°C). These conclusions were drawn by studying the changes in the infrared spectrum of the solvent cast material at increasingly higher temperatures ("moving window" 2D correlation infrared spectroscopy) and by quenching samples that were annealed at high temperature to room temperature. The latter technique allowed determination of the "frozen-in" morphology present at high temperature by atomic force microscopy (AFM) imaging at room temperature [117]. Even above T_{ODT}, local fluctuations in composition are expected to occur for SEBS. In spite of the saturated backbone in high-molecular-weight SEBS, polymer degradation occurs before the T_{ODT} is reached, and hence it is difficult to measure this temperature experimentally [114]. Lower-molecular-weight SEBS polymers could be readily processed under normal polyolefin plastic processing conditions (200°C–250°C) but cannot provide the necessary price–performance balance to become a product of commerce as an elastomer. A SEBS polymer with a PS end block MW of 3,400 (31.8 wt%) and a PEB midblock MW of 14,600 (68.2 wt%) exhibits a T_{ODT} of 142°C [118]. See next section for more on the T_{ODT} of SEBS.

A. Effect of Paraffinic Oil on T_{ODT} of SBCs

Since paraffinic oil is miscible with the rubber midblock, and relatively immiscible with the PS end blocks of SEBS, the addition of oil to this material should not change the solubility parameter difference between the rubber and PS phase, and hence the T_{ODT} of SEBS should not change with oil extension. However, it is generally known that the melt processing of "crystal" PS is facilitated by the addition of 1.5–2.0 wt% of mineral oil, and, on solidification of the PS melt, the oil does not bleed out, although the morphology of this product is unknown. PS is known to have limited melt miscibility with low-MW paraffinic oil, and hence SEBS T_{ODT} can be reduced by the addition of paraffinic oil, depending upon the amount and MW of the added oil. Thus, the T_{ODT} of certain low-molecular-weight SEBS grades, although too high to be experimentally measured due to thermooxidative decomposition, can be lowered sufficiently by the addition of paraffinic oil to allow measurement. For SEBS/paraffinic oil melts, the amount of oil partitioning between the PS and rubber phase depends upon the miscibility of oil with the respective phases, at the processing temperature.

There is a recent publication by Aoki et al. [119] that delineates the fundamental experimental aspects of SBC T_{ODT} measurement, using a low-MW styrene/ethylene–ethylene–butylene/styrene (SEEPS) triblock copolymer (Septon 4033, Kuraray, $M_w \approx M_n = 84,000$, $T_{ODT} > 300°C$, 30 wt% PS) and SEEPS/paraffinic oil melt blends. SEEPS is produced by the hydrogenation of SBIS, where the rubber block consists of about 50 wt% butadiene and about 50 wt% isoprene [120]. It is believed that the rubber midblock in SEEPS is more compatible with the PS end blocks than that in conventional SEBS, although details of the SEEPS rubber structure are unavailable. Rheological measurements were made in the linear viscoelastic region where material melt morphology is preserved, and the physical response of the melt to deformation is constant over the range of imposed strain. G' and G'' was measured at various temperatures as a function of strain rate ω. Measurements were made at 150°C, 170°C, 190°C, 200°C, 210°C, 220°C, and 230°C, for the SEEPS/50 wt% paraffinic oil (MW = 750) melt blend. At 220°C, the plot of log G' or log G'' versus log ω exhibited a sudden displacement to lower values at low ω (10–100 rad/s) compared with the plots at lower temperatures. For the linear plots obtained at 220°C, log G' or log G'' versus log ω had a slope of 2 and 1 respectively, typical of homopolymer melts in the terminal zone (Newtonian viscosity at low ω), where the narrow-molecular-weight distribution homopolymer has a MW that is several times the MW between entanglements. The plot of log G' versus log G'' was linear, with a slope of 2 at 220°C, characteristic of homopolymer melts just described. The plot was also invariant at 230°C, and the threshold temperature at which the plot of log G' versus log G'' became invariant with temperature was between 210°C and 220°C, which is system T_{ODT}. The invariance of G' versus G'' with temperature is characteristic of single-phase homopolymer melts of the previous description. For SEEPS and SEEPS/25 wt% paraffinic oil (MW = 750) melt blend, T_{ODT} was too high to measure experimentally. For a melt blend of SEEPS/75 wt% paraffinic oil (MW = 410), rheological experiments were conducted between 60°C and 140°C. Although G' dropped considerably with increasing temperature, no sudden drop in the log G' versus log ω plots was observed with increasing temperature. The plot of log G' versus log G'' was linear (slope $= 2$) and invariant with temperature, beginning at about 120°C. The change in rheological behavior of high oil (low-MW paraffinic oil) content compared with lower oil (higher-MW paraffinic oil) content SEEPS may be due to the increased volume of the oil-swollen PS domains in the former case. Thus, order–order transitions and domain dissolution over a wide temperature range may be occurring in the case of high oil content SEEPS. The rheological measurements may also be complicated by the presence of a separate oil phase, due to reduced miscibility of the SEEPS rubber phase with low-molecular-weight oil at higher temperatures.

B. SBCs as Compounded Materials

In elastomer applications, SEBS is never used alone; it is always compounded to improve product processability and performance and to lower product cost. PP, paraffinic oil, and fillers make up the bulk of a SEBS elastomer compound. In elastomer applications, high-molecular-weight SEBS is extended with 200 to over 400 phr of paraffinic oil. In certain oil-gel applications, the concentration of SEBS is as low as 5 wt% [121]. The oil contributes to compound processability and lowers cost without sacrificing the elastomer upper service temperature. Paraffinic oil is chosen to selectively swell the continuous rubber phase, leaving the discrete PS domains relatively unplasticized, thereby maintaining the integrity of these virtual cross-links at the elastomer upper service temperature. The MW of the PS end blocks is high enough to prevent significant plasticization by paraffinic oil and to provide sufficient incompatibility with the rubber phase for balancing TPE elastic properties (better with increased phase incompatibility) with processability (better with increased phase compatibility). Because SEBS is produced by the selective hydrogenation in the solution of the high vinyl butadiene rubber midblocks in SBS [122], the narrow-MWD distribution of the plastic and rubber blocks in SBS [123] (synthesized by anionic polymerization) is maintained in SEBS. Thus, a truly uniform rubber network structure swollen in paraffinic oil can be expected for SEBS due to the reorganization possible (at elevated temperature) in the PS domains.

The presence of a uniformly entangled rubber network and the low interphase volume (due to PS and rubber phase incompatibility—the interphase would hold less oil than the rubber phase) expected in high-molecular-weight SEBS would explain the large oil-holding capacity of this material. The rubber polymer chains can be viewed as being surrounded by a "tube" of oil, where the oil molecules are generally restricted to move within the tube but can cross over between tubes. The absorption of oil by SEBS is driven by the configurational entropy gained by the oil, which overcomes the conformational entropy lost on stretching of the rubber segments. There may be some lowering of system internal energy due to the adoption of low-energy conformations by the rubber segments. There also may be a limited enthalpic attraction between the oil and rubber. The rubber and the oil are nonpolar; therefore, no preferred orientation around the rubber molecule is expected for the oil in order to maintain the expected enthalpic attraction, thus minimizing the loss in entropy of the oil.

The viscosity of SEBS drops when it is plasticized by oil, but there is no increased phase mixing in the "melt." The apparent viscosity is reduced owing to the reduction in frictional forces between the rubber phase (when swollen in oil) and the wall of the capillary rheometer. This friction is not affected much by shear rate or temperature, so the apparent viscosity varies inversely with shear rate and is almost independent of temperature [112,113]. The flow of SEBS is best described by plug flow resulting from wall slip.

The presence of both PP and paraffinic oil is required for a dramatic improvement in the processability of a SEBS compound. Molten PP forms the viscous medium that allows ready transport of the biphasic SEBS during processing. The oil in the SEBS partitions between the SEBS and PP phases [124] (molten PP is miscible with paraffinic oil), thus reducing the viscosity of the molten PP and increasing its volume, which translates into improved SEBS compound melt processability. On cooling, the molten PP crystallizes, and the oil rejected from the crystalline phase partitions between the SEBS and amorphous PP phases. On cooling, the SEBS compound "hardens" rapidly due to the crystallization of the PP phase, thereby allowing rapid cycle time in end product manufacture.

C. SBC MORPHOLOGY

An example of typical SEBS morphology is represented in Figure 5.4. The morphology of a SEBS compound is dependent upon the relative proportions of PP, SEBS, and oil and upon processing conditions. Owing to the flow properties of SEBS already discussed, equilibrium morphology is not achieved in a typical compounding operation, where the residence time in the extruder is less than 3 min. At a low level of SEBS and oil, particulate SEBS has been found to be dispersed in PP when compounded in a twin screw extruder (75 wt% isotactic homo-PP, MI = 5.5; 13.3 wt% Kraton G1651 [33.3 wt% PS, MW = 29,000; 68 wt% PEB, MW = 116,000]; 11.7 wt% paraffinic oil). PP is the dispersed phase at very low PP levels, but a cocontinuous SEBS and PP phase is present in a composition range from about 10 to 55 wt% PP (11.6 wt% PP, 46.5 wt% Kraton G1651, 41.9 wt% paraffinic oil). The morphological assessments were based on extraction of injection-molded bars of SEBS compounds with xylene, at room temperature. Xylene extraction removed all oil in the sample. All SEBS present in the specimen was extracted in case of a continuous SEBS phase. In cases where all the SEBS was extracted, the residual PP formed a discreet phase if the xylene-extracted bar was reduced to a powder, or a continuous phase if the specimen retained its shape. The interdomain movement of the PS segments at the compound processing temperature causes the high-molecular-weight SEBS (which is a powder at room temperature) to "knit" together and form a continuous phase, especially when SEBS forms the bulk of the polymer blend. The paraffinic oil partitions between the SEBS rubber phase (the PS domains would also absorb a small quantity of oil) and the molten PP. It has been demonstrated that part of the molten PP, oil, and the PEB rubber phase of SEBS are miscible, allowing PP to form a continuous phase even at a very low PP level. The rubber and PP molecules are then entangled, and, on cooling, the trapped entanglements allow good adhesion between the phases and PP is nucleated across the phase boundary so that cocontinuity is maintained between the phases [106,123,125]. It is conceivable that the entanglement with a rubber molecule of an amorphous PP tie chain is anchored if the tie molecule is trapped within the same or different PP lamellae as it emerges from the rubber phase. Even at high elongations, the cocontinuous blends show a stress–strain behavior similar to that of rubber, with no sign of the typical necking phenomenon normally associated with PP at large deformation. It seems reasonable to propose that PP is present as thin coiled sheets and ligaments that simply uncoil during deformation, so that the PP phase itself is not subject to much stress and most of the deformation occurs in the SEBS. The PP phase reinforces the SEBS compound. In thermoset rubber, carbon black is used as rubber reinforcement. The presence of significant amounts of carbon black in the SEBS phase would considerably reduce SEBS compound melt processability due to increased SEBS melt viscosity.

From the foregoing discussion, it can also be understood how SEBS compounds with a hardness of about 0 Shore A can readily be produced. SEBS can absorb large quantities of paraffinic oil to yield a soft rubber, and the SEBS compound processability is enhanced by the oil together with a limited amount of the PP, which forms a cocontinuous hard phase alongside the cocontinuous SEBS rubber phase. During processing, the SEBS compound melt is simply slipping along the processing conduits on a thin film of a molten PP solution in oil. High nonfunctional filler (such as calcined clay that is contained mainly in the low viscosity solution of PP in oil during melt processing) and oil loading allows the production of low-cost SEBS compounds.

Because the oil-holding capacity of SEBS has been discussed, it is worth mentioning the oil-holding characteristics of commercially available SBS in connection with the wrist rest application, an example of which is a pad that is placed along the length of a computer keyboard, as wrist support. The highest-molecular-weight linear triblock SBS (Kraton D1101) is medium in MW compared to the highest-molecular-weight SEBS that is commercially available (Kraton G1651) [126]. Kraton D can probably hold only 100–150 phr of paraffinic oil without oil bleed. However, the oil gel of the wrist rest presumably contains low-molecular-weight SBS extended with perhaps 200 phr or more paraffinic oil. The SBS then increases oil viscosity, and oil bleed from the gel is prevented by encapsulation of the oil in an oil- and abrasion-resistant polyurethane cover. The oil-extended low-molecular-weight SBS can be readily processed (poured into a mold in the wrist rest application) at about 150°C, because of its lower (than SEBS) T_{ODT}. Moreover, the slight miscibility of the low-molecular-weight PS endblocks with paraffinic oil would further reduce hard- and soft-phase incompatibility, thereby improving gel processability. The high damping characteristics [127] of this gel (perhaps due to the large interphase volume created by the low molecular weight of the polymer and the mixing of small quantities of paraffinic oil into the styrene nano domains) may not be important in the wrist rest application. The lower cost of SBS compared to SEBS and the high oil loading (which also lowers cost) allowable without bleed due to the oil-resistant polyurethane cover makes SBS competitive in this low-end application where the product UV or thermooxidative stability requirement is minimal [128,129]. Moreover, intellectual property concerning SEBS gels, and the large number of SBS manufacturers compared to SEBS manufacturers, also allow the entry of SBS oil gels into this market.

D. SEBS Compound Upper Service Temperature Improvement

Even though the T_g of PS is about 95°C, under stress, the PS will flow at a temperature lower than its T_g. SBS loses most of its strength at 60°C–70°C [130,131]. In this case, the additional barrier to flow due to phase incompatibility between the rubber and the plastic is not sufficient to allow the PS nano domains to be good enough anchors to prevent disentanglement of the rubber chains and thus prevent viscous flow. Viscous flow occurs in low-molecular-weight SEBS (Kraton G1652, Table 5.4) at 65°C [131], in spite of the increased incompatibility between the rubber and plastic phases compared with SBS (elevated temperature stress–strain data). In SEBS compounds, the presence of PP helps to improve elastic recovery at elevated temperature. Nevertheless, because of the permanent plastic deformation of the styrenic domains and their reorganization as discussed earlier, the continuous use temperature of compounds containing high-molecular-weight SEBS is limited to 70°C, with a 100°C use temperature possible in applications where there is a limited load on the product.

Table 5.4 lists the physical properties of paraffinic oil blends of SEBS products Kraton G1651, G1650, and G1652 prepared by mixing in a laboratory Brabender. The SEBS materials have approximately the same PS content and are listed in order of

TABLE 5.4
SEBS Block Copolymers: Characterization and Properties

	PS (wt%)	PS ($M_w \simeq M_n$)	PEB (wt%)	PEB ($M_w \simeq M_n$)	Experimental[c] T_{ODT} (°C)
Kraton G1651[a]	33.3	29,000	66.7	116,000	
Kraton G1650[b]	29.0	13,500	71.0	66,400	350
Kraton G1652[a]	28.6	7,500	71.4	37,500	
Experimental [11]	31.8	3,400	68.2	14,600	142
	G1651	**G1651[e]**	**G1650**	**G1652**	
SEBS properties (50% SEBS, 50% paraffinic oil)[d]					
Hardness (Shore A)	16	17	25	25	
UTS (psi)	1148	1,721	1222	500	
UE (%)	904	1,119	791	560	
M100 (psi)	57	66	76	77	
CS (%), 22 h at 70°C	30	34	100	100	
CS (%), 22 h at 40°C	—	—	50	93	
TS (%), 10 min at RT	13	6	6	8	
Mixer removal	Crumbly	Crumbly	Sticky	Viscous oil	

[a] Ref. [126].
[b] Ref. [114].
[c] Ref. [118].
[d] All samples mixed under N_2 at 200°C and molded at 210°C except as noted. Mixing time approximately 19 min.
[e] Mixed under N_2 at 250°C and molded at 260°C.

decreasing MW. Note that for SEBS triblock copolymers, reducing PS MW while keeping the same weight percent of PS would require a reduction in rubber MW. The compression set increase with decreased SEBS MW can be related to permanent deformation of the PS microdomains and to the reorganization of these discrete PS domains by interdomain movement of the PS chain ends through the continuous rubber phase. Permanent plastic deformation of the PS phase may contribute only minimally to the compression set, because the volume of this phase is only 13% for a 30 wt% PS SEBS, assuming that all the added oil is present in the rubber phase and the interphase is neglected (PS density 1.05 g/mL; plasticized rubber density 0.86 g/mL). Hence, lowered SEBS MW must facilitate increased interphase movement at the molecular level, as discussed, due to the increase in phase compatibility. Increased phase compatibility is reflected in increased phase mixing for the lower-molecular-weight SEBS (higher hardness and higher M100; M100=modulus at 100% elongation) in comparison with the higher-molecular-weight materials (lower hardness and lower M100). Note that a continuous rubber phase is expected for these SEBS compositions. The increased T_g expected for the rubber phase of the low-molecular-weight SEBS would contribute to the increased set observed, but it is believed that the bulk of the set observed is due to interdomain movement of the PS segments. The property changes observed when high-molecular-weight SEBS is processed at different temperatures reflect the difficulty in achieving a stable morphology even after long processing times (compare columns 1 and 2 in Table 5.4).

The elastic recovery of SEBS compounds at elevated temperature can be improved by increasing the hard-phase T_g while maintaining or exceeding the incompatibility between the rubber and plastic phases over that of SEBS. The T_g of the hard phase can be increased by chemical modification of the PS end blocks in SEBS, by synthesis of triblock copolymers where the PS blocks are replaced by higher T_g hard blocks, and by compounding with a high T_g polymer that is miscible with the PS domains of SEBS. Alkylation of the PS phase of SEBS increases the hard-phase T_g but reduces the compatibility difference between the rubber and hard phases [113].

A recent publication [131] reviewed the methodology to enhance the high-temperature properties of SEBS by chemical modification, which increases the T_g of the PS glassy phase.

Poly(α-methylstyrene) (PαMS) has a T_g of 165°C, and α-methylstyrene (α-MS) can be polymerized by anionic, cationic, and free radical polymerization. Triblock copolymers with a polyisoprene midblock and PαMS end blocks have been produced by anionic polymerization, using a difunctional initiator, although the temperature had to be reduced to −78°C (instead of about −40°C for SBS synthesis), due to the low "ceiling" temperature for αMS polymerization [132]. An unsuccessful attempt to synthesize, by cationic polymerization, an α-MS/isobutylene/α-MS triblock copolymer has been reported, because of low "crossover" efficiency from the living cationic poly(isobutylene) termini to α-MS monomer [133]. TPEs based on PαMS are not expected to be of commercial value because of reversion of the polymer to monomer at elevated temperature [134]. Hence, the use of PαMS hard segments is unsuitable for improving the high-temperature compression set of SBCs.

Poly(2,6-dimethyl-1,4-phenylene oxide) (PPO) is the high T_g additive of choice for increasing SEBS upper service temperature. Paul and coworkers [135] showed that the solubility of PPO is much greater than that of PS itself in the PS nano domains of SEBS, owing to the exothermic heat of mixing in the case of PPO. The PPO MW should be less than or equal to that of the PS MW of the SEBS for miscibility, since limited conformations are available to it in this confined geometry [135–138]. Also, PPO of greater MW would lose much of its configurational entropy and gain only a small amount of translational entropy upon mixing [139]. The exothermic heat of mixing partially compensates for the unfavorable entropic effects associated with PPO confinement in the case of PPO–SEBS mixing.

Baetzold and Korberstein [118] studied solvent-blended low-molecular-weight PPOs (M_n/T_g [°C]: 1600/116, 700/161, 6000/196) with low-molecular-weight SEBS (PS [31.8 wt%], MW = 3,400, PEB: MW = 14,600 PEB = poly(ethylene/butene) SEBS rubber midblock). The PPOs raised the T_g (DMTA) of the PS domains to over the expected T_g (as per the Fox equation), for the case where these additives are uniformly distributed within and across all PS domains. At a low concentration (5–10 wt%) of the lower MW PPOs (M_n ~2700, ~1600), no separate PPO phase T_g could be detected by DMTA. Hence, the PPO distribution in PS may be macroscopically heterogeneous, that is, although the concentration of PPO in PS is locally homogeneous, there is a PPO concentration difference from PS domain to PS domain. However, electron microscopy did not support this interpretation. Hence, PPO is thought to be concentrated in the center of the PS domains in order to explain the observed T_g (DMTA, tan δ maximum, tan δ vs. T). The periphery of the PS domains would hold a reduced PPO concentration as this material is highly incompatible with PEB. For 5, 10, and 20 wt% concentration of PPO (M_n ~2700) in SEBS, a shoulder (DMTA, tan δ vs. T) that moves to progressively higher temperatures is seen at a higher temperature than the T_g of PS, reflecting the reduced PPO concentration at the PS domain periphery. At 20–30 wt% PPO, the T_g of PPO is also detected due to the formation of a separate PPO phase. For the higher MW PPO (M_n ~ 6000), even at 5 wt% concentration in SEBS, as expected, the T_g for a separate PPO phase is observed, in addition to the two T_gs from the PS phase. Unexpectedly, the T_g of PS is increased most by the lowest M_w PPO (M_n ~ 1600) and drops a few degrees with increased PPO MW. Presumably, the enthalpy of mixing favors increased concentration of the lower MW PPO at the center of the PS domains. As PPO MW increases, the concentration of PPO within the PS domains becomes more diffuse, resulting in a lower T_g. However, the breadth of the tan δ peak in the tan δ versus T curve, in the T_g region, suggests a concentration gradient of PPO within each PS domain.

In pure SEBS, the end group PS segments must extend to the center of the domain in order to maintain a constant density. Hence, it seems reasonable to think that even addition of PS homopolymer (instead of PPO) will swell the spherical domains of SEBS by occupying the domain core. Prior to Koberstein, Paul and coworkers had studied the dissolution of higher M_w PPO (M_n ~ 22,600) into SBS (Kraton D1101, 28.8 wt% PS) (M_n ~ 14,500 for PS; M_n ~ 67,500 for PBD; 16 wt% diblock content) [135]. PPO (M_n from 15,500 to 29,400) dissolution into the PS (M_n from 5,300 to 29,000) domains of higher MW SEBS was also studied by DSC and DMTA. The broad tan δ peaks in the T_g region of the solution blended products suggested a PPO concentration gradient within each PS domain, although no distinct transitions corresponding to PPO-poor and PPO-rich regions within the individual PS domains was identified, as in the Koberstein case [126].

TPEs with improved elevated temperature compression set have been produced by melt blending SEBS, PPO, PP, and paraffinic oil [140]. Gel compositions with softening points above 100°C were achieved by solution blending of PPO and SEBS and subsequent plasticization of the product isolated with paraffinic oil [141]. The replacement of paraffinic oil in this system with very-low-molecular-weight EPR as plasticizer (to prevent evaporative losses) results in soft SEBS gel compositions (hardness < 30 Shore A) with good elastic recovery at 150°C [142].

Low-molecular-weight SEBS is readily processable but has a limited upper service temperature (see earlier discussion). High-molecular-weight SEBS or SEBS with a high T_{ODT} has an increased upper service temperature at the expense of reduced processability. PP that is useful in SEBS compounding because of its partial melt miscibility with oil and PEB represents another limit beyond, which the upper service temperature of SEBS cannot be raised. PEB is not compatible with the commercially available, higher melting ($T_m = 240$°C), isotactic poly(4-methyl-1-pentene). Hence, the latter plastic cannot be used to improve the upper service temperature of SEBS compounds.

The increased T_g of the hard-phase modified SEBS would necessitate an increase in compound manufacturing temperature, time, and mixing intensity to achieve a suitable polymer morphology that would be stable at the TPE service temperature. Polymer thermooxidative and mechanical degradation may preclude these aggressive manufacturing conditions.

With the additional expense of material and compounding, SEBS compounds (presumably modified with PPO) have matched the 70°C temperature elastic recovery of PP/EPDM TPVs (Table 5.5) [143].

TABLE 5.5
Property Comparison of SEBS Compounds versus PP/EPDM TPVs

	Manufacturer, Trade Name	
	Multibase, Inc., Low Compression Set SEBS Compound	Advanced Elastomer Systems, PP/EPDM TPV
Property	Multiflex TPE A 5001 E LC	Santoprene® 101-55W185
Hardness (Shore A)	46	55
Tensile strength (psi)	800 at yield	640
Ultimate elongation (%)	600	330
Compression set (%) (70°C, 22 h, ASTM D395B, constant deflection)	19	19
Tear strength (pli)[a]	110	108
Specific gravity	1.06	0.97

[a] pli, pounds per linear inch.

E. SELECTED NEW DEVELOPMENTS IN SBCs

High-molecular-weight SEEPS triblock copolymers, similar to SEBS, are available from Kuraray. SEEPS is produced by the hydrogenation of styrene/~50 wt% butadiene– ~50 wt% isoprene/styrene triblock copolymers. The styrene content of these powdery products that vary in MW is around 30 wt%. None of the products flow at 230°C (2.16 kg). Although consisting of noncrystalline rubber blocks, SEEPS exhibits improved tensile strength over comparative SEBS due to the development of crystallinity on stretching. Improved tensile strength over SEBS compounds is also noted for PP/paraffinic oil/SEEPS melt blends [120]. Highly elastic ozone-resistant films can readily be extruded from SEEPS compounds that contain mostly SEEPS and oil. Judicious amounts of PS, or PP, or polyethylene are required for film melt processability and excellent physical properties [144].

When PS is the added plastic, it should be contained within the SEEPS styrene domains for film processability and good film physical properties. It is difficult to prepare such films from SEBS compounds. The lower "gel" content observed for SEEPS may be the reason for this observed difference [120]. "Gels" are partially swollen or overswollen regions, observed on the surface of an extruded film, in the SBC oil-gel network. Details of SEEPS morphology or rubber block structure has not been published. Another hallmark of SEEPS is its increased paraffinic oil-holding capacity over SEBS. "Oil-holding" capacity of an SBC refers to the amount of oil that can swell the polymer, without the formation of a separate oil phase. Thus, evaporation of oil and oil migration into oil-free polyolefinic materials in contact with SEEPS compounds (PP/SEEPS/oil) is limited, in comparison with SEBS compounds [120].

High- and low-molecular-weight SEBS, with varying butylene content in the rubber block, but higher than the 40 wt% butylene content in conventional SEBS, is commercially available as Kraton ERS ("enhanced rubber segment") polymers [145]. The styrene content in these materials varies from 13 to 33 wt%, with a rubber block T_g of −35°C (DSC) for many of the grades. Kraton ERS polymers have a lower T_{ODT} and are more miscible with PP than conventional SEBS. Lower-molecular-weight Kraton ERS grades can be readily melt blended with PP, without oil, for the preparation of clear, transparent compounds that are suitable for medical applications. Some low-molecular-weight ERS polymers have a low-enough T_{ODT} to be used in hot-melt adhesive applications as thermooxidatively stable SBS or SIS replacement. A high-flow ERS polymer is being touted for the preparation of breathable elastic films from melt blown fibers and in the impact modification of long glass fiber–reinforced 130 MFR homopolypropylene (Kraton MD 1648, 20 wt% styrene, 220 MFR [230°C, 2.16 kg], rubber $T_g = −13$°C [DMTA, 11 Hz], T_{ODT} ~200°C).

Asahi Kasei has commercialized styrene/butadiene-butylene/styrene (SBBS) triblock copolymers. These materials are obtained, presumably, by the partial hydrogenation of conventional SBS used for SEBS preparation. The pendent vinyl groups of the rubber backbone that are readily accessible to the hydrogenation catalyst may be almost all fully hydrogenated, and some residual unsaturation is retained in the rubber backbone. Hence, the thermooxidative stability of these polymers is better than that of SBS, but poorer than that of SEBS. These relatively low-molecular-weight products (MI [190°C, 2–16 kg] ~ 3 for all grades, lower T_{ODT} expected than comparative SEBS) are readily processable impact modifiers for PP, polyethylene, PS, and engineering plastics. They are also more thermooxidatively stable replacements for SBS and SIS in hot-melt adhesive applications. SBBS grades (Tuftec™ P) with 20, 30, 47, and 67 wt% styrene content are available.

Hydrogenated and unhydrogenated SIS triblock copolymers with high vinyl content from 3,4 polymerization of isoprene in the rubber midblock are available as Hybrar™ polymers from Kuraray. These high T_g materials (8°C to –32°C [DSC, 10°C/min]) with PS content ranging from 12 to 33 wt% have excellent damping characteristics [146,147]. The hydrogenated lower-molecular-weight materials have excellent compatibility with PP and can be used in the preparation of transparent oil-free compounds.

Thus, these compounds are useful for the manufacture of medical tubes, bags, and films, and food wrap, since they also have better air and moisture barrier over SEBS compounds. Furthermore, tubing produced from Hybrar compounds have excellent kind resistance. The high damping characteristics of Hybrar makes it useful in the formulation of hot-melt adhesives, athletic equipment grips, and in automotive sound damping applications. Steric hindrance precludes the complete hydrogenation of the rubber midblock of Hybrar, in a cost-effective manner. Thus, the rubber block in all Hybrar grades can be vulcanized. Hydrogenated high-molecular-weight Hybrar can be vulcanized during foaming (KL-7135, 33 wt% PS, rubber T_g = –15°C [DSC, 10°C/min], no flow at 230°C, 2.16 kg). Melt blends of unhydrogenated Hybrar and PS have greatly increased damping properties over unmodified PS.

Styroflex 2G66 is a styrene/styrene-ran-butadiene/styrene triblock copolymer, available from BASF. This SBC has a total PS end block content of 30 wt%, with about an additional 35 wt% of styrene copolymerized within the random SBR (T_g = –40°C, DMTA [148]) midblock. This 87 Shore A [35] product with narrow-MWD has an MW of about 130,000. The fuzzy borders around the PS nanodomains in the TEM of Styroflex is indicative of an extended interphase due to the presence of styrene in the continuous rubber phase [149,150]. Indeed, the softness of this product in spite of high total copolymerized styrene content is related to the extended interphase [151]. The stress–strain properties of the higher MW (~130,000) Styroflex 2G66 are comparable to that of lower MW (~70,000) SBS with 30 wt% endblock PS content. However, low T_{ODT} (145°C), and greater melt shear thinning behavior, allows Styroflex to be much more readily melt processable than the comparative SBS, at normal compound processing temperatures (170°C–210°C).

The rubber block in Styroflex is initially synthesized using a highly active difunctional initiator for anionic polymerization, where solvent combination and monomers feed ratio are adjusted to ensure a near random SB rubber, in spite of the higher reactivity of butadiene over styrene in anionic polymerization. Also, the chosen solvent combination limits the rubber vinyl content for improved product thermooxidative stability. The "living" ends of the SB rubber block are then used to initiate the polymerization of styrene to complete the synthesis of Styroflex. Styroflex is more thermooxidatively stable than comparative SBS. Styroflex TPE can be used as a PVC replacement, and the rubber phase of this material can be selectively softened over the thermoreversible PS domain polymer entanglement trappers for the continuous rubber matrix, with paraffinic oil, to yield very soft TPEs.

Transparent Styroflex can be used to impact and modify styrenic polymers such as general-purpose "crystal" PS (GPPS), high-impact PS (HIPS), ABS, styrene/ acrylonitrile (SAN), and expanded PS foam. The environmental stress cracking resistance of styrenic polymers is also improved by melt blending Styroflex into these materials. It is also compatible with polar polymers such as TPVs, nylons, and polyesters. The very low melt viscosity of Styroflex 2G66 allows it to be sheared into nanoparticles in the impact modification of polyethylene for film applications. Styroflex can be used to compatibilize polyolefins (PP, PE), polar plastics, and polyolefins with polar plastics. Presumably, compatibilization occurs by the prevention of coalescence of particles generated on subjecting an incompatible polymer melt blend to high shear, due to the presence of a large number of nanoparticles generated from a relatively small amount of added Styroflex.

Good elastic recovery, transparency, and excellent oxygen and water vapor permeability may allow food wrap application for Styroflex in multilayer films.

Hydrogenated styrene/SB/styrene triblock copolymers are available as Kraton™ A and Dynaron™ grades from Kraton Polymers and JSR, respectively. Hydrogenation increases both the thermooxidative stability and T_{ODT} of these unsaturated SBCs.

Unlike Styroflex, the rubber block segment near the PS end groups in Kraton A and Dynaron may be rich in styrene for T_{ODT} reduction. The rubber block structure of these materials are, however, not disclosed. High-styrene-content SBCs of this class with varied MW and rubber T_g are commercially available.

Styrene/isobutylene/styrene triblock copolymers, produced by cationic polymerization, are available from Kaneka (Sibstar™). Excellent gas impermeability, high damping characteristics, and enhanced thermooxidative stability over SEBS (which always contains a low level of unhydrogenated unsaturation) are the unique characteristics of these materials.

V. THERMOPLASTIC VULCANIZATES

A significant advance in polyolefin-based TPEs resulted from the discovery that EPDM rubber, when selectively cross-linked under shear (dynamic vulcanization) during melt blending with a compatible plastic, namely, isotactic homopolypropylene, results in a TPE with mechanical properties and fabricability far superior to those obtained from a simple blend of the elastic and plastic materials [152–159]. Indeed, the performance, price, and environmental impact of PP/EPDM TPVs have provided impetus for replacement of thermoset rubber by these TPEs. Penetration of the thermoset rubber market by PP/EPDM TPVs has been made possible by the breadth of the product service temperature (–40°C to 135°C), hardness range (25 Shore A to 50 Shore D), excellent fabricability and fabrication economics, and the ability of product scrap to be reprocessed, among other desirable environmental characteristics. PP/EPDM TPVs are products of commerce in thermoset rubber replacements through finished part cost savings realized by fabrication, design, and material economics. Compared to SEBS compounds, PP/EPDM TPVs exhibit better elastic recovery at a higher service temperature (100°C vs. 70°C). In "static" applications, PP/EPDM TPVs can provide service at 135°C versus 100°C for SEBS compounds. Very soft TPEs (0–5 Shore A) are based on SEBS; PP/EPDM TPVs with hardness lower than 25 Shore A are not commercially available.

A. DEFINITION OF DYNAMIC VULCANIZATION

Dynamic vulcanization is the process of producing a TPE by selective cross-linking of the rubber phase during mixing of a technologically compatible or compatibilized rubber and plastic blend of high rubber content, while minimally affecting the plastic phase. Rubber cross-linking is accomplished only after a well-mixed molten polymer blend is formed, and intensive blend mixing is continued during the curing process. The elastomeric TPV thus formed should ideally consist of a plastic matrix that is filled with 1–5 μm cross-linked rubber particles.

B. DEVELOPMENT OF DYNAMIC VULCANIZATES: HISTORICAL PERSPECTIVE

Dynamic vulcanization has its origin in the work of Gessler and Haslett [160] at Esso, where they demonstrated that carbon black–filled blends of isotactic PP and chlorobutyl rubber with good tensile strength could be obtained by curing the rubber (with zinc oxide) after the components were blended on a mill at room temperature, by further milling the blend at the curing temperature and above the melting point of PP. Tensile strength was lower when the blend was "statically" cured. Both "dynamic" and "static" cure resulted in thermoplastic compositions. The first patent claim, limited to chlorobutyl rubber, included blends with up to 50 wt% rubber in PP that could be cured with any curative that did not degrade PP. The incorporation of plasticizer oils into the blends is one of the items outlined in the second claim. The goal of this work may have been the impact modification of PP, because the rubber content of the blend was limited to 50 wt% in the patent claims. Captured in this work were the essential attributes of dynamic vulcanization as practiced today, including the use of rubber plasticizer oils, except for recognition of the importance of curing the rubber phase only after the formation of an intimate plastic and rubber blend, and the value of compositions containing a high rubber content.

   Gessler and Haslett did not continue their research on the dynamic vulcanization of rubber and plastic blends; the work of Fischer represented the next advance in this technology. It was shown that the properties of polyolefin blends with EPM or EPDM rubber (TPOs) could be dramatically improved if the rubber was first either statically or dynamically (on a mill, Banbury, or extruder) cured to a gel content of up to 90% before melt blending with the plastic. The gelled rubber was still processable (could be "banded" on a mill) prior to melt blending with the plastic. The TPO rubber content could be as high as 90%. The improvement in TPO properties was quantified by an increased "performance factor" (tensile strength [psi] × elongation at break [%] divided by elongation set at break) [161,162]. Another route to TPOs with improved properties was achieved by the use of EPM or EPDM rubber that was branched in a controlled manner during rubber production [163].

Fischer's work on TPOs culminated in the dynamic vulcanization (in a Banbury) of molten blends of PP/EPM or PP/EPDM with peroxide [164]. To maintain thermoplastic processability ("banding" of the final product on a mill or extrudability as a measure of product processability), the rubber cure state had to be limited. In the patent, the maximum cure state claimed for the rubber is 90% gel. Dynamic vulcanization tremendously increases melt viscosity over that of the TPO melt, which increases with increased rubber cure state, thus reducing TPV processability. TPV physical properties, however, improve with increased rubber cure state. Increased tensile strength, improved compression and tension set, lower swell in hydrocarbon oils, and improved flex fatigue and abrasion resistance are manifested in a TPV in comparison with the corresponding TPO. Nevertheless, because of PP plastic breakdown by the peroxide rubber curative employed by Fischer, the TPVs of the illustrative patent examples did not achieve their full property potential. To improve TPV melt processability, Fischer indicates the use of very limited quantities (~1 part per 100 parts of rubber) of process aids (e.g., epoxidized soybean oil, polymeric slip aids). Uniroyal's polyolefin thermoplastic rubber (TPR®, commercialized in 1972) is based on Fischer's work on dynamic vulcanization.

About the time Fischer was pursuing his studies on dynamic vulcanization, Paul Hartman at Allied Chemical Corporation claimed that butyl rubber could be grafted onto polyethylene by using a difunctional resole-type phenolic resin as rubber curative while the rubber and plastic were melt blended on a mill. The grafting was thought to occur via the end olefinic functionality in PE. The grafted material exhibited superior physical properties and had a greater capacity to disperse fillers such as carbon black and talc than the simple melt-blended product. Rubber crosslinking was avoided by using judicious amounts of the low functional phenolic resin curative. Grafted products of butyl rubber, EPDM, or diene rubbers such as SBR and NBR onto PP, PE, or poly(1-butene) were claimed [165–167]. The complete solubility of the products of this invention in hot xylene was taken as proof of grafting of the rubber onto plastic and the absence of cross-linked rubber. In all probability, the expected grafting reaction was minimal, and the rubber simply underwent chain extension in the presence of limited amounts of curative during dynamic vulcanization.

Monsanto entered the field of dynamic vulcanization with a patent by Coran et al. [168] that extended the work of Fischer to dynamic vulcanization of diene rubbers in a polyolefin matrix. The inventors demonstrated that thermoplastic compositions could be obtained even when the rubber was cured to a high cure state as opposed to Fischer's finding that the gel content of the rubber obtained in dynamic vulcanization should be less than 100% in order to maintain product thermoplastic processability. In addition, the Monsanto inventors realized that TPV physical properties improve when dynamic vulcanization is continued to achieve a high rubber cure state. TPV physical property improvement was also attributed to the presence of small rubber particles (less than 50 μm in diameter) dispersed in a plastic matrix. Subsequently, high-rubber-content TPVs based on butyl rubber were claimed by Monsanto [169]. The patent claims included compositions containing high levels of plasticizer oils.

In a coup de grace [170], Coran et al. demonstrated that, contrary to the Fischer partial cure requirement, PP/EPDM TPVs with both excellent physical properties and processability can be obtained when the rubber is cured to a high cure state. Sulfur, which does not degrade PP, was the curative of choice in the patent examples. Processable TPVs with a high rubber cure state could also be obtained by peroxide cure in the presence of a bismaleimide coagent that undoubtedly limited PP breakdown in addition to allowing the achievement of a fully cured rubber phase. It was also recognized that the presence of a high level of paraffinic oil allowed the preparation of soft, processable TPVs with excellent elastic recovery. On TPV plastic phase melting, the oil in the product partitions between the oil-swollen, cross-linked particulate rubber, and molten PP. The increased volume of plastic phase (solution of PP in paraffinic oil) in the molten TPV allows excellent TPV processability. When the TPV melt is cooled, some of the oil rejected from the plastic crystallites is absorbed into the rubber phase, and the rest of the oil is present as submicron pools in the amorphous plastic phase. At the same time, Gessler and Kresge [171] disclosed that PP/EPDM or EPM TPOs that were produced with high-molecular-weight rubber had desirable physical properties but were not processable. The TPOs exhibited both good physical properties and processability if paraffinic oil was added to the compositions.

Monsanto began its effort to commercialize TPVs with PP/EPDM/paraffinic oil compositions that were cured with sulfur. TPV morphology consisted of an oil-swollen, continuous PP matrix that was filled with cross-linked, micrometer-sized (1–5 μm) oil-swollen rubber particles. TPV mechanical properties and fabricability were dependent upon rubber particle size, with the size just indicated being preferred. During TPV processing, however, the rubber particles increased in size, presumably due to the breakage and re-formation of the polysulfidic cross-links that occur in the melt during particle collision. This unstable melt morphology ("melt stagnation" or phase growth of the dispersed rubber) resulted in poor and variable product fabricability and mechanical properties [172]. Evolution of gases with an unpleasant odor during manufacturing is another serious drawback of sulfur-cured TPVs.

This final obstacle to PP/EPDM TPV commercialization was overcome by Abdou-Sabet and Fath [173] by the use of a resole-type phenolic resin as curative. Melt stagnation was avoided by the formation of thermooxidatively stable cross-links in the rubber particle. Improved TPV resistance to hydrocarbon oils and compression set also resulted from the use of the phenolic resin curative. The advantages of using the phenolic resin in diene rubber-based TPVs were simultaneously recognized by Coran and Patel [174].

The granting of the Monsanto TPV patents led to heavy but unsuccessful opposition by several corporations. Uniroyal had entered the TPV market with a product for which the rubber phase was partially cured, but Santoprene rubber by Monsanto had superior physical properties due to its fully cured rubber phase that was achieved without plastic phase breakdown. Leading TPV suppliers today include ExxonMobil (Santoprene), Mitsui (Milastomer®), Sumitomo (Sumitomo TPE), and Teknor Apex (Sarlink®).

C. Principles of Dynamic Vulcanization

TPVs are complex systems that, when formulated and processed correctly, result in materials that show significant fabrication advantages over thermoset rubber. Six key requirements have been identified in the preparation of polyolefinic TPVs:

Principles of Dynamic Vulcanization
I.   Rubber and plastic compatibility
II.   Interphase structure
III.   Plastic phase crystallinity
IV.   Rubber vulcanization
V.   Morphology control
VI.   Melt viscosity control

1. Principle I: Rubber and Plastic Compatibility

An incompatible polymer blend would have poor physical properties due to poor adhesion between the phases. A miscible polymer blend would offer average properties of the blend components. In a blend with polymers of the "right" compatibility, the technologically useful properties of each of the blend components can be realized. The first principle of dynamic vulcanization is that the rubber and plastic should have the "right" compatibility, but should not be melt-miscible. If the plastic were miscible with the rubber, cross-linking of the rubber would result in the inclusion of large portions of plastic into the rubber particles, reducing the amount of plastic in the continuous phase. The resultant product would lose thermoplastic processability; the product of dynamic vulcanization would be a powder. Moreover, because the plastic's T_g is usually much higher than that of rubber, if the plastic is miscible with the rubber over a broad temperature range, the blend would not be elastomeric owing to a high average T_g. The extent of compatibility between the rubber and plastic required for TPVs with good physical properties and processability is difficult to quantify. Even between polymers that are considered technologically incompatible, such as PS and poly(methyl methacrylate), an interphase thickness of 50 Å has been measured [175]. Presumably, the high entropic penalty for demixing results in entanglement between different polymers in the interphase of incompatible polymer blends.

The more compatible the polymers, the greater the interphase thickness and interpolymer entanglements, and the better the mechanical properties of the polymer blend [176]. The difference between the "critical surface tension for wetting" of polymers is considered to be a rough measure of polymer interfacial tension [177,178]. Based on surface tension values, PP/EPDM blends (EPDM not characterized) are among the most compatible of the rubber and plastic blends evaluated by Coran and Patel in dynamic vulcanization [177].

The excellent compatibility of PP and EPR is reflected in the fine blend morphology that can be generated by melt mixing of the components [171,179–183]. Indeed, EPR is the impact modifier of choice for PP because of the good interfacial adhesion between these components [184,185]. Datta and Lohse [186] showed that in 80:20 by weight PP/EPR blends, EPR rubber particle size could be further reduced by the addition of an iPP-*g*-EP copolymer referred to as a compatibilizer. There should be a distinction drawn between emulsification and compatibilization. A diblock polymer acting as a true compatibilizer should have its block segments anchored in the phases being compatibilized. This would also imply an increase in interphase thickness over the uncompatibilized blend, which would also result in improved mechanical properties for the blend. A true compatibilizer is more likely to be formed by the in situ reaction of components that are already present in the phases being compatibilized. An example would be the impact modification of nylon by the melt blending of this plastic with a miscible blend of EPR and maleated EPR to generate fine particulate EPR in a nylon matrix [187]. The amine end groups of the nylon would then react with the grafted maleic anhydride moiety in the EPR to generate in situ a block copolymer in which the blocks are truly anchored in the phases being compatibilized.

In emulsification, as in the emulsification of oil in water by soap, the emulsifier is adsorbed only onto the surface of the particulate dispersed phase [188] and may also be present as a separate phase that is an additional barrier in preventing coalescence of the dispersed phase. These phenomena have been observed in the case where block copolymers have been melt mixed with incompatible polymer blends [189,190]. There is a recent report of the inability of PP-*b*-EPR to function as a compatibilizer in PP/EPR blends due to cocrystallization of the PP segment of the compatibilizer with the PP phase and rejection of the EP block into the interlamellar space of the PP crystals [191].

Key aspects pertaining to the interfacial adhesion of immiscible polymers are briefly reviewed by Adedeji and Jamieson [192], who studied particle size reduction due to emulsification and compatibilization in solution blends of SAN copolymers and PS produced with and without added poly(methyl methacrylate)-*block*-PS (PMMA-*b*-PS). Compatibilization or emulsification of the dispersed PS phase in SAN depended upon the polymer molecular masses, including the PMMA-*b*-PS compatibilizer segmental molecular mass. These determinations were made by studying crazes generated by the mechanical deformation of blended polymer films and observing preserved (due to emulsification) or fractured PS domains caused by efficient stress transfer across the SAN/PS interphase (due to compatibilization).

Chun and Han [114] also pointed out the importance of distinguishing between emulsification and compatibilization of polymer blends. The previously mentioned [186,193] observation of improved mechanical properties on size reduction of the dispersed EPR phase in the iPP matrix, by means of the iPP-*g*-EPR additive, may not be due to improved interfacial adhesion. The mechanical property improvement in impact-modified (by polyolefin rubber) isotactic PP has been shown to be dependent upon rubber MW [181,194–196], structure [181], and composition [181,195,197,198], rubber particle size and particle size distribution [181, 194,199–207], the ability of the rubber particle to cavitate [195,201,203,204,208–210], optimal interparticle distance [194,211–214] (and therefore PP ligament thickness), and perhaps modification of the PP crystal phase by the rubber particles [182,206,210,215,216], a compatibilizer [217], or PP nucleating additives [218,219].

Many of the aspects discussed in connection with the mechanical property improvement in impact-modified iPP are interrelated. For example, the rubber particle size obtained in an iPP/polyolefin rubber blend would depend upon rubber melt viscosity (and hence rubber MW) and processing conditions. Adhesion of the rubber to the iPP matrix would depend upon rubber MW, which can be related to the extent of miscibility of the low-molecular-weight ends of the two components [220]. However, if the rubber phase is cross-linked—for example, to stabilize the melt morphology—interphase adhesion would be affected. Particulate high MW or cross-linked rubber will not cavitate. The iPP crystal structure obtained on cooling iPP/polyolefin molten blends would also depend upon rubber particle size and cure state.

The impact strength of plastics would depend upon the morphology-dependent deformation behavior (shear yielding, crazing, and rubber particle cavitation) that ensues under the impact conditions. Impact strength is also dependent upon processing conditions [221,222], which would affect rubber particle size and shape and cause built-in stresses due to "frozen in" rubber particle shape and differential shrinkage between the plastic and rubber. During the cooling of molten impact–modified (with particulate polyolefin rubber) isotactic PP, for example, any voids created by differential shrinkage between the rubber and plastic phase will be filled in as long as the phases are mobile. Although there is considerable shrinkage in the plastic phase from the melting point to the crystallization temperature, once the morphology is frozen (corresponding to the crystallization temperature), the rubber particles shrink more than the plastic on cooling to room temperature. This differential shrinkage would have caused the rubber particles to shrink away from the plastic/rubber interface were it not for the good adhesion between the phases, but imposes triaxial tension on the rubber phase [206,223].

On bending, a piece of impact-modified PP sometimes results in stress whitening ("blush") due to scattering of light at the crease line, presumably owing to rubber particle cavitation. PP "blush" can be reduced by compounding in polyethylene, which is more compatible with the rubber than plastic, and hence, resides within the rubber particles. This allows a reduction in the rubber amount and hence limits rubber shrinkage and changes the stress distribution across the rubber particles. Rubber particle cavitation is prevented due to the excellent adhesion of the rubber to both the PP and polyethylene phases. The impact properties of the system are then also improved while minimizing stiffness decrease due to rubber particle incorporation into PP [224]. Fibrillar rubber morphology is undesirable in impact-modified plastics, because these types of rubber particles presumably act as inefficient craze initiators [196,225]. Finally, the behavior of plastics on impact would depend upon temperature and test speed, and at high test speeds, adiabatic heating in the deformation zone has to be taken into account [226,227].

A review of the structure and properties of PP/elastomer blends has been published [196,225,228]. Modification of iPP with particulate rubber offers improved material impact strength at reduced stiffness. It is desirable to achieve good material impact properties while maintaining stiffness. HIPS is filled with appropriately sized rubber particles, which allows good impact properties, but these rubber particles contain trapped particulate PS ("salami" structure), resulting in the material having a reduced rubber content, which limits the stiffness decrease on impact modification of PS [229]. This technique has also been applied to balancing iPP rigidity with toughness [230,231].

Compatibility and miscibility in PP/EPDM molten blends would be dependent upon polymer MW, the ethylene/propylene (E/P) ratio and E/P compositional distribution in the rubber molecules, and the length of sequences of the structural units in the rubber. Although iPP and atactic PP are miscible in the melt [232,233], iPP and sPP are not melt-miscible in all proportions, and hence the tacticity of iPP may have an influence on the compatibility and miscibility in PP/EPDM molten blends. Blends of PP and EPR copolymers have been shown by neutron scattering to be immiscible in the melt, even when the ethylene content of the EP copolymer is as low as 8 wt% [233]. Kyu and coworkers [234] found that iPP with an M_n of 247,000 and EPDM (70% ethylene, 5% ENB, ML 1+4 at 125°C=55) are miscible below the melting point but above the crystallization temperature of iPP, a lower critical solution temperature was observed [234]. Some miscibility of low-MW fractions between iPP and EPDM is indicated in the solid state because of a T_g increase of the rubber in the melt-blended product [222]. In PP/ethylene-1-octene plastomer blends, the lower rubber T_g observed has been attributed to trapped stresses in the rubber phase caused by a mismatch in the thermal contraction characteristics of the plastic and rubber [223].

It is well established that iPP/EPDM blends are immiscible at temperatures well above the iPP melting point at which dynamic vulcanization is conducted. The partitioning of paraffinic oil between molten PP and EPDM does not change the system's immiscibility characteristics [106].

Table 5.6 lists the formulations and properties of PP/EPDM/paraffinic oil melt-blended products (TPOs) and the corresponding TPVs. Paraffinic oil is completely miscible with EPDM and molten PP. The formulations were mixed in a laboratory

TABLE 5.6
Dynamic Vulcanization of PP/EPDM/Oil Blends: Compression-Molded Plaques from Brabender Preparations

| | | Increased Plastic Content and Decreased Oil Level | | | |
| | | | Sample | | |
	1 (TPO)	2 (TPV)	3 (TPO)	4 (TPV)	5 (TPO)	6 (TPV)
Hardness (Shore A)	10	37	27	49	89	90
UTS (psi)	48	404	132	720	1098	1757
UE%	112	239	361	318	389	412
M100 (psi)	46	172	120	256	956	1090
Compression set (%) ASTM D395B, constant deflection, plied discs	100	16	100	18	81	50
Tension set (%)	40	3	37	3	53	25
% Extractable rubber (cyclohexane) (based on total rubber)		3.7		4.2		1.9
Volume% (rubber + oil)		91		88		61

Brabender at 180°C, followed by curing of the molten blend under shear for TPV preparations, and subsequently, compression molded at 200°C to produce plaques for testing. The morphology of the products as probed by AFM is presented in Figure 5.5, with product numbers as in Table 5.6. Rubber is represented by the dark areas, and plastic domains by the lighter color. Products 1–4 have a continuous rubber phase and a discrete plastic phase, whereas the phase morphology is reversed in products 5 and 6. Oil was extracted from the TPV samples by Soxhlet using the cyclohexane/acetone azeotrope, following which the dried samples were subjected to extraction by cyclohexane at room temperature (24 h) to determine the amount of soluble rubber present. The data indicate that all the TPV samples had a high level of cross-linked rubber. The volume percent rubber in the formulations was calculated assuming that all the oil is in the rubber phase, with none in the amorphous portion of the plastic phase.

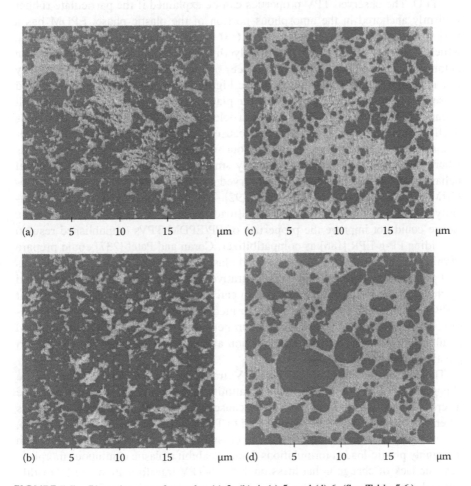

FIGURE 5.5   Phase images of samples (a) 3, (b) 4, (c) 5, and (d) 6. (See Table 5.6.)

Despite the high rubber cure state and rubber continuous/discrete plastic solid-state morphology, products 2 and 4 are still thermoplastic. Perhaps in the melt state, the rubber and plastic phases are cocontinuous or there is a continuous plastic phase. In both cases, thermoplastic processability is expected. For the TPVs in question, particulate rubber (of undetermined size) could form a continuous phase by impingement and entanglement of polymer chains at the points of contact of the rubber particles. The TPV morphology is dependent upon the level of shear introduced into the melt during dynamic vulcanization.

a.   Transformation of Thermoplastic Olefins to Thermoplastic Vulcanizates

On transformation of TPO to TPV there is a substantial improvement in the desirable elastomeric physical properties such as elastic recovery (compression set, tension set) and tensile strength (Table 5.6). This is counterintuitive, because compatibility between the rubber and plastic should decrease on cross-linking the rubber and result in poorer TPV properties compared with the corresponding TPO. The observed TPV properties can be explained if the particulate rubber is firmly anchored in the amorphous portion of the plastic phase. EPDM has a very low MW between entanglements (1660 for EP rubber with 46 wt% E) [206], which is not much higher than that of polyethylene (1250) [235]. The MW between entanglements for PP is 7000 [236]. Then, owing to the PP/EPDM compatibility already discussed, the two polymers are highly entangled in the interphase, and these entanglements are trapped when the plastic crystallizes. It is conceivable that a trapped PP segment on the rubber particle surface is a tie molecule that finds itself anchored to the rubber particle because it is part of one or more PP crystalline segments. This thinking is not without some precedent. In PS/PMMA blends where the interfacial tension is relatively small, no difference in the mechanical behavior of the interface could be observed in the presence or absence of PS-*b*-PMMA copolymers of different MWs [192]. This has been attributed to mechanically effective entanglements that are already present in the PS/PMMA blends. Lohse could not improve the properties of PP/EPDM TPVs (unpublished results) by adding PP-*g*-EPR [186] as compatibilizer. Coran and Patel [237] could prepare PP/nitrile rubber (NBR) TPVs with good elastic properties only if the plastic was first pretreated with the phenolic resin curative. The PP was thought to be modified by the phenolic resin (presumably due to end unsaturation in the plastic) such that a PP-*g*-NBR compatibilizer could be formed in situ during dynamic vulcanization. This approach did not improve the properties of PP/EPDM TPVs (unpublished results), which suggests that good adhesion already exists between the rubber and plastic phases in this system.

The hardness change on TPO-to-TPV transformation is dependent upon the change in product morphology and crystallinity in the plastic phase and is also due to cross-linking of the rubber phase. Because the rubber phase is still continuous when TPO Samples 1 and 3 are converted to TPVs 2 and 4, respectively (Table 5.6), the change in hardness can be related to cross-linking of the rubber. Because both the highly plastic-loaded formulations 5 and 6 exhibit a plastic continuous morphology, the lack of change in hardness on TPO-to-TPV transformation can be readily rationalized.

Inoue and Suzuki [238,239] studied the impact properties of PP (70 wt%)/EPDM (30 wt%) melt blends before and after dynamic vulcanization with a curative that is not expected to affect PP (N,N-m-phenylenebismaleimide/2,2,4-trimethyl-1,2-dihydroquinoline). No change in rubber particle size results from dynamic vulcanization. The considerable improvement in impact properties observed after dynamic vulcanization was related to increased interfacial adhesion between the rubber and the plastic (due to the in situ formation of a compatibilizer during cure by grafting of rubber onto the plastic), which caused increased shear yielding and crazing in the damaged plastic zone created by impact testing. It was also noted that the cross-linked EPDM particles acted as nucleating agents and decreased PP spherulite dimensions. The extent of the improved impact strength observed on dynamic vulcanization due to the change in plastic morphology was not established. The uncross-linked rubber particles in the iPP were shown to cavitate, and energy dissipation was thought to occur by craze initiation at the rubber particle sites subsequent to cavitation, with the cavitation process itself contributing to only a modest improvement in impact strength. Rubber particle cavitation was not observed on impact testing of the modified iPP containing cross-linked rubber particles as suggested by Ishikawa et al. [240]. Instead, plastic shear yielding was thought to occur at the rubber particle sites, which led to the formation of stronger crazes (compared with the case where iPP contained uncross-linked particulate rubber) due to stronger cold-drawn fibrils spanning the craze. These strong fibrils presumably resulted from the in situ rubber-to-plastic grafting that occurred on cross-linking of the rubber particles, as mentioned earlier. Similar results have been reported by Kruliš et al. [241] for dynamically vulcanized PP (80 wt%)/EPDM (20 wt%) blends (sulfur cure).

Differential shrinkage on cooling of the blends in question would cause the rubber particles to be in triaxial tension (see previous discussion) in both TPO and TPV, but the built-in strain should be detrimental to impact strength. The cross-linked rubber particles of the TPV are less likely to dissipate energy by cavitation compared to the TPO, which contains gel-free particulate rubber. Because iPP is the major component of the impact-modified plastic and impact modification of iPP by cross-linked rubber particles has a dramatic effect on the plastic crystal structure, it is reasonable to propose that the increase in impact strength observed on TPO-to-TPV transformation is due to modification of the plastic phase crystal structure by the cross-linked rubber particles according to our understanding of the impact behavior of plastics due to Argon and coworkers [242–244]. This work suggests that when the plastic crystalline structure that is nucleated around a rubber particle percolates throughout the specimen, impact properties are optimized. Therefore, the plastic ligament thickness must be small enough to allow overlap of the crystalline structures that are nucleated around the rubber particles. However, it should be noted that thin plastic ligaments would contribute greatly to product impact resistance by shear yielding on loading due to plane stress conditions, as opposed to lower energy absorption by crazing that would be the case for thick plastic ligaments due to plane strain loading.

Semicrystalline polymers such as nylon-6 and PP that are normally ductile fail in a brittle manner under impact loading. The impact toughness of these materials can be improved by the incorporation of rubber particles that are bonded to the matrix, at the expense of a reduction in material stiffness. The data of Argon and coworkers indicate that there is a dramatic increase in material impact toughness when the interparticle matrix ligament thickness is below a critical dimension. This phenomenon has generally been attributed to the overlapping of stress fields created by the appropriately spaced rubber particles of the "right" dimensions. Certain melt-blended PP/calcium carbonate composites are stiffer and have greater Izod impact strength than the unfilled plastic [244]. This effect was not observed for nylon-6 [243]. The key requirement for plastic toughening in these cases is rubber particle cavitation or debonding of the inorganic particle from the matrix prior to initiation of matrix plastic flow. The presence of rubber particles or particulate inorganic material induces the formation of a layer of oriented matrix crystals of well-defined thickness around the particles upon cooling after melt mixing. When the matrix ligament thickness is small enough to allow these oriented crystal structures to percolate throughout the specimen, a dramatic increase in material toughness is achieved. Nylon-6 could not be toughened with calcium carbonate particles using this concept, owing to only partial debonding of the rigid particles from the matrix, thereby causing stress concentration in the early phases of the impact response that caused a reduction in fracture toughness.

Extending this concept to the TPO transformation, it is known that the impact behavior of materials is controlled by the structure-dependent molecular response that is initiated under the specific impact test conditions (temperature, strain rate, adiabatic heating). In a PP/EPDM TPO system, the impact response has been demonstrated to proceed from a brittle response at low rubber volume with cavitation of rubber particles on impact to a ductile response characterized by matrix shear yielding without rubber particle cavitation or debonding of the rubber particles from the matrix at a higher rubber volume fraction. Increasing rubber particle volume in the PP matrix from 0% to about 40% resulted in a step increase in impact strength when the rubber volume consisting of the "right"-sized rubber particles established a PP ligament thickness of 0.1 μm, presumably due to establishment of the "percolated" PP crystal structure already discussed. Increasing the rubber particle volume beyond a certain level in the PP/EPDM compositions resulted in a drop in impact strength due to the decrease in matrix phase volume. Fracture toughness decreased with increasing strain rate as expected, but in some cases, on further increase in strain rate, PP/EPDM blend fracture toughness increased due to adiabatic heating of the impact damage zone and the surrounding volume. Also, in the TPO ductile region, fracture energy initially increases, then decreases, as temperature increases, due to the decrease in matrix yield strength with temperature [245]. As already discussed, physical property improvement on TPO-to-TPV transformation can be rationalized by presuming that an improvement in adhesion between the rubber particle and matrix accompanies this transformation. This improved adhesion would assist interphase stress transfer during impact loading of the Inoue and Suzuki TPVs [238,239], but the unexpected improvement in TPV impact resistance over that of TPO was probably due to PP crystal structure modification.

Coran and Patel [157] studied the properties of dynamic vulcanizates obtained from 99 rubber and plastic combinations. Only a very limited number of TPVs of this study were technologically useful, with the properties (physical properties, processability, and cost) of PP/EPDM-based products far surpassing those of the other materials. Although a wide variety of thermoplastics and elastomers could be combined to form TPVs, the best results, based on tensile strength and compression set, were attributed to the compatibility of PP and EPDM.

2. Principle II: Interphase Structure

The best adhesion (required for excellent TPV physical properties) between the cross-linked particulate rubber and the PP matrix is obtained via a "mechanical lock" of the rubber particles on to the amorphous phase of the PP matrix. Owing to excellent PP/EPDM melt compatibility, the molten PP molecules are entangled with the EPDM molecules on the surface of the cross-linked rubber particles. It is known that cross-linked rubber particles nucleate the crystallization of PP (higher PP crystallization temperature, DSC) in a TPV much more efficiently than the corresponding TPO. It is conceivable, then, that a trapped PP segment on the rubber particle surface is a tie molecule that finds itself anchored on to the rubber particle because it is part of one or more PP crystalline segments. The "mechanical lock" thus formed provides better interphase adhesion than added compatibilizer. In fact, increased addition of iPP-g-EPDM as compatibilizer to an iPP/EPDM TPV resulted in progressive loss of product tensile strength and elongation, presumably due to disruption of the "mechanical lock" between the rubber particles and the plastic matrix [246].

When a crystalline ethylene/α-olefin copolymer was added to a PP/EPR melt blend, the solid-state physical properties of the blend improved only when the additive of choice was incorporated into the particulate EPR and only if uniformly distributed crystallites ("cross-link" junctions) were formed within the particulate EPR [247].

Matrix failure was observed in the solid-state fracture of a iPP/30 wt% metallocene polyethylene (mPE) melt blend. The observed interfacial failure was attributed to the iPP crystallite anchoring discussed earlier. When Ziegler–Natta PE (Z-N PE) was used instead of mPE, the low level of completely amorphous material present in Z-N PE disrupted the "mechanical lock" under discussion, and interfacial failure was observed. Addition of an ethylene/ethylene–butene/ethylene triblock copolymer to the iPP/mPE melt blend resulted in the change of failure mode from matrix to interfacial [248,249].

Thus, there is precedent in the literature for the proposed TPV interphase structure.

3. Principle III: Plastic Phase Crystallinity

The third principle of dynamic vulcanization is that the plastic phase should be highly crystalline so as to provide sufficient cross-links (thermoreversible) for good elastic recovery. The plastic crystal melting point should be high enough to provide the desired elastomer upper service temperature. The plastic T_g should be low for improved TPV processability and low-temperature properties. There should be no plastic phase transitions between T_g and T_m because that would be detrimental to TPV high-temperature elastic recovery. Note that iPP exhibits an α-crystal transition between 30°C and 80°C [250]. TPV processability is enhanced by plastic materials with a broad MWD.

The continuous plastic phase controls both the upper service temperature and melt processability of the TPV. In the case of iPP/EPDM TPVs, the 165°C peak DSC melting point (which is lowered to about 155°C in commercially available 60 Shore A TPVs due to plastic crystal structure consisting of fragmented lamellae [see Section VII.D] and due to the presence of paraffinic oil [small effect for kinetic reasons]) of the isotactic PP allows a 100°C upper service temperature, which can be extended to 135°C in no-load applications. In polyolefinic systems, where polar interactions are absent, plastic "hydrodynamic volume" matching with the rubber will allow maximum compatibility with the rubber from the melt to the solid state by maximizing the entropy gain in the interphase region. For polyolefins, this can be related to the T_g of the plastic, which should be as close as possible to that of the rubber. PP has a T_g of 0°C–10°C (depending upon rate and type [DSC, DMA] of measurement), which is relatively close, for a plastic material, to the T_g of EPDM (–50°C for a 60 wt% ethylene EPDM with 4.5 wt% copolymerized ethylidene norbornene as cure site). Polyethylene with a T_g of –80°C is more compatible with the EPDM rubber under consideration, but this increased compatibility raises TPV melt viscosity beyond the fabrication capability of conventional plastics-processing equipment [251]. The relatively low T_g of iPP allows a low melt viscosity for this plastic, because on melting at 165°C, the melt is already 165°C above the T_g. This low PP melt viscosity is critical to TPV processability, because this viscosity is raised considerably on being filled with cross-linked EPDM particles.

It is now readily recognized that a completely amorphous polymer may not be suitable as the TPV plastic phase from the standpoint of processability. For example, commercially available, highly amorphous, bisphenol A-based PC with a T_g of 150°C will allow a use temperature well below 150°C. However, the melt viscosity of polymers decreases gradually above T_g, and PC is economically processable only just above 300°C. Filling PC with cross-linked rubber particles would raise the melt viscosity such that the material would be processable only above the plastic and rubber decomposition temperatures. Therefore, a completely amorphous plastic is unsuitable for use as the TPV plastic phase component.

Because the plastic material will become part of the TPV elastomeric system, it is reasonable to choose a ductile, as opposed to a brittle, material for this purpose. The mode of failure of a plastic material under, say, uniaxial tension will depend upon the temperature, the strain rate, adiabatic heating, the flaws present in the sample, and sample size. A brittle material will fail after a "small" deformation, and the damage zone around the failure surface will be "limited." For brittle material, failure generally starts with the development, perpendicular to the sample tensile direction, of microcracks (crazes) that are spanned by load-bearing and cold-drawn fibrils. The crazes rapidly coalesce into a crack that rapidly propagates perpendicular to the tensile direction, leading to material failure. "Crystal" PS (so called because of its high transparency) is a high T_g (~100°C), completely amorphous brittle material whose failure in tension begins with the formation of crazes that can be 0.1–2 μm thick and 50–1000 μm long, with fibril diameters varying from 4 to 10 nm [252]. Preceding craze formulation, there is an expansion in sample volume due to an increase in polymer hydrodynamic volume as the system attempts to reduce the applied stress. Argon and Hannoosh [253] showed that for small, highly perfect samples of PS, deformation by shear yielding precedes crazing. Sample preparation is important, because dust particles have been observed to act as craze nuclei in a thin film of PS, as observed by TEM. Impact-modified PS has been shown to deform by simultaneous rubber particle cavitation, crazing, and shear yielding [229].

If a bar of material is placed in uniaxial tension, the principal tensile stress acts on the area of the bar that is perpendicular to the tensile direction. If the material is isotropic, the maximum shear (or "sidewise") force acts on an area inclined at 45° to the tensile axis [254]. Macromolecules will align and flow past each other (shear yielding or plastic flow) wherever the stress in the slip plane exceeds the material yield stress. Deformation is initiated at a site of stress concentration created by a structural imperfection in the material, which may be caused by energetically unfavorable molecular arrangements ("built-in" stresses), voids, or the presence of foreign matter. The stress concentration is created by the "bending" of the stress field around the imperfection. It is preferable, of course, that the plastic material elongate to high strain by shear yielding rather than crazing before rupture, if it is to be chosen as the TPV plastic phase.

Once a damage zone is created around a site of stress concentration, more material is drawn from the sides toward the damage zone as the test specimen "necks" in the tensile direction. If the stress field generated is such that material from the sides cannot be drawn in fast enough (depending upon test conditions and molecular characteristics), polymer chains may slip past each other or rupture under load in the damage zone, creating a void that can cause an increase in stresses across the now reduced area of the test specimen, leading to brittle fracture. If the thickness of the specimen is small in comparison to the damage zone (plane stress conditions, Figure 5.6), one can expect shear deformation, and in the opposite case (plane strain conditions, Figure 5.7) for the same applied strain rate, brittle failure. As the strain rate is increased, the failure mode of a material can change from ductile to brittle. At very low strain rates, a material may show brittle behavior, as in the stress cracking of polyethylene [255]. A material is considered to be brittle if, under the test conditions in question, failure occurs largely due to crazing. A ductile material fails largely by shear yielding.

For elastomeric applications as in TPVs, a plastic that can deform at a relatively low stress is more desirable than other commercially available plastic materials; iPP meets this criterion [256]. Moreover, for small strains and short deformation time, iPP exhibits Hookean elasticity (storage of energy without any dissipation of the input energy as heat) and linear viscoelastic behavior (complete recovery of deformation on release of the deforming force, although part of the input energy is lost as heat) for small strains beyond the Hookean limit before undergoing yielding at about 8% strain [256,257].

a. iPP Morphology and Mechanical Properties

The most commonly occurring crystalline form of iPP is the α (monoclinic) form, which displays spherulitic morphology, as does the β (hexagonal) crystal modification but not the γ (orthorhombic) or smectic forms. These spherulites are composed of radial lamellae in which the c (long) crystalline axis is perpendicular to the lamellar growth direction (Figure 5.8). iPP α spherulites display a unique "cross-hatched"

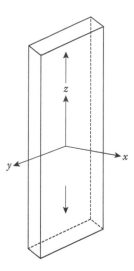

FIGURE 5.6   Plane stress condition (stress in yz plane only). Sample width y is large in comparison with the damage zone that is centered at the crossing of the axes and created by a tensile force in the z direction. Material contained in the width y restricts reduction in this dimension, generating stresses pointing away from the center in the y directions in response to the tensile force in the z direction. Sample thickness in the x direction is much smaller than the width, and the material from the x direction feeds the sample neck as it elongates in the z direction (low stress in the x direction). Note that since the sample dimension will decrease minimally in the y direction, plane strain conditions prevail in the xz plane.

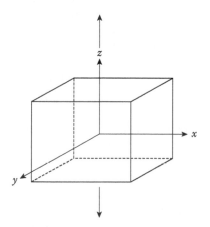

FIGURE 5.7   Plane strain condition (strain in the xz and yz planes). Sample width y and thickness x is large in comparison with the damage zone that is centered at the crossing of the axes and created by a tensile force in the z direction. Reduction of dimension in the xy plane is restricted due to large sample width and thickness, thus creating stresses pointing away from the center in the xy plane, resulting in a triaxial or dilatational stress field. Most of the strain occurs in the z direction, creating a large stress concentration in the damage zone that is unable to draw in much material from the xy plane. Chain breakage and/or slippage of the macromolecules in the damage zone leads to transfer of stress to other sites of stress concentration in the sample, and the test specimen breaks in a brittle manner with a small strain in the z direction.

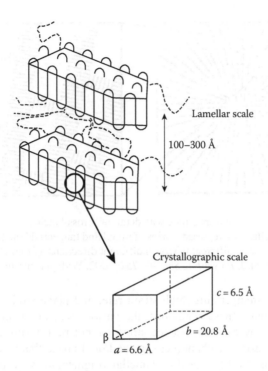

FIGURE 5.8 Chain folding in polypropylene lamellae. (From Zhang, L. et al., *J. Appl. Polym. Sci.*, 86, 2085, 2002. With permission.)

structure due to the presence of tangential lamellae (Figure 5.9), which is not observed for the lower density and lower melting β and γ forms. On heating or under mechanical stress, the β and γ forms can be converted to the α form. A lower density, lower melting mesomorphic crystal form that appears in iPP when it is fabricated under rapid cooling conditions also converts to the α form on heating [258].

iPP lamellae and spherulites are, of course, connected by tie molecules that are responsible for material continuity and therefore the mechanical integrity of the plastic. The initial Hookean response of PP observed at room temperature at low strain can be related to elastic deformation of the low T_g amorphous tie molecules. The linear viscoelastic region that is observed for low strain beyond the Hookean limit reflects enthalpic elasticity due to reversible interlamellar shear or reversible intralamellar fine chain slip (Figure 5.10) [256,257].

The fluctuation in amorphous layer thickness and the number and length of the tie molecules throughout the sample can produce a high stress concentration and cause material fracture at an average stress that is much lower than expected for the tie chain density present [259]. As PP crystallization conditions are adjusted to yield larger spherulite size, the material becomes more brittle due to the reduction in interspherulite tie molecules and the fracture mode changes from intra- to inter-spherulitic rupture; a transition from spherulitic yield to boundary yield is also observed [259–265].

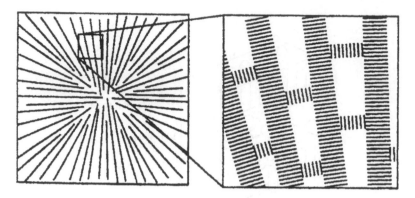

FIGURE 5.9 Schematic of α spherulite with detail of "cross-hatched" structure due to radial and tangential lamellae. ---- represent *c* axes of radial and tangential lamella. The *c* axis may be of the same size, or be different in size in different directions. (From Dijkstra, P.T.S., van Dijk, D.J., and Huétink J. *Polym Eng Sci., 42, 152,* 2002. With permission.)

In general, low-temperature, high strain rate, and plane strain conditions favor crazing of iPP. Under these conditions, the various relaxation times in the amorphous PP layers may be such that the stress cannot be transmitted effectively to the spherulites, causing tie chain breakage and/or slippage that results in craze formation. Voids can also be created by lamellar fragmentation. At high temperature and/or low strain rate, plastic deformation occurs through unraveling of the folded lamellar PP chains. Radial lamellar orientation results in anisotropic spherulite deformation. Depending upon the orientation to the tensile direction, lamella can be separated, sheared, or compressed. Voids can be formed within the spherulite through formation of a fibrillar structure containing small pieces of lamellae due to coarse chain slip (Figure 5.10) [257,260,261,266]. A transition from ductile to brittle behavior of PP was observed at room temperature as the strain rate was varied from 10^{-4} to 90 s^{-1} [267].

The low α-crystal transition temperature (30°C–80°C) [250,268] of PP allows chain pullout from the crystalline segments, delaying the onset of brittle fracture. At very low strain rates, as in the stress cracking of polyethylene, the tie molecules holding the lamellae together untangle and pull out of the lamellae to cause interlamellar separation and brittle fracture [255]. Although PP does not undergo stress cracking at ambient temperature, as PE does, when both materials are compared at the same temperature difference above T_g, their stress cracking behavior is comparable [269]. The low α-crystal transition temperature observed for both PP and PE lowers the energy barrier to stress cracking. This result also confirms the role of the amorphous component of the plastic in the stress cracking process.

4. Principle IV: Rubber Vulcanization

The fourth principle is that the rubber should be selectively cured to a high cure state for improved TPV elastic recovery, subsequent to the formation of an intimate rubber and plastic melt blend. The plastic should be minimally affected by the rubber curative. Some rubber and plastic grafting may provide the beneficial effects of

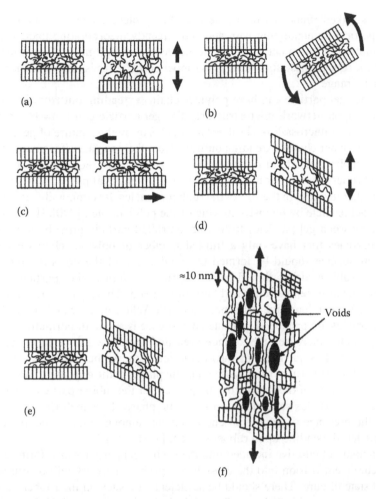

FIGURE 5.10 Polypropylene spherulite deformation mechanisms. Schematic diagram showing the variety of deformation mechanisms operative in a semicrystalline polymer, (a) interlamellar separation, (b) lamellar stack rotation, (c) interlamellar shear, (d) intracrystalline shear ("fine chain slip"), (e) intracrystalline shear ("coarse chain slip"), and (f) fibrillated shear. In bulk samples, these mechanisms coexist. (From Chen, C.Y. et al., *Polymer*, 38, 4433, 1997. With permission.)

improved phase compatibility at the expense of reduced TPV melt processability. High-molecular-weight rubber with a narrow-MWD would allow a high rubber cure rate and cure state. A short time (~30 s) is available for dynamic rubber vulcanization in a twin screw process.

The curative that is added after formation of the desired rubber and plastic blend should not, or only minimally, affect the plastic phase and should rapidly diffuse into the rubber phase. In the initial stage of the rubber cross-linking process, a branched molecule is formed, altering the viscosity characteristics in the surrounding rubber domain. The branched molecule is also a nucleus for the subsequent formation of a cross-linked rubber particle. The branched molecule can grow into a gelled rubber particle by entangling with and subsequently incorporating uncross-linked rubber molecules into a network. It is desirable that several nuclei form simultaneously in the rubber phase so that the final rubber particle size is limited to the desired micrometer range.

A rubber gel particle can have polymer chains extending out from its surface, depending upon network microstructure. The gel particle can grow by entanglement capture of uncross-linked rubber molecules or by the capture of gel particles that have polymer chains extending outward from their surface. It is undesirable to allow the rubber particle to grow beyond the 1–5 μm size required for good TPV physical properties in order to avoid the decline in physical properties that occurs with a further increase in the size of the rubber particles. It is impossible to break up a rubber gel particle by intensive mixing of the polymer melt [270]. Hence, before collision between gel particles fosters uncontrolled particle growth, cross-linked rubber particles that have only a limited number of polymer chains extending from their surface should be formed by rapid action of the cross-linking agent. Therefore, rubber curing should be fast enough to limit rubber particle growth, which is related to temperature and "mixing" speed. These parameters also control the diffusion of curative in the polymer melt. When appropriately cross-linked rubber particles are produced, particle coalescence is avoided. Formation of small rubber particles increases the surface area and therefore the surface energy of the rubber, which is quenched by the interaction of the rubber surface with the linear molecules of the rubber-compatible plastic phase, at the expense of an increase in the viscous drag of the molten plastic phase over the rubber particles. Thus, the cured rubber particles separate into the plastic phase. TPV melt rheological data suggest the presence of long-lived entanglements among almost touching elastomer particles, depending upon rubber content [271].

The amount of curative included into the rubber gel particles as it forms and the rate of curative diffusion into the cross-linking rubber network will determine the rate and state of cure. There should be sufficient cure sites on the rubber molecule to form a network with "adequate" cross-link density for "good" TPV properties such as compression set. A "lightly" cross-linked rubber particle can still have "low" extractables (in the standard test used to determine the amount of rubber molecules excluded from the rubber network) but may yield a TPV with poorer compression set than a TPV containing more densely cross-linked rubber particles.

5.   Principle V: Morphology Control

PP/EPDM TPVs are produced by melt blending of the rubber and plastic followed by selective curing of the rubber phase while the well-mixed blend continues to be intensively sheared. The fifth principle is that the mixing intensity during TPV preparation should cause fragmentation of the rubber phase into small (1–5 μm in diameter) cross-linked rubber particles and also allow plastic phase inversion. That is, the discrete or cocontinuous molten plastic phase should become continuous as rapidly as the continuous rubber phase is cured and broken up into cross-linked rubber particles. The process results in 1–5 μm diameter rubber particles filling a molten PP matrix. Depending upon the rubber-to-plastic ratio and the melt viscosity (therefore MW) of the individual phases and the blending conditions, the molten blend morphology prior to cure could consist of (1) particulate rubber in a molten plastic matrix, (2) cocontinuous rubber and plastic phases, or (3) molten plastic particles in a continuous rubber phase [200,215,272]. After dynamic vulcanization, it is desirable that irrespective of the initial blend morphology, the final product should consist of fine, fully cross-linked rubber particles in a plastic matrix. The variation in blend morphology of PP/EPR blends at 200°C and at a shear rate of 5.5 s⁻¹ due to variations in the weight fraction and viscosity of the components has been reported [273]. Owing to polymer melt shear thinning, the phase diagram varies with shear rate. Energy transfer between the phases during melt mixing is expected to be maximized by phase viscosity matching. Phase melt viscosities should also have a significant influence on the rubber particle size produced during dynamic vulcanization.

Note, however, that material shear thinning narrows the considerable viscosity difference observed between the plastic and rubber melt viscosities at low shear rates [271]. Melt viscosity differences observed at low shear rates would narrow considerably under the much higher shear rates for PPs with different melt flow rates used in dynamic vulcanization. Also, thermal, thermooxidative, and mechanochemical degradation will lower material melt viscosity [207,274]. Phase coalescence has been observed when the mixing intensity of uncured PP/EPDM melt blends is reduced [200,221,275,276]. Perhaps the factors just discussed are responsible for the formation of rubber particles of similar size by the dynamic vulcanization of PP/EPDM blends with different degrees of mismatch between rubber and plastic melt viscosity at low shear rates [277].

It is reasonable to conclude that the mixing intensity should be at least sufficient to maintain blend morphology and to facilitate curative diffusion into the rubber. Also, the finer the blend morphology, the easier it is to achieve the desirable TPV morphology. It is also evident that in the dynamic vulcanization of blends 2 and 3 (see first paragraph in this section), it is harder to achieve the desired TPV morphology (compared with dynamic vulcanization beginning with blend 1), with case 3 expected to yield TPVs with the largest rubber particle size. In the case of blend 3 (continuous rubber phase), the curative should not cross-link the rubber almost instantly. If this happened, the mixing process would have to grind the thermoset rubber generated into 1–2 μm particles, which is not possible [270]. However, the rubber curing should be fast enough to rapidly generate several rubber nuclei so as to limit rubber particle size, and the mixing should be intense enough to transport the gelled rubber particles into the plastic phase (phase inversion) before the undesirable agglomeration of rubber particles occurs. As the rubber cross-links (the plastic phase should not be affected by the curative), it becomes less compatible with the plastic, and hence it seems reasonable to propose that dangling chain ends and any loosely cross-linked network reside on the outside of the rubber particle, which should have a more tightly cross-linked core. As the curative diffuses to the interface between the molten rubber and plastic, curing at the rubber surface will create a viscosity mismatch between internal and surface rubber. The shearing and elongational forces during dynamic vulcanization should be intense enough to dislodge the cured rubber section from the surface of the rubber phase for particle size control.

If the rubber cure rate is too high and/or the mixing intensity too low, "skin" formation due to curing of rubber at the molten rubber and plastic interface will result in large rubber particles.

The most efficient packing for monodisperse hard spheres is a face-centered cubic arrangement with the spheres taking up 74 vol% of space, and therefore a minimum of 26 vol% plastic for a continuous phase is required if TPVs could be modeled in that way [278]. Of course, a TPV would have a distribution of rubber particle sizes

FIGURE 5.11 Scanning electron microscopic image of the morphology of Santoprene Rubber 8211-35 at increasing magnification from top to bottom. Note length of scale bar representing the dimension in micrometers..

that are not spherical and are deformable. Hence, a continuous plastic phase can be established in the solid state in a TPV with rubber volume fraction greater than 74%. Nevertheless, very soft TPVs (35 Shore A) appear to have a rubber continuous and discrete plastic phase morphology in the solid state (Santoprene Rubber 8211-35) (Figure 5.11). Of course, it can be argued that "winding" continuous plastic ligaments can appear to be discrete in a two-dimensional scanning electron micrograph, but the high rubber and oil content of this product would favor a discrete plastic phase over a continuous one. The rubber continuous phase in soft TPVs is probably due to closely packed cross-linked rubber particles held together by thermoreversible entanglements and "spot-welded" together by the discrete plastic phase. It can readily be seen how increasing the oil and rubber levels while decreasing the plastic content of very soft TPV formulations can produce a softer product with only limited processability and mechanical integrity. In the case of example 6 (Table 5.6), the higher level of plastic in the formulation permits continuous plastic phase morphology in both TPO and TPV.

The solid-state morphology of commercially available PP/EPDM TPVs with hardness greater than 35 Shore A is best described as consisting of particulate rubber in a continuous plastic matrix. There is an example of a 70 Shore A TPV product (Vegaprene®) that has a rubber continuous morphology [279], perhaps because it has a high rubber-to-plastic ratio and only a minimal amount of oil added for TPV processability and hardness control. The high percent rubber content (low oil and low filler content) may be responsible for the touted improved elastic recovery of Vegaprene over other TPVs of equivalent hardness [280].

Interpretation of 2D projections (TEM, AFM) of a 3D structure has been shown to be inadequate in assigning TPV morphology. For example, in a thin section cut for TEM, overlaying rubber particles may suggest a continuous rubber phase. Even if discreet rubber particles are observed, the rubber phase may still be continuous, as morphological information in the third dimension is not available. In electron tomography, 3D morphology of TPVs can be elucidated from 2D TEM images of the sample viewed from different directions [281].

6. Principle VI: Melt Viscosity Control

The molten rubber and plastic interfacial area increases tremendously on TPO-to-TPV transformation due to the presence of small cross-linked rubber particles that fill the plastic matrix. The viscous drag of the molten plastic over the rubber particles and rubber particle interactions would make a polyolefinic TPV melt almost unfabricatable without viscosity reduction by the process oil added.

The importance of the TPV interphase formation and adhesion between phases has already been clearly established. Even in the presence of paraffinic oil, which is necessary to control TPV melt viscosity, the trapped entanglements between the rubber particles and the PP amorphous phase are maintained. The sixth principle is that a plasticizer that minimally affects the adhesion between the rubber and plastic phases is necessary for the control of TPV melt viscosity. The use of a judicious amount of paraffinic oil will limit the decrease in adhesion between the rubber and plastic phase (disentanglement of rubber and plastic molecules in the interphase). The addition of excessive amounts of oil to a PP/EPDM TPV formulations will result in poor mechanical properties. Completion of dynamic vulcanization results in a solution of PP in oil that is filled with cross-linked, oil-swollen particulate rubber. Since cross-linked EPDM has a high affinity toward paraffinic oil, and since molten PP is miscible with paraffinic oil, it seems reasonable to propose that there is no difference in oil concentration between the plastic and rubber phase in a molten TPV, given the mobility of the oil molecules at the TPV melt processing temperature (200°C–240°C). Note that the room temperature specific gravity of amorphous EPDM rubber that is typically used in TPV manufacturing is about 0.86 and that of the amorphous phase of PP about 0.855 [282]. At the TPV production temperature, both molten materials are expected to have the same specific gravity and oil concentration, in spite of the higher cross-link and entanglement density in EPDM, in comparison with the entanglement density in PP. However, this thinking is not in keeping with the interpretation of published experimental results [283,284]. On cooling the TPV melt, the rubber particles nucleate PP crystallization, and the oil rejected from the crystallites are absorbed into the amorphous PP phase and also into the oil-swollen rubber particles. The oil distribution will, of course, depend upon TPV composition. Solid-state TPV morphology depends upon product composition and production conditions. The TPV morphology of many commercial products consists of oil-swollen particulate rubber, with most of the rubber particles having a diameter between 1 and 5 µm, contained in an oil-swollen, semicrystalline PP matrix. It is generally accepted that paraffinic oil in a TPV is present as a miscible component in both the rubber and amorphous plastic phase. Indeed, the reduction in T_g, as measured by DMTA, of the amorphous plastic phase and rubber, due to the presence of oil, has been used to quantify the oil distribution in TPVs, in the solid state [283,285–288].

The miscibility of paraffinic oil in the amorphous phase of PP is questionable, since microporous membranes can be produced by the crystallization of PP by cooling an oil solution of PP. Voids are created on solvent extraction of oil from the solid solution cast membrane [289]. Hence, in a TPV, the oil may be present as submicron pools in the PP amorphous phase, while being miscible in the rubber phase. Miscibility of oil in the PP amorphous phase would swell this phase uniformly, thereby deforming the PP crystallites via the PP chain segments of the crystallite-tethered rigid amorphous phase. Thus, the high rigid amorphous phase content [290], in comparison with the mobile amorphous phase, may preclude oil miscibility in semicrystalline PP. Obviously, oil must occupy volume in the PP amorphous phase without uniformly swelling this phase. Oil may be pooled in the mobile amorphous phase. On melting a TPV in an NMR tube, about a third of the oil in the formulation was neither mixed in with the EPDM rubber nor molten PP [291]. Although oil is completely miscible in the molten TPV, without the benefit of mixing, the submicron pools of oil in the amorphous phase of the solid TPV would remain as a separate phase in the melt due to slow diffusion of oil into the polymeric materials.

The T_g of neat iPP is observed at 0°C (DMTA, 0.1 Hz, 0.1% strain, tan δ maximum, tan δ vs. T). The shift of this maximum to lower temperatures in PP/EPDM TPVs is perhaps not due to miscibility of a paraffinic oil in the PP amorphous phase, but rather originates from dissipation of energy caused by the oil-filled microporous PP structure that results by crystallization of PP from the oil solution, in the TPV melt.

D. Rationalization of PP/EPDM TPV Elastic Recovery

Given that TPVs by definition almost always exhibit a continuous plastic phase, the mechanism of recovery has been an important subject of study and debate. It is clear that in order for TPVs to show good elastic recovery, the following conditions are necessary: First, within the two-phase structure, there must be an interphase that provides a high degree of adhesion between the phases. Second, the rubber phase must be sufficiently cured to behave elastically. The rubber particle size must be small enough that it can be deformed by the relatively thin PP ligaments. And finally, the plastic crystal structure should be modified by the rubber particles to produce an elastic plastic phase.

For excellent TPV elastic recovery, the iPP crystallites should be uniform in size and be uniformly distributed throughout the sample, as should be the particulate rubber dispersed phase. There should also be an appropriate balance between tie molecules and crystallites in the TPV plastic phase. A large number of tie molecules will maximize the load-bearing capacity of the plastic phase and allow maximum extensibility before rupture, provided the tie molecule distribution and length are uniform, which would minimize local stress concentrations [292]. However, the smaller the lamellae due to loss of molecular mass to tie molecules, the more readily will permanent deformation in the plastic phase occur, especially at elevated temperature or under creep conditions.

Thus, the nucleating effect of cross-linked rubber particles on the plastic phase [238–240,293] is beneficial for elastic recovery in PP/EPDM TPVs. In injection-molded PP/EPDM TPV plaques, the plastic α-crystal nucleation density was too high for the observation of individual spherulites by polarized light microscopy except at the specimen core where the spherulites were sparsely distributed. In the center of the specimen, where the melt cooling rate was the lowest, a few β spherulites were observed [293]. This is, of course, undesirable because the melting point of these crystals is lower than that of the α form and the β crystals are converted to the α form by heat or mechanical stress. These factors would be detrimental to TPV elastic recovery. However, it should be mentioned that β spherulites deform in a more ductile manner than the α form due to the absence of the "cross-hatched" structure in the former case [294]. In fact, β-nucleated PP is marketed for high-impact-strength applications (Borealis).

For a 73 Shore A PP/EPDM TPV, x-ray data suggest that not much spherulitic structure is present and that the PP crystallites consist of separated and/or fragmented lamellae [295]. The percent crystallinity of the plastic phase in a PP/EPDM TPV is not much lower than that of injection-molded iPP. The absence of a large number of spherulites in the 73 Shore A TPV suggests a uniform crystallite distribution in the plastic phase. A considerable part of the plastic phase network then consists of elastic (T_g 0°C) tie molecules.

Our understanding of TPV elastic recovery has been advanced by the pioneering work of Inoue et al. [295–298]. When 73 Shore A hardness PP/EPDM TPV was stretched to 200% elongation in the path of an x-ray beam, not much change in the iPP crystallite orientation was observed, and the orientation was completely recovered when the deforming force was released immediately after the 200% elongation.

On TPO-to-TPV transformation, particularly when the material rubber content is high, as already discussed, there is a considerable increase in interfacial area between the rubber and plastic phases due to breakup of the TPO continuous rubber phase into discrete, micrometer-sized, cross-linked rubber particles. This causes the continuous plastic phase to form a "cobweb" structure consisting of thin ligaments and thin coiled sheets, with good adhesion to the particulate rubber phase dispersed therein. There may be some thick "islands" of plastic in the structure, but when a tensile force is applied to a TPV sample, it can be surmised that most of the deformation will occur by "uncoiling" of the coiled thin ligaments and sheets, resulting in little actual deformation in the plastic phase at the molecular level. The plastic ligaments in the tensile direction are stretched, causing a compression of the rubber particles in between ligaments and causing movement of rubber particles in the stretch direction. The rubber particles should be small enough (~5 µm in diameter) to be readily deformed by the thin PP ligaments. Generally, the larger the rubber particles, the poorer the TPV elastic recovery.

Plastic ligaments that are perpendicular to the pull direction would be in compression. In continuous cyclic loading of soft PP/EPDM TPVs at room temperature, where not much time is available for strain recovery, the TPV takes a permanent set. However, after several cycles, the stress–strain "hysteresis loop" shows little change whether the cycling is conducted above or below the elastomer yield stress [277]. This suggests that some of the plastic ligaments have shear yielded under the prevailing plane stress conditions (see previous discussion). Plastic and rubber adhesion is not lost during this process, and, on release of the deforming force, the TPV recovers by rubber particle springback (release of equatorial compression), the recoiling of uncoiled PP ligaments, and the decompression of polar PP ligaments.

Huy et al. [299] studied the deformation of PP/EPDM TPVs by using polarized infrared spectroscopy. The plane of the polarized light was adjusted to be alternately parallel or perpendicular to the sample draw direction. The change in the intensity of the IR absorption bands peculiar to PP or EPDM rubber only, caused by molecular orientation, was studied. The study was conducted on iPP, thermoset EPDM rubber, and iPP/EPDM TPVs under tensile deformation. The Herman "orientation function" increased considerably during the deformation of pure iPP to high strain, more so than the thermoset EPDM. However, the PP crystal phase orientation was less than that of the EPDM in the TPV. At a low deformation (50% elongation) in a one-cycle loading–unloading test, the orientation in both TPV phases was completely recoverable. In a TPV stress relaxation test to 1000 min at 200% constant elongation, the orientation of the PP phase increased whereas that of the EPDM rubber phase decreased linearly. When the modulus of the TPV rubber phase was increased by increasing the rubber cure state, the orientation in the PP crystal phase increased as the PP ligaments probably experienced increased stresses while the harder rubber particles were "squeezed" on tensile loading.

The work of Huy et al. [299] and that of others supports the emerging mechanistic picture of the ability of TPVs to recovery elastically due to the recoverable deformation of highly cured, highly dispersed rubber particles and their influence on the crystalline structure of a continuous plastic phase of appropriate ligament thickness. There is continued interest in elucidating the mechanistic details of TPV elastic recovery [299–304].

E. THERMOPLASTIC VULCANIZATE PROCESSABILITY

PP/EPDM TPVs can be fabricated by any thermoplastic process such as extrusion, injection molding, and blow molding. The presence of paraffinic oil in the compound can increase the volume of the continuous molten PP phase by partitioning into the iPP melt from the rubber phase under the processing conditions, thereby allowing excellent melt flow. The oil rejected on crystallization of the molten PP is reabsorbed by the rubber phase, with some of the oil forming submicron pools in the amorphous PP phase. TPV melts are shear thinning, and a considerable lowering of viscosity can be achieved by increasing the melt shear rate over that achievable by increasing the melt temperature alone. In extrusion, the die swell of TPVs is much lower than that of molten PP itself, with softer TPVs (higher rubber content) exhibiting less die swell than harder TPVs (higher iPP content) [305]. This is due to the TPV melt plug flow that occurs due to wall slip, where the melt slips as it is pushed through the die, assisted by a thin film of molten PP in oil that lubricates the die surface. The energy supplied to the melt leads to little melt deformation (and therefore low die swell) owing to the low frictional forces between the melt and the die surface, and increased energy input to the melt simply results in more of the melt being pushed out of the die. Crystallization of the iPP on cooling of the extrudate "freezes" the smooth morphological characteristics of the extrudate. TPV processability is enhanced by the use of a plastic phase with a broad MWD.

F. THERMOPLASTIC VULCANIZATE HARDNESS CONTROL

In thermoset EPDM rubber, a relatively soft material in comparison with PP, carbon black is used as reinforcement and paraffinic oil as compound softener. A balance of carbon black and paraffinic oil is needed to adjust compound processability and hardness.

The use of carbon black–reinforced EPDM in dynamic vulcanization would increase product melt viscosity to an unacceptable level for processability, due to enhanced rubber particle interactions via surface carbon black. Hence, TPV hardness is controlled by the amount of PP in the formulation. A balance of PP/EPDM/ oil is required for controlling TPV processability and hardness. Harder TPV grades, because of higher PP content, are poorer in elastic recovery than softer grades. TPVs are colored by the addition of pigments as PP concentrates.

VI. FINAL COMMENTS

TPEs have the favorable fabrication economics and environmental impact of plastics processing over that of thermoset rubber. In comparison with thermoset rubber, TPEs allow superior fabricated part appearance, and rapid fabricability (short cycle time), particularly by injection and blow molding. Also, there is no need for the post fabrication operation of trimming flash from molded parts.

Oil-swollen, particulate (~1–5 μm in diameter) cross-linked EPDM rubber, contained in an oil-swollen PP matrix, best describes PP/EPDM TPV morphology. These materials continue to be at the forefront of thermoset rubber replacement, ever since the commercialization of Santoprene thermoplastic rubber over 33 years ago. Although the high-temperature elastic recovery of PP/EPDM TPVs cannot match that of thermoset EPDM rubber, TPVs best address this major drawback of other nonpolar and polar TPEs.

This article attempts to explain, on a molecular basis, the exceptional elasticity of TPVs and how only the PP/EPDM combination is ideal for TPV preparation, due to the "right" compatibility between two components. The melt miscibility of PP with paraffinic oil, and the capacity of cross-linked EPDM to become highly swollen in paraffinic oil, is key in allowing excellent product melt processability and in lowering product cost. The key raw material physical property requirements for commercially successful TPVs are discussed in detail. Also discussed in detail are the structure/property relationships of other TPEs such as SBCs, TPUs, and the latest developments in polyolefinic TPEs.

ACKNOWLEDGMENTS

I am indebted to Dr. Garth Wilkes, Distinguished Professor of Chemical Engineering, Virginia Polytechnic and State University, for his thoughtful review of the manuscript, which has enhanced its accuracy and clarity. Discussions with the following individuals are gratefully acknowledged: Dr. Geoffrey Holden, Holden Polymer Consulting; Dr. Jeffrey Koberstein, Professor of Polymer Science, Columbia University; Dr. Ed Kresge, Polymers Consultant; Dr. Yona Eckstein, Noveon; Dr. Gerald Robbins, Bay State Polymers; Dr. Steve Manning, Bayer; Drs. Dale Handlin, Kathryn Wright, Bing Yang, Kraton Polymers; and Dr. Brian Chapman, Kuraray. I also thank the late Marc Payne, formerly chief technology officer, Advanced Elastomer Systems, L.P., for review of this work. Cathy Parker and Brian Gray cheerfully helped me with literature searches and in obtaining paper copies of the numerous references cited in this work. I would also like to acknowledge the contributions of the late Norm Barber and Dr. Lili Johnson. I am also grateful to S. Rebecca Rose for manuscript preparation and for her patience with me during this process. Linda Penson has taken great pains with manuscript revisions. Finally, I thank ExxonMobil Chemical for allowing me to publish this work.

REFERENCES

1. Gowariker VR, Viswanathan NV, Sreedhar J. *Polymer Science*. New Delhi, India: Wiley, 1987, p. 2.
2. *Encyclopaedia Britannica*, Micropaedia, Vol. 10. Chicago, IL, 1988, p. 223.
3. *Encyclopaedia Britannica*, Macropaedia, Vol. 21. Chicago, IL, 1988, p. 291.
4. Mark JE. *Physical Properties of Polymers*. Washington, DC: American Chemical Society, 1993, pp. 5–7.
5. Chum SP, Kao CI, Knight GW. In: Scheirs J, Kaminsky W, eds., *Metallocene-Based Polyolefins*. Chichester, U.K.: Wiley, 2000.
6. Kresge EN. *Thermoplastic Elastomers*. Cincinnati, OH: Gardner, 1996, p. 103.

7. (a) Karjala TP, Walther BW, Hill AS, Wevers R. *SPE ANTEC Conference Proceedings* 1999, p. 2139; (b) Karjala TP, Cheung YW, Guest MJ. *SPE ANTEC Conference Proceedings* 1999, p. 2127; (c) Guest MJ, Cheung YW, Martin JM. *ACS Polymeric Materials Science and Engineering Conference Proceedings* 1999, p. 371; (d) Diehl CF, Guest MJ, Chaudhary BI, Cheung YW, van Volkenburgh WR, Walther BW. *SPE ANTEC Conference Proceedings* 1999, p. 2149; (e) Park CP. *SPE ANTEC Conference Proceedings* 1999, p. 2134; (f) Guest MJ, Cheung YW, Besto SR, Karjala TP. *SPE ANTEC Conference Proceedings* 1999, p. 2116; (g) Coran AY. Thermoplastic Rubber-Plastic Blends, In: Bhowmick AK, Stephens, HL, eds., *Handbook of Elastomers*, New York: Marcel Dekker, 2001, Chapter 8.
8. Stephens CH, Poon BC, Ansems P, Chum SP, Hiltner A, Baer E. *J Appl Poly Sci* 2006;100:1651.
9. Gilbert M. *J Macromol Sci Rev Macromol Chem Phys C* 1994;34(1):77.
10. Marshall RA. *J Vinyl Technol* 1994;161(1):35.
11. Barendswaard W, Litvinov VM, Souren F, Scherrenberg RI, Gondard C, Colemonts C. *Macromolecules* 1999;32:167.
12. Brookman RS. *J Vinyl Technol* 1988;10(1):33.
13. Witsiepe WK. *ACS Adv Chem Ser* 1973;29:39.
14. Adams RK, Hoeschele GK, Witsiepe WK. In: Holden G, Legge NR, Quirk R, Schroeder HE, eds., *Thermoplastic Elastomers*. Cincinnati, OH: Hanser/Gardner, 1996, Chapter 8.
15. Meckel W, Goyert W, Wieder W. In: Holden G, Legge NR, Quirk R, Schroeder HE, eds., *Thermoplastic Elastomers*. Cincinnati, OH: Hanser/Gardner, 1996, Chapter 2.
16. Petrovicé ZS, Budinski-Simendic J. *Rubber Chem Technol* 1985;58:685.
17. Bonart R. *Angew Makromol Chem* 1977;58/59:259.
18. Leung LM, Koberstein JT. *Macromolecules* 1986;19:706.
19. Peebles LH Jr. *Macromolecules* 1976;9(1):58.
20. Peebles LH Jr. *Macromolecules* 1974;7(6):872.
21. Miller JA, Lin SB, Hwang KKS, Wu KS, Gibson PE, Cooper SL. *Macromolecules* 1985;18(1):32.
22. Harrell LL. *Macromolecules* 1969;216:607.
23. Gogolewski S. *Collord Polym Sci* 1989;267:757.
24. Seymor RW, Estes GM, Cooper SL. *Macromolecules* 1970;3(5):579.
25. Lilaonitkul A, Cooper SL. In: Frisch KC, Reegen SL, eds., *Advances in Urethane Science and Technology*, Vol. 7. Wesport, CT: Technomic, 1979, p. 163.
26. Clough SB, Schneider NS, King AO. *J Macromol Sci Phys* 1968;B2:641.
27. Leung LM, Koberstein JT. *J Polym Sci Polym Phys Ed* 1985;23:1883.
28. Abouzahr S, Wilkes GL, Ophir Z. *Polymer* 1982;23:1077.
29. Van Bogart JWC, Gibson PE, Cooper SL. *J Polym Sci Polym Phys Ed* 1983;21:65.
30. Ferguson J, Patsavoudis D. *Eur Polym J* 1972;8:385.
31. Zhrahala RJ, Critchfield FE, Gerkin RM, Hager SL. *J Elast Plast* 1980;12:184.
32. Chang AL, Thomas EL. In: Cooper SL, Estes GM, eds., *Multiphase Polymers*, ACS Adv Chem Ser 176. Washington, DC: American Chemical Society, 1979, p. 31.
33. Wilkes GL, Samuels SL, Crystal R. *J Macromol Sci Phys* 1974;B10(2):203.
34. Wilkes GL. Private communication.
35. Koberstein JT, Russell TP. *Macromolecules* 1986;19:714.
36. Hu W, Koberstein JT. *J Polym Sci* 1994;B32:437.
37. Seymour RW, Cooper SL. *Macromolecules* 1973;6(1):48.
38. Van Bogart JWC, Bluemke DA, Cooper SL. *Polymer* 1981;22:1428.
39. Sung CSP, Schneider NS. *Macromolecules* 1975;8:68.
40. Yang WP, Macosko CW, Wellinghoft ST. *Polymer* 1986;27:1235.
41. Wilkes GL, Bagrodia S, Humphries W, Wildnauer R. *J Polym Sci Polym Lett Ed* 1975;13:321.

42. Wilkes GL, Wildnauer R. *J Appl Phys* 1975;46(10):4148.
43. Wilkes GL, Emerson JA. *J Appl Phys* 1976;47(10):4261.
44. Ophir ZH, Wilkes GL. In: Cooper SL, Estes GM, eds., *Multiphase Polymers*, ACS Adv Chem Ser 176. Washington, DC: American Chemical Society, 1979, Chapter 3.
45. Kong ESW, Wilkes GL. *J Polym Sci Polym Lett Ed* 1980;18:369.
46. Van Bogart JWC, Lilaonitkul A, Cooper SL. In: Cooper SL, Estes GM, eds., *Multiphase Polymers*, ACS Adv Chem Ser 176. Washington, DC: American Chemical Society, 1979, Chapter 1.
47. Heskett TR, Van Bogart JWC, Cooper SL. *Polym Eng Sci* 1980;20(3):190.
48. Martin DJ, Meijs GF, Gunatillake PA, McCarthy SJ, Renwick GM. *J Appl Polym Sci* 1997;64:803.
49. Saiani A, Daunch WA, Verbeke H, Leenslag JW, Higginis JS. *Macromolecules* 2001;34:9059.
50. Saiani A, Roches C, Eeckhaut G, Daunch WA, Leenslag J-W, Higging, JS. *Macromolecules* 2004;37:1411.
51. Saiani A, Novak A, Rodier L, Eeckhaut G, Leenslag J-W, Higging JS. *Macromolecules* 2007;40:7252.
52. Xu M, MacKnight WJ, Chen CHY, Thomas EL. *Polymer* 1983;24:1327.
53. Chen CHY, Briber RM, Thomas EL, Xu M, MacKnight WJ. *Polymer* 1983;24:1333.
54. Xu M, MacKnight WJ, Chen-Tsai CHY, Thomas EL. *Polymer* 1987;28:2183.
55. Chen TK, Shieh TS, Chui JY. *Macromolecules* 1998;31:1312.
56. Pytela J, Sufcak M. Novel polybutadiene diols for thermoplastic polyurethanes. *PU Latin America 2001*; Technical Update 4: Footwear and CASE Paper 1, Crain Communications, Ltd., London, UK, August 28–30, 2001.
57. Chao H, Pytela J, Tian N, Murphy J. Progress in chain extender evaluation for polyurethanes derived from hydroxyl-terminated polybutadiene resins. *API 2005 Polyurethanes Technical Conference and Trade Fair*, Houston, TX, October 17–19, 2005.
58. Henning S, Chao H. *Rubber World* 2006;235(1):25.
59. Weaver LB, Batra A, Ansems P. U.S. Patent 2010/0028568A1, February 4, 2010.
60. Weaver LB, Batra A, Ansems P. U.S. Patent 2010/0055358A1, March 4, 2010.
61. Sendijarevic A, Sendijarevic V, Frisch KC, Cenens JLR, Handlin DL. U.S. Patent 6,111,049, August 29, 2000.
62. Aneja A, Wilkes GL, Yurtsever E, Yilgor I. *Polymer* 2003;44:757.
63. Anjea A, Wilkes GL. *J Appl Polym Sci* 2002;85:2956.
64. Aneja, A, Wilkes GL. *Polymer* 2002;43:5551.
65. Russell DD, Moore CS. New high performance aromatic TPU elastomers. *ACS Rubber Division 158th Fall Technical Meeting*, Cincinnati, OH, October 17–20, 2000.
66. Russell DD. *Automotive Elastomers Conference*, Dearborn, MI, June 15, 2004.
67. Frick A, Mikoszek M. *Macromiol Symp* 2005;311:57.
68. Prisacariu C. *Polyurethane Elastomers*. New York: Springer-Verlag, 2011.
69. Kroschwitz JI, Howe-Grant M, eds. *Kirk-Othmer Encyclopedia of Chemical Technology*, Vol. 10. New York: Wiley, 1993, p. 624.
70. Bagdi K. Role of interactions on the structure And properties of segmented Polyurethane elastomers, PhD thesis, Department of Physical Chemistry and Materials Science, Budapest University of Technology and Economics, Budapest, Hungary, 2010.
71. Petrovic ZS, Ferguson J. *Prog Polym Sci* 1991;16:695.
72. Di Battista G, Peerlings HWI, Kaufhold W. *Rubber World* 2003;227:39.
73. Nelb RG, Chen AT. In: Holden G, Legge NR, Quirk R, Schroeder HE, eds., *Thermoplastic Elastomers*. Cincinnati, OH: Hanser/Gardner, 1996, Chapter 9.
74. Jakeways R, Ward IM, Wilding MA. *J Polym Sci Polym Phys Ed* 1975;13:799.
75. Nitzche SA, Wang YK, Hsu SL. *Macromolecules* 1992;25:2397.
76. Muramatsu S, Lando JB. *Macromolecules* 1998;31:1866.

77. Gabriëlse W, Soliman M, Dijkstra K. *Macromolecules* 2001;34:1685.
78. Gabriëlse W, Guldener V, Schmalz H, Abetz V, Lange R. *Macromolecules* 2002;35:6946.
79. Chen, K, Tang X. *J Appl Polym Sci* 2004;91:1967.
80. Kim KJ, Bae JH, Kim YH. *Polymer* 2001;42:1023.
81. Dow engage product selection guide. http://www.dow.com.
82. Dow specialty elastomers for automotive TPO compounds. http://www.dow.com.
83. Bensason S, Mimck J, Moet A, Chum S, Hiltner A, Baer E. *J Poly Sci B: Polym Phys* 1996;34:1301.
84. Bensason S, Stepanov EV, Chuna S, Hiltner A, Baer E. *Macromolecules* 1997;30:2436.
85. Chien JCW, Iwamoto Y, Rausch MD, Wedler W, Winter HH. *Macromolecules* 1997;30:3447.
86. Lieber S, Brentzinger H-H. *Macromolecules* 2000;33:9192.
87. Rieger B, Cobzan C, Troel C, Hild S. In: Baugh LS, Camich JM, eds., *Stereoselective Polymerization with Single-Site Catalysts*. Boca Raton, FL: CRC Press, 2007, Chapter 9.
88. Giller C, Gururajan G, Wei J, Zhang W, Hwang W, Chase DB, Rabolt JF, Sita LR *Macromolecules* 2011;44:471.
89. Cherian AE, Rose JM, Lobkovsky EB, Coates GW. *J Am Chem Soc* 2005;127:13770.
90. Anderson-Wile AM, Coates GW, Alamo RG. *Macromolecules* 2011;44:3436.
91. Mason AF, Coates GW. *J Am Chem Soc* 2004;126:16326.
92. Buscio V, Cipallo R, Friederichs N, Ronca S, Togron M. *Macromolecules* 2003;36:3806.
93. Buscio V, Cipullo R, Friederichs N, Ronca S, Talerico G, Togron M, Wang B. *Macromolecules* 2004;37:8201.
94. Ruokolainen J, Mezzenga R, Fredrickson GH, Kramer EJ. *Macromolecules* 2005;38:851.
95. Kramer EJ et al. *J Polym Sci B: Polym Phys* 2010;48:1428.
96. Sakuma A, Weiser M-S, Fujita T. *Polym J* 2007;39(3):193.
97. Weiser M-S, Thomann Y, Heinz L-C, Pasch H, Mülhaupt R. *Polymer* 2006;47:4505.
98. Furuyama R, Mitani M, Mohri J-I, Mori R, Tanaka H, Fujita T. *Macromolecules* 2005;38:1546.
99. Shen CLP, Hazlitt LG. *Macromol Symp* 2007;257:80.
100. Wang HP, Khariwala DU, Cheung W, Chum SP, Hiltner A, Baer E. *Macromolecules* 2007;40:2852.
101. Ansems P, Soediono M, Tahe A. Oil extension of olefin block copolymers, In: *SPE ANTEC 2007*, Cincinnati, OH, 2007.
102. Li S, Register RA, Weinhold JD, Lendes, BG. *Macromolecules* 2012;45:5773.
103. Chen HY, Poon B, Chum SP, Dies P, Hiltner A, Baer E. Olefin block copolymers as polyolefin blend compatibilizer, In: *SPE ANTEC 2007*, Cincinnati, OH, 2007.
104. Nakashima N. Compatabilization of PP and polymer of ethylene series by a new hydrogenated polybutadiene block copolymer. In: *"New Opportunities for Thermoplastic Elastomers 2000" RAPRA Conference Proceedings*, Amsterdam, Paper 8, March 6–7, 2000.
105. Sierra CA, Galan C, Fatou JG, Parelada MD, Barrio JA. *Polymer* 1997;38:4325.
106. Ohlsson B, Tornell B. *Polym Eng Sci* 1996;36:1547.
107. Gergen WP, Lutz RG, Davison S. In: Holden G, Legge NR, Quirk R, Schroeder HE, eds., *Thermoplastic Elastomers*. Cincinnati, OH: Hanser/Gardner, 1996, p. 301.
108. Holden G. *Understanding Thermoplastic Elastomers*. Cincinnati, OH: Hanser/Gardner, 2000, Chapter 3.
109. Barton AFM. *Handbook of Polymer: Liquid Interaction Parameters and Solubility Parameters*. Baco Raton, FL: CRC Press, 1990.
110. Ryan AJ, Stanford JL, Still RH. *Polym Commun* 1988;29:196.
111. Holden G, Bishop ET, Legge NR. *J Polym Sci C* 1969;26:37.
112. Ghosh S, Bhowmick AK, Roychowdhury N, Holden G. *J Appl Polym Sci* 2000;77:1621.

113. Gergen WP, Lutz RG, Davison S. In: Holden G, Legge NR, Quirk R, Schroeder HE, eds., *Thermoplastic Elastomers*. Cincinnati, OH: Hanser/Gardner, 1996, p. 304.

114. Chun SB, Han CD. *Macromolecules* 1999;32:4030.

115. Kim JK, Lee HH, Sakurai S, Aida S, Masemoto J, Nomura S, Kitagawa Y, Suda Y. *Macromolecules* 1999;32:6707.

116. Han CD, Baek DM, Kim JK. *Macromolecules* 1990;23:561.

117. Zhan T, Wu Z, Li Y, Luo J, Chen Z, Xia J, Liang H, Zhank A. *Polymer* 2010;51:4249.

118. Baetzold JP, Koberstein JT. *Macromolecules* 2001;34:8986.

119. Takaheshi R, Ishigai A, Suzuki T, Nishitsuji, Koda T, Aoki Y. *Nihon Reoroji Gakkaishi* 2012;40(4):171 (*Journal of the Society of Rheology*, Japan, article in English).

120. Chapman BK, Bell TE. The advantages of SEEPS styrenic block copolymers In: *SPE TOPCON*, Akron, OH, Conference Paper 2005.

121. Handlin DL, Jr.. U.S. Patent 4,798,853, January 17, 1989.

122. Holden G, Legge NR. In: Holden G, Legge NR, Quirk R, Schroeder HE, eds., *Thermoplastic Elastomers*. Cincinnati, OH: Hanser/Gardner, 1996, p. 54.

123. Quirk RP, Morton M. In: Holden G, Legge NR, Quirk R, Schroeder HE, eds., *Thermoplastic Elastomers*. Cincinnati, OH: Hanser/Gardner, 1996, Chapter 4.

124. Ohlsson B, Hassander H, Tornell B. *Polym Eng Sci* 1996;36:501.

125. Gergen WP, Lutz RG, Davison S. In: Holden G, Legge NR, Quirk R, Schroeder HE, eds., *Thermoplastic Elastomers*. Cincinnati, OH: Hanser/Gardner, 1996, pp. 315–322.

126. Tucker PS, Barlow JW, Paul DR. *Macromolecules* 1988;21:2794.

127. Lindlof JA. U.S. Patent 3,676,387, July 11, 1972.

128. Wolf RJ, Barnidge TJ, Darvell WK, Kirchhoff KJ. U.S. Patent 5,713,544, February 3, 1998.

129. Yates PM. U.S. Patent 6,506,271, January 14, 2003.

130. Fetters LJ, Morton M. *Macromolecules* 1969;2:453.

131. Wright T, Jones AS, Harwood HJ. *J Appl Polym Sci* 2002;86:1203.

132. Quirk RP, Morton M. In: Holden G, Legge NR, Quirk R, Schroeder HE, eds., *Thermoplastic Elastomers*. Cincinnati, OH: Hanser/Gardner, 1996, p. 87.

133. Faust R. In: Salamone JC, ed., *Polymeric Materials Encyclopedia*, Vol. 5. Boca Raton, FL: CRC Press, 1996, p. 3820.

134. Ohtani H, Tsuge S. In: Salamone JC, ed., *Polymeric Materials Encyclopedia*, Vol. 3. Boca Raton, FL: CRC Press, 1996, p. 1791.

135. Tucker PS, Barlow JW, Paul DR. *Macromolecules* 1988;21:1678.

136. Aggarwal SL, Livigni RA. *Polym Eng Sci* 1977;17:498.

137. Mayes AM, Russell TP, Satija SK, Majkrzak CF. *Macromolecules* 1992;25:6523.

138. Winey KI, Thomas EL, Fetters LJ. *Macromolecules* 1991;24:6182.

139. Hong KM, Noolandi J. *Macromolecules* 1983;16:1083.

140. Akiyama Y, Kishimoto Y, Mizushiro K. U.S. Patent 4,772,657, September 20, 1988.

141. Francis JV, Overbergh NMM, Lodewijk J, Vansant MFG. U.S. Patent 5,710,206, January 20, 1998.

142. Hall JE, Wang X, Takeichi H, Mashita N. U.S. Patent 6,184,292, February 6, 2001.

143. Shell Technical Bulletins: Kraton® G7821X, G7702X, and G7722X. Multibase and GLS Corporation websites.

144. Chapman BK, Foster HA. U.S. Patent Application Publication 2012/0207996A1, August 16, 2012.

145. Flood JE, Wright KJ. A new low viscosity elastic SEBS block copolymer for compounding and elastic fabric applications. *SPE TOPCON*, Akron, OH, 2014, Conference Paper.

146. Kilian D, Kishiis, Jogo Y, Shachi K. *TPE Magazine* 2010;4:220.

147. Chapman BK, Kilian D. *TPE Magazine* 2012;1:28.

148. Huber R. SBS block copolymer with unique structure and properties. *TPE 2005 Proceedings*, RAPRA Technology, Berlin, Germany, September 14–15, 2005.

149. Knoll K, Niebner N. *Macromol Symp* 1998;132:231.
150. Knoll K, Niebner N. Styroflex: A new transparent styrene-butadiene copolymer with high flexibility. In: Ouirk R, ed., *Application of Anionic Polymerization Research*. ACS Symposium Series. Washington, DC, 1998, pp. 112–128.
151. Ganb M, Standinger V, Thunga M, Knoll K, Schneider K, Stamm M, Weidisch R. *Polymer* 2012;53:2085.
152. Naskar K. Dynamically vulcanized PP/EPDM thermoplastic elastomers: Exploring new routes for crosslinking with peroxides, PhD Thesis, University of Twente, Enschedes, The Netherlands, 2004.
153. Kresge EN. *Rubber Chem Technol* 1991;64(3):469.
154. Payne MT, Rader CP. In: Cheremisinoff NP, ed., *Elastomer Technology Handbook*. Boca Raton, FL: CRC Press, 1993, Chapter 14.
155. Coran AY, Patel RP. In: Karger-Kocsis J, ed., *Polypropylene*, Vol. 2: *Copolymers and Blends*. London, U.K.: Chapman & Hall, 1995, Chapter 6.
156. Abdou-Sabet S, Puydak RC, Rader CP. *Rubber Chem Technol* 1996;69:476.
157. Coran AY, Patel RP. In: Holden G, Legge NR, Quirk R, Schroeder HE, eds., *Thermoplastic Elastomers*. Cincinnati, OH: Hanser/Gardner, 1996, Chapter 7.
158. Karger-Kocsis J. In: Shonaike GO, Simon GP, eds., *Polymer Blends and Alloys*. New York: Marcel Dekker, 1999, Chapter 5.
159. Abdou-Sabet S, Datta S. In: Paul DR, Bucknall CB, eds., *Polymer Blends*, Vol. 2. New York: Wiley, 2000, Chapter 35.
160. Gessler AM, Haslett WH. U.S. Patent 3,037,954, June 5, 1962.
161. Fischer WK. U.S. Patent 3,758,643, September 11, 1973.
162. Fischer WK. U.S. Patent 3,862,106, January 21, 1975.
163. Fischer WK. U.S. Patent 3,835,201, September 10, 1974.
164. Fischer WK. U.S. Patent 3,806,558, April 23, 1974.
165. Hartman PF, Eddy CL, Koo GP. *SPE J* 1970;26:62.
166. Hartman PF, Eddy CL, Koo GP. *Rubber World* 1970;163(1):59.
167. Hartman PF. U.S. Patent 3,909,463, September 30, 1975.
168. Coran AY, Das B, Patel RP. U.S. Patent 4,104,210, August 1, 1978.
169. Coran AY, Das B, Patel RP. U.S. Patent 4,130,535, December 19, 1978.
170. Coran AY, Patel RP. U.S. Patent 4,130,534, December 19, 1978.
171. Gessler AM, Kresge EN. U.S. Patent 4,132,698, January 2, 1979.
172. Abdou-Sabet S, Datta S. In: Paul DR, Bucknall CB, eds., *Polymer Blends*, Vol. 2. New York: Wiley, 2000, p. 524.
173. Abdou-Sabet S, Fath MA. U.S. Patent 4,311,628, January 19, 1982.
174. Coran AY, Patel R. U.S. Patent 4,271,049, June 2, 1981.
175. Sperling LH. *Introduction to Physical Polymer Science*. New York: Wiley, 2000, p. 545.
176. Sperling LH. *Introduction to Physical Polymer Science*. New York: Wiley, 2001, pp. 509–516.
177. Coran AY, Patel RP. In: Holden G, Legge NR, Quirk RP, Schroeder H, eds., *Thermoplastic Elastomers*. Cincinnati, OH: Hanser/Gardner, 1996, p. 175.
178. Sperling LH. *Introduction to Physical Polymer Science*. New York: Wiley, 2001, Chapter 12.
179. Phillips RA, Wolkowicz MD. In: Moore EP Jr, ed., *Polypropylene Handbook*. Cincinnati, OH: Hanser/Gardner, 1996, Chapter 3, Section 5.
180. Chung O, Coran AY. *Rubber Chem Technol* 1997;70(5):781.
181. D'Orazio L, Mancarella C, Martuscelli E, Sticotti G, Cecchin G. *J Appl Polym Sci* 1999;71:701.
182. D'Orazio L, Mancarella C, Martuscelli E, Sticotti G, Ghisellini R. *J Appl Polym Sci* 1994;53:387.

183. Martuscelli E. In: Karger-Kocsis J, ed., *Polypropylene*, Vol. 2: *Copolymers and Blends*. London, U.K.: Chapman & Hall, 1995, Chapter 4.

184. Maier C, Calafut T. *Polypropylene: The Definitive User's Guide and Databook*. Norwich, NY: Plastics Design Library, 1998, pp. 21–24.

185. Liang JZ, Li RKY. *J Appl Polym Sci* 2000;77:409.

186. Datta S, Lohse DL. *Polymeric Compatibilizers*. Cincinnati, OH: Hanser/Gardner, 1996, pp. 86–90.

187. Keskkula H, Paul DR. In: Kohan MI, ed., *Nylon Plastics Handbook*. Cincinnati, OH: Hanser/Gardner, 1995, p. 415.

188. Sperling LH. *Introduction to Physical Polymer Science*. New York: Wiley, 2001, Chapter 12, Section 5.

189. Li J, Favis BD. *Polymer* 2002;43:4935.

190. Kim JR, Jamieson AM, Hudson SD, Manas-Zloczower I, Ishida H. *Macromolecules* 1999;32:4582.

191. Nitta K, Kawada T, Yamohiro M, Mori H, Terano M. *Polymer* 2000;41:6765.

192. Adedeji A, Jamieson AM. *Polymer* 1993;34:5038.

193. Lynch JC, Nauman EB. *Polym Mater Sci Eng* 1994;71:609.

194. Premphet K, Paecharoenchai W. *J Appl Polym Sci* 2002;85:2412.

195. Yokoyama Y, Ricco T. *Polymer* 1998;39:3675.

196. Ling Z, Rui H, Liangbin L, Gang W. *J Appl Polym Sci* 2002;83:1870.

197. Mäder D, Thomann Y, Suhn J, Mülhaupt R. *J Appl Polym Sci* 1999;74:838.

198. Naiki M, Matsumura T, Matsuda M. *J Appl Polym Sci* 2002;83:46.

199. Jang BZ, Uhlmann DR, Vander Sande JB. *Polym Eng Sci* 1985;25:643.

200. Pukánszky B, Fortelný I, Kovár J, Tüdös F. *Plastics Rubber Compos Process Appl* 1991;15(1):31.

201. Dompas D, Groeninckx G. *Polymer* 1994;35:4743.

202. Mehrabzadeh M, Nia KH. *J Appl Polym Sci* 1999;72:1257.

203. van der Wal A, Verheul AJJ, Gaymans RJ. *Polymer* 1999;40:6057.

204. van der Wal A, Gaymans RJ. *Polymer* 1999;40:6067.

205. Karger-Kocsis J, Kalló A, Szafner A, Bodor G, Sényei Z. *Polymer* 1979;20:37.

206. Jang BZ, Uhlmann DR, Vander Sande JB. *J Appl Polym Sci* 1985;30:2485.

207. Dao KC. *Polymer* 1984;25:1527.

208. Chou CJ, Vijayan K, Kirby D, Hiltner A, Baer E. *J Mater Sci* 1988;23:2521.

209. Lu J, Wei GX, Sue HJ, Chu J. *J Appl Polym Sci* 2000;76:311.

210. Bedia EL, Astrini N, Sudarisman A, Sumera F, Kashiro Y. *J Appl Polym Sci* 2000;78:1200.

211. Grellmann W, Seidler S, Jung K, Kotter I. *J Appl Polym Sci* 2001;79:2317.

212. Wu S. *J Appl Polym Sci* 1988;35:549.

213. Wu S. *ACS Polym Mater Sci Eng* 1990;Fall:220.

214. Wu S. *Polym Eng Sci* 1990;30(13):753.

215. Pukánszky B, Tüdös F, Kalló A, Bodor G. *Polymer* 1989;30:1399.

216. Da Silva ALN, Rocha MCG, Lopes L, Chagas BS, Coutinho FMB. *J Appl Polym Sci* 2001;81:3530.

217. Kim BC, Kwang SS, Lim KY, Yoon KJ. *J Appl Polym Sci* 2000;78:1267.

218. Macauley NJ, Harkin-Jones EMA, Murphy WR. *Polym Eng Sci* 1998;38:516.

219. Sherman LM. *Plastics Technol* 2002;48(7):44.

220. Barlow JW, Paul DR. *Polym Eng Sci* 1987;24(8):241.

221. Karger-Kocis J, Csikai I. *Polym Eng Sci* 1987;27(4):241.

222. Xiao HW, Huang SQ, Jiang T, Cheng SY. *J Appl Polym Sci* 2002;83:315.

223. Mäder D, Bruch M, Maier RD, Stricker F, Mülhaupt R. *Macromolecules* 1999;32:1252.

224. Dwyer SM, Boutni OM, Shu C. In: Moore EP Jr, ed., *Polypropylene Handbook*. Cincinnati, OH: Hanser/Gardner, 1996, p. 220.

225. Zhang L, Huang R, Wang G, Li L, Ni N, Zhang X. *J Appl Polym Sci* 2002;86:2085.
226. van der Wal A, Nijhof R, Gaymans RJ. *Polymer* 1999;40:6031.
227. van der Wal A, Gaymans RJ. *Polymer* 1999;40:6045.
228. Martuscelli E. In: Karger-Kocsis J, ed., *Polypropylene*, Vol. 2: *Copolymers and Blends*. London, U.K.: Chapman and Hall, 1995, Chapter 4.
229. Rios-Guerrero L, Keskkula H, Paul DR. *Polymer* 2000;41:5415.
230. Zhang H, Wang J, Li J, Cao S, Shen A. *J Appl Polym Sci* 2001;79:1345.
231. Zhang H, Wang J, Cao S, Wang Y. *J Appl Polym Sci* 2001;79:1351.
232. Lohse DJ, Wissler GE. *J Mater Sci* 1991;26:743.
233. Lohse DJ. *Polym Eng Sci* 1986;26:1500.
234. Chen CY, Yunus WMd.ZW, Chiu HW, Kyu T. *Polymer* 1997;38:4433.
235. Ferry JD. *Viscoelastic Properties of Polymers*. New York: Wiley, 1980, p. 374.
236. Eckstein A, Suhn J, Friedrich C, Maier RD, Sassmannshausen, Bochmann M, Mülhaupt R. *Macromolecules* 1998;31:1335.
237. Coran AY, Patel R. *Rubber Chem Technol* 1983;56:1045.
238. Inoue T, Suzuki T. *J Appl Polym Sci* 1995;56:1113.
239. Inoue T. *J Appl Polym Sci* 1994;54:723.
240. Ishikawa M, Sugimoto M, Inoue T. *J Appl Polym Sci* 1996;62:1495.
241. Kruliš Z, Fortelney I, Kovar J. *Collect Czech Chem Commun* 1993;58:2642.
242. Argon AS, Bartczak Z, Cohen RE, Muratoglu OK. In: Pearson RA, Sue HJ, Yee AF, eds., *Toughening of Plastics: Advances in Modeling and Experiments*. ACS Symposium Series 759. Washington, DC: American Chemical Society, 2000, Chapter 7.
243. Wilbrink MWL, Argon AS, Cohen RE, Weinberg M. *Polymer* 2001;42:10155.
244. Thio YS, Argon AS, Cohen RE, Weinberg M. *Polymer* 2002;43:3661.
245. Gaymans RJ, van der Wal A. In: Pearson RA, Sue HJ, Yee AF, eds., *Toughening of Plastics: Advances in Modeling and Experiments*. ACS Symposium Series 759. Washington, DC: American Chemical Society, 2000, Chapter 8.
246. Naskar K. Dynamically vulcanized PP/EPDM thermoplastic elastomers: Exploring new routes for crosslinking with peroxides PhD thesis, University of Twente, Enschede, the Netherlands, 2004, Chapter 6.
247. Nomura T, Nishio T, Iwanamik, Yokomizo K, Kitano K, Toki S. *J Appl Polym Sci* 1995;55:1307.
248. Chaffin KA, Bates FS, Brant P, Brown GM. *J Polym Sci B: Polym Phys* 2000;38:108.
249. Poon BC, Chum SP, Hiltner A, Baer E. *Polymer* 2004;45:893.
250. Brandrup J, Immergut EH, Grulke EA, eds. *Polymer Handbook*. New York: Wiley, 1999, V/15&V/26.
251. Abraham T, Barber N. U.S. Patent 6,667,364, December 23, 2003.
252. Kramer EJ. *Microscopic and Molecular Fundamentals of Crazing. Advanced Polymer Science*. Berlin, Germany: Springer-Verlag, 1983, pp. 52–53.
253. Argon AS, Hannoosh JG. *Philos Mag* 1977;36:1195.
254. Ebewele RO. *Polymer Science and Technology*. Boca Raton, FL: CRC, 2000, p. 361.
255. Lustiger A, Markham RL. *Polymer* 1983;24:1647.
256. Ehrenstein GW. *Polymeric Materials*. Cincinnati, OH: Hanser/Gardner, 2001, pp. 175, 253.
257. Coulon G, Castelein G, G'Sell C. *Polymer* 1998;40:95.
258. Phillips RA, Wolkowicz MD. In: Moore EP Jr, ed., *Polypropylene Handbook*. Cincinnati, OH: Hanser/Gardner, 1996, Chapter 3.
259. Peterlin A. *J Polym Sci A-2* 1969;7:1151.
260. Way JL, Atkinson JR, Nutting J. *J Mater Sci* 1974;9:293.
261. Freidrich K. *Prog Colloid Polym Sci* 1978;64:103.
262. Ishikawa M, Ushui K, Kondo Y, Hatada K, Gima S. *Polymer* 1996;37:5375.
263. van der Wal A, Mulder JJ, Thijs HA, Gaymans RJ. *Polymer* 1998;39:5467.

264. van der Wal A, Mulder J, Gaymans RJ. *Polymer* 1998;39:5477.
265. Sugimoto M, Ishikawa M, Hatada K. *Polymer* 1995;36:3675.
266. Zhang XC, Butler MF, Cameron RE. *Polymer* 2000;41:3797.
267. Dijkstra PTS, van Dijk DJ, Huétink J. *Polym Eng Sci* 2002;42:152.
268. Hu WG, Schmidt-Rohr K. *Acta Polym* 1999;50:271.
269. Lustiger A. Private communication.
270. Coran AY, Patel RP. In: Holden G, Legge NR, Quirk, R, Schroeder HE, eds., *Thermoplastic Elastomers*. Cincinnati, OH: Hanser/Gardner, 1996, p. 159.
271. Kruliš Z, Fortelný I. *Eur Polym J* 1997;33:513.
272. Manchado MAL, Biagiotti J, Kenny JM. *J Appl Polym Sci* 2001;81:1.
273. Romanini D, Garagnani E, Marchetti E. Reactive blending: Structure and properties of cross-linked olefinic thermoplastic elastomers, Paper presented at the *International Symposium on New Polymeric Materials*, European Physical Society (Macromolecular Section), Naples, Italy, June 9–13, 1986.
274. Hatanaka T, Mori H, Terano M. *Polym Degrad Stability* 1999;64:313.
275. Jain AK, Gupta NK, Nagpal AK. *J Appl Polym Sci* 2000;77:1488.
276. Goharpey F, Katbab AA, Nazockdast H. *J Appl Polym Sci* 2001;81:2531.
277. Wright KJ, Lesser AJ. *Rubber Chem Technol* 2001;74:677.
278. Leutwyler K. *Sci Am*, September 14, 1998.
279. Cretin JL. (Hutchinson Worldwide) Presentation at the *Automotive Elastomers Conference*, Detroit, MI, June 18–19, 2002.
280. Garois N. U.S. Patent 6,028,142, February 22, 2000.
281. Sengupta P, Noordermeer JWM. *Macromol Rapid Commun* 2005;26:542.
282. Phillips RA Wolkowicz MD. In: Moore Jr EP, ed., *Polypropylene Handbook*. Cincinnati, OH: Hanser/Gardner, 1996, p. 126.
283. Jayaraman K, Kolli V, Kang S-Y, Kumar S, Ellul MD. *J Appl Polym Sci* 2004;93:113.
284. Sangers W, Sengupta P, Noordermeer J, Picken S, Gotsis A. *Polymer* 2004;45:8881.
285. Abraham T, Barber N, Mallamaci M. *Rubber Chem Technol* 2007;80:324.
286. Ellul MD. *Rubber Chem Technol* 1998;71:244.
287. Sangers W, Wubbenhorst M, Picken S, Gotsis A. *Polymer* 2005;46:6391.
288. Caihong L, Xiembo H, Fangzhong M. *Polym Adv Technol* 2007;18:999.
289. Lloyd DR, Barlow JW, Kinzer KE. *AIChE Symp Ser* 1989;84(261):28.
290. Zia Q, Milena D, Androsch R. *Macromolecules* 2008;41:8095.
291. Winters R, Lugtenburg J, Litvinov VM, van Duin M, de Groot HJM. *Polymer* 2001;42:9745.
292. Lustiger A, Marzinsky CN, Mueller RR. *J Polym Sci: Polym Phys* 1998;36:2047.
293. Cakmak M, Cronin SW. *Rubber Chem Technol* 2000;73:753.
294. Aboulfaraj M, G'Sell C, Ulrich B, Dahoun A. *Polymer* 1995;36:731.
295. Yang Y, Chiba T, Saito H, Inoue T. *Polymer* 1998;39:3365.
296. Kikuchi Y, Fujui T, Okada T, Inoue T. *Polym Eng Sci* 1991;31:1029.
297 Kikuchi Y, Fujui T, Okada T, Inoue T. *J Appl Polym Sci* 1992;50:261.
298. Aoyama T, Carlos AJ, Saito H, Inoue T, Niitsu Y. *Polymer* 1999;40:3657.
299. Huy TA, Luepke T, Radusch JJ. *J Appl Polym Sci* 2001;80:148.
300. Okamoto M, Shiomi K. *Polymer* 1994;35:4618.
301. Oderkerk J, Groeninckx G, Soliman M. *Macromolecules* 2002;35:3946.
302. Oderkerk J, de Schaetzen G, Goodeis B, Hellemans L, Groeninckx G. *Macro molecules* 2002;35:6623.
303. Boyce MC, Socrate S, Kear K, Yeh O, Shaw K. *J Mech Phys Solids* 2001;49:1323.
304. Boyce MC, Yeh O, Socrate S, Kear K, Shaw K. *J Mech Phys Solids* 2001;49:1343.
305. Abdou-Sabet S, Datta S. In: Pual DR, Bucknall CB, eds., *Polymer Blends*, Vol. 2: *Performance*. New York: Wiley, 2000, p. 550.

6 Carbon Black

Wesley A. Wampler, Leszek Nikiel,
and Erika N. Evans

CONTENTS

I. Introduction 210
II. Definitions.. 210
III. Carbon Black Manufacturing Process... 211
    A. Reaction.. 212
        1. Furnace Process 212
        2. Thermal Process.. 214
        3. Reactor Conditions versus Properties 215
    B. Filtration/Separation 215
    C. Pelletizing... 216
    D. Drying 216
IV. Controlling the Quality of Carbon Black 217
    A. Specific Surface Area.. 217
    B. Structure... 220
    C. Tint Strength 221
    D. Pellet Properties 222
    E. Impurities 223
    F. In-Rubber Tests 224
V. Effect of Carbon Black on Rubber Properties 225
    A. Mixing and Dispersion.. 225
    B. Uncured Rubber Properties... 228
    C. Cured Properties 228
VI. Carbon Black Classification and Various Grades... 232
    A. Examples of N100 Series ($N_2SA = 121$–150)...................................... 233
    B. Examples of N200 Series ($N_2SA = 100$–120)...................................... 233
    C. Examples of N300 Series ($N_2SA = 70$–99)... 236
    D. Examples of Carcass or Semireinforcing Grades 236
    E. Thermal Grades 237
VII. Surface Chemistry and Modifications to Carbon Black................................ 237
    A. Carbon Black Surface Chemistry 238
    B. Carbon Black Modification... 239
        1. Postprocess Modification 240
        2. In-Process Modification 243
VIII. Conclusions.. 245
References.. 246

I. INTRODUCTION

Carbon black is produced by the incomplete combustion of organic substances, probably first noted in ancient times by observing the deposits of a black substance on objects close to a burning material. Its first applications were no doubt as a black pigment, and the first reported use was a colorant in inks by the Chinese and Hindus in the third century A.D. [1]. It was not until the early twentieth century when carbon black was first mixed into rubber that its possible usefulness in this area was explored. The fact that carbon black has the ability to significantly improve the physical properties of rubber (often referred to as reinforcement) has provided its largest market today, that is, the tire industry. Currently about 8.5 million metric tons of carbon black is used worldwide in tires annually of the 11.6 worldwide demand [2]. A typical tire contains 30%–35% carbon black, and there are normally several grades of carbon black in the tire, depending on the reinforcement requirements of the particular component of the tire. Of course, carbon black is also used in many nontire rubber applications owing to its ability to reinforce the rubber and to its use as a cost reduction diluent in the compound. Nontire rubber products currently require about 2.3 million metric tons of carbon black annually on a worldwide basis [2].

This chapter brings the reader up to date on how carbon black is manufactured, how its quality is controlled, how the carbon black characteristics influence rubber properties, and how the different grades of carbon black are classified and used, then finally presents a review of carbon black surface chemistry and how the modification of these surfaces holds substantial promise for future developments.

II. DEFINITIONS

Before beginning, there is merit in reviewing some basic definitions in carbon black technology. Although it is not attempted to present a comprehensive list of definitions, several important ones will be given, and the reader is referred to American Society for Testing and Materials (ASTM) D3053 for additional carbon black terminology [3]:

Carbon black: An engineered material, primarily composed of elemental carbon, obtained from the partial combustion or thermal decomposition of hydrocarbons, existing as aggregates of acinoform morphology, which are composed of spheroidal primary particles and turbostratic layering within the primary particles.

Carbon black particle: A small spheroidal-shaped (paracrystalline, nondiscrete) component of a carbon black aggregate; it is separable from the aggregate only by fracturing. Particle diameters can range from less than 20 nm in some furnace grades to a few hundred nanometers in thermal blacks.

Carbon black aggregate: A discrete, rigid, colloidal mass of extensively coalesced particles; it is the smallest dispersible unit of carbon black. Aggregate dimensions can range from as about 50 to 300 nm for furnace blacks. Figure 6.1 shows the distinction between a particle and an aggregate in carbon black.

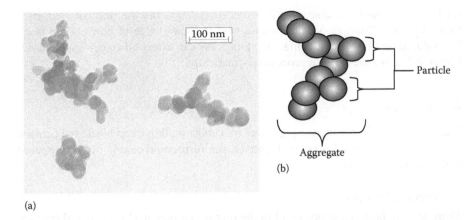

(a)

(b)

FIGURE 6.1   (a) Carbon black aggregate as viewed by transmission electron microscopy and (b) a schematic showing the distinction between carbon black particles and the aggregate. (Photograph by David Roberts.)

Carbon black agglomerate: A cluster of physically bound and entangled aggregates. Agglomerates can vary widely in size from less than a micrometer to
a few millimeters in the pellet.

Carbon black pellet: A relatively large agglomerate that has been densified
in spheroidal form to facilitate handling and processing. Pellets range in
diameter from tenths of a millimeter to 2–3 mm.

Carbon black structure: The degree of irregularity and deviation from sphe-
ricity of the shape of a carbon black aggregate. It is typically evaluated
by absorption measurements that determine the voids between the aggregates and agglomerates and thus indirectly the branching and complexity of
shape of the carbon black aggregates and agglomerates.

Carbon black specific surface area: The available surface area in square
meters per unit mass of carbon black in grams. Typically the adsorption
of molecules such as iodine or nitrogen is measured and then either the
amount adsorbed per unit mass is reported or a specific surface area is calculated based on current adsorption theories.

III.   CARBON BLACK MANUFACTURING PROCESS

The carbon black manufacturing process consists of several distinct segments. Each
segment is important for ensuring economical production and for meeting customer
expectations:

1. Reaction
2. Filtration/separation
3. Pelletizing
4. Drying

Each segment could be discussed in exhaustive detail, but the purpose here is to furnish a short description that allows a working knowledge of how carbon black is produced and how the manufacturing process can affect customer applications. Figure 6.2 shows the furnace process schematically.

A.   REACTION

There are two main production processes for rubber grade carbon black: the furnace process and the thermal process. However, the furnace process is by far the more dominant process today.

1.   Furnace Process

There are two broad categories within the furnace carbon blacks: tread and carcass. The processes for manufacturing the two are very similar in most respects, the main differences being that carcass carbon black (used mainly in tire carcasses, sidewalls, and other semireinforcing applications) is made at lower temperatures and lower reaction velocities and with longer residence times than tread carbon blacks. Tread blacks are used in tire treads and in areas where higher levels of reinforcement are needed. Because of these differences in reaction kinetics, carcass carbon blacks are lower in specific surface area than tread blacks.

Carbon black is formed very quickly and at very high temperatures typically generated from the combustion of natural gas with air but with insufficient oxygen to reach the stoichiometric ratio and corresponding temperature. The reaction occurs in refractory-lined vessels that are required to sufficiently contain the high-temperature reactor gas stream. The refractory lining presents a problem because of constant erosion at high velocities. The erosion contributes to contamination of the carbon product, which is not good for any customer product application. The erosion of refractory can also significantly change the cross-sectional area of the "choke" in tread grade furnace reactors, affecting several carbon black properties, most significantly surface area, structure, and tint levels. The "choke" is a narrowing section of the furnace reactor (on tread but not carcass reactors) that is necessary to attain the velocities required to produce the high levels of surface area desired. Velocities can approach supersonic levels at the choke and temperatures approach 3400°F (1870°C).

In the first stage of the process, hydrocarbon fuels are used to generate temperatures via combustion that create an exothermic reaction with temperatures ranging from 2400°F (1315°C) to 3400°F (1870°C). This high temperature is necessary to supply the energy required to "crack" or "split" the carbon–hydrogen bond of the raw material feedstock. The specific surface area of carbon black, which is probably the most important quality parameter, is directly proportional to the reaction temperature. This means that because more fuel is used to attain higher reaction temperatures for the higher-surface-area carbon blacks, there is a resulting higher production cost.

An endothermic reaction ("cracking") proceeds concurrently with the exothermic reaction. A hydrocarbon (feedstock) is injected into the reactor for the production of

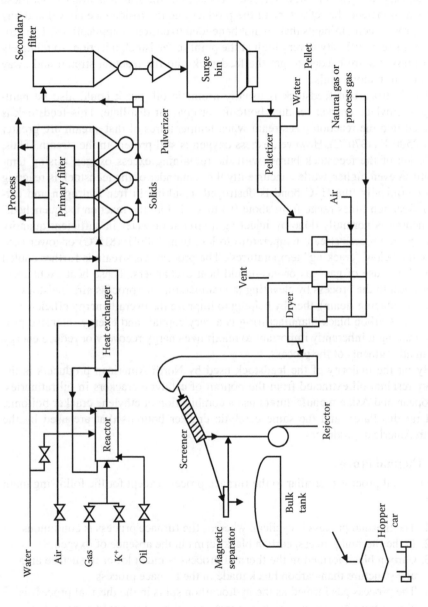

FIGURE 6.2 Schematic of the furnace carbon black process.

carbon black at elevated pressures and temperatures. High feedstock injection pressures and temperatures are necessary to attain good economics and minimize coke formation. Coke is formed from rapid cooling of the oil droplets or from oil droplet impingement on the reactor refractory walls. This coke is sometimes referred to in the industry as "grit" or "sieve residue" (because of the way it is tested), but these terms also include the refractory in the product due to erosion (see above) and any other process contaminants that are not beneficial to customer applications. The process gas stream velocity is very high at the point of feedstock injection, so relatively high pressures are needed to get the feedstock into the reaction stream and away from the refractory walls.

The hydrocarbon feedstock is usually aromatic oil, but it could also be natural gas, ethylene cracker residual bottoms, or coal tar distillate. This feedstock is injected into the reaction gas stream when temperatures of that stream are greater than 2500°F (1370°C). However, excess oxygen is still present in the stream. Thus, a portion of the feedstock burns, with the remaining excess oxygen raising temperatures even higher, while concurrently the remainder of the feedstock is reacting endothermically (the H–C bond is destroyed, resulting in free hydrogen and carbon). Reaction times range from about 0.3 to 1 s before the reaction is "quenched." Quenching is normally done by injecting a stream of water in sufficient quantity to drop the process stream temperature to less than 1500°F (815°C) or lower (i.e., dropping below "cracking" temperatures). The process gas stream is further cooled through the use of gas–gas or gas–liquid heat exchangers. These heat exchangers return heat to the process by elevating the temperature for process air, feedstock, or water (producing steam), thereby helping to improve the overall energy efficiency of the plant. Carbon black manufacturing is a very capital- and energy-intensive process, making it inherently important to maximize energy recovery or reduce energy use in all segments of the process.

By far the majority of the feedstock used by North American producers is the heavy residual oil extracted from the bottom of catalytic crackers in oil refineries. European and Asian manufacturers use a combination of ethylene cracker bottoms, coal tar distillates, and the same catalytic cracker bottoms that are used by the North American producers.

2. Thermal Process

The thermal process is similar to the furnace process except for the following main areas:

1. The thermal process is cyclical, whereas the furnace process is continuous.
2. In the thermal process, carbon black forms in the absence of oxygen.
3. Carbon black formed in the thermal process is much lower in surface area and structure than carbon black made in the furnace process.
4. The process gas formed as the hydrocarbon splits in the thermal process is almost pure hydrogen, which requires special handling processes and procedures, whereas the process gas formed in the furnace process is mostly N_2 and H_2O, with smaller amounts of CO, H_2, CO_2, C_2H_2, and CH_4.

The feedstock for thermal black can be natural gas or catalytic cracker bottoms. Thermal carbon blacks are not as reinforcing as furnace black, can have lower levels of hydrocarbon residuals on the surface, and are lower in tint or blackness. There are some areas where these properties are beneficial, but by far the vast majority of carbon black (>90%) production in the world uses the furnace process.

As a side note, the thermal process was developed in the United Kingdom in the early 1900s as a method to produce hydrogen gas for use in cities to augment or replace coal burning. Carbon black was a secondary product in this H_2-producing process.

3. Reactor Conditions versus Properties

Carbon black has two primary properties (surface area and structure) that are important to the majority of end users and are controlled predominantly in the reaction area. Specific surface area is manipulated by controlling reaction temperature, reaction time, and reaction velocity. Structure (or branching) is manipulated by increasing or decreasing the amount of turbulence at the point of feedstock injection in the reaction forming zone or by the addition of metallic salts (potassium salts being by far the most prevalent) to prevent the formation of carbon black particulate structure.

B. FILTRATION/SEPARATION

Carbon black is formed in a reactor with less oxygen present than would be required for complete combustion, resulting in many species of gas components in the process gas stream. Gas species present include H_2O, N_2, CO, H_2, CO_2, CH_4, C_2H_2, and trace amounts of other compounds such as SO_2 and H_2S. The carbon black formed in the reaction section must be separated from these gaseous components. This is accomplished through the use of various types of commercially available cloth filter bags. At this stage of the process, the carbon black is in a "loose" or "fluffy" state at about 500°F (260°C). The surface area of the carbon black being very high (25–150 m²/g), the loose product is unmanageable for most customers. Carbon black in this state is extremely light, and a few grams can easily obscure most of the light in a 4000 ft³ room. The gas, often referred to as tail gas, does contain combustible components (H_2, CO, CH_4), but the heat content is very low because of the high quantities of nitrogen and water present, 45–75 Btu/ft³ (1676–2794 kJ/m³). Natural gas, by comparison, averages around 950–1000 Btu/ft³. Even though the heat content is quite low, most carbon black manufacturers have developed technology that allows combustion of this process gas to supply heat to the process or to generate steam and/or electricity. The tail gas that is not utilized within the process is either incinerated or sent to a flare. In an effort to become more environmentally friendly, the carbon black industry is exploring options to control and lessen SO_x/NO_x emission concentrations. Energy recovery is essential to maintain energy efficiency and meet environmental compliance requirements.

After separation, the carbon black is conveyed (pneumatically or mechanically) to the next segment of the process, where it is pelleted and dried for ease of shipment and handling by the customers.

C. PELLETIZING

Most customers need carbon black delivered in bulk quantities in a form that is easy to convey and also easy to disperse into their compound (rubber, plastic, ink, paint, etc). To get the loose carbon black into a pelleted form that meets these needs, the carbon black producers are obliged to use mechanical pin mixers, chemical pelleting aids (such as molasses or lignosulfonate), water, and equipment of high capital and continuous operating costs. Because carbon black is formed from a hydrocarbon raw material (which does not mix naturally with water) and has high surface area and structure, large amounts of water are needed to form the pellets, normally with a pelleting aid added to facilitate "wetting." Water content of the product leaving the pelleting area ranges from 35% to 65% by weight. Water is used extensively in the carbon black process—about five times more water than feedstock.

Customers expect to receive uniform pellets capable of withstanding the rigors of being shipped hundreds to thousands of miles but not so hard as to impede incorporation with a minimum of mixing energy and time. It is also highly desirable to minimize the unpelleted carbon black (or minimize pellet breakdown) so as to mitigate customer concerns about fugitive carbon black in their plants.

D. DRYING

The wet pellets, having a high concentration of water, are not a desirable final product form. Therefore, carbon black producers are obliged to use large amounts of energy (with significant capital investment) to drive the water from the wet pellet. It is necessary to lower the moisture content from approximately 50% by weight as it leaves the pelletizer to less than 1% for shipment to customers. Most producers use the process gas, sometimes called tail gas, separated from the carbon black in the filtration section of the process to supply the fuel needed to dry the wet pellets. Although this is an inexpensive fuel, the capital involved to collect, direct, and support combustion of this low-Btu gas is relatively high.

After drying, the pellets are conveyed to bulk storage tanks for packaging into bags (ranging from 50 to 2000 lb), bulk trucks (45,000 lb), or railcars (100,000 lb).

A small number of customers prefer the final product in different forms for one reason or another. But the wet-pelleted furnace-type products dominate the industry in terms of volume.

Other forms of final product are

1. *Dry pellets*: Using a rotating drum and recycling some carbon black pellets, the loose carbon black is rolled into pellets via mechanical tumbling action. Dry pellets are softer than the wet pellets and are used in

applications where the product must disperse in a vehicle with lower energy than wet pellets.

2. *Powder carbon black*: The carbon black can be directly packaged before going through the pelleting and drying stage. Typically the customers for this kind of product are looking for carbon black that is very easy to disperse uniformly with minimum energy. Freight costs and packaging costs are naturally higher than for wet-pelleted carbon black because of the lower density.

A process that has virtually disappeared because of environmental concerns is the channel black process in which natural gas is burned and the resulting carbon black is collected on channel irons that are continuously scraped to obtain the product. It is a highly inefficient process that releases much of the carbon black to the environment. Due to the highly oxidative environment in which the carbon black is produced, it has a high oxygen content (3%–5%), which results in slow curing characteristics in rubber.

IV. CONTROLLING THE QUALITY OF CARBON BLACK

To control the quality of carbon black during production, it must be tested for the characteristic properties that can be related to its performance in rubber. Before discussing carbon black characterization and the various quality control tests, it is worthwhile to point out that the carbon black industry has done numerous things to standardize and improve the product received by customers. Examples of this would be the establishment of industry-wide target properties for each grade of carbon black [4], standard practices for calculation of process indices from process control data [5], standard methods for sampling packaged and bulk shipments [6,7], standard practices for reducing and blending samples [8], standardized test methods for every quality parameter and establishment of standard reference blacks with accepted values to ensure uniformity of test data from any lab [9], and a laboratory proficiency program that cross-checks data between over 60 labs worldwide on a semiannual basis.

It is only appropriate that a more detailed discussion of the characterization properties used for quality control purposes is now undertaken in some detail. Table 6.1 briefly summarizes the quality control tests, what they measure, and how they should be used.

A. SPECIFIC SURFACE AREA

The specific surface area is by definition the available surface area in square meters per unit mass of carbon black in grams. This parameter is evaluated through the use of adsorption measurements. In the absence of significant microporosity, which includes almost all rubber grade carbon blacks, the measure of specific surface area exhibits an inverse correlation with the size of the carbon black particles [10]. In theory, the calculation of the amount of surface in square meters is

TABLE 6.1

Brief Summary of the Quality Control Tests for Carbon Black, What They Measure, and How They Should Be Employed

Test	Measures	Use[a]
Oil absorption no. (OAN)	Structure	A
Compressed oil absorption no. (COAN)	Structure after compression	B
Iodine adsorption no.	Surface area	A
Nitrogen surface area	Total surface area	B
STSA	External surface area	B
Tinting strength	Fineness/color	B
Pellet hardness	Strength of pellets	A
Fines content	Dustiness level	A
Pour density	Bulk density	B
Mass strength	Resistance to packing	C
Pellet size distribution	Pellet sizes	C
Toluene discoloration	Extractables	C
Ash content	Inorganics from water	B
Heating loss	Moisture	A
Sieve residue	Contaminants	A
Natural rubber mix	300% modulus, tensile strength	B

[a] A, typical specification property; B, specified only if application is critical to this measurement; C, needs to be used only for process control.

$$S\left(m^2\right) = \frac{W_m NA}{M} \tag{6.1}$$

where
  S is the surface area
  W_m is the weight of the adsorbate monolayer (g)
  N is Avogadro's number (6.023×10^{23} mol^{-1})
  A is the cross-sectional area of adsorbate (m^2)
  M is the molecular weight of the adsorbate (g/mol)

Thus, the specific surface area, in square meters per gram, can be determined by dividing S by the mass of the unknown sample. However, because of the energetically heterogeneous surface of carbon black [11], no molecules adsorb in a monolayer, and even theories that account for multilayer adsorption assume an energetically homogeneous surface [12]. Nonetheless, adsorption tests still provide the best available technique for quality control of carbon black specific surface area, and the most widely used is the adsorption of iodine from aqueous solution. Other methods are also used to assess this property, and each will subsequently be reviewed. Regardless of the technique, it is clear that this is a property that greatly influences the final properties of compounds that contain the carbon black. Increasing only the specific surface area of the carbon black used in a rubber compound will typically increase such attributes as the compound's blackness, stiffness, hysteresis, and wear resistance.

The iodine number test is a well-defined procedure [13] in which a sample of carbon black is added to a 0.0473 N solution of iodine, whereupon it is shaken, then centrifuged to separate the solid. The resulting solution is titrated with 0.0394 N sodium thiosulfate to an endpoint. From this titration, the amount of iodine that adsorbed to the carbon black surface can be calculated, and the result is reported as the grams of iodine adsorbed per kilogram of carbon black (g/kg). Note that these units are not in terms of surface area per unit mass despite the fact that this is what it attempts to assess and monitor. The measurement does have some drawbacks because it can be affected by any entities on the surface that may react chemically with iodine, due to such things as excessive residual oil or oxidation of the carbon black surface. However, under normal conditions (i.e., with no process changes occurring to produce such surface entities), the method provides a reliable, precise, and simple technique for assessing and monitoring the specific surface area.

Nitrogen adsorption measurements are made on carbon black by exposing the carbon black to various partial pressures of nitrogen with the sample at liquid nitrogen temperatures and then applying the ideal gas laws to determine the number of nitrogen molecules that adsorbed. The measurements are made using a multipoint static–volumetric automated gas adsorption apparatus according to standard procedures [14]. From earlier experiments it was determined that the nitrogen molecule had a cross-sectional area of 16.2 Å^2, and by using this value and the Brunauer–Emmett–Teller (BET) method [12] or the de Boer method modified by MaGee known as statistical thickness surface area (STSA) [15], a total specific surface area or an external specific surface area in square meters per gram, respectively, is calculated. Although, like the iodine number method, these give good relative determinations to changes in process conditions that are believed to change this parameter, there is some question as to whether the adsorption process gives us a true measure of the specific surface area or is significantly affected by the nature of the surface, because in both methods there is an assumption that the surface is energetically homogeneous and it has been demonstrated that this is not the case with carbon black [11]. A simple reporting of the amount of nitrogen adsorbed per gram of carbon black would avoid this conflict in interpretation.

It is also to be noted that the STSA method is carried out at higher partial pressures of nitrogen than the BET method and uses the de Boer model to try to remove influences of adsorption into micropores in order to calculate an external surface area. This calculation was derived empirically from experiments in which an N762 carbon black was tested and assumed to have no micropores. The STSA test indicates that there is microporosity in relatively low-specific-surface-area tread blacks that by other methods have not shown microporosity, and this apparent discrepancy has not been resolved. STSA has, however, been demonstrated to be more insensitive to heat and oxidative treatments than any other specific surface area measurement and thus provides a superior measurement of external surface area [15]. The STSA method has clearly been shown to be a better alternative to evaluating the external surface area than its predecessor, which was cetyltrimethylammonium bromide surface area measurements [16], which are no longer in practice primarily due to poor precision but other issues with the methodology [17].

B. STRUCTURE

"Structure" is a term that has been used for many years in the carbon black industry to describe the other main quality parameter of carbon black. It is basically a measure of the complexity in shape of the carbon black aggregates within a sample. Carbon black aggregates vary quite widely in morphology (size and shape factors), from the large individual spheres found in some thermal blacks to small highly complicated, branched aggregates in high-structure, high-surface-area carbon blacks. The concept of structure is used in an attempt to assess this aggregate shape parameter. Figure 6.3 shows the difference between a high-structure and a low-structure carbon black as observed under a transmission electron microscope. The complex and varied shapes of the carbon black aggregates lead to the creation of voids between the aggregates in any samples of carbon black that are greater than the voids that would be created if the aggregates were simple spheres of equivalent size. It is this fact that has led to the commonly used techniques of measuring internal void volumes as a means of indirectly assessing the shape, or "structure," of aggregates within a carbon black sample. In general, the greater the measured internal void volume, the more complex, open, and branched the aggregates within a sample are and the greater the structure. The measurements are made using either volumetric measurements under specific pressures or, more commonly for quality control, oil absorption measurements. In either case it is clear that this is a parameter of carbon black that has a significant influence on the compound in which the carbon black is dispersed. Increasing only the structure of the carbon black used in a rubber compound will typically increase the compound's hardness, viscosity, stress at high strain, and wear resistance.

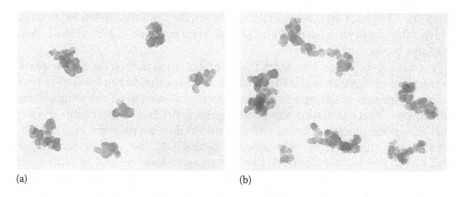

(a)   (b)

FIGURE 6.3   (a) N326 (low structure) and (b) N358 (high structure) carbon blacks as viewed by transmission electron microscopy. (Photograph by David Roberts.)

Oil absorption is the method of choice for quality control purposes for assessing the structure of carbon black by applying the techniques in ASTM D2414 [18]. The test is simply a vehicle demand test where the oil (either dibutyl phthalate [DBP], paraffin oil, or epoxidized sunflower oil) is added dropwise through an automated buret to a sample of carbon black that is being rotated by blades in a chamber much like an internal mixer, and when enough oil is added to fill all the voids between the aggregates, there is a change in the mixture from a free-flowing powder to a semiplastic agglomeration, which raises the torque on the rotating blades to a preset torque endpoint, or alternatively, the entire torque curve is recorded and the endpoint is a certain percent (typically 70%) of the maximum torque. Most commonly it is reported as the oil absorption number (OAN) in units of milliliters of oil per 100 g of carbon black. Paraffinic oil and epoxidized sunflower oil were added to the procedure as a means for companies to move away from the more environmentally unfriendly DBP. It was observed many years ago that this measurement was greatly influenced by the amount of work that is needed to be exerted on the carbon black sample for it to be easily manipulated and that it was not always in alignment with the amount of "structure" that was influencing compound properties. Thus, an alternative method was developed and adopted for oil absorption wherein the sample is compressed at 24,000 psi four times (24M4) before the oil absorption is measured [19]. Thus, this alternative test, referred to as compressed oil absorption number (COAN), seeks to approximate the level of structure present in a carbon black after it is mechanically mixed into rubber. The difference between a typical oil absorption value and a compressed oil absorption value can vary anywhere from 3 to almost 50 units depending on the grade. Although the COAN has proved itself to be a useful tool, one is cautioned to consider that the breakdown of structure may vary considerably according to the parameters of the polymer into which the carbon black is mixed.

It was proposed many years ago that volumetric measurements of the carbon black are made under specified pressures to assess the structure parameter, but the equipment of the times were not very sophisticated and oil absorption became the method of choice for structure measurements. This "void volume" test was revived in the 1990s when improved technology made the test much more accurate and precise. Compressed void volumes are now an ASTM test and are obtained by measuring the compressed volume of a weighed sample in a cylindrical chamber as a function of the geometric mean pressure exerted by a movable piston [20,23]. A profile of void volume as a function of the mean pressure provides a means to assess carbon black structure at varying levels of density and aggregate reduction. Due to the nature of this test, it is more closely aligned with the COAN values than the OAN values obtained from liquid absorption measurements. To date the test has not gained acceptance for quality control purposes but may do so in the future because it is much faster than oil absorption and appears to be just as accurate and precise.

C. TINT STRENGTH

For the tint strength test, a sample of carbon black is mixed into a paste with a white powder (zinc oxide) and plasticizer, the paste is thinly spread on a smooth surface, and the reflectance of the paste is measured [21]. Each time the test is performed, a reference N330 carbon black is likewise tested, and the tint strength is the ratio of the reflectance of the standard to that of the sample. In this way, a carbon black sample that causes the paste to be blacker in color than the standard and thus have less reflectance than the standard will have a higher tint strength (>100) than the standard.

The tint strength test has obvious applications to customer applications where color is critical. However, for other applications, there is some debate about its usefulness because it is highly correlated to other carbon black properties. The tint strength results are correlated directly with the carbon black specific surface area (the smaller the carbon black entities, the more dispersed these black bodies are in the paste, leading to higher tint strength) and are inversely correlated with the carbon black structure (the more highly branched the aggregates, the more voids and the less coverage of the whiteness of the zinc oxide, meaning lower tint strength). Tint strength ultimately measures the degree of dispersion of the carbon entities in the zinc oxide–containing paste. Higher tints indicate more highly dispersible carbon.

D.   PELLET PROPERTIES

As discussed in Section III, carbon black must typically be densified in the form of pellets to facilitate transport and handling. These pellets must be hard enough to withstand the transportation, unloading, and handling needed for the customer, yet must be soft enough to not have difficulty in breaking down and subsequently dispersing in the polymer into which they are mixed. Thus, several tests have been developed to assess the quality of the pellets produced. Without doubt the two most important tests developed for evaluating the quality of the pellets and predicting whether the customer will encounter difficulties in handling or mixing are the determination of fines content and pellet hardness. Other pellet quality tests for carbon black include pellet size distribution, bulk density, and mass strength.

The "fines" content of carbon black pellets is determined by placing a 25 g sample onto a 125 μm screen and shaking for 5 min, with the material passing through the screen being considered the fines [22]. The instrument used for the shaking, called a Ro-Tap, performs a rotary shaking motion and has a hammer that taps the top screen. Depending on the type of unloading and transportation system at the receiving location of the carbon black, the maximum amount of the 5 min fine that can be tolerated is a typical specification property. Excessive fines can lead to problems with unloading, dustiness, and/or flowability. The test can also be done using a 20 min shake, and the difference between the 20 and 5 min fines tests is known as the attrition [22]. The attrition is a good indication of the amount of pellet breakdown that might occur as the pellets are handled through conveying systems. It is also a property that is typically monitored in the process, because high attrition values give production personnel an indication that there are problems with the pelletizer. In either test the sample should be riffle split (blended) before testing to ensure uniformity of the fines in the sample.

Pellet hardness testing is typically done on pellets that are between 1.4 and 1.7 mm in diameter, which are obtained by sieving the samples through a U.S. No. 14

screen and collecting the pellets retained by that screen on a U.S. No. 12 screen in a 1 min shake. An automated tester employing a piston then brings one pellet at a time against a load cell until it fractures [24]. Normally, only 20–50 pellets are tested on a sample for reasonable testing time considerations, but the container may actually contain millions of pellets of many different sizes, and thus it is not surprising that the statistical reliability of the test is notoriously poor. In spite of this fact, the test has still proved to be an invaluable tool for assessing the quality in regard to whether the pellets will be too hard to disperse or too soft to maintain integrity.

Pellet size distribution is tested by production personnel to monitor their pelletization processes. Sieve analysis is done to determine the relative amounts of pellets in six size intervals: <0.125, 0.125–0.25, 0.25–0.50, 0.50–1.0, 1.0–2.0, and >2.0 mm [25]. Bulk density, or pour density, is a simple test wherein a sample is poured into a container of known volume and the mass is measured in order to calculate a density [26]. Bulk density varies appreciably between grades and is needed for converting between mass and volume in shipping, handling, and compounding on the commercial scale. Not surprisingly, the bulk density can be correlated inversely with the oil absorption values, because higher oil absorption leads to aggregates and agglomerates that will not pack as closely in the pellet and thus have a lower observed pellet density. The mass strength test [27], once called the pack point test, measures the minimum force required to compact a relatively large sample of pellets into a coherent mass. An excessively low value indicates that the sample may tend to dust or pack during unloading or conveying. The test is relatively simple and fast and is used by process personnel as a quick measure of pellet quality.

E.   IMPURITIES

Carbon black is basically elemental carbon. Because of the feedstock and manufacturing process, it does, however, contain a small but significant amount of noncarbon constituents. The main heteroatoms incorporated into the carbon structure are hydrogen, oxygen, and sulfur. Thermal blacks typically contain less than 1% of these heteroatoms and furnace grades less than 2%–3%. None of these heteroatoms have been determined to affect the quality of the rubber product in which the carbon black is mixed, and thus their measures have not been developed into quality control tests. Many people have questioned whether the sulfur in the carbon black affects the vulcanization in sulfur-based curing systems, but it appears that the sulfur is tightly bound in the carbon black structure and is thus unavailable as free sulfur [28]. Oxygen in high amounts such as are found in channel blacks and some treated carbon blacks can cause the cure rate to slow in an amine-based sulfur vulcanization system because there can be enough acidic oxygen surface complexes (such as carboxylic groups) to appreciably react with the amine-based accelerator and make it unavailable for curing reactions [29,30]. Other noncarbon constituents, which are most frequently process contaminants, can adversely affect quality; these include moisture, ash, extractables, and the various impurities sometimes found from water wash sieve residue analysis. Moisture is a parameter typically found on customer specifications and is determined by measuring the mass loss at 125°C. Ash content of carbon black arises primarily from the salts and minerals in process water and is measured to ensure satisfactory purity of the carbon black in applications where purity is critical. Ash is determined by measuring the residue remaining after the combustion of the carbon black in an air atmosphere, normally at a temperature of 550°C [31].

Extractables are the oily residues remaining on the sample during carbon black formation and result from the reaction being quenched in the furnace before the decomposition of the oil has reached completion. The test for extractables, typically important only for process control, is done semiquantitatively by determining the amount of discoloration (by measuring the percent transmittance at 425 nm wavelength) of the toluene used to extract the carbon black sample [32]. Note that the lower the value of percent transmittance, the greater the amount of oily residue remaining on the carbon black.

Other impurities are found by determining the amount of material (often called sieve residue or grit) that resists passage through screens of a specified size after washing with water and the application of gentle mechanical rubbing [33]. The material found can be from many origins such as refractory failure, coke formation, and metal degradation of process equipment. Typical screen size openings are 45 Am (U.S. No. 325) and 0.5 mm (U.S. No. 35). Other screen sizes may be used, because the purpose is to ensure that these impurities are limited to small amounts and do not cause problems such as surface blemishes or degradation of any performance properties in the products in which the carbon black is used. Similar to the sieve residue test [33] is a new relatively automated method for mechanical flushing of a carbon black sample with water through a sieve to determine the amount of nondispersible matter in the sample [109].

Manufacturers of mechanical rubber goods (MRG) whose applications are very sensitive to defects due to impurities worked with the carbon black industry to develop grades of carbon black that are extremely clean (very low ash and sieve residue) to minimize the defects in their products. Carbon black manufacturers took several actions to accomplish this objective of new, cleaner grades of carbon black, including special units dedicated to producing this less contaminated carbon black. Other actions included developing reactors that minimized coke formation, using filtered or reverse osmosis water for the process, filtration of the feedstock oil, and replacement of carbon steel in the process with stainless steel. Despite the fact that these carbon blacks cost more to produce, they were viewed favorably by the specialized MRG customers because the reduction in scrap cost would often easily offset the increase in carbon black cost.

F. IN-RUBBER TESTS

ASTM has developed two rubber recipes specifically for evaluating carbon black in rubber. One formula is for natural rubber [34] and the other for styrene–butadiene rubber [35]. The formulations are shown in Table 6.2. Normally when any test is to be done in these recipes, one also mixes and tests the current Industry Reference Black (IRB) and reports the data as differences from the IRB in order to minimize fluctuations in data due to mixing differences. The values for the current IRB are found in ASTM D1765 [4]. Years ago, customers commonly specified requirements

TABLE 6.2

ASTM Formulations D3192 (Natural Rubber) and

D3191 (Styrene–Butadiene Rubber)

Ingredient	D3192 (NR), phr	D3191 (SBR), phr
SBR-1500		100.00
Natural rubber, SMRL	100.00	
Carbon black	50.00	50.00
Zinc oxide	5.00	3.00
Stearic acid	3.00	1.00
Sulfur	2.50	1.75
TBBS (accelerator)		1.00
MBTS (accelerator)	0.60	
Total	161.75	156.75

on stress–strain properties in the natural rubber recipe, but their use has been declining because most customers did not observe much usefulness from these data (as opposed to the usefulness of physicochemical properties of carbon black discussed earlier) and it has been gradually removed from customer specifications.

V. EFFECT OF CARBON BLACK ON RUBBER PROPERTIES

The physical properties imparted to a given rubber compound by carbon black are dominated by three factors: (1) the loading of the carbon black, (2) the specific surface area of the carbon black, and (3) the structure of the carbon black. Table 6.3 shows a generalization of how these factors influence the rubber properties, but the reader is cautioned that there are many exceptions to these relationships and that the type of polymer, presence or absence of oil, type of cure system, and many other factors may also alter those relationships. The more detailed discussion that follows is divided into three categories: (1) the mixing and dispersion processes that occur initially, (2) the processing properties of the uncured compound, and (3) the physical properties of the cured compound.

A. Mixing and Dispersion

Carbon black is incorporated into rubber through shear forces generated by adding the carbon black to rubber in an internal mixer or open mill. The addition of the carbon black causes the torque developed in an internal mixer to rise to a maximum before slowly dropping while the temperature of the mixed stock continuously rises. The temperatures generated during mixing generally increase as the loading of carbon black, the specific surface area of the carbon black used, or the structure of the carbon black used is increased. The initial rise to a maximum torque is generally referred to as the incorporation stage because the polymer is filling the voids between the carbon black aggregates and agglomerates, generally to a point

TABLE 6.3
Effect of Carbon Black on Rubber Properties

Rubber Property	Effect of Carbon Black on Rubber Properties		
	Surface Area	Structure	Loading
Uncured properties			
Mixing temperature	Increases	Increases	Increases
Die swell	Decreases	Decreases	Decreases
Mooney viscosity	Increases	Increases	Increases
Dispersion	Decreases	Increases	Decreases
Loading capacity	Decreases	Decreases	—
Cured properties			
300% modulus	Insignificant	Increases	Increases
Tensile strength	Increases	Insignificant	Increases[a]
Elongation	Insignificant	Decreases	Decreases
Hardness	Increases	Increases	Increases
Tear resistance	Increases	Decreases	Increases[a]
Hysteresis	Increases	Insignificant	Increases
Abrasion resistance	Increases	Insignificant	Increases[a]
Low-strain dynamic modulus	Increases	Insignificant	Increases
High-strain dynamic modulus	Insignificant	Increases	Increases

[a] Increases to an optimum and then decreases.

at which the mixture becomes a coherent rubbery composite. Subsequently, this process continues as the torque decreases and processes such as deagglomeration (reduction of agglomerate sizes through breakdown of the agglomerates into aggregates) and distribution (movement of the aggregates or agglomerates throughout the matrix and sometimes more preferentially into one polymer if it is a polymer blend) take place. Depending on the mixing conditions, carbon black type, polymer type(s), etc., there is a final dispersion of the carbon black aggregates throughout the polymeric medium. This dispersion of the carbon black in the polymer is critical, and in general, the better the dispersion, the better the performance properties of the carbon black–filled rubber compound. It has been recognized to be of such importance that it has been the subject of many research studies [36–39]. One aspect worth noting is that it has been observed that carbon blacks with higher structure generally give shorter incorporation times, and this can be postulated to be due to the fact that the voids between the aggregates are greater owing to the higher degree of branching in the aggregates (they cannot pack as closely), which would leave larger voids that could be more easily filled with rubber during mixing. Another aspect of mixing is the loading capacity (limit to the amount of carbon black that can be incorporated into the rubber while still maintaining a rubbery composite), which normally decreases as the surface area and/or structure of the carbon black increases.

It is clear that the assessment of the level of dispersion in a carbon black–filled rubber compound is a key parameter for predicting performance. The ASTM standard test method [40] for evaluating dispersion of carbon black in rubber uses four techniques:

Method A is a fast qualitative visual comparison of a torn or cut specimen versus reference photographs at 10–20× magnification to give the sample a rating from 1 (worst) to 5 (best).

Method B is a time-consuming and laborious quantitative test done by measuring with a light microscope the percentage of area covered by black agglomerates in microtomed sections of the compound.

Method C is a relatively fast quantitative test wherein the cut surface of a rubber specimen is traced with a stylus that measures the amount of roughness caused by the carbon black agglomerates but requires a laborious calibration for each system studied.

Method D is a quantitative test in which the surface roughness caused by undispersed carbon black agglomerates is assessed on a cut rubber specimen using an interference microscope.

Additional techniques for assessing dispersion besides the ASTM methods are quite numerous. Some are just extensions of the ASTM methods such as the disperGRADER, which essentially duplicates method A but with more reference photographs, software for additional analysis, and the ability to test uncured rubber [41]. Another example is surface roughness measurements with a stylus as in method C, but by scanning in an X–Y plane (rather than using a single line scan), reconstruction of a 3D surface is possible [42]. Slight variations of method D are available using light interferometric techniques.

One problem with all the aforementioned methods is that they address only the macrodispersion of the carbon black as opposed to the microdispersion. In general, microdispersion is at scales of nanometers to fractions of a micrometer, whereas macrodispersion is at scales of several micrometers to millimeters. Problems with macrodispersion refer to poorly dispersed carbon black that may present itself as lumps of filler that for some reason was not fully deagglomerated. Poor macrodispersion can often be related to problems with failure properties and appearance.

Microdispersion refers to the degree to which the aggregates and agglomerates have been dispersed at the submicrometer level, which influences such factors as the amount of interfacial area between the carbon black and polymer (important for the degree of interaction that will take place) and the extent to which the filler–filler network, held together by van der Waals forces, has formed. The filler–filler network plays a dominant role in the low-strain dynamic properties of the compound, which will be discussed in more detail later. The level of microdispersion can be observed qualitatively in a 2D mode using a microtomed section of rubber under a transmission electron microscope but does not lend itself well to reasonable quantification. Electrical resistivity measures microdispersion in the bulk sample, but it is important to note that measurements must be evaluated as relative comparisons to samples of identical composition in order to restrict the influence on resistivity to dispersion differences. When comparing these "like" compounds, an improved dispersion results in higher resistivity due to increased coverage of the carbon black aggregates with the insulating polymer.

B. Uncured Rubber Properties

Once carbon black is mixed into rubber, the resulting filled rubber compound is subjected to processes such as calendering, extrusion, and molding before it is cured to make the finished rubber good. As would be expected, the addition of carbon black changes the properties of the uncured rubber significantly. The addition of carbon black increases the viscosity of the compound, and these increases in viscosity can be correlated with increasing loading of the carbon black, with increasing structure of the carbon black used, and, to a lesser extent, with increasing surface area of the carbon black. These increases in viscosity with carbon black additions obviously change the flow characteristics of the filled compound. It is noted that the typical polymer by itself, when made to flow at low shear rates, will exhibit a shear stress proportional to the shear rate (Newtonian flow), whereas the carbon black–filled polymer results in highly non-Newtonian flow. In most processes there is an extrusion step, and carbon black is well known to influence the amount of swelling the rubber compound experiences when passing through a die. This die swell is the ratio of the cross-sectional area of the extrudate to that of the die and is greater than 1 with rubber compounds. The incorporation of carbon black into the compound reduces the amount of swelling that will occur from passing through a die, and this improvement (or reduction in swelling) can be increased by increasing the loading of carbon black, increasing the structure of the carbon black used, and/or increasing the surface area of the carbon black used.

C. Cured Properties

Once the carbon black–filled rubber compound has been molded, it is cured into a finished product. In general for the tire industry, accelerated sulfur vulcanization systems are used to cure the rubber at high temperature, and the simple presence of any grade of carbon black, even in low amounts, causes a significant reduction of the time before curing starts (induction time). This observation has led to the hypothesis that carbon black may play a catalytic role in the vulcanization process [43]. The physical properties of the final cured rubber product are highly influenced by the type and amount of carbon black. Higher-specific-surface-area carbon blacks tend to give better wear resistance to the rubber as well as greater heat loss (hysteresis) in a tire tread application than their lower-specific-surface-area counterparts. As the filled compound is subjected to higher strains (>10%), the physical properties become less influenced by the specific surface area of the carbon black and increasingly influenced by the structure of the carbon black. Carbon black structure appears to play only a small role in performance at low strains. Thus, higher-structure carbon blacks tend to give greater reinforcement as observed by higher modulus at high strains in cured rubber. Increasing the loading of carbon black, whatever grade, tends to also increase the strength of the

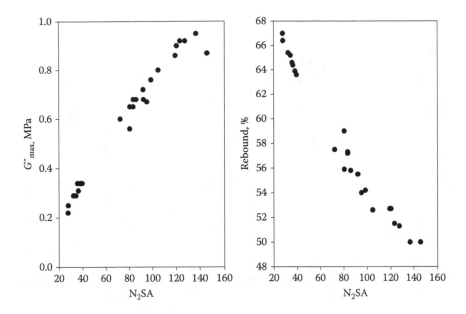

FIGURE 6.4 Relationship of carbon black nitrogen surface area to selected rubber properties.

rubber, but some properties, such as tensile strength and abrasion resistance, tend to decrease after a certain loading. Figures 6.4 through 6.6 demonstrate some of the relationships just described.

It is worthwhile to discuss the current theories on how and why carbon black reinforces rubber. Rubber is a material that has found utilization because it can be deformed and then recover from the deformation. These deformations can be characterized by three parameters: strain amplitude, frequency of deformation, and temperature. Regarding the reinforcing role of carbon black, it has been demonstrated that the strain dependence is the most important of the three parameters [44,45], so further discussion will concentrate in this area. Considerable research has been done on the dynamic mechanical properties of filled compounds [46–48], which forms the basis for the following discussion. It has been shown that the behavior of the polymer/carbon black composite is different in two domains: low strain (<10%) and high strain (>10%). Figure 6.7 shows the response of the elastic or storage modulus (G') and the viscous or loss modulus (G'') from very low strains (0.1%) to 10% strain for a typical carbon black compound and for the corresponding unfilled polymer. It is clear that the response is quite different for the carbon black–filled compound and that the filler is the main contributor to the reinforcement. It is theorized [47] that the carbon black aggregates and agglomerates dispersed throughout the polymer matrix form a network that is held together by van der Waals–type forces. Because of the nature of the forces holding the network together, this network is very sensitive to even small changes in strain and continues to separate as the strain increases, which decreases the stiffness of the composite, leading to the observed decrease in

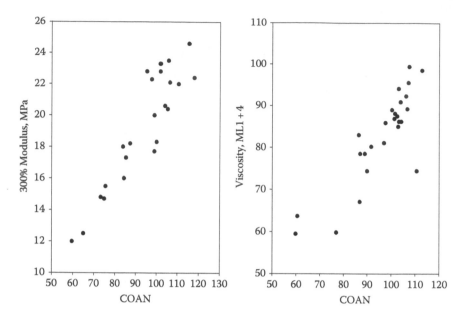

FIGURE 6.5 Relationship of carbon black structure to selected rubber properties.

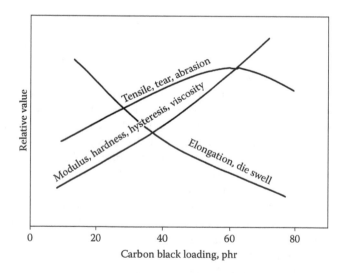

FIGURE 6.6 Generalized relationships between carbon black loading and selected rubber properties.

G' (the elastic component of the modulus). As the network breaks, energy is dissipated as heat, which leads to the observed rise in G'' (the viscous or loss component of the modulus) until it reaches a maximum before decreasing. This maximum in viscous modulus at low strain (G''_{max}) is correlated with hysteresis (energy loss) characteristics of the finished rubber good, most notably the rolling resistance behavior

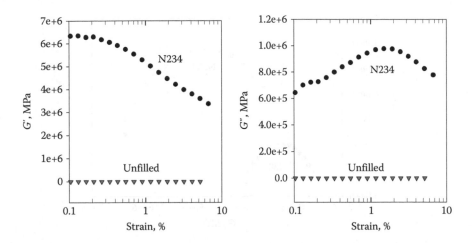

FIGURE 6.7 Relationship of G' and G'' with strain for N234-filled SBR (D3191) and unfilled D3191.

of tires. Because these low-strain properties are highly dependent on the strength of the carbon black network, which is held by weak van der Waals forces, it is not surprising that the specific surface area (which is inversely proportional to the size of the particles and aggregates) plays a dominant role. It is known that the smaller the object, the greater the attractive forces due to either more or stronger van der Waals bonds, because comparisons are made at the same mass of carbon black. It is, of course, observed that the high-surface-area blacks give higher G'_{max} and G''_{max}; however, the structure appears to play little or no role at low strain. As a side note, many in the industry have also used the maximum in tan δ (the ratio of G'' to G') at 60°C for the correlation to energy loss in the compound instead of G''_{max}, but the problem with this is one of the mathematics of the relationship that demonstrates that the energy dissipated per strain cycle is directly related to the G'' value at constant strain amplitude, and thus in order to make comparisons of tan δ to evaluate energy loss, the G' values must be equivalent, which is typically not the case. An excellent approach for making comparisons and understanding behavior regarding carbon black reinforcement in low-strain dynamic properties is the <G-plot> representation, where G' is plotted versus G'' as shown in Figure 6.8. In this plot, first considered by Payne and Whitaker [48] and popularized by Gerspacher [47], the lowest strain is on the right and strain increases as the curve moves to the left.

The other domain of carbon black reinforcement is that of high-strain properties. It is in this region that the surface area of the carbon black begins to play only a small role yet the structure of the carbon black has a very significant influence. As noted earlier, compound properties such as 300% modulus (stress at 300% strain) and dynamic properties above 10% strain are highly correlated to the structure of the carbon black. Once again, structure is a measure of the complexity, shape, and irregularity of the carbon black aggregate owing to factors such as the degree of branching and the number of particles per aggregate. It stands to reason that the higher its structure, the more the carbon black would perturb the polymer movement

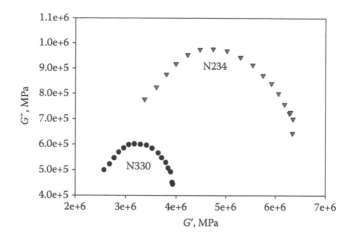

FIGURE 6.8 <*G*-plot> of N234 and N330 in D3191 (SBR).

occurring rapidly at high strains, causing increased stress at equivalent strain compared to a lower-structure carbon black.

VI.   CARBON BLACK CLASSIFICATION AND VARIOUS GRADES

Committee D24 of the ASTM is devoted to carbon black. This committee recognized years ago the value of having a system for designating the grades and having industry-wide accepted test values for the grades of carbon black produced. A standard classification system for carbon black used in rubber products [3] was developed wherein each carbon black grade is assigned a four-character designation. In this classification system, the first character is a letter indicating the relative effect of the carbon black on the cure rate of a typical rubber compound containing the carbon black. The letter *N* indicates a normal curing rate typical of furnace black that has not been modified, and *S* indicates a relatively slower curing rate typical of channel blacks that have many oxygen surface groups or of furnace blacks that were modified during production in such a way as to reduce the curing rate. These are the only two letters presently used for the first character. The second character in the nomenclature system is a number that indicates the average specific surface area as measured by nitrogen adsorption according to D6556 [14]. Table 6.4 shows the 10 arbitrary groups of average surface area, assigned the numbers 0–9. The last two characters are numbers that are arbitrarily assigned.

Each carbon black grade with an ASTM designation then has two target values: the OAN and the iodine adsorption number agreed to between the carbon black suppliers. This is found in Table 6.1 of ASTM D1765 [4] and given in this text as Table 6.5. Also in this table are typical properties for each grade for COAN, N_2SA, STSA, tint, and pour density.

It is not the purpose of this chapter to review all the grades of carbon black, but a few examples will be given in order for the reader to understand some of the differences between grades.

TABLE 6.4

Classification of Carbon Blacks by Nitrogen Surface Area

Group No.	Avg. N_2 Surface Area (m²/g)
0	>150
1	121–150
2	100–120
3	70–99
4	50–69
5	40–49
6	33–39
7	21–32
8	11–20
9	0–10

A. EXAMPLES OF N100 SERIES ($N_2SA = 121–150$)

N110—$N_2SA = 127$ (Iodine No. = 145), OAN = 113. This grade gives superior abrasion resistance, high reinforcement, and high tensile strength. It is recommended for tire tread rubber, bridge pads, and conveyor belts.

N134—$N_2SA = 143$ (Iodine No. = 142), OAN = 127. This high-structure N100 series rubber gives higher abrasion resistance than N110 and high tensile strength. It is well suited for truck and passenger tire treads, especially for heavily loaded truck tires.

In the classification system before ASTM D1765, the N100 series would have typically been classified as super abrasion furnace (SAF).

B. EXAMPLES OF N200 SERIES ($N_2SA = 100–120$)

N220—$N_2SA = 119$ (Iodine No. = 121), OAN = 114. This provides excellent abrasion resistance, high tensile strength, and good tear properties. It is recommended for passenger and, especially, truck tires as well as MRG. In the pre-D 1765 system, it was classified as intermediate SAF.

N234—$N_2SA = 119$ (Iodine No. = 120), OAN = 125. N234 provides superior abrasion resistance in comparison with N220 as well as excellent wear and extrusion properties typical of the 1970s' improved process high-structure blacks. It is used in tire treads including high-performance tires, retread rubber, tank pads, and conveyor belt covers. In the old classification system, it could have been a SAF despite being in the N200 series.

N231—$N_2SA = 111$ (Iodine No. = 121), OAN = 92. This grade has low structure, high abrasion resistance, and excellent tear resistance. It is used mainly in tires where resistance to tear is of primary importance such as many off-road applications.

TABLE 6.5
ASTM Grades of Carbon Black

ASTM Class	Iodine Adsorption No. D1510 (g/kg)	Oil Absorption Number (OAN) D2414 (10^{-5} m^3/kg)	Oil Absorption Number Compressed (COAN), D3493 (10^{-5} m^3/kg)	NSA Multipoint, D6556 (10^3 m^2/kg) (m^2/g)	STSA, D6556 (10^3 m^2/kg) (m^2/g)	Tint Strength, D3265
N110	145	113	97	127	115	123
N115	160	113	97	137	124	123
N120	122	114	99	126	113	129
N121	121	132	111	122	114	119
N125	117	104	89	122	121	125
N134	142	127	103	143	137	131
N135	151	135	117	141	—	119
S212	—	85	82	120	107	115
N220	121	114	98	119	106	116
N231	121	92	86	111	107	120
N234	120	125	102	119	112	123
N293	145	100	88	122	111	120
N299	108	124	104	104	97	113
S315	—	79	77	89	86	117
N326	82	72	68	78	76	111
N330	82	102	88	78	75	104
N335	92	110	94	85	85	110
N339	90	120	99	91	88	111
N343	92	130	104	96	92	112
N347	90	124	99	85	83	105
N351	68	120	95	71	70	100

(Continued)

TABLE 6.5 (Continued)
ASTM Grades of Carbon Black

ASTM Class	Iodine Adsorption No. D1510 (g/kg)	Oil Absorption Number (OAN) D2414 (10^{-5} m³/kg)	Oil Absorption Number Compressed (COAN), D3493 (10^{-5} m³/kg)	NSA Multipoint, D6556 (10^3 m²/kg) (m²/g)	STSA, D6556 (10^3 m²/kg) (m²/g)	Tint Strength, D3265
N356	92	154	112	91	87	106
N358	84	150	108	80	78	98
N375	90	114	96	93	91	114
N539	43	111	81	39	38	—
N550	43	121	85	40	39	—
N582	100	180	114	80	—	67
N630	36	78	62	32	32	—
N642	36	64	62	39	—	—
N650	36	122	84	36	35	—
N660	36	90	74	35	34	—
N683	35	133	85	36	34	—
N754	24	58	57	25	24	—
N762	27	65	59	29	28	—
N765	31	115	81	34	32	—
N772	30	65	59	32	30	—
N774	29	72	63	30	29	—
N787	30	80	70	32	32	—
N907	—	34	—	9	9	—
N908	—	34	—	9	9	—
N990	—	38	37	8	8	—
N991	—	35	37	8	8	—

C. Examples of N300 Series ($N_2SA = 70-99$)

N330—$N_2SA = 78$ (Iodine No. = 82), OAN = 102. This is one of the most important basic blacks in the industry. It provides good economic abrasion resistance with high resilience, easy processing, and relatively good tensile and tear properties. It has a wide range of applications in both tires and MRG for high-severity applications. The old system classified it as high abrasion furnace (HAF).

N326—$N_2SA = 78$ (Iodine No. = 82), OAN = 72. This is a carbon black with significantly lower structure than N330. It has good reinforcement and processability like N330 but with better tensile and tear properties. It is used in tires for carcass compounds, belt skim, and steel cord adhesion compounds. It also finds uses in MRG for high-severity applications.

N339—$N_2SA = 91$ (Iodine No. = 90), OAN = 120. This is another of the 1970s' "improved process" carbon blacks that gives superior abrasion resistance, extrusion properties, and dynamic properties. It has found use in tire treads and highly stressed MRG.

All the aforementioned grades and any of the N100, N200, and N300 series carbon blacks are known broadly as either "tread blacks" because most but not all find application in tire treads or "hard blacks" because they give much higher durometer hardness to the compounds they are mixed in than other broad groups known as either the "carcass blacks" (again because many but not all find application in tire carcass compounds) or "soft blacks" (because they give relatively lower durometer hardness in compounds they are mixed into). Another distinction that one might come across is the tread grades being referred to as reinforcing carbon blacks and the carcass grades as semireinforcing carbon blacks. The carcass, or soft, blacks are typically the N500, N600, and N700 series carbon blacks, and some examples are given as follows. There are no N400 series blacks listed in ASTM D1765.

D. Examples of Carcass or Semireinforcing Grades

Carcass or semireinforcing grades include N500 ($N_2SA = 40-49$), N600 ($N_2SA = 33-39$), and N700 Series ($N_2SA = 21-32$):

N550—$N_2SA = 40$ (Iodine No. = 43), OAN = 121. In the old system, this black was known as fast extrusion furnace (FEF) because it provides fast, smooth extrusions and gives a smooth surface to the extruded product. It also imparts medium abrasion resistance, high strength, and low shrinkage and die swell. It is used in tire carcasses, cushion gum, tubing, cable jacketing, and any extruded goods that require excellent dimensional stability.

N660—$N_2SA = 35$ (Iodine No. = 36), OAN = 90. This black was originally known as general-purpose furnace (GPF) because of its wide applications in tire carcasses, tubes, belts, hose, and many other industrial products. The lower structure gives lower modulus and viscosity than, for example, N550.

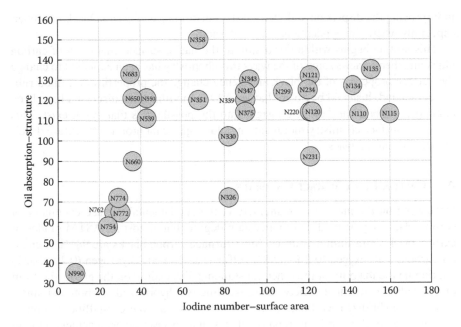

FIGURE 6.9   Plot of structure versus surface area for various carbon black grades.

N762—$N_2SA = 29$ (Iodine No. = 27), OAN = 65. Originally known as semireinforcing furnace (SRF) owing to its good mechanical processing efficiency
and its ability to be highly loaded in rubber, which makes it useful in such
applications as hoses, belts, tire bead insulation, and plastics.

E. THERMAL GRADES

Thermal grades are very low in surface area and structure. For example, N990 has a
nitrogen surface area of 8 and an OAN of 38. It can be highly loaded in rubber and
finds applications in belts, hoses, various extruded products, and plastics. Figure 6.9
shows graphically how the various grades relate to each other in terms of structure
and surface area.

VII. SURFACE CHEMISTRY AND MODIFICATIONS TO CARBON BLACK

The standard ASTM grades of carbon black filler remain satisfactory for most current rubber applications. Nevertheless, there exists a long continuing history of
research into ways to modify the basic properties of carbon black. In some cases, the
aim has been to address certain problems in the furnace during production, such as
coke formation [49,50]. In others, ways to modify carbon black to obviate the addition of other compound chemicals (i.e., antioxidants) were sought [51]. But by far
the greatest interest in modifying the properties of rubber grade carbon black lies

in improving the dispersion and viscoelastic performance of carbon black in rubber, especially where tire applications are concerned.

This section begins with a discussion of the native surface chemistry of carbon black and then reviews some of the ways this chemistry has been exploited to change the surface properties of the black and hence change the properties of a filled rubber compound. It should be noted that not all the modifications discussed have been applied to rubber, but they give insight into the possibilities that might exist. This avenue of surface chemistry modification holds significant promise for future developments in carbon black.

A.   CARBON BLACK SURFACE CHEMISTRY

Depending on the manner of manufacture, the fundamental surface chemistry of carbon black may be modified to a certain degree, but almost all ASTM grades of carbon black share similar features. Most furnace carbon blacks are 98+% elemental carbon, with oxygen (0.2%–0.5%), sulfur (1%–2%), and hydrogen (0.2%–0.4%) making up the majority of the other constituents. Depending on the grade, the form of the carbon is distributed between crystalline (sp^2 hybridized) carbon and amorphous (sp^3 hybridized) carbon. X-ray diffraction reveals the crystalline carbon to be composed of three or four graphitic planes in a turbostratic arrangement with an interplanar distance slightly greater than that of graphite [10].

As mentioned previously, the most common constituents of carbon black aside from elemental carbon are oxygen, hydrogen, and, depending on the feedstock, a significant amount of sulfur (1%–2%), but it is not clear whether this is confined to the surface or is distributed throughout the carbon black aggregates. Gas chromatography performed on samples heated at 1°C/min–2.5°C/min up to 1500°C indicates that the major decomposition products are carbon monoxide, carbon dioxide, hydrogen gas, and, to a lesser extent, hydrogen disulfide. The relative amounts of decomposition products vary with the type and grade of carbon black evaluated. For example, channel blacks, known to have much higher oxygen content than furnace blacks, give proportionately more carbon oxides [52]. It is worthwhile pointing out at this juncture that although oxygen groups are present on furnace blacks as produced, they are relatively small in number and thus play little role "as is" in reinforcement.

From pyrolysis and simple titration studies, several types of functional groups on carbon black have been proposed [53]. Carbon dioxide appears to be derived from functional groups containing two oxygens, such as lactones and carboxylic acid. On the other hand, groups such as quinones and phenols produce carbon monoxide upon decomposition. Likewise, hydrogen gas is most likely produced from the reduction of –CH or –OH surface groups. Similarly, H_2S could be produced from free or bound sulfur and hydrogen radicals.

Adsorption studies have been made to identify the regions of carbon black having the highest activity [54], and these appear to be at the crystallite edges, where the density of available π electrons from the aromatic system would be greatest. In fact, Wang and Wolff [54] identified a difference in activity between crystallite edges, the planar surface, and the amorphous region, the apparent order of activity being edges > amorphous > planar surface. This work has been confirmed in other adsorption studies, which also proposed an even higher-energy region for the slit-shaped cavities between crystallite edges [11]. It is this heterogeneity in the arrangement and ordering of carbon atoms that may actually be the most important aspect of the surface activity.

This lends support to the idea that interaction of carbon black with rubber during compounding increases with increasing availability of crystallite edges or surface defects. It is believed that the π electrons in high density in carbon black crystallite edges interact with π electrons in the unsaturated polymer bonds through van der Waals attraction. The reduction in certain carbon black reinforcing properties when the black is graphitized (thereby reducing the density of crystallite edges) is another indication that crystallite edges play a key role in the interaction of carbon black with rubber [54]. But the high activity of these regions also means that the carbon black aggregates are strongly attracted to each other through the same van der Waals forces [47]. This strong attraction between carbon aggregates could be described as carbophilicity.

This has a dual effect. On the one hand, carbophilicity contributes greatly to the relative strength of the filler–filler network, and this has a significant effect on the low-strain dynamic properties (i.e., hysteresis) of cured rubber as mentioned earlier. In the bulk, the interaction of carbon black aggregates with one another adds a strength to rubber in much the same way as a bundle of threads forms a strong cord or rope. On the other hand, the carbophilic attraction makes dispersion of the agglomerates or aggregates all the more difficult. Surface characteristics of the carbon black will also undoubtedly influence the degree to which the polymer and filler will interact. So a question arises as to which is more important: dispersion or polymer–filler interaction. In either case it appears clear that the activity of the carbon black surface plays an important role.

The energy crises of the 1970s focused research efforts of all involved in manufacturing tires and other rubber products to reduce the heat loss (hysteresis) of rubber compounds. Thus, one of the major goals of tire, rubber, and carbon black manufacturers is to reduce hysteresis in tires without sacrificing treadwear or traction, through the improvement of rubber properties via modified polymers or carbon black. Because carbon black remains the filler of choice for rubber compounders, ways to improve the hysteresis of rubber by lessening carbophilicity or increasing the interaction of carbon black with polymers continue to be pursued. This leads to the idea of carbon black modification.

B. CARBON BLACK MODIFICATION

Numerous papers have been published over the years examining the myriad of ways in which researchers have attempted to modify the surface of carbon black in order to change its basic properties or the properties of materials incorporating the modified black. This section will highlight the history and some recent developments in this field; however, an exhaustive review of the subject is beyond the scope of this text.

In general, two approaches have been taken in modifying the surface chemistry of carbon black: postprocess modification and in-process modification. Much work has been done to explore the reactions of carbon black with various gases at elevated temperatures [55]. These early adsorption studies formed the basis for other modification methods, in particular, wet chemical and plasma treatments. Running almost parallel with this work has been the in situ modification of carbon black in a furnace, usually by injecting additives in conjunction with carbon black oil (CBO) or at another location.

1. Postprocess Modification

a. Gas Adsorption

The chemical modification of the surface of carbon black began when scientists first discovered that simple inorganic molecules could be adsorbed onto the surface of the carbon black substrate [53]. Initial investigations were not intended to modify the surface in such a way as to change the basic properties of rubber compounds in which the carbon was incorporated. Instead, these studies attempted to employ adsorbents as means to probe the basic nature and chemistry of this substrate.

As part of a series on carbon black and compounding published in 1957, Studebaker [55–57] reported on the interaction of such gases as H_2, NH_3, and H_2S with channel blacks at elevated temperatures. He noted striking changes in rubber properties as a result of chemical treatment. Specifically, the rates of cure, moduli, abrasion resistance, and electrical conductivity all appeared to increase dramatically. Although he did not offer any definite reason for this effect, he did note that the reaction of these molecules with carbon black appeared to be similar to reactions with simple aromatic compounds.

Using the observation that carbon black might follow classical organic chemistry in adsorbing or reacting with various chemicals, investigators such as Bansal, Puri, and Donnet explored the interaction of carbon black with Cl_2, I_2, Br_2, and H_2S. Their work was intended primarily to elucidate the nature of chemisorption on carbon black, but they also noted changes in cure rates and other properties [53]. These and other investigations have shown that the order of reactivity of halogens to carbon black follows the sequence chlorine > bromine > iodine, in accordance with their respective nucleophilicities [53]. Desorption studies at elevated temperatures further indicate that some of the halogen is retained, probably due to C–X bonding on the surface.

Other work of a similar nature has continued, although gas reactions were determined to be impractical for any serious production scale treatment of carbon black. However, investigators have recently focused on treating channel-type (high O_2) black with ammonia gas at elevated temperatures with the intent of substituting ether-type oxygen atoms with nitrogen [58]. The aim is to change the nature of the crystalline edges so that pyrrole or pyridinic groups are formed on these edges. It has been found that such a black improves the coordination of certain metal atoms and may provide a means of making carbon black more effective as a catalyst for certain nonrubber applications.

b. Oxidation

Some of the earliest methods of treating carbon blacks to improve or change their basic properties involved posttreatment in aqueous or organic media. The most common commercial example of this is the use of oxidizers on the black after the aggregates have been produced. The early channel blacks were found to have interesting properties, most notably wettability, that were attributed to the presence of oxygen groups on the surface of these blacks. Currently, only one company in Asia makes true channel black, environmental concerns having led to its demise in North America and Europe. Other advantages of oxidation include an increased nitrogen adsorption value, possibly due to an increase in the number of active sites on the surface of the black but most probably due to induced microporosity, which allows a high level of conductivity with smaller loadings. Polymer–filler interaction might be enhanced as well, although the increased level of oxygen groups may offset this. Numerous patents have been filed over the years covering various methods of oxidizing the surface [59–68].

Commercially, several manufacturers use such chemicals as ozone, hydrogen peroxide, or nitric acid to remove part of the carbon by cutting micropores into the surface. This produces a less dense black, the so-called conductive black. Other oxidants tried include chromic acid and permanganates [69]. Electron microscopic images of such blacks even show the interior of the aggregates to be hollow. The advantage is that high levels of conductivity may be achieved at relatively low loadings (e.g., 2–20 parts of carbon black per 100 parts of polymer, phr). Nonporous carbon blacks require much higher loadings (e.g., 40–60 phr) to reach the same level of conductivity. These oxidized blacks are especially useful in applications such as electrical wire sheathing or antistatic rubber mats, where a flexible yet conductive material is required.

Recent work has been presented on reactive in situ mixing of oxidized carbon blacks (ozone treated) in combination with in-chain functionalized solution-polymerized styrene–butadiene rubber to improve the hysteresis performance of the filled rubber [110]. It is proposed that these compounds are a close approximation of silica-filled compounds in regard to hysteresis and traction performance.

c. Reaction of Carbon Black with Diazonium Salts

Diazonium salts are formed from aromatic amines and have the general formula Ar–$N_2^+X^-$. They undergo two main classes of reactions: replacement, in which the nitrogen is lost as N_2, while some other group becomes attached to the aromatic ring, and coupling, in which the nitrogen is retained in the product. Diazonium salts are valuable in organic synthesis owing to their ability to form so many classes of compounds [70]. The coupled diazo compounds have found use as radical polymerization initiators, along with various peroxides. These work by eliminating nitrogen or oxygen gas during decomposition, leaving carbon radicals on the corresponding organic groups. Depending on reaction conditions, the decomposition may be triggered by ultraviolet light or heat. Examples include 2,2'-azobisisobutyronitrile (AIBN) and benzoyl peroxide (BPO). In theory, any organic group could be incorporated into a diazo or peroxide compound, although the diazo class is generally more stable and easier to synthesize [71]. It seems reasonable, then, to try to bond organic groups to the electron-rich active sites of carbon black through a radical intermediate, and in fact, one company has patented the addition of diazonium salts to carbon black in a pelletizer in order to achieve this [72–74]. A carbon black bearing such groups may interact better with polymer during mixing, especially if the polymer has been modified so that a secondary reaction between functional groups on the polymer and carbon black might occur. Applications to date in this area are for the most part nonrubber.

d. Plasma Treatment

Put simply, a plasma is an ionized gas capable of conducting an electric current. Examples include lightning, the *aurora borealis*, and neon or fluorescent light. Commercially, it is generally produced in a chamber containing a gas at low pressure that is in contact with two electrodes. If a material such as carbon black is allowed to come in contact with the plasma field, several reactions may occur simultaneously, most notably gas ionization, surface ablation, and radical formation of both substrate and gas. If the plasma energy is pulsed (on/off cycles), bond formation from radical intermediates is observed [75].

Although this has become an area of increasing interest recently, plasma modification of carbon black has been discussed in the literature since the mid-1960s. Early interest in high-temperature plasmas focused on using the plasma energy to produce carbon black from a hydrocarbon stream passing through the plasma region [76,77]. Although this was feasible in principle, it was not economically advantageous under the state of the art at that time. However, advances in plasma technology sparked renewed interest in its use as a heat source in manufacturing furnace blacks. It has been suggested that better control of structure and even microstructure may be possible.

In this case, the treatment should be classified as in-process modification, although numerous papers have appeared describing the use of plasma for postprocess modification as well. The ability of a low-temperature plasma to form radical sites on both the carbon black substrate and the ionized gas makes it feasible to bond almost any molecule directly to the carbon black surface. Indeed, any material that can be ionized in the gaseous state could be used for surface treatment [78,79]. This includes not only simple molecules such as halides, ammonia, nitrates, hydroxyl groups, and carboxyl groups but also polymer monomers such as styrene [80]. It is already well known that plasma treatment can impart hydrophilicity to carbon black, but it should also be a quick, efficient way to tailor the surface of the black for preferential bonding to compatible polymers, catalysts, or other materials.

Currently, one major drawback to full-scale plasma treatment is the fact that the black must be treated as a batch, and no large-scale batch plasma processing equipment yet exists with this capability. Another disadvantage is that the black must be treated in the fluffy state for maximum effectiveness. Finally, because of insufficient demand for specialty blacks (other than oxidized blacks), carbon black producers are generally unwilling to make the capital expenditures necessary for plasma treatment. Still, research into ways to use plasma to treat loose black in situ (as part of the reactor) is ongoing, and this technology may soon open up entirely new vistas for carbon black, especially in advanced materials.

e. Polymer Grafting

Because it has long been recognized that better filler–polymer interaction should result in better dispersion during mixing, another approach to treating the carbon black surface involves grafting polymers to the surface. As mentioned earlier, the active sites on carbon black readily adsorb many different molecules. Further, under the right conditions, the black also forms free radicals and may be used to initiate radical polymerization in solution, making it possible to grow polymers directly on the surface [81]. This has been tried on a number of occasions using, for example, benzene [82], latex [83], polyaniline [84], dendrimers [85], and conductive polymers such as polypyrrole [86]. Although several of these were claimed to improve dispersion during mixing, the properties of cured rubber do not seem to be affected as much as was hoped, perhaps because the sites needed for strong filler–filler interaction are blocked by the grafted polymers. Yet again, research into this area remains strong.

2. In-Process Modification

a. Chemical Addition

Addition of various chemicals has been the subject of many patents over the years. Many attempts have been made at just using an improved pelletizing aid to improve dispersion or some other performance parameter. Additions of chemicals can also be done to physically adsorb that chemical such that it can react with the medium that it is being formulated into. One such recent development is the addition of alkali metal polysulfides that adsorb onto the carbon black surface strongly enough that during mixing with rubber, they can continue to stay adsorbed and then interact with the polymer molecules [111]. With polysulfides, there is a strong indication of increased interaction as evidenced by increased bound rubber and increased cross-link density of the polymer, probably from sulfur linkages formed during the vulcanization process. In addition, the adsorbed molecules decrease the filler–filler interaction, thereby decreasing the hysteresis that leads to improved rolling resistance in tire compounds. The increased filler–polymer interaction helps prevent loss in abrasion resistance that typically occurs with the greater filler–filler interaction from the adsorbed molecule.

b. Metal Addition

As mentioned previously, the addition of certain alkali metals, in particular potassium, to a carbon black reactor is used as a means of structure control. Over the past 40 years, most other metals in the periodic table have also been explored for this and other purposes. Metals such as chromium [87], calcium, strontium, and magnesium [88] were added, oddly enough, to increase surface area without necessarily decreasing structure. Rare earth metals were added in an attempt to form carbon–metal hybrids [89], presumably to reduce the amount of pure metal needed in certain applications such as catalysis. Iron, nickel, and cobalt were added to the quench section of a reactor to form so-called magnetic carbon blacks [90]. The same author also added these and barium or aluminum salts to the feedstock to achieve the same purpose [91] and recently investigated the incorporation of iron and nickel posttreated carbon black in filled rubber vulcanizates [91a] to make a smart material. In other studies, salts of almost every metal in the periodic table were added to feedstocks in order to form carbon black–metal hybrids for use in elastomeric compounds [92,93].

It is anticipated that these hybrids will form metal bonds to diazo coupling agents and give carbon black dispersions equal to or better than the improved dispersions of silica in polymer with coupling agents. Carbon black so treated might also find such nonrubber uses as low-cost catalysts or other smart materials applications.

c. Wide Aggregate Size Distribution Blacks

Several techniques can be employed to widen the aggregate size distribution during the manufacturing process. These modifications have been improved upon over the years resulting in improvements in the hysteresis performance of the rubber compounds that they are incorporated into. It is proposed that by widening the distribution, there is an increased amount of smaller and larger aggregates that are able to reach the polymer interface more easily leading to improved polymer–filler interaction. At the same time, the difference in aggregate sizes creates more voids and less contact between aggregate causing decreased filler–filler interaction. Both of these lead to better compound hysteresis.

d. Carbon Black–Silica Dual-Phase Filler

This hybrid filler is manufactured by the blending of carbon black with silica during the formation process of each. Indeed, the use of silica as an alternative filler for rubber compounds has been considered almost as long as carbon black has been in use. There are several reasons for this. First, it resembles carbon black in its morphology; that is, it is an aggregate of spheroidal particles, though of a much smaller particle size, on the order of 10 nm. Second, the chemical nature of the surface is quite different from that of carbon black, being covered with a large number of hydroxyl groups. The combination of silica's small particle size and unique surface chemistry, without any coupling agents, leads to rubber properties fairly comparable to those of carbon black–filled rubber, but with some drawbacks such as inferior treadwear [94]. The presence of so many hydroxyl groups results in strong hydrogen bonding between aggregates. For this reason, early attempts to disperse silica in a rubber compound were not very successful, and large agglomerates resulted. Still, when silica was compounded with carbon black, these large agglomerates improved rubber tear properties and helped prevent crack propagation [95].

It was not until the early 1970s that the use of organosilane coupling reagents was found to improve silica dispersion in rubber. Essentially, an organosilane has a dual functionality, with one end terminated by an alkoxide (e.g., an ethoxide in the case of SI-69) and the other end terminated by sulfur atoms. During mixing, the alkoxides react with the hydroxyl groups on silica and liberate the free alcohol as a Si–O–Si bond is formed between silica and the silane. The sulfur atoms on the other end appear to be drawn to the polymer during mixing and may participate in cross-linking during curing [96]. The use of an organosilane appears to improve dispersion of silica in rubber by reducing the amount of hydrogen bonding between aggregates while at the same time improving the polymer–filler interaction. The use of organosilanes has led to significant improvements in the use of silica for tire treads in regard to performance properties, especially rolling resistance and traction, without significantly sacrificing treadwear. Their use in tires increased significantly during the 1990s.

Despite the apparent superiority of silica as a filler in tires, its use comes with a cost. The only silica found to perform adequately in a rubber compound is precipitated silica. The cost of making this material has historically been higher than that of carbon black, although continuous improvements in manufacturing efficiency have steadily brought this cost more in line with carbon black. The organosilane coupling reagent remains a significant cost barrier to the use of silica on a large scale. In addition, silica is more difficult to compound than carbon black, leading to increased processing costs for tire manufacturers. Finally, the ethanolic fumes generated during compounding raise certain health and environmental issues. Still, the potential threat silica may pose to carbon black in the future has spurred research efforts to find inexpensive ways to make carbon black perform in a similar manner in tire applications.

Some research into producing carbon black with silica groups involves post-treating the carbon black with various organosilicon compounds [97], for example, substituting polyethylene glycol with dimethyl silicon groups [98] during pelletization. However, more common efforts involve forming the carbon black in the presence of silica, resulting in a hybrid blend, the so-called dual-phase filler. In an early embodiment of the art, a patent was published that described the production of a carbon black–silica pigment made by passing carbon black exhaust gas through a slurry of precipitated silica [99]. It was claimed that the resulting black had about a 28% silica content. Since then, numerous patents have appeared discussing this type of approach [100–105]. Each of these methods has a subtle difference, but all are generally similar. Basically, silicon-containing species are coinjected or reacted with CBO as the oil is pyrolyzed to produce carbon black. These may include such compounds as organosilanes, organochlorosilanes, siloxanes, organosilicates, silazanes, and even silicon polymers [106]. The resulting material has regions of silica and carbon in the same aggregate. The hybrid aggregates can be easily incorporated in a rubber compound by coupling with organosilanes or organosulfides [107]. This is designed to improve the trade-off between wear resistance and rolling resistance. It is claimed that the hybrid silica–carbon black imparts less hysteresis and better wet traction than carbon black alone yet has better wear resistance than silica alone.

Early hybrids of this type had the silica randomly dispersed in the carbon black aggregate, which gave improved performance in truck tires with heavy filler loading but was not as advantageous in passenger tires. However, by increasing the amount of silicates in the reactor and optimizing the distribution of the silica domain on the carbon black aggregate (using multistage cofuming technology), a dual-phase filler has been produced that is claimed to give improved performance in passenger tires [108]. Some of these dual-phase fillers have been commercialized, but none have yet reached widespread use, possibly due to the greater production costs associated with the silica additives.

VIII. CONCLUSIONS

Although the carbon black industry is quite mature, carbon black remains the filler of choice for most rubber applications. The advantages of relatively low production costs, well-established testing and quality control methods, and considerable user experience by rubber compounders would indicate that carbon black is not likely to disappear anytime soon. Rather, continuous R&D efforts into improved production techniques, in situ property modification, and surface treatment will help carry this material into the future as new applications are discovered.

REFERENCES

1. Donnet JB, Voet A. *Carbon Black Physics, Chemistry, and Elastomer Reinforcement.* New York: Marcel Dekker, 1976, pp. 1–33.
2. Notch Consulting 4Q2013 Carbon Black Quarterly Newsletter, Table 7, page 12.
3. ASTM D3053-13a. Annual Book of ASTM Standards, Section 9 Rubber Volume 09.01, 2014 edition, American Society of Testing and Materials, West Conshohocken, PA.
4. ASTM D1765-13. Annual Book of ASTM Standards, Section 9 Rubber Volume 09.01, 2014 edition, publisher American Society of Testing and Materials, 100 Barr Harbor Drive, West Conshohocken, PA 19428-2959.
5. ASTM D4583-95 (2009). Annual Book of ASTM Standards, Section 9 Rubber Volume 09.01, 2014 edition, publisher American Society of Testing and Materials, 100 Barr Harbor Drive, West Conshohocken, PA 19428-2959.
6. ASTM D1799-03a (2008). Annual Book of ASTM Standards, Section 9 Rubber Volume 09.01, 2014 edition, publisher American Society of Testing and Materials, 100 Barr Harbor Drive, West Conshohocken, PA 19428-2959.
7. ASTM D1900-06 (2011). Annual Book of ASTM Standards, Section 9 Rubber Volume 09.01, 2014 edition, publisher American Society of Testing and Materials, 100 Barr Harbor Drive, West Conshohocken, PA 19428-2959.
8. ASTM D5817-03a (2008). Annual Book of ASTM Standards, Section 9 Rubber Volume 09.01, 2014 edition, publisher American Society of Testing and Materials, 100 Barr Harbor Drive, West Conshohocken, PA 19428-2959.
9. ASTM D4821-14. Annual Book of ASTM Standards, Section 9 Rubber Volume 09.01, 2014 edition, publisher American Society of Testing and Materials, 100 Barr Harbor Drive, West Conshohocken, PA 19428-2959.
10. Hess WM, Herd CR. Microstructure, morphology and general physical properties. In: Donnet JB, Bansal RC, Wang MJ, eds., *Carbon Black Science and Technology.* New York: Marcel Dekker, 1993, pp. 89–173.
11. Schroder A, Kluppel M, Schuster RH, Heidberg J. Energetic surface heterogeneity of carbon black. *Kautsch Gummi Kunst* 2001;54(5):260–266.
12. Brunauer S, Emmett PH, Teller E. Adsorption of gases in multimolecular layers. *J Am Chem Soc* 1938;60:309–319.
13. ASTM D1510-13. Annual Book of ASTM Standards, Section 9 Rubber Volume 09.01, 2014 edition, publisher American Society of Testing and Materials, 100 Barr Harbor Drive, West Conshohocken, PA 19428-2959.
14. ASTM D6556-14. Annual Book of ASTM Standards, Section 9 Rubber Volume 09.01, 2014 edition, publisher American Society of Testing and Materials, 100 Barr Harbor Drive, West Conshohocken, PA 19428-2959.
15. MaGee R. Evaluation of the external surface area of carbon black by nitrogen adsorption. *Rubber Chem Technol* 1995;68(4):590–600.
16. Saleeb FZ, Kitchener JA. The effect of graphitization on the adsorption of surfactants by carbon blacks. *J Chem Soc* 1965;10:911–917.
17. Gerspacher M, O'Farrell CP, Wampler WA. An alternate approach to study carbon black. *Rubber World* 1995;212(3):26–29.

18. ASTM D2414-13a. Annual Book of ASTM Standards, Section 9 Rubber Volume 09.01, 2014 edition, publisher American Society of Testing and Materials, 100 Barr Harbor Drive, West Conshohocken, PA 19428-2959.

19. ASTM D3493-13. Annual Book of ASTM Standards, Section 9 Rubber Volume 09.01, 2014 edition, publisher American Society of Testing and Materials, 100 Barr Harbor Drive, West Conshohocken, PA 19428-2959.

20. ASTM D7854-13. Annual Book of ASTM Standards, Section 9 Rubber Volume 09.01, 2014 edition, publisher American Society of Testing and Materials, 100 Barr Harbor Drive, West Conshohocken, PA 19428-2959.

21. ASTM D3265-14. Annual Book of ASTM Standards, Section 9 Rubber Volume 09.01, 2014 edition, publisher American Society of Testing and Materials, 100 Barr Harbor Drive, West Conshohocken, PA 19428-2959.

22. ASTM D1508-12. Annual Book of ASTM Standards, Section 9 Rubber Volume 09.01, 2014 edition, publisher American Society of Testing and Materials, 100 Barr Harbor Drive, West Conshohocken, PA 19428-2959.

23. Joyce GA, Henry WM. Modeling the equilibrium compressed void volume of carbon black. *Rubber Chem Technol* 2006;79(5):735–764.

24. ASTM D5230-13. Annual Book of ASTM Standards, Section 9 Rubber Volume 09.01, 2014 edition, publisher American Society of Testing and Materials, 100 Barr Harbor Drive, West Conshohocken, PA 19428-2959.

25. ASTM D1511-12. Annual Book of ASTM Standards, Section 9 Rubber Volume 09.01, 2014 edition, publisher American Society of Testing and Materials, 100 Barr Harbor Drive, West Conshohocken, PA 19428-2959.

26. ASTM D1513-05. Annual Book of ASTM Standards, Section 9 Rubber Volume 09.01, 2014 edition, publisher American Society of Testing and Materials, 100 Barr Harbor Drive, West Conshohocken, PA 19428-2959.

27. ASTM D1937-13. Annual Book of ASTM Standards, Section 9 Rubber Volume 09.01, 2014 edition, publisher American Society of Testing and Materials, 100 Barr Harbor Drive, West Conshohocken, PA 19428-2959.

28. Lewis JE, Devinney ML, McNabb CF. Radiotracer studies of carbon black surface interactions with organic systems. II. Use of sulfur-35 in initial studies on the reactivity of "bound" surface sulfur. *Rubber Chem Technol* 1970;43:449–462.

29. Cotton GR. The effect of carbon black surface properties and structure on rheometer cure behavior. *Rubber Chem Technol* 1972;45:129–144.

30. Haws JR, Cooper WT, Ross EF. Another look at S-type carbon blacks. *Rubber Chem Technol* 1977;50:211–216.

31. ASTM D1506-99 (2013). Annual Book of ASTM Standards, Section 9 Rubber Volume 09.01, 2014 edition, publisher American Society of Testing and Materials, 100 Barr Harbor Drive, West Conshohocken, PA 19428-2959.

32. ASTM D1618-99 (2011). Annual Book of ASTM Standards, Section 9 Rubber Volume 09.01, 2014 edition, publisher American Society of Testing and Materials, 100 Barr Harbor Drive, West Conshohocken, PA 19428-2959.

33. ASTM D1514-04 (2011). Annual Book of ASTM Standards, Section 9 Rubber Volume 09.01, 2014 edition, publisher American Society of Testing and Materials, 100 Barr Harbor Drive, West Conshohocken, PA 19428-2959.

34. ASTM D3192-09. Annual Book of ASTM Standards, Section 9 Rubber Volume 09.01, 2014 edition, publisher American Society of Testing and Materials, 100 Barr Harbor Drive, West Conshohocken, PA 19428-2959.

35. ASTM D3191-10. Annual Book of ASTM Standards, Section 9 Rubber Volume 09.01, 2014 edition, publisher American Society of Testing and Materials, 100 Barr Harbor Drive, West Conshohocken, PA 19428-2959.

36. Hess WM. Characterization of dispersions. *Rubber Chem Technol* 1991;64:386–449.
37. Medalia AI. Electrical conductance in carbon black composites. *Rubber Chem Technol* 1986;59:432–454.
38. Coran AY, Donnet JB. The dispersion of carbon black in rubber. *Rubber Chem Technol* 1992;65:973–997.
39. Gerspacher M, O'Farrell CP. Carbon black, microdispersion and treadwear. *Rubber Plastic News* 1993;23(6):85–87.
40. ASTM D2663. Annual Book of ASTM Standards, Section 9 Rubber Volume 09.01, 2002, 429–439.
41. Putman JB. An improved method for measuring filler dispersion of uncured rubber. In: *ACS Rubber Division Fall Meeting*, Cleveland, OH, 2001, Paper No. 19, pp. 1–10.
42. Gerspacher M, O'Farrell CP, Yang HH. Better filler dispersion impacts essential compound characteristics. In: *ITEC'96 Conference Proceedings*, Akron, OH, pp. 43–47.
43. Wampler WA, Gerspacher M, Yang HH. Carbon black's role in compound curing behavior. *Rubber World* 1994;210(1):39–43.
44. Gerspacher M, Lansinger CM. Carbon black characterization and application to study of polymer/filler interaction. In: *ACS Rubber Division Spring 1988 Conference*, Dallas, TX, April 19–22, 1988, Paper No. 7, p. 23.
45. Gerspacher M, O'Farrell CP, Yang HH. Carbon black network responds to dynamic strains: Methods and results. *Elastomerics* 1990;122(11):23–30.
46. Medalia AI. Effect of carbon black on dynamic properties of rubber vulcanizates. *Rubber Chem Technol* 1978;51:437–523.
47. Gerspacher M. Dynamic viscoelastic properties of loaded elastomers. In: Donnet JB, Bansal RC, Wang MJ, eds., *Carbon Black Science and Technology*. New York: Marcel Dekker, 1993, pp. 377–387.
48. Payne AR, Whitaker RE. Low strain dynamic properties of filled rubbers. *Rubber Chem Technol* 1971;44:440–478.
49. Speck ME. Production of carbon black. U.S. Patent 3,494,740, 1970.
50. Speck ME. Production of carbon black. U.S. Patent 3,512,934, 1970.
51. Aboytes P, Iannicelli J. Antioxidant carbon black. U.S. Patent 3,323,932, 1967.
52. Riven D. Surface properties of carbon. *Rubber Chem Technol* 1971;43:307–343.
53. Bansal RC, Donnet JB. Surface groups on carbon blacks. In: Donnet J-P, Bansal RC, Wang M-J, eds., *Carbon Black*, 2nd edn. New York: Marcel Dekker, 1993, pp. 175–220.
54. Wang M-J, Wolff S. Surface energy of carbon black. In: Donnet J-P, Bansal RC, Wang M-J, eds., *Carbon Black*, 2nd edn. New York: Marcel Dekker, 1993, pp. 229–243.
55. Studebaker ML. The Chemistry of reinforcement l–Some reactions between carbon black and simple inorganic molecules. *Rubber Age* 1957;81:661–671.
56. Studebaker ML. The chemistry of reinforcement ll–Scorch in natural rubber stocks containing carbon black. *Rubber Age* 1957;81:837–842.
57. Studebaker ML. The chemistry of reinforcement lll–model systems containing carbon black and squalene. *Rubber Age* 1957;81:1027–1083.
58. Stöhr B, Boehm HP, Schlögl R. Enhancement of the catalytic activity of activated carbons in oxidation reactions by thermal treatment with ammonia or hydrogen cyanide and observation of a superoxide species as a possible intermediate. *Carbon* 1991;29:707–720.
59. Heller GL. Treatment of carbon black. U.S. Patent 3,216,843, 1965.
60. Johnson PH. Oxidation of carbon black. U.S. Patent 3,318,720, 1967.
61. Gunnell TJ. Nitric acid treatment of carbon black. U.S. Patent 3,336,148, 1967.
62. Johnson PH. Process of modifying furnace carbon black. U.S. Patent 3,353,980, 1967.
63. May CC. Treatment of powdered oil furnace carbon black. U.S. Patent 3,364,048, 1968.
64. Deery HJ. Oxidation treatment of carbon black. U.S. Patent 3,398,009, 1968.
65. Hagopian E. Process for aftertreating carbon black. U.S. Patent 3,495,999, 1970.

66. Henderson EW. Preoxidation and pelleting of carbon black. U.S. Patent 3,870,785, 1975.
67. Hunt HR. Oxidizing carbon black and use of it in elastomers. U.S. Patent 4,075,140, 1978.
68. Wilder CR. Carbon black and process for preparing same. U.S. Patent 4,518,434, 1985.
69. Suetsugu K, Maruyama E, Same K. Black coloring agent. U.S. Patent 3,992,218, 1976.
70. Morrison RT, Boyd RN. *Organic Chemistry*, 6th edn. Englewood Cliffs, NJ: Prentice-Hall, 1992, pp. 866–875.
71. Seymour RB, Carraher CE Jr. *Polymer Chemistry: An Introduction*. New York: Marcel Dekker, 1981, pp. 281–319.
72. Belmont JA, Amici RM, Galloway CP. Reaction of carbon black with diazonium salts, resultant carbon black products and their uses. U.S. Patent 5,900,029, 1999.
73. Belmont JA, Amici RM, Galloway CP. Reaction of carbon black with diazonium salts, resultant carbon black products and their uses. U.S. Patent 5,851,280, 1998.
74. Belmont JA, Amici RM, Galloway CP. Reaction of carbon black with diazonium salts, resultant carbon black products and their uses. U.S. Patent 6,042,643, 2000.
75. Lieberman MA, Lichtenberg AJ. *Principles of Plasma Discharges and Materials Processing*. New York: Wiley, 1994, pp. 191–216, 265–300.
76. Jordan ME. Method for producing carbon black. U.S. Patent 3,331,664, 1967.
77. Bjornson G. Plasma preparation of carbon black. U.S. Patent 3,409,403, 1968.
78. Johnson MM. Plasma treatment of carbon black. U.S. Patent 3,408,164, 1968.
79. Binder M, Mammone RJ. Method of treating carbon black and carbon black so treated. U.S. Patent 5,500,201, 1996.
80. Akovali G, Ulkem I. Some performance characteristics of plasma surface modified carbon black in the (SBR) matrix. *Polymer* 1999;40:7417–7422.
81. Vidal A, Riess G, Donnet JB. Process for grafting polymers on carbon black through free radical mechanism. U.S. Patent 4,014,844, 1977.
82. Aboytes P, Iannicelli J. Modified carbon blacks. U.S. Patent 3,335,020, 1967.
83. Mabry MA, Rumpf FH, Westveer SA, Morgan AC, Podobnik IZ, Chung B, Andrews MJ. Methods for producing elastomeric compositions. U.S. Patent 6,040,364, 2000.
84. Maruyama T, Ishikawa K, Kirino Y. A new carbon black to reduce rolling resistance. In: *ACS Rubber Division Spring Meeting*, Savannah, GA, 2002, Paper No. 38, pp. 1–11.
85. Tsubokawa N, Satoh T, Murota M, Sato S, Shimizu H. Grafting of hyperbranched poly(amidoamine) onto carbon black surfaces using dendrimer synthesis methodology. *Polym Adv Technol* 2001;12:596–602.
86. Wampler WA. Conducting polymer–carbon black composites. PhD dissertation, The University of Texas at Arlington, Arlington, TX, 1995.
87. Johnson PH. Carbon black process. U.S. Patent 3,206,285, 1965.
88. Hinson FA Jr. Production of carbon black. U.S. Patent 3,408,165, 1968.
89. Jordan ME, Burbine WG. Production of carbon black. U.S. Patent 3,383,175, 1968.
90. Otto WKF. Process for production of metal bearing carbon black. U.S. Patent 3,431,205, 1969.
91. Otto WKF. Magnetic carbon blacks. U.S. Patent 3,448,052, 1969.
91a. Probst N, Grivei E, Fockedey E. New ferromagnetic carbon based functional filler. *Kautsch Gummi Kunst* 2003;56:595–599.
92. Mahmud K, Wang MJ, Kutsovsky YE. Elastomeric compounds incorporating metal treated carbon blacks. U.S. Patent 6,150,453, 2000.
93. Wang M-J, Mahmud K. Elastomeric compounds incorporating metal treated carbon blacks. U.S. Patent 6,017,980, 2000.
94. Wagner MP. Reinforcing silicas and silicates. *Rubber Chem Technol* 1976;49:703–774.
95. Cruse RW, Hofstetter MH, Panzer LM, Pickwell RJ. Effect of polysulfidic silane sulfur on rolling resistance. *Rubber Plastics News* 1997;26(18):14–17.

96. Hunsche A, Goerl U, Mueller A, Knaack M, Goebel T. Investigations concerning the reaction silica/organosilane and organosilane/polymer. *Kautsch Gummi Kunst* 1997;50:881–889.
97. Wolff S, Gorl U. Carbon blacks modified with organosilicon compounds, method of their production and their use in rubber mixtures. U.S. Patent 5,159,009, 1992.
98. Murray LK. Wet-pelleting of carbon black. U.S. Patent 3,844,809, 1974.
99. Scott OT. Method of producing a carbon black silica pigment. U.S. Patent 4,211,578, 1980.
100. Mahmud K, Wang MJ, Francis RA. Elastomeric compounds incorporating silicon-treated carbon blacks and coupling agents. U.S. Patent 5,877,238, 1999.
101. Mahmud K, Wang MJ. Method of making a multi-phase aggregate using a multi-stage process. U.S. Patent 5,904,762, 1999.
102. Labauze G. Rubber composition based on carbon black having silica fixed to its surface and on diene polymer functionalized with alkoxysilane. U.S. Patent 5,977,238, 1999.
103. Kawazura T, Ishikawa K. Process for the production of surface-treated carbon black for the reinforcement of rubbers. U.S. Patent 6,020,068, 2000.
104. Mahmud K, Wang MJ, Method of making a multi-phase aggregate using a multi-stage process. U.S. Patent 6,057,387, 2000.
105. Mahmud K, Wang MJ. Method of making a multi-phase aggregate using a multi-stage process. U.S. Patent 6,364,944, 2002.
106. Kari A, Freund B, Vogel K. Carbon black and processes for manufacturing. U.S. Patent 5,859,120, 1999.
107. Brown TA, Wang MJ. Elastomer compositions with dual phase aggregates and pre-vulcanization modifier. U.S. Patent 6,172,154, 2001.
108. Wang MJ, Kutsovsky Y, Zhang P, Hehos G, Murphy LJ, Mahmud K. Using carbon-silica dual phase filler. *Kautsch Gummi Kunst* 2002;55:33–40.
109. ASTM D7724-11. Annual Book of ASTM Standards, Section 9 Rubber Volume 09.01, 2014 edition, publisher American Society of Testing and Materials, 100 Barr Harbor Drive, West Conshohocken, PA 19428-2959.
110. Herd C, Edwards C, Curtis J, Crossley S, Schomberg KC, Gross T, Steinhauser N, Kloppenberg H, Hardy D, Lucassen A. Use of surface treated carbon blacks in an elastomer to reduce compound hysteresis and tire rolling resistance and improve wet traction. U.S. Patent Application 20130046064, publication date February 21, 2013.
111. Jacobsson M, Cameron P, Neilsen J, Nikiel L, Wampler W. Improving hysteresis through filler modifications and smart compounding. *Rubber World* 247(5):22–26, February 2013.

7 Silica and Silanes

Anke Blume, Louis Gatti, Hans-Detlef Luginsland,
Dominik Maschke, Ralph Moser, J.C Nian,
Caren Röben and André Wehmeier

CONTENTS

I. Introduction 252
II. Silica 253
   A. General Considerations and Basic Information 253
   B. Essentials of Rubber Silica 255
   C. Characterization of Silica (Analytical Properties) 255
      1. Methods for Characterizing the Morphology of Silica 256
      2. Characterization of Surface Chemistry 261
      3. Chemical Bulk Analyses 264
   D. Process and Technology 265
      1. Production Process 265
      2. Influence of Process Parameters on Product Properties 265
      3. Dispersibility and Surface Activity 266
      4. Typical Process for Rubber Silica 266
   E. Product Overview and Future Trends 266
      1. Commercial Products 266
III. Silanes 267
   A. Basic Considerations 267
      1. Definitions: Primer, Adhesive, and Modifier 267
      2. Function of Silanes as Adhesion Promoters and Fields of
          Application 268
      3. Overview of Reactions of Silane Coupling Agents 268
   B. Essentials of Rubber Silanes 269
      1. General Structure of Rubber Silanes 269
      2. History of Rubber Silanes 269
      3. Types of Rubber Silanes and General Applications 269
   C. Process and Technology of Rubber Silanes 270
   D. Characterization of Silanes and Reactions 271
      1. Methods of Analysis 271
      2. Silica–Silane Coupling 273
      3. Silane–Rubber Coupling 274
   E. Product Overview and Applications 275

IV.   Silica and Silanes in Rubber 276
      A.   General Considerations 276
           1.   Carbon Black as Filler in Rubber Applications 277
           2.   White Fillers in Rubber Applications 277
      B.   Rubber Reinforcement 277
           1.   General Considerations 277
           2.   Types of Interactions 279
           3.   In-Rubber Performance 283
      C.   Future Trends 287
           1.   Tire Label and Silane Coupling Efficiency 287
           2.   Reduced Ethanol Emission 290
           3.   Improved Processing 290
V.   Applications of Silica and Silane.. 291
      A.   History... 291
      B.   Silica in Shoe Soles.. 292
      C.   Silica in Industrial Rubber Goods.. 295
           1.   Seals, Cables, Profiles, and Hoses.. 295
           2.   Rice Hulling Rollers... 299
           3.   Soft Rollers... 300
           4.   Conveyor Belts and Power Transmission Belts 301
           5.   Engine Mounts, Belts, and Air Springs....................................... 303
           6.   Boots and Bellows.. 303
           7.   Floor Coverings.. 303
           8.   Golf Balls 305
      D.   Silica in Tires 305
           1.   Passenger Car Tire Treads.. 305
           2.   Winter Tire Treads 313
           3.   Truck Tires 316
           4.   Earth Mover and Off-the-Road Tires 317
           5.   Solid Tires 318
           6.   Tire Body.. 319
           7.   Motorcycle Tires... 321
           8.   Bicycle Tires... 322
VI.   Summary 323
Acknowledgments... 324
References... 324

I.   INTRODUCTION

The invention of synthetic amorphous silica, in the late 1940s, also marked the birth of the development of new rubber compounds with many more options and improved performance features. Silica's progress from a simple filler used for producing colored rubber compound to an important physical performance additive for substantially improving rubber compounds, especially in tire applications, took technically about 20 years. But it did not take long to learn that silica-containing compounds achieve significantly better tensile strength and tear strength. Therefore, silica became an important physical reinforcement ingredient in all kinds of rubber goods in the 1950s and 1960s. Silica manufacturers started to develop and offer specific silica for different requirements.

After the first success with commercially available mercaptopropyltrimethoxysilane in the late 1960s, the polysulfidic silane Si 69®* from Evonik Industries AG (formerly Degussa AG) was introduced in 1972 and set the benchmark for significantly improved silica-containing compounds. Silanes turned silica into an active chemical reactant in the rubber compound. Additional 20 years were necessary to commercially develop the silica and silane success story for the high-volume tire application that began in Europe.

In the early stages, the use of silica–silane compounds grew slowly. The consumption of silanes in 1990 was still below 3000 metric tons per year. The introduction of the "Green Tire" by Michelin, with a tread compound based on silane from Evonik and silica from Solvay S.A. (formerly Rhodia S.A. or Rhône Poulenc S.A.), was a real breakthrough in tire innovation marking the beginning of a new area for the entire industry.

Silica had become an active, high-performance ingredient for rubber compounds. Only homogeneously and well-dispersed silica particles are able to provide improved reinforcement in the rubber compound. As a consequence and in order to meet market demands, all major silica manufacturers were developing highly dispersible (HD) silica.

With the importance of silanes in rubber steadily increasing, professional research and development became even more important in order to respond to new requirements and set new trends, to investigate and discover the complex chemistry that is involved with silanes in silica-filled rubber compounds. Therefore, this chapter emphasizes the chemistry in rubber compounding based on Evonik long dedication and experience in silica and silane research and development.

II. SILICA

A. General Considerations and Basic Information

A glance at the occurrence of the homologous chemical elements silicon and carbon in a variety of locations (Table 7.1) reveals that although life is based on carbon, the accessible outer layer of the earth's lithosphere is primarily of a siliceous nature [1,2]. Silicon as a carrier of "inorganic life" occurs in nature almost exclusively in the form of crystalline solids in approximately 800 different siliceous minerals. Even in the evolutionary history of humans, there are numerous indications of the omnipresence of silicates in our natural habitat. Siliceous minerals were utilized either by processing natural silicate deposits (e.g., clay, kaoline, chalk, or talc) or by means of chemical conversion (silica, silicones or ceramics). The properties of pure elemental silicon are now of pivotal importance in the manufacture of integrated switching circuits and therefore also form the basis for the age of electronics.

Evonik began dealing with siliceous chemicals because of the company's involvement in the use of carbon black (CB). In addition to their classic application as black

---

* Si 69® is a registered trade name of Evonik Industries AG.

TABLE 7.1

Occurrence of the Elements Silicon and Carbon

Location	Occurrence of Silicon	Occurrence of Carbon	Most Common Elements	
Outer space	0.003%	0.005%	H+He	99%
Earth's crust	27.7%	0.1%	O	51%
Human organism	0.001%	9.5%	H+O	89%

Source: Greenwood, N.N. and Earnshaw, A., *Chemistry of the Elements*, Pergamon Press, Oxford, U.K., 1984.

pigment, CBs were increasingly used as active fillers in the rubber industry, in particular in the manufacture of automobile tires. Because the starting material for CB at the time—natural gas—was not available in sufficient quantities in Germany, a substitute was sought that could be prepared from indigenous starting materials. The concept of "white CB" was born, and research into the manufacture of siliceous fillers commenced. The idea to apply the manufacturing method for CB to volatile silicon compounds (pyrogenic silica, AEROSIL®) also originates from that time (one of the main inventor was Evonik chemist Harry Kloepfer [3]):

> Because the significance of a white active filler for the rubber industry is very evident, I performed several experiments on the manufacture of a "white carbon black" parallel to the first active carbon black experiments. Even during the initial experiments, I endeavored to imitate the manufacturing conditions for gas black. The important aspect in the manufacture of gas black seemed to me to be the direct separation of ultra-fine carbon particles from the gas phase, i.e., the production of aerosols.

As described earlier, fine silica can be produced by a pyrogenic reaction or by precipitation. Silicates are manufactured by a precipitation process. It is an interesting task to precipitate silica and silicates from aqueous media. The methods are so variable that various products customized for the respective applications are attainable. Manufacturers and users of fine silica and silicates require analytical characteristic values in order to specify and compare products [4]. Owing to the differences in chemical composition of silica, silicates, and other natural substances such as chalk or siliceous minerals, the characterization usually commences with physical/chemical analysis, which provides important information on the product composition, the main constituents, and, of particular importance, the secondary constituents. One significant constituent of precipitated silica and silicates is water, which occurs in various quantities either chemically bonded or adsorbed. The specific surface area, mean particle size, particle-size distribution (PaSD), pH, and absorption of certain oils all play significant roles in the final products. These somewhat "classical" methods of analysis and a number of new methods for characterizing silica and silicates will be dealt with in greater detail in Section II.C.

The versatility of fine silica and silicates is attributable to a variety of fundamental properties. In chemical terms, they are largely inactive and exhibit high thermostability; they influence the viscosity of liquids and act as anti-sedimentation agents; they are capable of producing matting effects and of preventing adhesion between foils and the caking of powders. Because of their high absorbing power, they also serve as carriers for feed and pesticides. Targeted organic modification of the silica surface converts "water-friendly," or hydrophilic, silica into "water-repelling," or hydrophobic, products. The hydrophobic properties have proven to be very useful for certain applications, for example, in silicone rubber or for defoaming liquids. The oldest application of "white" silica and silicates, their use in shoe soles and technical rubber articles, and the use of silica in modern automobile tires will be discussed later.

B. ESSENTIALS OF RUBBER SILICA

The oldest use for fine silica and silicates is in shoe soles. The high demands with regard to wear can be met with ease by such reinforced vulcanized materials. The great advantages of light-colored reinforcements vis-à-vis CB are the possibility of satisfying practical, not to mention fashion-related, requirements, whether they be that shoe soles be nonmarking, transparent, or of a particular color or whether it be necessary to stamp cable sheaths with colored lettering.

If the fineness and therefore the high specific surface of a substance are considered to be the main contributing factors to the reinforcing effect in rubber, then fine silica and silicates must act similarly to rubber black. This is indeed the case, but clear distinctions must be made [5]. In simple terms, the cross-linking of rubber during vulcanization proceeds in a different manner in the presence of CB than in the presence of silica or silicates. The reasons for this are adsorption of cure ingredients, the lack of interaction between silica and the polymer, and the immiscibility of hydrophilic silica with the hydrophobic rubber. Consequently, silica is used mostly in conjunction with bifunctional organosilanes, such as Si 69®. Organosilanes undergo chemical bonding to the silica surface and offer a functional group that binds to the rubber, thereby providing the potential for cross-linking to the rubber. The behavior of silica and organosilanes will be described in detail in Section III.

C. CHARACTERIZATION OF SILICA (ANALYTICAL PROPERTIES)

Unlike other chemicals, silica is usually not only characterized by its chemical composition–it is mostly SiO_2–but by its physical properties, such as the ratio between the "inner" and "outer" surfaces, the size and shape of pores, the absorption capacity, the surface roughness, the primary particle size, the formation of aggregates from these primary particles with the development of siloxane bonds, and the combination of aggregates to form agglomerates held together by hydrogen bondings. Other methods attempt to characterize the *surface chemistry*; that is, they describe the number of silanol groups and their arrangement, surroundings, and reactivity and the chemical composition and degradation behavior of the silica surface. Furthermore, there are a number of methods that can collectively be termed bulk chemical analysis.

This review attempts to summarize physical and chemical analytical methods for characterizing silica [6–8] but does not lay claim to completeness. The "traditional" characterization methods are only briefly described because an comprehensive overview is given in the relevant literature [9–16, 24–26]. The main emphasis here is on the new methods—those that have been specially developed or adapted for the purpose of silica analysis. Moreover, some analytical methods are more of scientific interest and are therefore not described in detail either. An extensive bibliography to each method is provided for more detailed information.

1.   Methods for Characterizing the Morphology of Silica

Specific surface. The specific surface of a silica is generally determined by using the Brunauer–Emmett–Teller (BET) adsorption method [9] or a modification thereof [10,11]. For measurement, the sample is cooled to the temperature of liquid nitrogen. At low temperatures, nitrogen is adsorbed on the silica surface. The quantity of adsorbed gas is a measure of the size of the surface. When performed under defined conditions, the BET method yields perfectly reproducible results. The BET method always provides the sum of the so-called outer geometrical surface and the inner surface, that is, the surface within the porous silica structure. If the value obtained for the surface by using the BET method is compared to that from electron microscopic images, which indicate only the outer surface, then it is at least possible to estimate the ratio of outer and inner surfaces. Evaluation of other sections of the BET isotherms provides additional information. The so-called C value calculated from the BET isotherms gives a qualitative indication of the magnitude of the interaction between the surface of the silica and the adsorbed material and hence of the chemical reactivity of the surface [12]. If it is ensured not only that the surface of a sample is covered with nitrogen but also that the pores are filled, then the distribution of mesopores (pores between 2 and 30 nm in size) can also be determined by using the Barrett–Joyner–Halenda method [13].

Hexadecyl-trimethyl-ammonium bromide (CTAB) surface area. This method is known from the area of CB technology and is also used in the case of silica. It is based on the adsorption of surface-active molecules from aqueous solutions. The adsorbed molecule is hexadecyl-trimethyl-ammonium bromide CTAB and was first used by Saleeb and Kitchener [14]. The preferred adsorption site for these large CTAB molecules is the outer, geometrical surface, which correlates quite well with the surface area accessible to the rubber [15,16]. Comparison with the BET surface, which is the sum of the outer and inner surfaces of a filler, provides an indication, with a certain margin of error, of the ratio between the inner surface and the total surface. This principle is pictured in Figure 7.1.

Sears determination of the specific surface. The Sears number is a measure of the number of silanol groups on the surface and therefore a measure of the specific surface of a silica [17,18]. The Sears number is equivalent to the quantity of 0.1 N NaOH required to titrate a suspension of silica from pH 6 to pH 9. The acidic silanol groups on the silica surface react with NaOH. The Sears number gives an indication of the number of reactive centers on the surface of a silica.

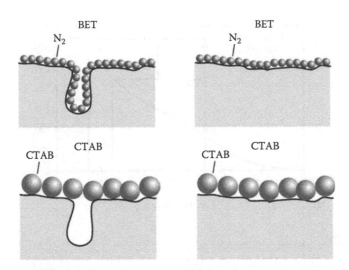

FIGURE 7.1   Difference of nitrogen and CTAB adsorption on porous and nonporous silica.

Pore volume and pore size distribution. The term "pore volume" and a specific evaluation method have not been described explicitly [19,20]. The pore volume of synthetic silica can be understood as (1) the surface roughness, (2) the micro- or sub-micropore volume within the particles or aggregates, or (3) the void volume. The most common method of determining the pore volume in a silica is mercury porosimetry [21,22]. The measured quantity in mercury porosimetry is the pressure, p, required to force the mercury into the pores of a sample. The necessary pressure is inversely proportional to the pore diameter. If the volume of mercury at this pressure is known, then the pore volume can be calculated. Comparison of the measured curves of different silica reveals distinct differences in intrusion and is indicative of very different structures of the measured products.

Estimation of intra-aggregate and interaggregate structure [23]. The sharp step at the end of the intrusion curve $V = f(R)$ (Figure 7.2a) corresponds to the intrusion of mercury inside the pores originating from the open shape of silica aggregates. The intrusion volume in this range of pore sizes is characteristic of the intra-aggregate structure of silica aggregates. This intra-aggregate structure can be described quantitatively by use of the structure index (IS), which is the mercury intrusion volume measured between R_{min} and R_{max}, pore radius values corresponding, respectively, to the beginning and the end of the step.

R_{min} and R_{max} can be easily determined by considering the derivative curve $dV/dR = f(R)$ (Figure 7.2b). R_{max} is chosen to match the following condition on the slope:

$$\left| \frac{E}{dR}\left(\frac{dV}{dR}\right)\bigg|_{E_{max}} \right| \leq 0004 \times \left(\frac{dV}{dR}\right)_{max}$$

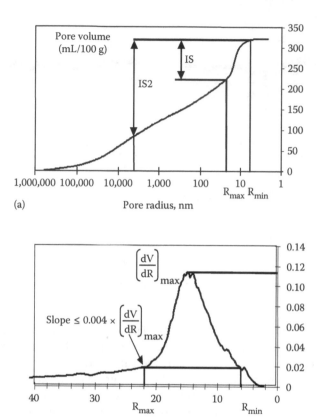

(a)   Pore radius, nm

(b)   Pore radius, nm

FIGURE 7.2 (a) Intrusion curve and (b) determination of R_{min} and R_{max} for the calculation of IS.

R_{min} is chosen to match the following condition:

$$\left(\frac{dV}{dR}\right)\bigg|_{R_{min}} = \left(\frac{dV}{dR}\right)\bigg|_{R_{max}}$$

The structure index IS depends only on the size and shape of individual aggregates.

The global structure of a silica, including both intra-aggregate and interaggregate pore volumes, can be estimated by using a second structure index called IS2, defined as the total intrusion volume corresponding to pore radius values smaller than 4000 nm. IS2 depends on both aggregate shape and aggregate organization resulting from the drying process.

Oil absorption number (formerly known as dibutylphtalate (DBP) number). The assessment of the liquid-absorbing capacity of synthetic silica and silicates may involve the absorption of dioctyladipate (DOA is used, since DBP is considered a "Carcinogenic, Mutagenic or Toxic for Reproduction" (CMR) substance) [24–26].

This largely automated measurement technique provides an indication of the total volume of liquid that can be absorbed by a silica sample. The magnitude of the DOA number gives an initial indication of the interaggregate structure and the processing and dispersion properties of a silica.

Void volume. The void volume [27], that is, the silica structure as a function of the pressure, is considerably more reliable than the DOA number. In the void volume method, the sample is subjected to a defined coarse crushing, and then a specified quantity of the sample is transferred to a cylindrical glass chamber with a volume scale. The chamber is sealed by means of a moving piston. For the measurement, a constant pressure is applied to the piston until the volume of the sample in the chamber remains constant. Then the measured value is read off. Subsequently, the pressure exerted on the sample is increased again. Figure 7.3 shows the decrease in structure (or the void volume) as the pressure increases.

The conventional ULTRASIL® VN 3 GR is compared to the HD-silica Zeosil®* 1165 MP and ULTRASIL® 7000 GR with the similar CTAB surface area. It can be seen that the HD silica have a higher void volume even at the start of the measurement. As the pressure is increased, the structure decreases yet always remains well above the level of the respective reference. In other words, the higher structure remains intact and can be penetrated more easily by polymers during mixing, which is indicative of the more advantageous dispersion behavior of the silica.

Microscopic methods. The only method allowing direct insight into the dimensions of interest in the case of silica is electron microscopy, which provides information on the size of primary particles and aggregates or agglomerates and, with certain limitations, on the PaSD of an examined sample [28–30]. Here, the advantage of electron microscopy–being able to directly observe a species–is also the disadvantage. The statistics are not very good. "Electron microscopic surfaces" can be

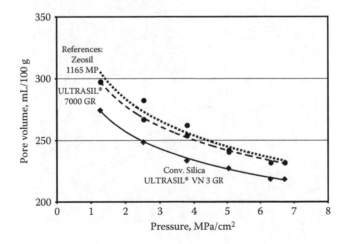

FIGURE 7.3 Decrease in structure of silica as pressure increases.

* Zeosil (Solvay SA) and ULTRASIL (Evonik Industries AG) are registered trade names.

calculated from the various PaSDs, and these can be compared with those from BET measurements or other investigations.

TEM. TEM [30] works in much the same manner as light microscopy: electrons are passed through a thin object and, following their interaction with the prepared sample, are used to produce an image. However, the resolution exceeds that of a light microscope by a factor of 1000; for TEM, resolution is 0.2–0.3 nm (for a light microscope, it is ~200 nm). TEM images with high resolution provide valuable information on the composition or structure of different silica samples. With a suitable imaging technique, even the crystalline short-range order of silica and silicates can be detected. The structure of the silica can also be recognized, as the manner in which the primary particles are connected to each other (Figure 7.4).

During the last decade, electron tomography has been transferred from the field of medical and biological studies into the characterization of finely divided materials in three dimensions at enhanced electron optical resolution [28].

For example, in order to characterize structural parameters of silica, an aggregate that is deposited on a carbon film, which is supported by TEM sample holder, is tilted relative to the incident beam in, for example, 1° steps, and a series of about 150 TEM images at changing tilt angle is recorded. This set of images is arranged to a film to visualize the 3D shape of an aggregate. Recorder, composer, and visualizer software is used to generate an electronic reconstruction of the object. This allows to determine the angle dependence of aggregate parameters such as the minimum and maximum diameter, the mean projected area, the circumference and convex circumference as a measure of structural complexity, and shape parameters.

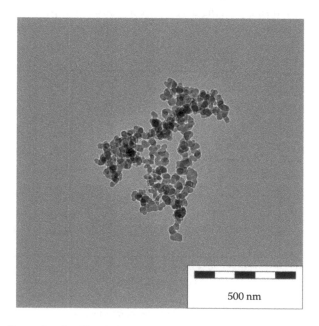

500 nm

FIGURE 7.4 Example of a silica aggregate.

Statistical evaluations of the size and structure of products can be enhanced from two dimensions—the projected TEM image of an aggregate on the detector plane—into the third dimension. Furthermore, also the dispersion of reinforcing fillers in polymers can be visualized by large-scale imaging on cryoultramicrotome sections, for example, of rubber [29].

Scanning electron microscopy (SEM). The scanning electron microscope [30] is not actually a microscope, in the sense that it uses electromagnetic lenses in order to magnify images, similar to the case with light optics (this comparison also applies to the transmission electron microscope). It merits the name "electron microscope" only because it produces a strongly magnified image with the help of electrons. As is also the case with TEM, SEM initially produces beams of electrons from an electron source. The extremely sharply focused beam of electrons produces very good resolution and depth of focus. It is this great depth of focus in particular that makes SEM superior to TEM for certain applications.

PaSD. Besides various scattering methods, one method of particular interest is the so-called disc centrifuge method (CPS Instruments Europe). The ultrasonically dispersed sample is injected into a fast-rotating disc, containing an aqueous density gradient. Separation of the different particle sizes occurs due to gravimetric acceleration inside the disc. The silica is detected optically and based on mathematic models, the particle size can be computed [31].

2. Characterization of Surface Chemistry

Chemical reactions of silanol groups. In the literature, a variety of methods are described that are suitable for determining silanol groups and their chemical reactivity on the surfaces of silica [32–34]. All of these methods are based on applying the experience gathered in small-molecule and low-molecular-weight chemistry to surface chemistry. However, the fact that this analogy is flawed is demonstrated by the following simple consideration. On the surfaces of solids, particularly in the interior of micropores, spatial inhibition, other equilibrium conditions, and other reaction possibilities may prevail. Chemical reactions are nonetheless an important tool for characterizing solid surfaces. The reactions generally yield easily reproducible characteristic values allowing different products to be compared. A small selection of the reactions described in the literature will be dealt with briefly here.

In the determination of silanol groups with lithium aluminum hydride ($LiAlH_4$), a silica sample is first degassed in a vacuum and then allowed to react with $LiAlH_4$ at room temperature, and the resultant hydrogen is determined volumetrically [35,36]. A further method is the reaction of a sample with an alkyl lithium or alkyl magnesium reagent [37] followed by volumetric determination of the resulting alkane. Further reactions may be performed with alcohols [38]; chlorosilanes [39,40]; hexamethyldisilazane, boron trichloride, aluminium chloride [41]; or boroethane [42–44].

NMR spectroscopy. With ^{29}Si NMR [19,45–48] examinations of solids, it is possible to detect different surroundings of the silicon atom on the basis of the oxygen atoms and hydroxy groups in the silica sample (Scheme 7.1). The ratios of the detected signal intensities correspond to the proportions of the various Si surroundings in

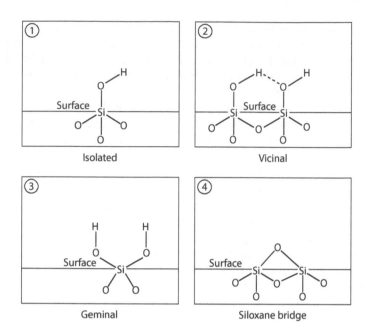

SCHEME 7.1 Surroundings of Si in silica: geminal, isolated, vicinal, and siloxane bridges.

the sample. With the solid nuclear magnetic resonance (NMR) method, it is possible to distinguish between three main groups around the silicon atom in silica:

1. Isolated, terminal SiOH (with a chemical shift of approximately −100 ppm)
2. Vicinal, adjacent mutually hydrogen-bonded SiOH groups (with a chemical shift of approximately −110 ppm)
3. Geminal SiOH (with a chemical shift of approximately −90 ppm)
4. Siloxane bridges (bulk) (with a chemical shift of approximately −110 ppm)

The relative content of isolated or geminal silanol groups and siloxane bridges is quantifiable, and the different silanol groups are assigned different reactivities. For example, with respect to organosilanes as coupling agents [49], geminal and isolated groups are considered to be the most reactive. It was shown in further studies, that both, the germinal and the isolated silanol groups react completely during the hydrophobation process [50].

Infrared (IR) spectroscopy. IR spectroscopy is another important method to differentiate between various silanol groups [51]. An overview of the detectable groups is given in Table 7.2 [52].

Variations of the IR technique contribute the further refinement or elucidation of the chemical structure on the surface of silica samples. In the near IR, for instance, the SiOH groups can be quantified by means of a combined oscillation band. With diffuse reflectance IR Fourier transform spectroscopy (DRIFT), an insight can be obtained how the SiOH groups change as the degree of chemical modification increases. Details on the implementation of other spectroscopic methods can be found in the

TABLE 7.2
Silanol Groups Detectable by IR Spectroscopy

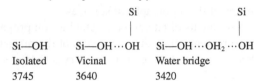

	Si—OH	Si—OH···OH	Si—OH···OH$_2$···OH
	Isolated	Vicinal	Water bridge
Position of IR band (cm^{-1})	3745	3640	3420

Silica powder, measurement in transmission

⇓ E (e.g., 3745/1870 cm^{-1})

Si—O combination band

Relative amount of different silanol groups

Source: Janik, M., Internal Report 05701, Degussa, AG, 2001.

literature [52–58]. These include x-ray photoelectron spectroscopy [59], incoherent neutron scattering, and inverse gas chromatography [60,61]. The latter can also be used to derive the morphology index [62]. The use of some of these methods for silica analysis is still in the incipient stage and thus will not be dealt with here in detail.

A special FT-IR spectrometer can be used for the acquisition of silica spectra recorded in the 400–5500 cm^{-1} range. The water that is adsorbed on the silica surface (under atmospheric condition) can be characterized by a δH_2O band near 1630 cm^{-1} and by a $(\nu+\delta)$ H$_2$O combination band at about 5260 cm^{-1} [63]. The overall concentration of SiOH groups can be calculated by using the area of the band at ~4570 cm^{-1} assigned to the $(\nu+\delta)$ SiOH band [63].

Thermoanalytical methods. Thermal analysis [64,65] is the term given to a group of methods that measure a physical property of a substance (and/or its reaction products) as a function of temperature or time while subjecting the substance to a controlled temperature program. With differential thermal analysis (DTA), it is possible to monitor the changes in enthalpy of a sample during the course of a temperature program. The precondition here is that the corresponding processes (e.g. chemical reactions, phase changes) are sufficiently short to allow a thermal effect to be observed. DTA is often combined with thermogravimetry (TG), which records the weight loss of the sample as a function of the temperature. From the mass loss, information can be obtained on resulting products and on the possible course of degradation reactions [66]. In the case of silica, rational information can be obtained only by means of a combined DTA–TG measurement, because the thermal effects observed in DTA are generally negligible on their own [67]. Although it is difficult to draw conclusions regarding the structure of a silica surface from the combination of DTA and TG alone, the results obtained can nonetheless be compared to those from other investigations. From DTA–TG measurements, it is possible to calculate surfaces, which can be compared with BET surfaces, for example. An indication is obtained as to the ratio between the "outer" and "inner" surfaces, and it is also possible to distinguish between different silanol groups, as is also the case with IR and NMR.

Atomic force microscopy (AFM). Since the beginning of the 1990s, AFM [68–71] has been used to characterize surfaces of amorphous and crystalline, synthetic and biological products including liquid crystals and films. AFM combines several analytical methods in one instrument:

1. Images of the topography of surfaces
2. Measurements of lateral forces of adhesion properties
3. Measurements of modulation forces of the material's "hardness"
4. Measurements of force–distance curves to characterize the "hardness" and "adhesion"

AFM can be used to characterize the microstructure of CB and silica surfaces. AFM images allow the analysis of filler dispersion in the rubber matrix. Depending on the resolution of the images, filler aggregates and agglomerates in the polymer matrix can be identified.

Small-angle scattering (SAS). The method of SAS is a well-known tool to characterize the structure of fine particles and has been used to investigate fillers for about 60 years [72,73]. Because precipitated silica behaves in most cases as fractal scatterer, the scattering of these materials can be described with fractal structural concepts. In a typical scattering experiment, it is possible to estimate primary particle size and distribution, mass fractal dimension, and aggregate size; all these parameters depend on the type of silica investigated. For precipitated silica samples, typical mass fractal dimensions of $d_m = 1.9 \pm 0.2$ can be found. A huge advantage of the SAS method is the examination of silica structure when the material is dispersed in various media, for example, in rubber [74,75,76].

3. Chemical Bulk Analyses

X-ray diffraction. Synthetic silicas and silicates are amorphous solids. That is, unlike crystalline solids, they do not possess an infinite 3D long-range order. Consequently, use of the classic x-ray diffraction method is not possible. Silicas, however, like glass, do have areas of short-range order that can be determined by appropriate evaluation of the diffuse x-ray diffraction bands. Silica from a variety of manufacturing processes differs from one another in terms of their x-ray diffraction bands. When the sample is tempered at temperatures as low as 200°C, changes in the short-range order can be detected by using x-ray diffraction [19].

Loss on drying and on ignition. Loss on drying [77] and loss on ignition [78] are significant characteristic parameters that can be used to characterize the differences between synthetic silica. Silicas prepared by means of precipitation exhibit a loss on ignition of more than 3% (typical values are in the region of 5%), provided they have not received special aftertreatment. In the case of pyrogenic silica, the loss on ignition is less than 3%.

Electrokinetic measurements. Several electrokinetic methods lend themselves to the determination of the surface activity of silica. The majority are based on measurement of the zeta (electrokinetic) potential. These include electrophoresis, flow potential, and electroacoustic measurements such as measurement of the ultrasound vibration potential or electrokinetic sonic amplitude. In 1991, a working group headed by Grundke [79] established the following correlation for silica partly modified with organosilanes: the lower the SiOH concentration on the silica surface, the lower the hydrophilic nature of the silica, the lower the surface activity, and the higher the zeta potential in potassium chloride solution. This observation is accompanied by an upward shift in the isoelectric point (point of zero charge). Because the isoelectric point is a measure of the acidity of a surface, this signifies that the modified surface is of lower acidity than the unmodified surface, because some of the SiOH groups have reacted with the organosilanes.

D.   PROCESS AND TECHNOLOGY

1.   Production Process

The various steps of manufacturing precipitated silica are displayed schematically in Scheme 7.2. In the first step (precipitation), the raw materials consisting of water glass (sodium silicate solution) and a mineral acid (normally sulfuric acid is used) are dosed into a stirred vessel containing water. In many cases, once a defined pH value has been set, the components are fed continuously to the reactor, this process taking place simultaneously over a certain time interval. Another possibility is to first supply a particular quantity of water glass and initially dose just the sulfuric acid. Normally, this is followed by a second stage in which water glass and sulfuric acid are added simultaneously under defined reaction conditions.

2.   Influence of Process Parameters on Product Properties

During the reaction time, primary particles are first formed in the reactor; later these particles react with each other, accompanied by dehydration, to form aggregates. Within the aggregates, the primary particles are linked together via siloxane bonds. During this process, the aggregates are deposited to form larger units, or agglomerates. In these agglomerates, the aggregates are held together by hydrogen bonding or van der Waals interactions that are considerably weaker than siloxane bonds. A state of equilibrium, which is dependent on the process conditions and can be easily influenced, is reached between the aggregates and agglomerates (Scheme 7.3). After that, the obtained suspension is filtered and the filter cake washed. It can then be resuspended and spray-dried or fed directly to a short-term drying process. Depending on the drying technology, the product can be optionally first milled and then granulated or granulated directly to convert it into a low-dust form.

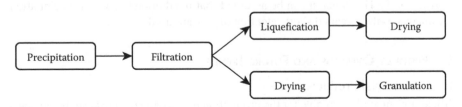

SCHEME 7.2   Schematic representation of the silica manufacturing process.

Precipitation:

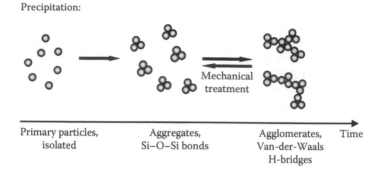

Primary particles,	Aggregates,	Agglomerates,	Time
isolated	Si–O–Si bonds	Van-der-Waals
		H-bridges

SCHEME 7.3   Particle formation during precipitation.

3.   Dispersibility and Surface Activity

An easy incorporation of the silica into the rubber mixture and its good dispersion are crucial because they have a considerable effect on the processing costs and on the rubber product's performance. By selecting appropriate process parameters, it is possible to influence the surface properties and therefore the reactivity of the silica toward the organosilane. The silanol groups on the silica surface should be present in a sufficient amount and be accessible for reaction with organosilanes to ensure a quantitative coupling (see Section III.A). Furthermore, the silane should not be adsorbed in silica pores, which are not accessible for the rubber.

4.   Typical Process for Rubber Silica

At present, a variety of manufacturers supply products that can be divided into three groups:

1. Products that can be described as first-generation, standard, or conventional silica.
2. Products belonging to the second generation of silica. Such products are termed easily dispersible silica or semi-HD silica.
3. The third and latest generation of silica comprises a product group charac- terized by excellent dispersion. These products are described as HD silica.

Depending on the process parameters and on the techniques used for manufacture, silicas are obtained that can be assigned to one of these three groups. With regard to group 3, the HD silica, it can be assumed that in addition to special precipitation parameters, only so-called short-retention dryers are used.

E.   PRODUCT OVERVIEW AND FUTURE TRENDS

1.   Commercial Products

Silicas are currently available for the production of tires and technical rubber articles from four main global manufacturers [80–83] and increasingly from local manufac- turers, especially in Asia.

In recent years, the developments in the area of rubber silica aimed to expand the application of these types of silica and also strove to improve the in-rubber performance toward higher abrasion resistance. Here are some current development targets:

- Expanding the application of precipitated silica and tailor-made silanes for *truck tires*. These types of tires use natural rubber (NR) instead of synthetic rubber types used commonly in passenger car tires.
- Usage of silica in *body compounds* of the tire. Common "rubber" silica is used mainly in the tread. The development needs new silica with low specific surface area.
- Development of silica that have a *tailor-made* PaSD. With these new types, it should be possible to improve the reinforcing behavior or adjust the compounding behavior to a new level.
- Application of silica with a *high specific surface area*. With these new types of silica, it should be possible to increase the in-rubber reinforcement.
- Improvement of the *microdispersion* of silica. Rubber compounds with an improved microdispersion are expected to have a superior abrasion behavior as compared to HD silica.
- Implementation of *"standard" grades*. Driven by the globalization of the tire industry, it is essential to standardize silica grades on a global basis.

III. SILANES

A. BASIC CONSIDERATIONS

Inorganic materials based on silicon or metals and the world of organic substances founded on carbon chemistry are as incompatible as oil and vinegar. To link these antipoles, the main task consists obviously in joining the two, with the silicon and the carbon in a single molecule. Silanes fulfill this requirement and are therefore predestined as adhesion promoters in composite materials. The best reference on the structure, chemistry, application, and history of silanes as coupling agents is probably the one of Plueddemann [84], who worked 35 years in the research department of Dow Corning, published numerous papers, and filed more than 90 patents. The most commonly used bifunctional silane coupling agents are based on the two structures I and II.

$$(RO)_3Si \diagdown \diagup \diagdown \qquad (RO)_3Si \diagup \diagdown \diagup \diagdown X$$

$$I \qquad\qquad II$$

$$R = CH_3, C_2H_5; X = \text{functional group}$$

1. Definitions: Primer, Adhesive, and Modifier

For the bonding of a polymer to an inorganic substrate, several kinds of adhesion promoters have to be used. Depending on the thickness and composition of the bonding layer (interface), the adhesion promoter may be defined as (1) a primer, (2) an adhesive, or (3) a finish or modifier. A *primer* mainly consists of a lubricant, a binder, and a coupling agent capable of bonding to the substrate.

The primer is applied from solution and forms a film with a thickness of 0.1–10 μm and a certain mechanical strength. The primer may be used for the direct adhesion of the matrix or as a preparation for a following top coating able to provide the coupling to polymer. Hydrolysable silanes are commonly used as coupling agents that condense to a polysiloxane structure on the surface.

An *adhesive* is a gap-filling composite consisting of, for example, polymers or resins, fillers, and an adhesion promoter and in some cases a silane.

In contrast to these adhesion formulations that provide a uniform and rather thick coverage of the surface, a *finish* or *surface modifier* forms a very thin layer, theoretically a monolayer. However, most of the hydrolysable silanes result in a several-monolayer-thick modification of the surface. In this regard, the coupling agents used for the silanization of a siliceous filler in rubber applications should be defined as surface modifiers.

2. Function of Silanes as Adhesion Promoters and Fields of Application

Bifunctional silanes can be used to chemically link an organic material to an inorganic substrate. Historically, one of the first main applications was the bonding of glass fibers to thermosetting and thermoplastic resins to increase the reinforcement, especially at high humidity. Depending on the type of polymer system, the functional group X can consist of an amino, glycidoxy, methacryloxy, or chloro group. The reinforcement of thermoplastics and rubbers by silane-modified siliceous fillers is another "big-volume" application. For this application, sulfur-functional silanes or vinylsilanes are often used. Silanes with an amino, epoxy, or mercapto group promote the improvement of the adhesion of sealants to an inorganic surface. Apart from these major applications, several others are summarized by Plueddemann [84] and Panster and coworkers [85]. A new focus of research with increasing significance is the investigation of the use of solvent- or waterborne silane formulations for metal adhesion, whose importance may also increase in the rubber industry [86]

3. Overview of Reactions of Silane Coupling Agents

In general, a silane adhesion promoter consists of a hydrolysable group such as trimethoxy or triethoxysilyl group that provides the coupling of the silane to the inorganic substrate (the use of the more reactive acetoxysilyl group is unusual for rubber applications). Most commonly used rubber silanes are based on the triethoxysilanes since the released ethanol is nontoxic. In the case of siliceous materials bearing silanol groups on the surface, Si–O–Si bonds are formed and alcohol is released. In the presence of moisture, an intermolecular condensation reaction between neighboring silanes is also possible. This condensation reaction can lead to the formation of multilayers and stabilization of the silane layer on the surface.

The organofunctional group X is responsible for the coupling reaction with the polymer matrix. To achieve optimal bonding to the matrix under cure conditions, the functional group has to be selected very carefully with regard to the chemical structure and reactivity of the polymer, the polymer precursor, or the resin [84]. Aminosilanes may be used as adhesion promoters in several composites such as phenolic, melamine, or furane resin, epoxy composites, and polyurethanes as well as coupling agents for polyesters and polyamides. Glycidoxy silanes and methacrylate silanes are also widely used for composite materials based on epoxy and acrylic resins, respectively. For polyolefins, vinyl- and methacryloxy-functional silanes are common, and for the coupling in sulfur-cross-linked elastomers, (blocked) mercaptosilanes and di- and polysulfide silanes are most suitable.

B. ESSENTIALS OF RUBBER SILANES

1. General Structure of Rubber Silanes

As mentioned earlier, for the coupling of siliceous fillers to unsaturated elastomers, silanes with one hydrolysable moiety to couple with the filler and one functional group to react with the rubber are optimal (structures I and II). The silica coupling is most often achieved by a trialkoxysilyl group, preferably triethoxysilyl moiety, reacting with the silanol groups on the filler surface during mixing. In the case of sulfur-cured compounds, sulfur-functional silanes are recommended, because they react with the rubber in the allyl position to the double bond during the vulcanization process. In peroxide-cured compounds, unsaturated silanes such as vinylsilanes perform best. With exception of the vinylsilane, the two functional groups in the silane are linked by a hydrocarbon spacer, preferably a propylene group, which also contributes to the hydrophobation of the polar silica surface.

2. History of Rubber Silanes

After the successful introduction of bifunctional silanes for glass fiber–reinforced composites in the early 1960s, the use of silanes in silica-filled rubber compounds started in the late 1960s. The first commercially available coupling agents were the highly reactive 3-mercaptopropyltrimethoxysilane for sulfur-cured compounds [87] and 3-methacryloxypropyltrimethoxysilane and vinyltrimethoxysilane for peroxide-cross-linked rubber compounds [88,89]. With the addition of these silanes, a remarkable increase in the reinforcement of white-filled compounds was achieved. The next major development step was the launch of the polysulfide silane Si 69® by Evonik in 1972; it showed good reinforcement properties and surmounted the short scorch behavior of the mercaptosilane [90,91]. The introduction of fuel-efficient tires based on precipitated silica replacing CB in tread compounds gave a considerable boost to Si 69®. Consumption has steadily increased ever since. Additional grades were developed and introduced to the market-based customers' requirements.

3. Types of Rubber Silanes and General Applications

For sulfur-cured rubber compounds, the three following types of sulfur-functional silanes are commonly used:

Di- and polysulfide silanes	$(RO)_3Si$ ⌒⌒ S_x ⌒⌒ $Si(OR)_3$	III
Mercaptosilanes	$(RO)_3Si$ ⌒⌒ SH	IV
Blocked mercaptosilanes	$(RO)_3Si$ ⌒⌒ S–PG	V

Where R = –CH$_3$, –C$_2$H$_5$, alkylpolyether; x = 2–8; PG = CN, octanoyl.

Whereas the mercapto- (IV, R = –CH$_3$ or –C$_2$H$_5$) and thiocyanatosilanes (V, R = –CH$_3$ or –C$_2$H$_5$, PG = CN) are used mainly for industrial rubber goods and for shoe soles, the main application of the di- and polysulfide silanes (III, R = –C$_2$H$_5$, x = 2–8) is in the tire industry. More recently, the demand for further improved overall tire performance and a need for higher throughput during compounding have led to the development of innovative silane solutions. Novel mercaptosilane Si 363® (IV, R = –C$_2$H$_5$ and alkyl polyether) [92] or blocked mercaptosilane NXT® (V, R = –C$_2$H$_5$, PG = octanoyl) [93] offer improved rolling resistance in comparison to Si 69®/TESPT. Compared to the high amounts of silane needed in tire tread applications, the consumption of silanes for industrial rubber goods is lower, but steadily increasing. In most cases, the mercaptosilane is based on a trimethoxysilyl group (MTMO, IV, R = –CH$_3$), whereas the thiocyanatosilane (TCPTEO, Si 264™, V, R = –C$_2$H$_5$, PG = CN) and the di- and polysulfide silanes (III, Si 69®/TESPT: R = –C$_2$H$_5$, x = 4; Si 266®/TESPD; R = –C$_2$H$_5$, x = 2) have a triethoxysilyl moiety. The trimethoxysilanes show a faster hydrolyzation speed, but the released methanol can cause serious health and safety problems in the mixing department. Therefore, the triethoxysilyl group is preferred for most rubber applications.

As mentioned earlier, vinylsilane I is best suited for the reinforcement in peroxide-cured compounds such as sealants, cables, and profiles. But owing to the low flash points of the trimethoxy- and triethoxysilanes, the trimethoxyethoxysilane (I, R = CH$_3$), with its higher flash point, or fillers premodified with a vinylsilane are also often used [94]. The latest development for the in situ modification of silica, avoiding the mutagenic and teratogenic methoxyethanol, is the oligomerized vinyltriethoxysilane (VTEO) Dynasylan® 6498, which has a high flash point, a reduced ethanol emission, and a high coupling efficiency [95,96]. Besides these rubber silanes, 3-chloropropyltriethoxysilane (CPTEO, II, R = –C$_2$H$_5$, X = Cl) for metal oxide-cured chloroprene rubber [97] and 3-aminopropyltriethoxysilane (AMEO, II, R = –C$_2$H$_5$, X = NH$_2$) for carboxylated nitrile rubber and halobutyl rubber can also be used [98,99]. An overview regarding the use of special silanes for special elastomers has been given by Klockmann et al. [100], and a summary of the commercial rubber silanes and their main applications is given in Section III.E.

C. Process and Technology of Rubber Silanes

As mentioned earlier, most functional silanes of structure II are based on 3-chloropropyltrimethoxysilane or CPTEO [85]. The starting materials can be either trichlorosilane VI, obtained from the reaction of silicon with hydrogen chloride, or trimethoxysilane VII, gained from the analogous reaction of silicon with methanol (Scheme 7.4). Both educts are extremely sensitive to moisture and are flammable; VI is also strongly corrosive and hazardous, and VII is toxic (T+). This demands very careful handling of these substances in the following hydrosilylation step. The addition of allyl chloride to the silanes VI and VII is carried out most often with a heterogeneous or homogeneous platinum catalyst, respectively [101], and results in the corresponding chloropropylsilanes VIII and IX. After the hydrosilylation reaction and separation of the by-products, an esterification step with an alcohol follows in the case of VIII, whereas IX can be functionalized directly by nucleophilic

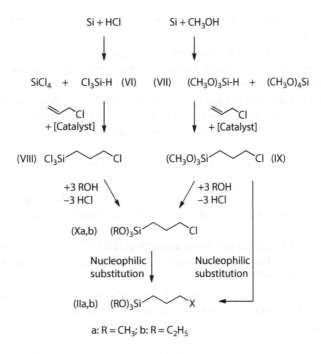

SCHEME 7.4 Possible production process for functionalized trialkoxysilylpropyl silanes.

substitution to the final product IIa with a trimethoxysilyl group. To produce the preferred triethoxysilyl silane, IX has to be cross esterified with ethanol followed by the final functionalization to product IIb.

An overview of the most common functionalizations is given by Panster and coworkers [85]. For the production of common sulfur silanes III, bis(triethoxysilylpropyl)disulfide (Si 266®/TESPD), and tetrasulfide (Si 69®/TESPT), respectively, the precursor silane Xb is reacted with a source of sulfur (Na_2S, Na_2S_x, and elemental sulfur) in aqueous or nonaqueous solutions [102,103].

D. CHARACTERIZATION OF SILANES AND REACTIONS

1. Methods of Analysis

For the characterization and quality control of silanes, several analytical methods based on the identification of the functional groups are well established. The different methods can be classified as (1) those analyzing a characteristic element or a functional group and (2) those identifying the silane component itself. The following are the most important methods used to analyze a characteristic functional group:

1. Gravimetric measurement of residue on ignition, which determines the amount of SiO_2 resulting from the content of silicon in the silane [104].
2. Elemental analysis to determine, for example, the amount of sulfur, carbon, chlorine, and nitrogen.

3. IR spectroscopy to identify the functional group by its characteristic vibration modes.
4. ^{13}C and ^{29}Si cross polarization nuclear magnetic resonance spectroscopy (CP/NMR), to analyze the alkoxysilyl group with regard to the substitution pattern at silicon as well as the degree of formation of polysiloxanes [105–107]. The silicon atoms of the organosilanes are detected in the ^{29}Si NMR spectrum between −70 and −40 ppm. The organofunctional group can be identified with ^{1}H NMR spectroscopy.

There are analytical methods to determine the silane component:

1. Gas chromatography (GC) for silanes that vaporize without decomposition.
2. Mass spectrometry (MS), which gives a characteristic "fingerprint" of each silane.
3. High-performance liquid chromatography (HPLC), which is most often used for the analysis of silane mixtures as in the case of the sulfur silanes TESPD and TESPT.
4. With the use of response factors, UV detection is capable of measuring the weight distribution of the sulfides from S2 to S10 and the average sulfur chain length [108].

In general, sulfur silanes are routinely characterized by methods including GC, NMR, or HPLC analysis. In Table 7.3, typical analytical data of the sulfur silanes Si 69® and Si 266® (Evonik) are compared. It is obvious that the sulfur content of the disulfide silane (Si 266®) is lower than that of the tetrasulfide, but with the correct sulfur adjustments in the compound, comparable vulcanizate data are achieved. As a rule of thumb, the replacement of Si 69® with Si 266 demands an additional amount

TABLE 7.3
Comparison of Typical Analytical Data for Si 69® and Si 266

	Si 69®	Si 266®
Residue of ignition, %	22.5	25.0
Sulfur content, %	22.6	13.9
HPLC analysis, UV 254, nm		
S2 (RF 31.3), %	17.1	85.2
S3 (RF 8.9), %	29.6	13.1
S4 (RF 4.9), %	23.7	1.2
S5 (RF 3.2), %	15.4	0.5
S6 (RF 2.4), %	7.7	—
S7 (RF 1.8), %	3.3	—
S8–S10, %	2.5	—
$\langle S_x \rangle$	3.8	2.15

Note: RF, Response factor.

of sulfur (9.5% of the amount of Si 69®), and owing to the lower molecular weight, the dosage of the disulfide silane is 10% lower.

2. Silica–Silane Coupling

For the performance of a silane, the reactions are of major importance. The alkoxysilyl group of the silane can react with silanol groups of the surface of glass fibers or siliceous fillers. This hydrophobation reaction has been studied intensively for a number of substrates and silanes under various conditions. For the rubber industry, it is of crucial importance to understand the coupling behavior of trialkoxysilanes with silica and silicates during the mixing process (in situ modification) and the presilanization of fillers (ex situ modification). The kinetics of the coupling reaction can be determined by using a model system of silica and an inert solvent to imitate the rubber matrix. The reactions of the silanes can be monitored by the analysis of the unreacted silane, the hydrolysis and condensation products (if possible to detect), and the released alcohol by, for example, HPLC and GC [109]. Another possibility to monitor the progress of the reaction is the use of ^{29}Si MAS-NMR to determine the change in the substitution pattern of the silane and the change in the silanol content of the silica [109]. ^{29}Si and ^{13}C MAS-NMR spectroscopy are also useful to analyze the degree and structure of the silane coupling of, for example, ex situ modified silica.

Taking the results of the earlier cited studies into account, it is reasonable to assume that prior to the silica–silane coupling, at least one alkoxy group of the silane is hydrolyzed to the more reactive silanol, which then reacts with a silanol group of the silica, releasing water (primary reaction). The hydrolysis reaction is the rate-determining step, and up to a certain point, an increase in water as well as an increase in temperature increases the reaction speed. Furthermore, the hydrolysis and condensation reactions are catalyzed by acids and bases. The reactivity of the silane decreases with more alkyl substituents bound to silicon and with increasing steric hindrance of the alkyl and alkoxy groups [110,111]:

$$H_3C(C{=}O)O \qquad\qquad H_3CO \qquad\qquad C_2H_5O \qquad\qquad C_2H_5O$$
$$H_3C(C{=}O)O{-}\underset{|}{Si}{-}R \;\; > \;\; H_3CO{-}\underset{|}{Si}{-}R \;\; > \;\; C_2H_5O{-}\underset{|}{Si}{-}R \;\; > \;\; H_3C{-}\underset{|}{Si}{-}R$$
$$H_3C(C{=}O)O \qquad\qquad H_3CO \qquad\qquad C_2H_5O \qquad\qquad C_2H_5O$$

Due to the fact that trialkoxysilanes can also undergo an intermolecular condensation reaction, oligo- and polysiloxanes can be formed on the silica surface (secondary reaction) and, as a side reaction, in the rubber matrix. This condensation reaction needs water and is slower than the primary reaction. Scheme 7.5 shows a simplified reaction model for the silica–silane coupling reaction [109].

Taking the surface coverage of the Si 69® molecule into account, a complete monolayer coverage of the silica surface ($\approx$150 m^2/g) under ideal conditions would be reached with 8% Si 69® [112]. This concentration is the common amount of Si 69® used to modify silica for tire applications (see Section IV.D).

A special technique, the IR operando technique, can be used to detect "online" the grafting of a silane, for example, propyltriethoxysilane on the surface of silica [113]. The IR spectrum of the ν (OH) region shows the behavior of OH groups

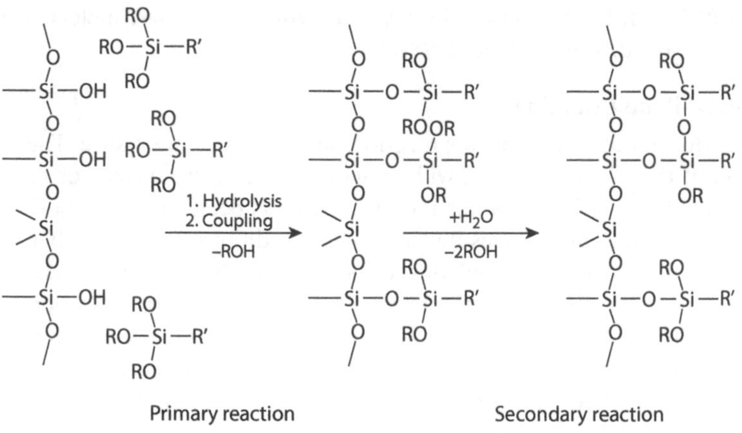

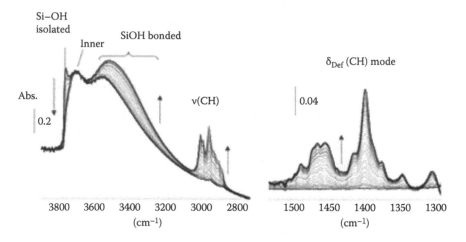

SCHEME 7.5 Simplified silica–silane coupling reaction.

FIGURE 7.5 IR spectra of silica during the reaction with PTEO at 383 K (spectra recorded for $t_{reaction}$ = 0–45 min, thick lines before reaction and after 45 min on stream).

during the reaction and enables to follow the ongoing process of the silica–silane reaction (Figure 7.5).

3. Silane–Rubber Coupling

The reaction of the organofunctional group with the rubber takes place during the vulcanization process. Therefore, it is reasonable to investigate the silane coupling reaction in commonly used model systems imitating the vulcanization mechanism. Suitable model olefins such as 2-methyl-2-pentene, 2,3-dimethyl-2-butene, and squalene, which imitate the diene rubber, and methods used to follow the vulcanization reaction were reviewed by Nieuwenhuizen et al. [114]. The existence of a silane–rubber bond after curing has been proven by the detection of a reaction

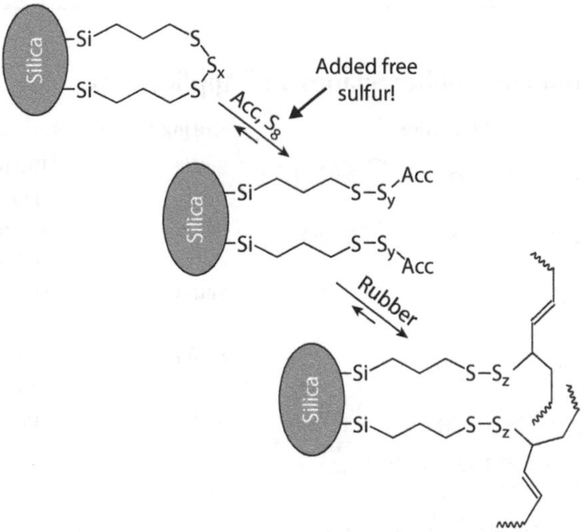

SCHEME 7.6 Simplified silica–rubber coupling reaction with TESPT (x = 1.7) and TESPD (x = 0).

product of TESPT with 2-methyl-2-pentene by GC–MS and the ^{29}Si NMR characterization of the reaction products with 5-trimethylsilyl-2-methyl-2-pentene [108]. Investigations regarding the mechanism and kinetics of coupling reactions of TESPD and TESPT with sulfur and an accelerator system have been performed using squalene as the model olefin [108,115,116]. It has been demonstrated that the sulfides with a sulfur bridge consisting of least two S-atoms can react with the rubber at high temperatures, but the addition of sulfur and accelerators speeds up the reaction strongly and increases coupling efficiency. Furthermore, it has been shown that the reactivity of the polysulfide increases with increasing sulfur chain length. According to the results, it is evident that the disulfide silane needs to incorporate sulfur for the succeeding coupling reaction, which is possible only in the presence of an accelerator system. Without sulfur and accelerator a coupling reaction is not possible [117]. In conventional and semiefficient cure systems, the tetrasulfide silane TESPT also acts as a sulfur acceptor, because the formation of di- and polysulfidic bridges requires more sulfur than the silane already contains. This sulfur incorporation for an effective coupling reaction is pictured in Scheme 7.6. The existence of the chemical silica–silane–rubber coupling in the rubber is demonstrated clearly by the significant increase in reinforcement, as demonstrated by works of Wagner [88], Wolff [90], and Thurn and Wolff [118]).

E. Product Overview and Applications

The most common rubber silanes can be purchased from various producers in the United States, Europe, and Asia, but owing to some specific differences, not all silanes are interchangeable. Table 7.4 gives an overview of the rubber silanes and their main applications.

TABLE 7.4

Overview of Common Rubber Silanes and Applications

Silane	Structure	Curing System	Application
TESPT	$(EtO)_3Si$ ⌒⌒ S_4 ⌒⌒ $Si(OEt)_3$	Sulfur	Tire treads, shoe soles, industrial rubber goods
TESPD	$(EtO)_3Si$ ⌒⌒ S_2 ⌒⌒ $Si(OEt)_3$	Sulfur	Tire treads, industrial rubber goods
MTMO	$(MeO)_3Si$ ⌒⌒ SH	Sulfur	Shoe soles, industrial rubber goods
MPTES	$(EtO)_3Si$ ⌒ SH	Sulfur	Shoe soles, industrial rubber goods
Si 363®	$(RO)_3Si$ ⌒⌒ SH $R = -C_2H_5$ and alkylpolyether	Sulfur	Tire treads
NXT®	$(EtO)_3Si$ ⌒⌒ S—C(=O)—⌒⌒⌒⌒	Sulfur	Tire treads
TCPTEO	$(EtO)_3Si$ ⌒⌒ SCN	Sulfur	Shoe soles, industrial rubber goods
VTEO	$(EtO)_3Si$ ⌒=	Peroxide	Industrial rubber goods
VTMOEO	$(MeOC_2H_4O)_3Si$ ⌒=	Peroxide	Industrial rubber goods
CPTEO	$(EtO)_3Si$ ⌒⌒ Cl	Metal oxide	Chloroprene rubber
MEMO	$(MeO)_3Si$ ⌒⌒ O—C(=O)—C(=CH$_2$)	Peroxide	Textile adhesion
AMEO	$(EtO)_3Si$ ⌒⌒ NH_2	Sulfur	Special polymers, metal adhesion
PTEO	$(EtO)_3Si$ ⌒⌒	Sulfur, peroxide	Industrial rubber goods, processing aid
OCTEO	$(EtO)_3Si$ ⌒⌒⌒⌒	Sulfur, peroxide	Industrial rubber goods, processing aid

IV. SILICA AND SILANES IN RUBBER

A. GENERAL CONSIDERATIONS

The phenomenon of the reinforcement of rubber by the addition of an active filler produces pronounced improvements in mechanical properties such as tensile strength, tear resistance, abrasion resistance, and modulus [119–121]. These improvements can be attributed to the inclusion of a solid dispersed phase, resulting in an internal stress that is higher than the external stress applied to the sample. This strain amplification is caused by the addition of the undeformable filler to the viscoelastic rubber matrix (hydrodynamic effect) and the partial immobilization of rubber on the filler surface and in the structure of the active filler. Without this reinforcement, most rubber products would be inconceivable. The fillers are generally classified into CB and white fillers.

1. Carbon Black as Filler in Rubber Applications

CB is not only the most widely used active filler but also the oldest one. Since the introduction of reinforcing blacks for use in rubber at the beginning of the twentieth century, considerable research work has been done to understand the mechanism of reinforcement. The two major characteristics of active blacks are their surface area and aggregate structure. These characteristics determine the static and dynamic in-rubber properties and hence make it possible to tailor-made the performance of rubber products. Besides their high reinforcement and broad product range, most CBs are relatively easy to process due to their physically induced reinforcing in contrast to the chemical based in the case of the silica/silane system.

2. White Fillers in Rubber Applications

Although the use of CB results in an outstanding high reinforcement, nonblack fillers such as clays, carbonates, silicates, and pyrogenic and precipitated silica are also needed. These fillers differ not only in their chemical structure but also in their particle size and shape. Compared to CB, these fillers show advantages in cut and flex resistance and heat buildup, they are nonconducting, and they are required for colored products, but the reinforcement they provide, with regard to modulus and abrasion resistance, is limited. Therefore, the less active clays, silicates, and carbonates are often used in rather high amounts for industrial rubber goods to reduce compound costs and to attain a given hardness level. Higher reinforcement is achieved by the use of pyrogenic or precipitated silica or a blend of the low-surface-area (LSA) fillers with these highly active silicas. To achieve higher reinforcement, silane coupling agents that provide a chemical link between the filler and the rubber are needed (see also Section III).

Detailed studies with regard to the reinforcing mechanism of the silica/silane system have been published [117,122,123].

B. Rubber Reinforcement

1. General Considerations

According to Payne, the modulus of filled rubber samples can be separated into strain-independent and strain-dependent contributions. The increase of modulus by the cross-linking of the rubber matrix, the aforementioned hydrodynamic effect due to the addition of a rigid filler, and the "in-rubber structure" (defined by the possibility of a filler preventing part of the rubber from being deformed) are strain independent. In addition to these contributions, the addition of an active filler results in the formation of a filler network, which also increases the modulus. This contribution of the filler network is strain dependent, because increasing strain leads to a successive breakdown of the filler network into subnetworks until, at high deformations, the network is reduced to a minimum or at nonfinite high strains is completely destroyed (Scheme 7.7). With the breakdown of the filler network, the rubber that is trapped in the network is released and can take part in the deformation [124]. This stress softening of filled rubber samples is well known as the Payne effect [125], which is defined as the difference between shear modulus values measured at low and high strains, ΔG^*. The various contributions to the modulus are visualized in Scheme 7.8.

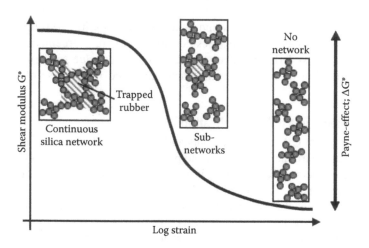

SCHEME 7.7   Strain-dependent breakdown of the filler network (Payne effect).

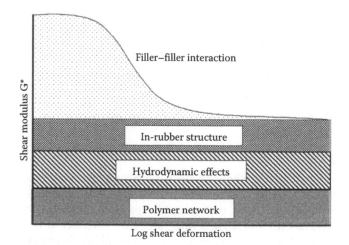

SCHEME 7.8   Contributions to the modulus of filled compounds according to Payne. (From Wang, M-J., *Rubber Chem. Technol.*, 71, 520, 1998.)

2. Types of Interactions

CB–CB and CB–polymer interactions. The surface of CB consists of small, disordered graphitic crystallites with only a few oxygen-containing groups at the edges of the basal planes. Apart from these functional groups, the CB surface is hydrophobic and therefore rather compatible with a hydrocarbon polymer. The sites with the highest energy, leading to a strong filler–polymer interaction, are at the edges (van der Waals bond) [121]. Because of strong interaction, a rubber shell is formed, which determines the bound rubber content. But the CB also builds a filler–filler network in the rubber matrix, which decreases with increasing strain. The filler–filler flocculation, resulting in the filler network, and the filler–polymer interaction, leading to the formation of a rubber shell, are competing processes. Therefore, a graphitization, eliminating the active sites, leads to far less filler–polymer interaction but a stronger filler–filler network. The effect of the filler flocculation on the dynamic behavior was reviewed by Wang [126].

Silica–silica and silica–rubber interaction. Wolf and coworkers intensively studied the adsorption energies and rubber interactions of CB and silica with unpolar and polar molecules [127–129] and rubber [130–134]. A major conclusion was that a strong reinforcement with silica is possible only in polar polymers such as acrylonitrile butadiene rubber, whereas for the reinforcement of unpolar rubbers, coupling agents are required. In Figure 7.6, the energies of adsorption of the polar acetonitrile, ΔG°_{MeCN}, are plotted against ΔG°_{C7}, the energy of adsorption of the unpolar heptane for both CBs and silica [128]. The results prove that the interaction of silica with acetonitrile is much higher than with heptane, whereas CB shows a stronger interaction with the alkane. According to these researchers, silica has a high specific component of the surface energy, γ_s^{sp}, which correlates with the filler–filler interaction, but a very low dispersive component, γ_s^{d}, responsible for the silica–polymer interaction.

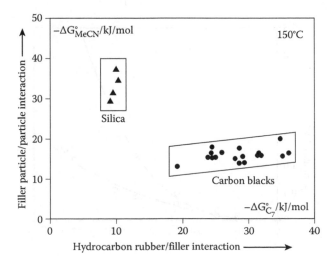

FIGURE 7.6 Energies of adsorption of acetonitrile versus those of heptane for carbon blacks and silicas.

Another conclusion of Wolff and coworkers was that the rather low interaction of silica with nonpolar polymers and the strong silica–silica interaction by hydrogen bonding resulted in pronounced flocculation of neighboring silica clusters, causing high viscosities and a high hardness. Figure 7.7 depicts the increase in the Mooney viscosity, normalized on CB, in NBR and NR plotted versus the volume fraction of the silica [133]. As expected, the increase in the viscosity in the nonpolar NR, with its low silica–polymer interaction, is much stronger than in the polar NBR, where CB and silica show comparable networking. Therefore, the reinforcement of NBR with unmodified silica is much higher than that of nonpolar rubbers.

The influence of the silica surface area and silica loading on the Payne effect has been investigated in [122]. It has been demonstrated that at a constant silica loading, the filler–filler network strongly increases with increasing surface area because of the decreasing interaggregate distance [135], but the increasing filler loading also causes a strong increase in both the Payne effect ΔG^* and the loss modulus. This results in a higher hysteresis loss under cyclic deformation. The two influences on the filler networking can be seen in Scheme 7.9 and are valid for both CB and silica. According to the kinetic cluster–cluster–aggregation model, the increase in ΔG^* above the percolation threshold is proportional to $\varphi \times 3.5$, where φ is the filler loading. Compared to CBs with a similar surface area, the filler network of an unmodified silica is significantly higher, resulting also in high loss moduli.

Considering the different contributions to the moduli (Scheme 7.8) and the network formation (Scheme 7.7), it is evident that the addition of LSA fillers such as clay and silicates mainly increases hardness and lowers compression set (hydrodynamic effect), but the filler network is very low, even at high loadings. With increasing surface area of the filler, the network increases, which leads to a higher compound hardness, higher moduli at low and moderate strains, and better tear

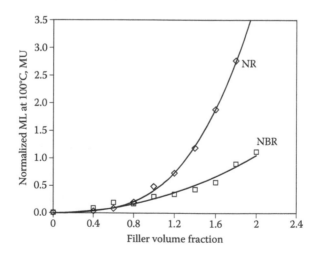

FIGURE 7.7 Normalized Mooney viscosity, ΔML, as a function of the filler volume fraction of NR and NBR.

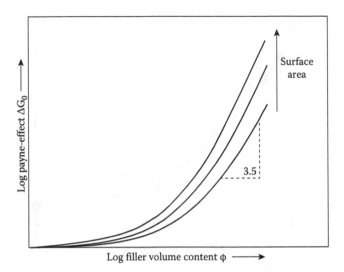

SCHEME 7.9   Increase of the filler network with increasing filler loading and surface area.

resistance but also to a strong rise in the viscosity and deterioration of the compression set due to the filler flocculation, leading to the formation of the network. Furthermore, the polar silica surface absorbs parts of the accelerator, which results in a poor cure rate and low cross-link density. Therefore, unmodified precipitated silicas are mainly used in small quantities (5–15 phr) to improve tear resistance in a blend with CB for the main reinforcement. The use of higher amounts requires a silane modification for a silica–rubber coupling or at least the addition of silica activators such as glycols and amines to establish good processability and a high curing rate.

Silica–silane–rubber interaction. The use of silanes allows a modification of the polar silica surface that makes the silica more compatible with the rubber matrix. The hydrophobation of the silica surface with monofunctional alkylalkoxysilanes, which couple to the silica but cannot react with rubber, leads to a strong reduction of the silica network. This results in a reduced Payne effect, lower viscosity (Figure 7.8), and lower hardness. As long as the hydrophobation of the silica surface is not complete, the formation of a silica network is still possible, but its strength will be lower than that of the unmodified silica. At moderate strains, these filler networks or subnetworks still contribute to the modulus. But at very high strains, the network is destroyed to a minimum or in dependency of the strength of shear forces applied completely destroyed. Owing to the lack of a polymer–filler interaction, the reinforcement or the high strain modulus is low [122]. This is in line with the results of Wolff et al. [134], who clearly demonstrated that the hydrophobation of the silica surface leads to a strong reduction in γ_s^{sp} without increasing γ_s^{d}.

The addition of alkylalkoxysilanes in highly filled compounds is beneficial if filler flocculation, viscosity, and stiffness have to be reduced. In combination with coupling agents, especially, these silanes act as specific silica processing aids, because the high strain modulus remains nearly unchanged [136].

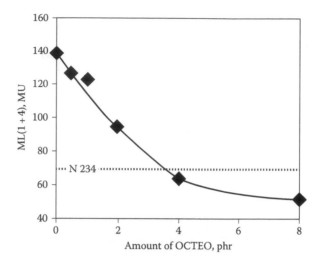

FIGURE 7.8 Decrease in the Mooney viscosity with increasing amount of the alkylsilane OCTEO in S-SBR/BR (80 phr ULTRASIL® 7000 GR), with 80 phr N 234 as reference.

A high silica–rubber interaction is achieved by chemical coupling using bifunctional silanes, as mentioned earlier. Figure 7.9 compares the shear moduli (Payne effects) and tan δ of an unmodified silica, a silica modified with a monofunctional hexadecyltrimethoxysilane and the coupling agent Si 69®. As expected, the silica modification reduces the filler network and the Payne effect, but the

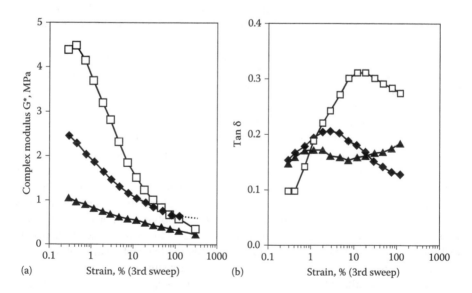

FIGURE 7.9 Course of (a) the complex modulus and (b) the hysteresis loss in the vulcanizate measured with the RPA at 60°C and 1.6 Hz. S-SBR/BR with 80 phr ULTRASIL® 7000 GR (■) plus 10 phr Si 69®, (▲) plus 8 phr HDTEO, and (□) without silane modification.

modulus at high strains is increased with Si 69® modification. This increase is due to the formation of an "in-rubber structure" by the chemical coupling [122]. Furthermore, the loss modulus is significantly lowered, which results in lower tan δ values.

There are three main reasons for this low hysteresis loss of the compound with the silica/silane filler system. On the one hand, the energy consumed for slippage of the polymer chains on the surface is low because the silica–rubber interaction is rather weak compared to the strong physical interaction of the rubber with CB. On the other hand, part of the rubber is chemically immobilized on the surface and therefore cannot be removed even under high stress, which is in agreement with findings of ten Brinke et al. [137]. Furthermore, it has to be considered that the strength of the silica–silica network after silanization is strongly reduced and therefore a breakdown under deformation is less energy consuming than in the case of an active CB or unmodified silica.

3. In-Rubber Performance

Green compound. One major concern in the preparation of rubber compounds is the incorporation, dispersion, and distribution of the active filler. For the application of the silica/silane filler system, prereacted silica/silane products can be used, but in situ modification of the silica with the silane, that is, the chemical reaction of silica and silane to a covalent bond during the mixing process, is more common. In contrast to mixing CB or a prereacted product, the in situ modification requires, in addition to optimal dispersion, precise control of the chemical reaction in the mixer. Like most chemical reactions, silanization is influenced by the temperature, the concentration of the starting materials and the final products remaining in the compound, the transportation and diffusion processes, and possible catalysis. The findings with regard to the in situ mixing process of the silanes TESPT and TESPD have been summarized in [123,138]. It has to be ensured that the mixing time is long enough and the mixing temperature sufficiently high to complete the silanization reaction and to remove the formed alcohol and the formed alcohol/water mixture, respectively, from the batch. Batch temperatures below 140°C require inadequate long mixing cycles, whereas above 160°C, the polysulfides in the TESPT split off sulfur and react with the rubber, causing a so-called prescorch. The disulfide TESPD exhibits a by far better prescorch safety as it is depicted in Figure 7.10.

Furthermore, the coupling reaction of the silane with the silica should occur simultaneously with the dispersion process. Hence, the silica and silane should be added simultaneously in the first mixing cycle. A good silanization is needed not only to guarantee the best reinforcement but also to reduce compound hardening during storage (flocculation) [139]. As previously stated, a certain amount of water speeds up the silanization reaction, which suggests that a moisture content of the silica in the range of 4%–6% is optimal. This range corresponds to the equilibrium moisture content of precipitated silica. Moisture levels below 3% lead to higher viscosities and stiffness, due to a deteriorated silanization reaction [123]. The effect can be shown by an increasing the Mooney viscosity with lower moisture contents of the silica as it is depicted in Figure 7.11.

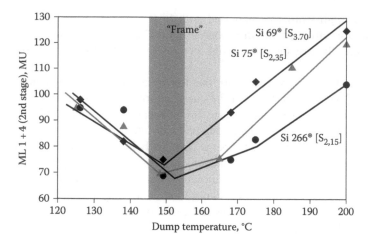

FIGURE 7.10 Scorch safety as function of the Mooney viscosity versus compounding temperature of TESPT = Si 69® and TESPD = Si 75® and TESPD = Si 266®.

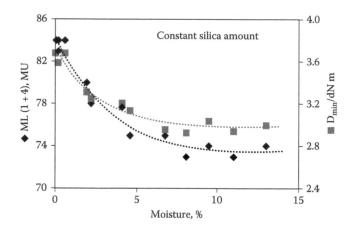

FIGURE 7.11 The Mooney viscosity and D_{min} as function of the silica moisture content.

The main influences on the silanization reaction are summarized in Scheme 7.10. As can be seen, the choice of the optimal mixing temperature is of crucial importance.

Another important point to be considered regarding the compounding of silica and silane is the fact that in compounds with a high amount of silica, all the silane can reach the silica surface during mixing. But the smaller the amount of silica in a blend with, for example, CB, the less probable it is that the added silane will reach the silica surface quantitatively in the given mixing time. Therefore, the mixing of blends containing small amounts of silica can require comparatively more silane than in compounds where silica is the main filler. In these cases the use of prereacted products is advisable.

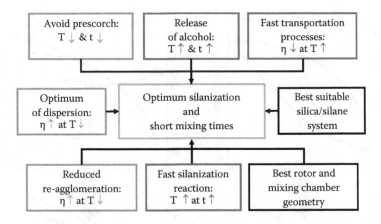

SCHEME 7.10 Main influences on the in situ silanization reaction.

Vulcanizate. The vulcanization mechanism of silica-filled compounds and the formation of the cross-linked polymer matrix and silica–silane–rubber network using TESPD and TESPT have been investigated in many different papers, for example [116,117]. It was demonstrated that during the vulcanization process, both networks, matrix cross-linking and silica–rubber coupling, are formed simultaneously and cannot be separated. Because the sulfur-functional silanes are sulfur acceptors (Scheme 7.6), the two cross-link reactions compete for the added free sulfur, so that the amounts of silane and sulfur determine the cross-linking structure and hence the reinforcement characteristic of the sample. This competition of the matrix and silica–silane–rubber coupling for the added sulfur is shown in Scheme 7.11. For example, an increasing amount of sulfur at a constant level of TESPT leads to an increase in the coupling efficiency of the silane until all the silane is activated. But because the activation of the silane also consumes free sulfur, the matrix cross-link density increases only slightly. When a high amount of TESPT has already been activated, the matrix cross-link increases more strongly. In contrast to this, an increase in the amount of TESPT at a constant amount of sulfur leads to a higher total number of silica–silane–rubber bonds. But owing to the incorporation of sulfur by the silane, an increase in the amount of TESPT reduces the matrix cross-link density, and, as stated earlier, in the case of the disulfidic silane TESPD, this effect is even more pronounced. Therefore, it is reasonable to assume that a variation in the amount of silane or the amount of sulfur changes the ratio of the matrix and silica–silane–rubber networks and that these two networks cannot be varied independently [117]. The silica–silane–rubber coupling efficiency can be shifted to higher values by the application of mercaptosilanes.

Keeping the earlier-stated influences regarding the amounts of silane and sulfur in mind, knowing the fact that silica–silane–rubber coupling contributes more to the modulus than matrix cross-linking [88,123], and considering that improved hydrophobation results in lower hardness values and reduced Payne effect, the in-rubber data performance of silica/silane-filled rubber compounds can be predicted qualitatively. For example, an increase in the amount of silane leads to a reduction of filler network and an increase in the moduli (in-rubber structure). This results

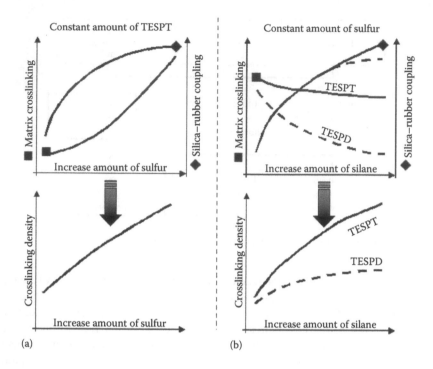

SCHEME 7.11   Influence of (a) the amount of sulfur and (b) the amount of silane on cross-linking densities.

in lower viscosities, higher static moduli, reduced elongation at break, and better abrasion resistance. With regard to the dynamic behavior, the stiffness increases with increasing overall cross-link density and decreases with improved hydrophobation. The hysteresis loss and compression set improve with increasing amounts of silane. Figure 7.12 pictures the increase in modulus 300% as the amounts of sulfur and Si 69® increase. It is obvious that an increase in the modulus with an increase in Si 69® could be compensated by a reduction in the amount of sulfur, but according to Scheme 7.11, this results in a shift from matrix coupling to silica–silane–rubber coupling. This has a certain importance for compounding if, for example, a higher surface area of a silica, resulting in a stronger silica network, has to be compensated by a higher silane dosage without increasing the modulus too strongly.

As previously mentioned, TESPT and TESPD differ in their average sulfur chain length x, and an adjustment of the difference in overall sulfur amount is recommended when TESPT is substituted by TESPD or a silane, which is not acting as a sulfur donor or vice versa. For instance, if 1 phr Si 69® (x = 3.7) is substituted by 0.90 phr Si 266® (x = 2.2), it is advisable to increase the elemental sulfur within the cure package by 0.10 phr.

As can be seen in Table 7.5, the silica coupling with TESPT and TESPD (compounds II and III) results in a strong increase of the static moduli, reduced elongation at break, and lower DIN* abrasion loss compared to the compounds without

---

* DIN: Deutsches Institut für Normung (German Institut for Standardization)

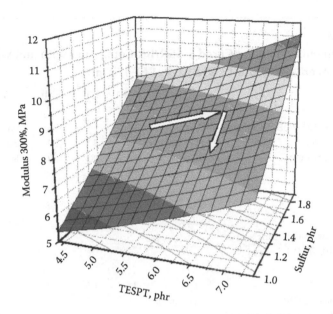

FIGURE 7.12 Influence of the amounts of TESPT (Si 69®) and sulfur on the modulus 300%.

silica–silane–rubber coupling (compounds I and IV). The reinforcement of these compounds is comparable to that of the N 234 CB compound V, but the hysteresis loss tan δ (60°C), which correlates with the rolling resistance, is significantly reduced.

Apart from the sulfur vulcanization, cross-linking with a peroxide initiator has crucial importance for mechanical rubber goods. In order to improve the mechanical properties of peroxide-cured compounds filled with white fillers, the addition of vinylsilanes is recommended. In contrast to the fairly high dosage of sulfur-functional silanes in products requiring high abrasion resistance, the addition of only 1 phr of Dynasylan® VTEO to 50 phr of a 160 m²/g silica already results in a strong improvement of static in-rubber data (Table 7.6) [96]. Higher cross-link densities can be achieved by increasing the amount of the radical initiator and adding activators such as triallyl cyanurate (TAC) or trimethylolpropane trimethacrylate (TRIM).

The silane amount needs also to be adjusted to the filler loading and the silica surface area by taking the specific surface area of the silica (measured with cetrimonium bromide [CTAB]) into consideration. For some applications, it is also recommended to adjust the accelerator content analogously to the silica loading and specific surface area (measured with N_2 according to the method of BET [Brunauer, Emmett, Teller]).

C. Future Trends

1. Tire Label and Silane Coupling Efficiency
One of the main driving forces for the development of new silica/silane technologies is the tire industry. Due to the introduction, respectively, and the announcement of

TABLE 7.5

Reinforcement of a Sulfur-Cured S-SBR/BR Compound with and without Coupling Agents; ULTRASIL® VN 3 GR, BET 175 m²/g

		I	II	III	IV	V
Formulation						
S-SBR, 25% S; 50% 1.2B; 37.5 phr oil	Polymer	96	96	96	96	96
BR, cis-1,4B>96%	Polymer	30	30	30	30	30
ULTRASIL® VN 3 GR	Silica	80	80	80	80	—
N 234	CB	—	—	—	—	80
Si 69®	Silane	—	6.4	—	—	—
Si 266®	Silane	—	—	5.8	—	—
PTEO	Silane	—	—	—	6.0	—
Other chemicals, ZnO₃; stearic acid 2; aromatic oil 10, 6PPD 1.5; wax 1						
Cure system						
Sulfur		1.5	1.5	2.1	1.5	1.5
Diphenyl guanidine (DPG) 2; Cyclohexylbenzothiazole sulfenamide (CBS) 1.5	Accelerator					
In-rubber properties						
ML(1+4) (MU)		170.0	59.0	59.0	80.0	73.0
$M_H - M_L$ (165°C), dN m		20.7	16.2	16.5	13.4	14.1
δ 10%, min		0.9	1.7	2.6	3.7	1.0
δ 90%, min		43.5	18.0	11.9	6.8	2.3
Tensile strength (ring), MPa		14.1	12.7	13.0	12.7	11.4
Modulus 100%, MPa		2.3	2.0	2.1	0.8	1.9
Modulus 300%, MPa		5.3	10.3	10.4	2.2	9.8
Elongation at break, %		710	340	350	860	330
Shore A hardness		81	64	65	55	68
DIN abrasion, mm³		136	83	83	304	122
MTS, freq. 16 Hz; preforce 50 N; amp. force 25 N						
Dyn. modulus E* (0°C), MPa		70.7	15.1	17.8	26.4	50.8
Dyn. modulus E* (60°C), MPa		46.4	7.3	8.0	7.4	9.7
Loss factor tan δ (0°C)		0.191	0.475	0.515	0.531	0.461
Loss factor tan δ (60°C)		0.063	0.128	0.130	0.194	0.275

tire labels around the world, for example, in Europe, Japan, South Korea, Brazil, and the United States, which classify the tire performance, the importance of fast implemented innovations has become most important. In all different tire labels, the rolling resistance is one of the most important criteria, since this is a major contributor to the fuel consumption of a car.

The application of mercaptosilanes is well known to lead to a significantly reduced rolling resistance at constant wet grip and abrasion (Scheme 7.12). The chemical background of the mercaptosilanes is an enhanced silane coupling efficiency without

TABLE 7.6

Reinforcement of a Peroxide-Cured EPM Compound with and without Dynasylan® VTEO

		I	II
Formulation			
EPM	Polymer	100	100
ULTRASIL® VN 3 P	Silica	50	50
Dynasylan® VTEO	Silane	—	1
Other chemicals, oil 30, PEG 2,			
Zn stearate 0.5			
Cure system DCP 40%	Accelerator	7	7
In-rubber properties			
ML(1+4), MU		115.0	112.0
$M_H - M_L$, 180°C, dN m		17.2	17.3
t 10%, min		0.6	0.5
t 90%, min		7.6	5.3
Tensile strength, MPa		13	12.6
Modulus 100%, MPa		1.3	1.8
Modulus 300%, MPa		2.8	4.9
Modulus 300%/100%		2.2	2.7
Elongation at break, %		790	530
Shore A hardness		62	65
Ball rebound, 60°C, %		62.7	63.2
Compression set, %		26.7	20.3
DIN abrasion, mm^3		146	92
E* 60°C, MPa		11.8	12.9
tan δ, 60°C		0.162	0.13

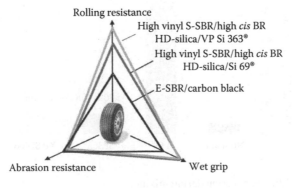

SCHEME 7.12 So-called magic triangle of tire performance.

changing the amount of sulfur. This results in a shift from matrix to silica/silane–rubber coupling, as described earlier.

2. Reduced Ethanol Emission

Depending on the mixing process, on average one to two of the three ethoxy groups per triethoxysilyl unit of a standard silane are split off as ethanol. Therefore, the ethanol emission occurs during the mixing and downstream process. This can cause problems in the mixing department if an optimal venting system is not ensured and if regulations limit the amount of volatiles. Furthermore, a classification into low–volatile organic compound (VOC) and VOC-free automobiles is under discussion in the United States [140]. Presilanized silica or white masterbatches, which have fewer or no reactive ethoxy groups, could be an alternative to the in situ reaction of silica and silane. Another approach is the use of silanes with less or no ethoxy units, for example, the mercaptosilane Si 363®, the silatrane XP Si 466 ,or the oligomerized protected mercaptosilane NXT*Z (Scheme 7.13).

3. Improved Processing

Compared to CB, the mixing of the two-component silica/silane system needs greater attention in the mixing department and special mixing devices with exact temperature control. Mixing is more time-consuming and also more expensive, and it has to be ensured that the silica/silane coupling is completed during the mixing. Porosity and even blisters formed by ethanol or ethanol/water mixtures not sufficiently removed from the batch can cause serious problems during downstream processing. Furthermore, optimal mixing control is needed to ensure batch-to-batch consistency comparable to that achieved with CB. Therefore, the processing has to be adjusted very precisely for each formulation. The compounders and process engineers have increased their knowledge and experience within recent years, but the raw material suppliers have also made a great effort to offer solutions. With the introduction of HD silica, the mixing time, and in some cases also the mixing cycles, could be reduced and the performance was also improved [5,141].

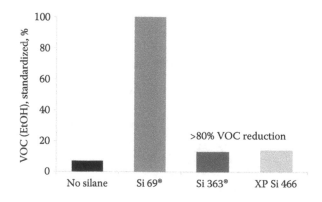

SCHEME 7.13 Ethanol emission during mixing.

The use of the more temperature-stable disulfide silanes has improved the tolerance toward temperature peaks, and the mixing temperature can be increased by ≈10°C [108,138,142,143]. This results in better batch-to-batch consistency and may allow shorter mixing times and a faster release of ethanol. With the introduction of several special processing aids and a completed silica/silane reaction, stickiness on the mills can be reduced. The construction of the mixers, especially the rotor geometry, to ensure optimal dispersion, distribution, and temperature control is also still improving [144].

Starting with large tangential mixers, the trend in the tire industry toward intermeshing mixers in sizes of 200–350 L (standard today 320 E) is dominant due to a better cooling efficiency and temperature control during mixing, important for the silica/silane technology. Also, the so-called tandem mixture technology where one intermeshing mixer is placed on the top of the next open, that is, without a ram, intermeshing mixer is one major development step [145]. Also the downstream processing that allows a fast cooldown of the batches is improved continuously.

The development of white- and black-filled powder rubbers (polymer–filler masterbatches) brought continuous mixing back into discussion [146–149]. The future will show whether this mixing concept is superior to the batch process.

V. APPLICATIONS OF SILICA AND SILANE

A. History

At the end of the nineteenth century, zinc oxide (ZnO) was discovered as a filler for rubber goods including tires, but between 1910 and 1920, there was a shift from the use of ZnO to gas blacks because of the shortage of zinc metal during World War I and the higher reinforcement afforded by CB [150]. In 1925 colloidal silicon dioxide (SiO_2) was investigated, and stress–strain experiments confirmed its reinforcing effect in NR [151]. In 1941, precipitated silica-containing magnesium oxide (MgO) proved to have a pronounced toughening and hardening effect [152] in rubber, and in 1942, high-temperature flame hydrolysis was introduced by Degussa (now named Evonik) for the production of pyrogenic silica [153,154]. The introduction of precipitated silica for commercial purposes by the Columbia Chemical Division of the Pittsburgh Plate Glass Co. followed in 1948 [82,155]. With this silica, significantly improved tensile and tear strengths of the vulcanizate were achieved. Due to the improved reinforcement, the white color and the low-conductivity silica was used in several industrial rubber goods.

The shoe sole industry rapidly became the most important consumer of white reinforcing fillers [156]. The use of silica resulted in a nonmarking "nuclear" sole with good abrasion, tear, and flex resistance (all premium-quality soles made in the United States contained silica in those times) [157]. Furthermore, it was used in belts, hoses, and rolls for flexibility and in nonmarking transparent and translucent compounds other application examples were wire and cable jackets and in special types of hoses, for which high abrasion resistance and a coloration for easy identification or visibility are desirable, silica was the best choice. Oil well parts based on chloroprene rubber and filled with silica showed exceptionally high resistance to gaseous and liquid petroleum fractions combined with the good durability needed in oil production. At that time, silica was also introduced in rubber adhesion compounds to improve rubber–metal bonding. Good overviews of the applications of silica in rubber goods are given by Wagner and coworkers [158,159].

In the late 1950s, further investigations came to the conclusion that the use of silica results in remarkably good reinforcing properties if a coupling agent, such as 3-methacryloxypropyltrimethoxysilane in peroxide-cured compounds and 3-mercaptopropyltrimethoxysilane in sulfur-cured compounds, is added [84,87,88,160,161] (see Section III.B). The use of high amounts of the highly reactive mercaptosilane was limited because the scorch time became too short. When Evonik introduced the more slowly curing coupling agent Si 69® (TESPT, bis(3-triethoxysilylpropyl)tetrasulfide) in 1972, the breakthrough of the silica–silane filler system in the rubber industry was achieved [90,91,118,162].

The filler system based on silica and the silane coupling agent Si 69® was first launched in winter tire tread compounds in 1974 and led to an improvement in winter performance never seen before [163,164]. This first big success came to an abrupt end when the compounds turned out to have a strong hardening effect accompanied by a loss in grip. After that, for a period of nearly 20 years, the silica alone or the silica–silane system was used in only small amounts in tire compounds to achieve special performance features such as improved cut and chip behavior and reduced heat build-up, with CB used as the main filler to ensure high reinforcement. In nontire applications, this filler system was used in nonmarking shoe soles, belts, engine mounts, and many colored industrial rubber goods [165]. For these applications, not only the sulfur-functional silanes but also silanes bearing reactive double bonds, such as vinyl- or methacrylsilanes, were in use. The use of silica for these niche applications changed in the early 1990s with the introduction of the "Green Tire" by Michelin in Europe [166]. Part of the concept was a tread compound filled with HD silica combined with an optimized amount of the sulfur-functional organosilane Si 69®, resulting in lower rolling resistance and better wet traction. Since then, the demand of the rubber industry for precipitated silica and silane has increased continuously, and now more than 80% of the original equipment tires in Europe contain this filler system in the tread compound. In the last few years, newly-designed mercaptosilanes are introduced as coupling agents in order to decrease the rolling resistance of passenger car tires further.

B. Silica in Shoe Soles

An important field of application for the silica–silane filler system is the segment of nonmarking colored shoe soles, which require good abrasion resistance, high stiffness, and high elasticity. Table 7.7 gives an example of a model shoe sole formulation and compares the vulcanizate data achieved with and without added silane (amounts of rubber ingredients are given in parts per hundred rubber, phr). As can be seen, the moduli, hardness, and resilience are higher for compound II, which contains 1.5 phr Si 69® as coupling agent. This silica–silane–rubber coupling results in improvements of 50% for the compression set value and 40% for the DIN abrasion resistance.

TABLE 7.7

Model Shoe Sole Compound with and without the Addition of Si 69® (TESPT); ULTRASIL® 7000 GR, BET 175 m²/g

		I	II
Formulation			
S-SBR, 25% styrene	Polymer	50	50
E-SBR, 23% styrene	Polymer	30	30
BR >96% *cis*-1,4	Polymer	20	20
ULTRASIL® 7000 GR	Silica	45	45
Si 69®	Silane	—	1.5
Other chemicals, oil 10, ZnO 3, stearic acid 1, Diethylene glycol (DEG) 1.5, antiaging 2, accelerator MBT 1.25, sulfur 1.8			
In-rubber properties			
ML(1+4), MU		95.0	90.0
$M_H - M_L$, 160°C, dN·m		4.6	5.5
Tensile strength, MPa		14.3	20.0
Modulus 100%, MPa		1.7	2.7
Modulus 300%, MPa		3.3	7.8
Modulus 500%, MPa		5.7	16.1
Elongation at break, %		810	570
Shore A hardness		67	72
Resilience, %		55.3	59.5
Compression set 22 h at 70°C, %		30.1	20.1
DIN abrasion loss, mm³		148	106

A widely used general shoe sole formulation based on 40 phr of isoprene rubber and 60 phr of butadiene rubber is shown in Table 7.8. Moduli and elasticity are at good levels, but further improvements in abrasion are achieved with the replacement of the conventional silica by the HD silica ULTRASIL® 7000 GR. Jogging shoe soles are often based on E-SBR* and BR† in a 50:50 blend filled with approximately 50 phr precipitated silica.

Table 7.9 shows another shoe sole formulation which leads to glassy transparency. The polymers are peroxide-cross-linked to prevent a yellowish color which is typical for a sulfur-cured system. A pyrogenic silica like AEROSIL® 200 V is used for the best transparency, and the oligomeric vinylsilane Dynasylan® 6498 is used as coupling agent for good abrasion resistance. By using the same formulation but one with 1 phr Dynasylan® 6498 and one without adding silane, the resulting Mooney viscosity is significantly reduced by the presence of silane (Figure 7.13a). This means that only a very small amount of Dynasylan® 6498 is necessary reduce the viscosity and therefore makes the process easier. At the same time, the very small amount

---

* ESBR: Emulsion-Solution Butadiene Rubber
† BR: Butadiene Rubber

TABLE 7.8
General-Purpose Shoe Sole Compound

		I	II
Formulation			
IR, 98% *cis*	Polymer	40	40
BR, 96% *cis*-1,4	Polymer	60	60
ULTRASIL® VN 3 GR	Silica	50	—
ULTRASIL® 7000 GR	Silica	—	50
Si 69®	Silane	2	2
ZnO		3	3
Stearic acid		2	2
Polyethylene glycol		2.5	2.5
Wax		1	1
Antioxidant		1.5	1.5
Naphthenic oil		5	5
Cure system: ZEPC 0.4, CBS 1, sulfur 1.7	Accelerator System		
In-rubber properties			
ML(1+4), MU		98	104
$M_H - M_L$, dNm		41.4	41.1
$t_{10\%}$, min		9.3	9.0
$t_{90\%}$, min		11.4	11.2
Mooney scorch t_5 (121 °C), min		42	38
Tensile strength dumbbell, MPa		14.9	16.5
Modulus 100%, MPa		2.9	2.9
Modulus 200%, MPa		5.7	5.9
Modulus 300%, MPa		9.4	9.6
Elongation at break, %		430	450
Shore A hardness		71	71
Tear resistance (Die C), N/mm		49	52
Tear resistance (trouser test), N/mm		17	15
Ball rebound, 60°C, %		62	62
DIN abrasion loss, mm³		83	64

of Dynasylan® 6498 significantly improves abrasion resistance (Figure 7.13b) and increases the moduli.

In addition to the aforementioned polymers, foamed urethanes are used in light-weight soling material for fashionable thick heels that have very good abrasion resistance properties but the big disadvantage of being slippery [167].

At this time, there are two main trends for new developments for shoe soles: sport shoe soles that are thinner and lighter and multifunctional soles with special compounds for high rebound, good wet traction, and good abrasion resistance. In addition, fashion shoes have been introduced with extra thick soles of low density (microcellular) [168].

TABLE 7.9

Transparent Shoe Sole Compound

Formulation

BR	Polymer	70
SBR	Polymer	10
IR	Polymer	20
Aerosil® 200 V	Silica	30
Dynasylan® 6498	Silane	1
DCP 99	Accelerator	0.3

In-rubber properties

ML(1 + 4), MU	104
$M_H - M_L$, dN m	32.3
$t_{10\%}$, min	1.1
$t_{90\%}$, min	19.5
Mooney scorch t_5 (121 °C), min	14
Tensile strength dumbbell, MPa	13.7
Modulus 100%, MPa	1.6
Modulus 300%, MPa	4.6
Elongation at break, %	650
Shore A hardness	60
Tear resistance (Die C), N/mm	44
Tear resistance (trouser test), N/mm	10
DIN abrasion loss, mm³	29
Akron abrasion, C.C.	0.051

C. SILICA IN INDUSTRIAL RUBBER GOODS

1. Seals, Cables, Profiles, and Hoses

The silica–silane filler system is used for industrial rubber goods that require high reinforcement combined with the possibility to manufacture white or colored products such as seals, hoses, and profiles. For belt applications, the silica is used to improve tear resistance. And in some dynamic applications, the silica–silane filler system is responsible for the reduction of the heat buildup. As an example, the benefits of reinforcement by the addition of 2 phr Si 69® (TESPT) to a sealing compound based on NBR are demonstrated in Table 7.10. The moduli and tear resistance are higher for compound II with Si 69® than for the silica compound without silane, I. The compression set is 25% lower, and the abrasion resistance is improved by nearly 60%.

Washing machine sealing compounds require good compression set, high flexibility, good tear and swelling resistance, and the ability to be colored. Therefore, silica in combination with a coupling agent is the filler system of choice. As an example, Table 7.11 shows washing machine sealing compounds that use silica with surface areas of 125 and 165 m²/g. Compound II with LSA silica (ULTRASIL® VN 2 GR) shows several advantages compared to compound I with medium-surface-area silica ULTRASIL® VN 3 GR: lower Mooney viscosity, higher moduli, and good

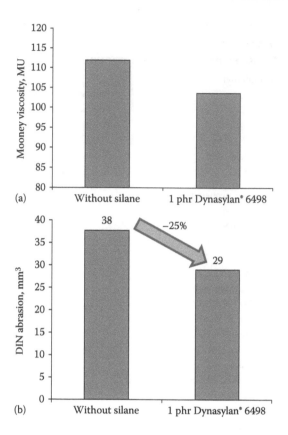

FIGURE 7.13 (a) Reduction of the viscosity by using only a small amount of silane and (b) reduction of the DIN abrasion by using only a small amount of silane.

compression set. Blends of natural fillers and silica can be used for a further adjustment of the flexibility, compression set and swelling resistance.

A typical formulation for oil seals is shown in Table 7.12. The basic acrylic rubber (ACM) was tested together with different silanes. PTEO as a monofunctional silane that can couple only to the silica but not to the polymer delivers a poor reinforcing behavior. AMEO shows the highest reactivity, the highest cross-linking density, but also a prescorch tendency. The use of glycidyloxypropyltriethoxysilane (GLYEO) or methacryloxypropyltrimethoxysilane (MEMO) leads to an increased cross-link density and to an efficient reinforcement (Figure 7.14).

In some cases, pre-reacted silicas are also used for industrial rubber goods, because the silanization reaction, commonly established during the mixing process, is already completed. Soft compounds mixed at rather low temperatures have particular need of such pre-reacted silica. Furthermore, processing is easier, mixing time may be reduced, and the required reinforcement is sufficiently high compared to an in situ modified silica [94]. Evonik offers several types of pre-reacted silica (brand name COUPSIL®) based on silica with 125 and 165 m^2/g and modified with vinylsilanes and sulfur-functional silanes [169].

TABLE 7.10

Comparison of the In-Rubber Data for a Sealing Compound Based on NBR with and without the Addition of Si 69® as Coupling Agent

Formulation		I	II
NBR, 35% ACN (acrylonitrile rubber)	Polymer	100	100
ULTRASIL® 7000 GR	Silica	45	45
Si 69®	Silane	—	2
Other chemicals, oil 2, ZnO₃, stearic acid 1, DEG 1.5, antiaging agent 3			
Cure system, CBS 1.8, DPG 1.2, ZBDC 0.15, sulfur 2.2	Accelerator system		
In-rubber properties			
ML (1+4), MU		28 €	24 €
$M_H - M_L$ (160°C), dN m		9.1	9.7
Tensile strength, MPa		19.8	19.5
Modulus 100%, MPa		15	19
Modulus 300%, MPa		32	7.1
Modulus 500%, MPa		72	16.4
Elongation at break, %		660 €	550 €
Shore A hardness		61 €	64 €
Tear resistance, Die C, N/mm		31 €	50 €
Resilience, %		39 €	42 €
Compression set 22 h at 70°C, %		35.6	28.2
DIN abrasion loss, mm³		187 €	119 €

Industrial rubber goods that require excellent compression set and high heat resistance are destined for peroxide curing, but their tear resistance is strongly deteriorated compared to that of sulfur-cured articles [170]. With the use of pure silica or silicates, the cross-link density and hardness can be adjusted precisely, but to improve the mechanical properties, vinylsilanes such as VTEO are needed to establish the filler-to-rubber coupling [96]. As in the case of CB-filled compounds, the addition of coagents such as TAC and TRIM can also be recommended for compounds with vinylsilanes. These activators strongly increase the cross-link density (see Section III.B.3). Table 7.13 demonstrates the advantage of Dynasylan® VTEO (Si 225) as coupling agent in an H-NBR formulation. The addition of only 1.5 phr Dynasylan® VTEO leads to much higher moduli, a strong reduction in the compression set, and lower swell values in toluene, commonly required for example for fuel and oil seals.

A peroxide-curing system is used together with white fillers and vinylsilanes in many cables, seals, profiles, and hose compounds based on ethylene propylene diene monomer rubber (EPDM), on ethylene propylene rubber (EPM), or ethylene vinyl acetate copolymer (EVA) [171]. Due to the low flash point of 38°C of the vinylsilane and the requirement of a one- or two-stage mixing process, pre-silanized natural and synthetic fillers are widely used for such applications. Calcined clays are often used

TABLE 7.11
Washing Machine Sealing Compounds

		I	II
Formulation			
EPDM oil ext (paraffin), 5% diene, 54% ethylene	Polymer	150	150
ULTRASIL® VN 3 GR	Silica	60	—
ULTRASIL® VN 2 GR	Silica	—	60
Si 69®	Silane	2	2
Other chemicals, oil 30, ZnO (active) 4, stearic acid 2, bisphenolic antioxidant 1.5			
Cure system, ZBPD 2, ZBED 0.8, MBT 1, DPG 2, sulfur 80% on bound polymer	Accelerator system		
In-rubber properties			
ML(1+4), MU		62	55
HITEC 170°C; 0.5°			
$M_H - M_L$, dN m		5.2	5.2
$t_{10\%}$, min		3.2	2.4
$t_{90\%}$, min		16.6	14.6
Tensile strength, MPa		14.6	11.9
Modulus 100%, MPa		1.5	1.4
Modulus 500%, MPa		9.8	11.1
Modulus 500%/100%		6.5	7.9
Elongation at break, %		620	520
Shore A hardness		53	52
Compression set, 22 h at 70°C, %		17	15
Compression set, 70 h at 100°C, %		48	43
DIN abrasion, 10 N, mm³		166	146
Surface roughness (topography) %		24.3	13.2

TABLE 7.12
Typical Formulation for Oil Seals Based on ACM Rubber

Formulation		
HyTemp AT 71	Polymer	100
ULTRASIL® VN 2 GR	Silica	50
Silane		Variable
Stearic acid		2
Struktol WB 222		2
Vulkanox OCD-SG		2
Vulkanol 81		5
Na Stearate 80		3.5
Sulfur		0.4

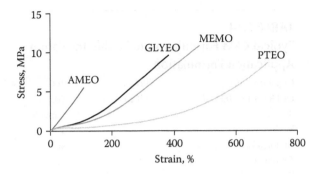

FIGURE 7.14 Stress–strain curves of ACM compounds with different silanes.

TABLE 7.13

Fuel Sealing Compound with and without the Addition of Dynasylan® VTEO; ULTRASIL® VN 2 P, BET 125 m²/g

		I	II
Formulation			
H-NBR	Polymer	100	100
ULTRASIL® VN 2 P	Silica	60	60
Dynasylan® (VTEO)	Silane	—	1.5
Fatty acid ester 10, TAC 3, initiator (40%) 7			
In-rubber properties			
ML(1+4), MU		112	103
$M_H - M_L$ (165°C), dN m		32.1	69.8
Tensile strength, MPa		18.1	21.2
Modulus 50%, MPa		2.5	4.2
Modulus 200%, MPa		5.1	10.8
Elongation at break, %		600	210
Shore A hardness		87	91
Swell (Toluene), %		150	100
Compression set, 24 h at 150°C, %		86.3	32.6

in cables for high-voltage applications because of their good insulating properties and ease of mixing and processing.

Another example for a cable insulation is a formulation based on chlorosulfonated polyethylene rubber (CSM). Table 7.14 shows a typical formulation where different silanes were tested. Significant reinforcement was achieved with Si 69®, AMEO, and GLYEO. AMEO shows a prescorch tendency. Therefore, the use of Si 69® and GLYEO is favored for well-balanced compound properties (Figure 7.15).

2. Rice Hulling Rollers

The important requirements for rice hulling rollers are nonstaining properties, good abrasion resistance, and high durability combined with an acceptable level of the

TABLE 7.14

Typical CSM Formulation for Cable Insulation Application Formulation

Hypalon® 4085	Polymer	100
ULTRASIL® 7000 GR	Silica	50
Silane		Variable
MgO		4
Naftolen ZD		40
Tetrone A		2.8
DEG		3

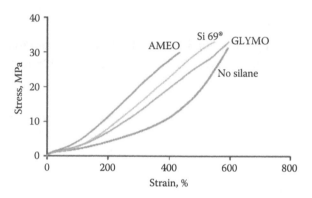

FIGURE 7.15   Stress–strain behavior of different silanes in a typical CSM formulation.

Mooney viscosity for production. The silica–silane filler system is best suited for this application, as demonstrated in Table 7.15 with a typical compound for Asian countries based on styrene-butadiene rubber (SBR) and filled with 90 phr silica. As can be seen, the addition of only 4 phr of the coupling agent Si 69® results in a reduced Mooney viscosity of more than 20% and remarkable improvements in abrasion resistance and modulus 200% (of ca. 50%). The use of the high-surface-area silica ULTRASIL® 9000 GR leads to the best reinforcing behavior.

Table 7.16 shows a typical formulation for the Indian market based on NBR. The increase of the surface area of the silica leads to an improved abrasion resistance. (Figure 7.16).

3.   Soft Rollers

Another example of a nonstaining industrial rubber good is a soft roller. There are two polymers commonly used: NBR, which is resistant to solvents such as ethers and gasoline, and EPDM, which is resistant to solvents such as ketones, alcohols, and esters. Table 7.17 shows a general formulation for a soft roller compound based on EPDM. The use of an HD silica, for example, ULTRASIL® 7000 GR, simplifies the mixing process of such soft compounds that generate low shear forces during mixing.

TABLE 7.15
Typical SBR Rice Hulling Roller Formulation and In-Rubber Properties

		I	II	III	IV	V	VI
Formulation							
SBR 1502	Polymer	100	100	100	100	100	100
Pliolite S6H	High-styrene resin	30	30	30	30	30	30
ULTRASIL® 7000 GR	Silica	90	—	—	90	—	—
ULTRASIL® VN 3 GR	Silica	—	90	—	—	90	—
ULTRASIL® 9000 GR	Silica	—	—	90	—	—	90
Si 69®	Silane	—	—	—	4	4	4
Other chemicals, ZnO 6; stearic acid 2; tackfier 8; naphthenic oil 8; PEG 4; BHT 1.2; wax 2							
Cure system							
Sulfur 4.5; DPG 1.1; CBS 1.2	Accelerator system						
In-rubber properties							
MS(1+4), (MU)		44	40	65	31	30	40
$M_H - M_L$ (150°C), dN m		24.4	24.9	28.7	24.4	23.6	29.8
$t_{10\%}$, min		4.1	4.7	3.6	4	4.7	3.7
$t_{90\%}$, min		24.8	23.8	27.6	20	20.8	23.5
Tensile strength (dumbbell), MPa		12.4	11.4	13.3	15.4	14.7	16.2
Modulus 100%, MPa		4.8	4.7	4.4	7.5	8	7.4
Modulus 200%, MPa		6.6	6.4	6.2	12	12.3	11.9
Modulus 300%, MPa		9.5	9.2	8.9	—	—	—
Elongation at break, %		390	370	420	280	260	290
Shore A hardness		93	93	93	93	94	94
DIN abrasion, mm³		248	258	239	196	207	185

As can be seen in Table 7.17, the better dispersion of ULTRASIL® 7000 GR results in a lower Mooney viscosity, a higher tensile strength, greater elongation at break, and better surface quality compared to the conventional silica ULTRASIL® VN 3 GR.

4.   Conveyor Belts and Power Transmission Belts

Conveyor or power transmission belts demand high abrasion resistance, good moduli, good fatigue resistance, and low compression set. The silica/silane system is mainly used in colored articles. The basis for conveyor belts is normally NR or SBR. For power transmission belts, chloroprene rubber (CR), EPDM, or HNBR is often used in combination with silica and the coupling agent CPTEO to ensure good abrasion behavior.

TABLE 7.16
Typical NBR Rice Hulling Roller Formulation and In-Rubber Properties

		I	II
Formulation			
NBR 34% ACN	Polymer	100	100
ULTRASIL® VN 3 GR	Silica	65	—
ULTRASIL® 7000 GR	Silica	—	65
Si 69®	Silane	5	5
Other chemicals, ZnO$_5$, stearic acid 2, DOP 5, DEG 0.7, bisphenolic antioxidant 1.5, phenolic resin 20			
Cure system, CBS 2.2, sulfur 2.1	Accelerator system		
In-rubber properties			
ML(1+4), MU		58	57
$t_{10\%}$; 150°C, min		14.6	12.9
$t_{95\%}$; 150°C, min		85.5	80.8
Tensile strength, MPa		19.4	19.2
Modulus 100%, MPa		6.7	6.6
Modulus 200%, MPa		12.9	13.0
Elongation at break, %		310	290
Shore A hardness		90	91
Die A tear, N/mm		44	43
Compression set, 70 h at 100°C, %		51.7	52.2
DIN abrasion loss, mm^3		93	88
Dispersion; surface roughness (topography), %		13.8	8.1

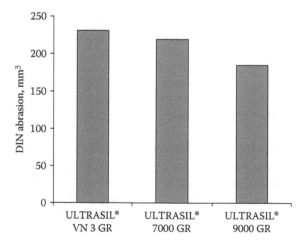

FIGURE 7.16   DIN abrasion of a typical NBR rice hulling roller formulation.

TABLE 7.17
Soft Roller Model Compound

		I	II
Formulation			
EPDM, 4% diene, 56% ethylene	Polymer	150	150
ULTRASIL® VN 3 GR	Silica	60	—
ULTRASIL® 7000 GR	Silica	—	60
Si 69®	Silane	4.8	4.8
Other chemicals, oil 50; ZnO 4; stearic acid 2	Accelerator system		
Cure system, MBT 1, DPG 1.2, ZMBT 1.5,			
ZBPD (50%) 2, ZDMC 0.8, sulfur 1.5			
In-rubber properties			
ML(1 + 4), MU		47	42
$t_{10\%}$; 150°C, min		3.6	3.5
$t_{95\%}$; 150°C, min		14.4	14.7
Tensile strength, MPa		6.3	9.3
Modulus 100%, MPa		1.0	1.0
Modulus 300%, MPa		3.7	3.9
Elongation at break, %		430	500
Shore A hardness		46	46
Tear resistance, N/mm		5	5
Compression set, 70 h at 100°C, %		14.7	15.5
Ball rebound, 23°, %		60	58
Dispersion; surface roughness (topography), %		24.9	11.0

5. Engine Mounts, Belts, and Air Springs

In engine mounts, driving belts, and air spring compounds [172], blends of CB and white fillers together with organosilanes (frequently pre-treated white fillers like COUPSIL®) are used to improve the dynamic properties because of the increasing demand for heat resistance. EPDM belt compounds filled with LSA silica have been developed [173] and are already in use.

6. Boots and Bellows

A lot of boots and bellows formulations are based on chloroprene polymer. Table 7.18 shows a typical CR formulation where different silanes were tested in comparison to the CB N 220. The highest reactivity is indicated by mercapto- and aminosilanes but with a tendency to prescorch behavior for aminosilanes. The use of the mercaptosilane Si 263 delivers the same reinforcing behavior than the CB N 220 (Figure 7.17).

7. Floor Coverings

Besides the common PVC floorings, there are also floorings based on elastomers such as E-SBR and EPDM. For the production of such colored floorings, aluminum

TABLE 7.18

Typical Boots and Bellows Formulation on the Basis of CR

Formulation		I	II
Baypren 210	Polymer	100	100
ULTRASIL® 7000 GR	Silica	50	—
CORAX® N 220	CB	—	50
Silane		2	—
Stearic acid		0.5	0.5
Maglite D		4	4
Disflamoll DPK		15	15
Naftolen ZD		10	10
Vaseline		2	2
Vulkanox 4010		1	1
ZnO		5	5
ETU-80		1	1

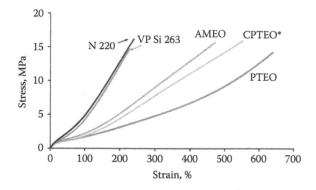

FIGURE 7.17   Stress–strain behavior of different silanes in a CR formulation.

silicates, pre-treated silicas, or conventional silicas are added to establish the required hardness (Shore A hardness 80–90) and good wear resistance. Furthermore, compression set needs to be low.

A typical flooring formulation contains approximately 85 phr E-SBR (or 65 phr E-SBR and 20 phr BR) in combination with 15 phr high-styrene resin masterbatch and is filled with 80–120 phr of non-reinforcing mineral fillers to increase hardness and 30 phr aluminum silicate (e.g., ULTRASIL® AS 7) or 20 phr of a silica (e.g., ULTRASIL® VN 3 GR) to guarantee the required abrasion resistance. The high cross-link density is achieved by curing with 2 phr accelerator and 3 phr sulfur. Because such compounds show high electrical resistance ($>10^{10}$ Ω) further additives are needed to establish antistatic behavior.

8. Golf Balls

In golf balls, normally high-*cis* BR and an acrylate such as butylene glycol dimethacrylate (BDMA) are used together with silica to ensure high hardness and elasticity. Additionally, BDMA acts as a softener in the uncured compound and as a hardener in the vulcanizate (Table 7.19). The cross-linking is usually carried out with dicumyl peroxide (DCP).

D. Silica in Tires

1. Passenger Car Tire Treads

The main application of the silica/silane filler system is currently the segment of passenger car tire treads (the so-called Green Tire, launched in 1992 by Michelin in Europe) [166,174], where it achieves a reduction in fuel consumption due to a considerable decrease in rolling resistance and improvement of wet traction. This extension of the "magic triangle of tire performance" (Scheme 7.12) was made possible with the use of an HD silica as the main filler, modified with a sulfur-functional silane (TESPT), and a combination of a high-T_g solution styrene butadiene copolymer (S-SBR) with a low-T_g 1,4-polybutadiene (BR). The high-*cis* 1,4-polybutadiene and an excellently dispersed silica, modified with a relatively high amount of TESPT (Si 69®), are needed to reach an abrasion resistance comparable to that of compounds with CB alone. In particular, the improved wet grip performance, as an important safety aspect, and the reduced fuel consumption, as an environmental aspect, made this tire concept a great success in Europe.

TABLE 7.19
Golf Ball Model Compounds

		I	II	III
Formulation				
BR >96% *cis*-1,4	Polymer	100	100	100
ULTRASIL® 7000 GR	Silica	30	50	60
Dynasylan® (VTEO)	Silane	1	1	1
BDMA DL 75		20	40	40
Initiator (40%)		10	10	10
In-rubber properties				
ML(1+4), MU		65	97	144
MDR 180°C; 2°				
$t_{10\%}$, min		0.3	0.2	0.2
$t_{90\%}$, min		3.9	4.4	3.4
Shore A hardness		96	97	98
Shore D hardness		61	78	79
Ball rebound, 60°C, %		65.4	53.4	55.5
Resilience, %		62.7	55.8	54.6

TABLE 7.20
Typical Green Tire Tread Formulation

		phr
Stage 1		
S-SBR[a]	Polymer	96
BR *cis*-1,4	Polymer	30
Silica		80
X 50-S® (50% Si 69® on CB)		12.8
ZnO		3
Stearic acid		2
Aromatic oil		10
Wax		1
6PPD		1.5
Stage 2		
Stage 3		
CBS	Accelerator	1.5
DPG	Accelerator	2
Sulfur		1.5

Table 7.20 gives an example of such a simplified tread compound, which is commonly mixed in a two- to four-stage mixing process, with one or two remill cycles, to establish optimum silica dispersion and finish the silane coupling reaction to the silica (see Section IV.B).

Figure 7.18 shows the improvements attained in rolling resistance (hysteresis loss tan δ at 60°C) and wet traction (measured by the Laboratory Abrasion Tester LAT 100 (system Grosch) [175–177] with the gradual replacement of the tread black N 234 by the silica/silane reinforcing system in an S-SBR/1,4-BR compound [178–180]. The hysteresis loss is improved by 50%, which results in about 20% reduction in

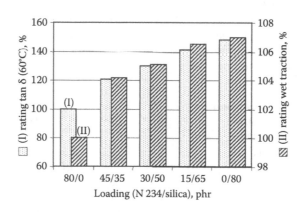

FIGURE 7.18 Improvement in hysteresis loss at 60°C (I) and wet traction (II) through gradual.

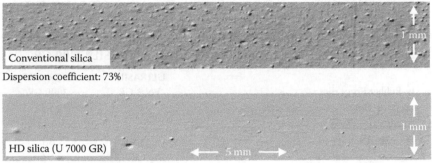

FIGURE 7.19 Dispersion of HD and conventional silica, measured by topography. (From Geisler, H., Bestimmung der Mischgüte, *Presented at the DIK-Workshop*, Hannover, Germany, November 27–28, 1997.)

rolling resistance and 3%–4% less fuel consumption [181]. According to the LAT 100 measurements, the wet traction is improved by 7%, resulting in shorter wet braking distances (ABS breaking).

To demonstrate the benefit of an HD silica in a tire tread compound, ULTRASIL® 7000 GR is compared to a conventional silica in the Green Tire tread formulation in Table 7.20. Figure 7.19 shows their dispersion behavior. The conventional silica ULTRASIL® VN 3 GR gives fair dispersion, with a peak area of 15% measured by the topography method (surface roughness analysis acc. to [182,183]) and a dispersion coefficient of 73% [184]. The HD silica ULTRASIL® 7000 GR shows excellent dispersion with a peak area of only 1% and a dispersion coefficient of 97%. Compared to ULTRASIL® VN 3 GR, ULTRASIL® 7000 GR is less compacted and has a higher void volume and a higher DBP and DOA number, due to a special precipitation and drying process. These characteristics explain the strongly improved dispersion behavior of this HD silica.

Table 7.21 shows the compound and vulcanizate data. The better the dispersion of the silica (within the same surface area range) and the weaker the filler–filler interaction, the lower is the Mooney viscosity. For the same reason, the heat buildup and tan δ at 60°C are also lower in the case of ULTRASIL® 7000 GR. The abrasion behavior of these two silica compounds was tested with the LAT 100 (Figure 7.20) and on the road (Figure 7.21). Both tests confirmed the expected improvement in abrasion behavior with improved dispersion. According to LAT 100 measurements, wet traction of these compounds is similar.

The use of silica with even higher surface areas, which means decreased primary particle sizes, and adjusted silane amounts further improves tread wear. However, an increase in the Payne effect, leading to a higher hysteresis loss, as well as higher adsorption of accelerators, has to be considered. Therefore, an exchange of silica with differences in surface area should be accompanied at least by an adjustment in the silane coupling and curing system.

TABLE 7.21

Comparison of Green Tire Tread Compounds with the Conventional Silica ULTRASIL® VN 3 GR and the HD Silica ULTRASIL® 7000 GR

In-Rubber Properties	ULTRASIL® VN 3 GR	ULTRASIL® 7000 GR
ML(1+4) (MU)	75	67
HITEC, 3.0°, 165°C		
$M_H - M_L$, dN m	8.1	8.1
$t_{10\%}$, min	6.0	6.1
$t_{90\%}$, min	12.0	11.6
Tensile strength, MPa	17.8	17.6
Modulus 100%, MPa	3.0	2.6
Modulus 300%, MPa	12.3	13.0
Modulus 300%/100%	4.1	5.0
Elongation at break, %	370	370
Shore A hardness	68	64
Ball rebound, 0°C, %	12.0	10.8
Ball rebound, 60°C, %	55.2	55.8
Goodrich flexometer, 20°C, 25 min, 0.225 in., 108 N		
Heat buildup,°C	112	105
Permanent set, %	4.2	3.5
MTS, 16 Hz, 50 N preload, 25 N amplitude force		
E*, 0°C, MPa	37.7	32.7
tan δ, 0°C	0.431	0.461
E*, 60°C, MPa	10.4	9.2
tan δ, 60°C	0.127	0.113
Dispersion, peak area (topography), %	15.2	1.0

In the following, three different silicas are investigated in a Green Tire tread compound with constant and on the surface area (CTAB) adjusted silane concentrations (Si 266® and Si 363®) (Tables 7.22 through 7.24) in order to demonstrate the possibilities given for compounding with different silica specific surface areas and silanes [185].

Si 266®:

EtO, OEt ... OEt, OEt
EtO-Si⌇⌇⌇$S_{2,15}$⌇⌇⌇Si-OEt

Si 363®:

OR
RO–Si–(CH$_2$)$_3$ –SH
OR

with R = C$_2$H$_5$ or alkylpolyether

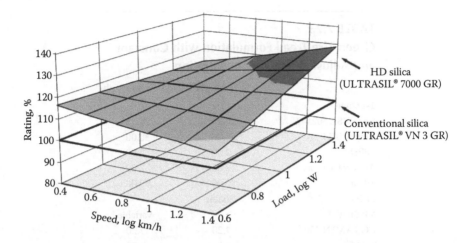

FIGURE 7.20 Abrasion behavior of HD and conventional silica, measured by the LAT 100 (system Grosch) under various severities.

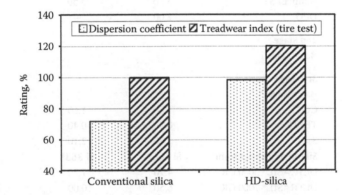

FIGURE 7.21 Improved tread wear on the basis of improved dispersion (calculation of the dispersion coefficient in accordance with Ref. [184]).

TABLE 7.22
Typical Analytical Properties of Three HD Silicas

		ULTRASIL® 5000 GR	ULTRASIL® 7000 GR	ULTRASIL® 9000 GR
CTAB$_{specific surface area}$	m²/g	110	160	200
BET$_{MP; specific surface area}$	m²/g	115	175	235
DOA$_{oil absorption number}$	g/100 g	180	200	225
Volatiles$_{105°c/2h}$	%	5.5	5.2	5.5
pH$_{5\% in water}$		6.2	6.5	6.5

TABLE 7.23

Green Tire Tread Formulation with Constant and Adjusted Silane Concentration

	Si 266®, phr	VP Si 363®, phr
1st stage		
Buna VSL 5025–2	96.25	96.25
Buna CB 24	30.00	30.00
6PPD	2.00	2.00
Antiozonant wax	2.00	2.00
Silica	90.00	90.00
Si 266®	Variable	
VP Si 363®		Variable
CORAX® N 330	7.20	7.20
TDAE	8.75	8.75
ZnO	2.00	2.00
Stearic acid	1.00	1.00
TMQ	1.50	1.50
Aktiplast ST	3.50	3.50
DPG	2.00	—
2nd stage		
Batch stage 1		
3rd stage		
Batch stage 2		
CBS	1.60	1.60
TBzTD	0.20	0.40
S_8	2.10	2.10
Silane Adjustment/phr	**Si 266®**	**Si 363®**
ULTRASIL® 5000 GR	4.70	7.20
ULTRASIL® 7000 GR	6.50	10.00
ULTRASIL® 9000 GR	8 10	12.50

The silanes in this test series have a different molecular weight (M_n (Si 363®) = 988 g/mol; M_n (Si 266®) = 480 g/mol). This means that 10.0 phr of Si 363® corresponds to less than 40% Si units as for 6.5 phr of Si 266®. The viscosity of the green compounds generally increases with increasing surface area of the silica (Figure 7.22). A higher surface area at the same filler volume leads to a shorter particle–particle distance. Therefore, a higher filler–filler network via Si–O⋯H–O–Si bonding exists (Figure 7.23). An adjusted silane content leads to a more comparable filler–filler network for all silicas.

The vulcanizates with the high-surface-area silica ULTRASIL® 9000 GR shows the highest tensile strength and elongation. Those with the LSA silica ULTRASIL® 5000 GR contains the highest moduli and lowest elongation at break. This is independent of the used silane. There are two possible explanations for this behavior:

TABLE 7.24

Mixing Procedure

1st	stage	GK 1.5 E, 80 rpm, chamber temp. 80°C; fill factor 0.56
0,0′	–0,5′	Polymers
0,5′	–1,0′	Vulkanox HS; Vulkanox 4020
1,0′	–2,0′	1/2 silica; silane; ZnO
2,0′		Raise ram and clean
2,0′	–3,0′	1/2 silica; premix: CB and oil; other ingredients
3,0′		Raise ram and clean
3,0′	–4,0′	Mix at 155°C (rpm adjustment if necessary)
4,0′		Dump; pass through a wide nip on an open mill one time
2nd	**stage**	GK 1.5 E, 80 rpm, chamber temp. 90°C; fill factor 0.53
0,0′	–1,0′	Batch stage 1
1,0′	–3,0′	Mix at 155°C (rpm adjustment if necessary)
3,0′		Dump; pass through a wide nip on an open mill one time
3rd	**stage**	GK 1.5 E, 40 rpm, chamber temp. 50°C
0,0′	–0,5′	Batch stage 2
0,5′	–2,0′	Ingredients 3rd stage
		Dump and homogenize on open mill
2,0′		Cut 3 * left; 3 * right; roll up and pass 3 * through a tight nip and 3 * through a wide nip; sheet off

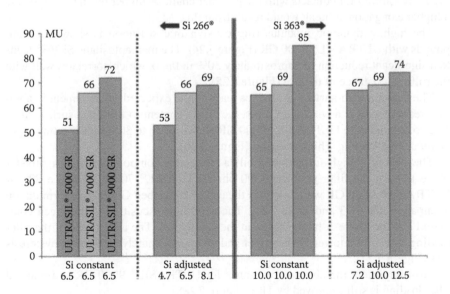

FIGURE 7.22 The Mooney viscosity depends on surface area of the silica (ML(1+4) at 100°C/MU, third mixing step [final mix]).

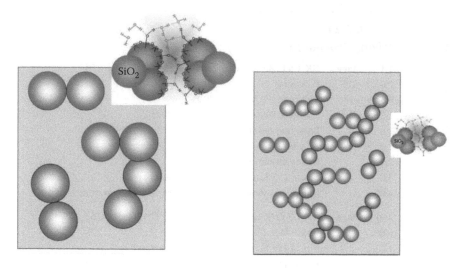

FIGURE 7.23 Model: a higher surface area at the same filler volume leads to shorter particle–particle distance.

One explanation is that the accelerators are adsorbed at the silica surface. Therefore, a higher-surface-area silica adsorbs more accelerator. This leads to a lower cross-linking density, a lower moduli, and a higher elongation at break. The other explanation is that the polymer interacts in a different way with the hydrophobic fillers. Higher-surface-area silica means smaller particles. As a result of this, there is a higher possibility for contacts with the polymer chain. A higher tensile strength and a higher elongation at break would result from this.

The highest dynamic modulus (highest dynamic stiffness) is shown in compounds with ULTRASIL® 9000 GR (Figure 7.24). The mercaptosilane Si 363® leads to a significant reduction of approximately 30% in tan δ, which correlates well with the rolling resistance of the tire (Figure 7.25).

The use of a high-surface-area silica leads to the expected improvement in abrasion resistance. It is advisable to adjust the silane content to the silica surface area. The compound with ULTRASIL® 9000 GR shows an up to 30% higher rating especially at high loadings (high severity) (Figure 7.26).

The use of a high-surface-area silica gives the compounder also a broader range to vary the filler content. A 90 phr ULTRASIL® 7000 GR and a 80 phr ULTRASIL® 9000 GR were tested in the earlier described Green Tire formulation (compare Table 7.23 and Table 7.24). The resulting viscosities of the green compound and the Shore A hardness are at the same level. The adjustment of the filler loading results in a lower Payne effect and an approximately 10% lower hysteresis loss (Figure 7.27).

Most important the abrasion resistance for ULTRASIL® 9000 GR at a reduced filler loading is still improved by 11% (Figure 7.28).

This example demonstrates the wide range of possibilities a compounder has to play with the silica/silane reinforcing system.

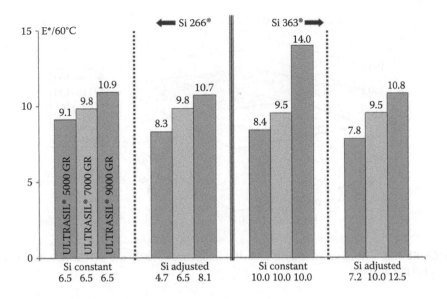

FIGURE 7.24 Dynamic modulus E* at 60°C in MPa (Zwick, 50 N ± 25 N; 16 Hz).

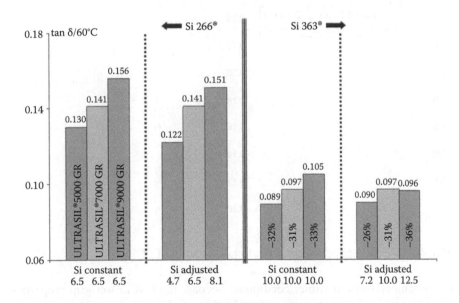

FIGURE 7.25 Loss factor tan δ at 60°C (Zwick, 50 N ± 25 N; 16 Hz).

2. Winter Tire Treads

The excellent wet grip performance and reduced hysteresis loss achievable with the silica/silane filler system offers the possibility to further improve winter tire tread compounds. First, the increase in the amount of polymers with a low or moderate glass transition temperature (T_g) results in a lower stiffness needed for a high grip

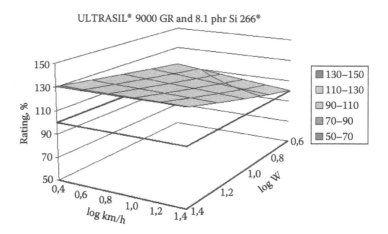

FIGURE 7.26 Abrasion resistance LAT 100 (reference = 100% = ULTRASIL® 5000 GR and 6.5 phr Si 266®).

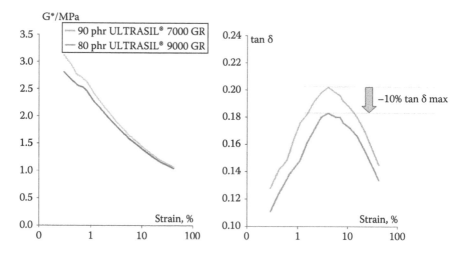

FIGURE 7.27 RPA measurement (second strain sweep [60°C; 1.6 Hz]).

at low temperatures in winter conditions. Second, the loss in wet grip properties known for such a T_g shift is compensated by the use of a medium-surface-area silica (CTAB, 165 m²/g). To further improve the wet traction, the T_g of the polymer has to be increased, which consequently also increases stiffness at low temperature and deteriorates rolling resistance. These drawbacks can be compensated by the use of a silica with a lower surface area (110–130 m²/g), leading to lower dynamic stiffness and reduced hysteresis loss. In addition to the optimized tread compound, a modern tread design is essential for the performance jump achieved within the last decades, as demonstrated in Table 7.25.

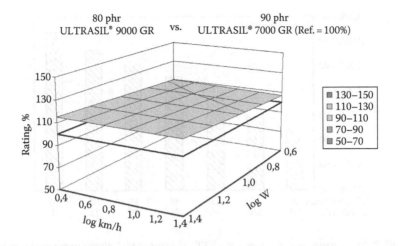

80 phr
ULTRASIL® 9000 GR   vs.   90 phr
ULTRASIL® 7000 GR (Ref. = 100%)

FIGURE 7.28   Abrasion resistance LAT 100 (reference = 100% = ULTRASIL® 7000 GR).

TABLE 7.25
Performance Comparison between Winter Tires of the 1980s and Modern Winter Tires, Speed-Rated *T* (192)

	Improvement, %
Dry handling	20
Dry braking	10
Wet handling (subjective)	30
Wet braking	30
Aquaplaning (transverse)	30
Snow properties	20
Ice properties	20
Rolling resistance	35
Abrasion resistance	20
Pass-by noise, dB (A)	–3–4

Skid resistance was identified as one critical property of a winter tire tread compound [186]. Skid resistance in the temperature range of 0°C to –15°C can be related to the compliance 1/E* of the tread material. The higher the compliance 1/E*, the better the skid resistance should be. Softer compounds with higher resilience were also found to have a higher friction on ice in the temperature range of 0°C to –20°C [187]. Both lab predictors of the behavior of wet traction are not as good as the predictor for rolling resistance, the tan δ at 60°C. However, both deliver a first hint.

As stated earlier, such a reduction in stiffness is achieved by the use of HD LSA silica [188]. The effect on compliance of a gradual replacement of the ULTRASIL® 7000 GR with a CTAB surface area of 160 m²/g by ULTRASIL® 5000 GR with a

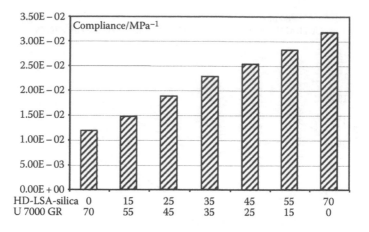

FIGURE 7.29 Change in compliance at –20°C, measured by Eplexor under constant force, as ULTRASIL® 7000 GR is replaced with an HD silica (see text).

TABLE 7.26
Winter Tire Tread Formulation

		phr
Formulation		
S-SBR (25% styrene)	Polymer	40
BR (96% *cis*-1,4)	Polymer	45
NR	Polymer	15
Silica		70
N 375	CB	20
X 50-S® (50% Si 69® on CB)		6
Other chemicals, oil 35, ZnO 3, stearic acid		
2, wax 1.5, 6PPD 1, TMQ 1		
Cure system, ZBEC 0.1, CBS 1.7, DPG	Accelerator	
1.7, sulfur 1.4	system	

surface area of 115 m^2/g is shown in Figure 7.29. The winter tire test formulation is depicted in Table 7.26. It is clearly demonstrated that the compliance at –20°C is gradually improved with increasing content of the LSA silica.

3. Truck Tires

The first use of silica (5–10 phr without silane) in truck tire treads based on NR had the goal of improving their tear properties (cut and chip behavior) [189,190], but the amounts were fairly low in order to avoid drawbacks in tread wear. Higher amounts of silica require TESPT like Si 69® as coupling agent. Investigations by Wolff [191] demonstrated that the partial replacement of CB with the silica/silane filler system resulted in a strong decrease in rolling resistance and only small drawbacks in wear

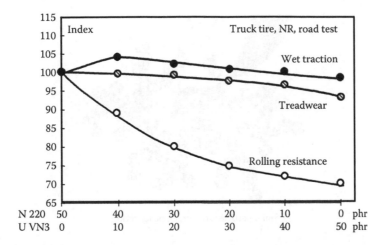

FIGURE 7.30 Truck tread test results as a function of the N 220/silica ratio.

(Figure 7.30). The full replacement led to a reduction of 30% in rolling resistance. The wet traction remained nearly stable, and the tread wear index was decreased by only 5% when a silane-modified silica was used to replace N 220 CB in a NR truck tread [192,193]. The use of a high structure black like N 121 in combination with an HD silica would be advantageous in this respect [194,195]. It was also observed that good overall performance was achieved with a blend of HD silica and CB and certain adjustments in silane content and the cross-link density [196].

One future trend for the achievement of excellent tread wear may be the use of a high-surface-area silica in the CTAB surface area of approximately 200–250 m^2/g [197]. Due to the pressure of the worldwide labeling also for truck tires over the last years, all major tire producers, raw material suppliers, and different universities are investigating the use of the silica/silane reinforcing system especially in NR in more detail. The focus is momentary given to reduce the rolling resistance, but abrasion resistance is still the major challenge when substituting highly active CBs by the silica/silane system in truck tire tread compounds.

4. Earth Mover and Off-the-Road Tires

For earth mover (EM) treads and in general for off-the-road (OTR) tread compounds, a partial replacement of CB with precipitated silica is a widely accepted practice in the tire industry for the improvement of properties such as tear resistance and cutting resistance [198]. The theory is that the small silica particles can stop the crack propagation (Figure 7.31).

On-the-road or OTR tires for higher-speed conditions and/or longer distances on the road also contain some silica, but in combination with a silane coupling agent, in the tread compound. Table 7.27 shows a model EM tread compound on the basis of NR. Silicas with different surface areas were tested. A silica with a BET surface area of approximately 160 m^2/g shows the best balance in molded groove tear resistance, heat build-up, and pico-abrasion index [199,200].

Small particles stop crack propagation

FIGURE 7.31   Theory about the crack propagation.

TABLE 7.27
Model EM Tread Compound

		phr
Formulation		
NR (SMR-5)	Polymer	100
N 220	CB	42
Silica		17
ZnO		5
Stearic acid		2
Aromatic oil		3
Cumarone indene resin		3
Antiozonant		1.5
Antioxidant		1.5
Polyethylene glycol		0.3
TBBS	Accelerator	1.1
Sulfur		2

The use of silica in sidewalls of OTRs for improved puncture resistance and sidewall reinforcement is a new development tendency. In combination with three plies of polyester, the silica compound should offer a three times higher tear resistance than a conventional CB compound together with good abrasion resistance and retention of sidewall flexibility [201].

5. Solid Tires

The market for solid tires, also called industrial tires, comprises a large group of products including tires for forklifts, lift trucks, transport carriages, construction machines, excavators, tanks, and low loaders, with the first two being the largest

TABLE 7.28
Heavy-Duty Solid Tire Treads

		I	II
Formulation, phr			
NR	Polymer	100	100
CORAX® N 375	CB	43	—
ULTRASIL® VN 3 GR	Silica	—	40
Si 69®	Silane	—	5
Other chemicals, ZnO_5; stearic acid 2;			
wax 1; IPPD 1			
DCBS	Accelerator	1	3
Sulfur		2	1.8
In-rubber properties			
MDR, 150°C (MPV), %		18.1	3.1
Tensile strength, MPa		19.3	24.8
Modulus 300%, MPa		10.8	10.7
Elongation at break, %		460	510
Shore A hardness		61	61
Goodrich flexometer (0.25 in., 23°C, 108 N preload)			
Temp change, 120 min,°C		124	30
Blowout, min		139	≫4320

application fields [202]. The advantage of a solid tire is a considerably higher load-bearing capacity and a nearly maintenance-free life in comparison to pneumatic tires. The demands for such tires are high elasticity, low rolling resistance, high resistance to outer destructive effects, and good abrasion resistance. Most of the tires that are produced as solid tires have a tread compound reinforced with active CBs. But especially in the case of battery-driven forklifts, a low rolling resistance is required to increase the service time. For indoor applications, nonmarking tires are also desired. The silica/silane filler system fulfills these demands best, so a small but growing percentage of solid tires are produced with silica in the tread. In the United States, the nonmarking tires already represent 15% of this market. Two examples of tread compound formulations are shown in Table 7.28. Compound I is a standard formulation for a heavy-duty solid tire tread with N 375, and compound II is the nonmarking version for the same application.

Newly designed solid tires for forklifts are characterized by higher tensile strength, further reduction in rolling resistance, improved grip, improved ability to withstand outer influences such as ozone, high durability, and better wear resistance at high loadings.

6. Tire Body

Significant improvements in wet traction, rolling resistance and service life of a whole tire can only be achieved by taking all parts of the tire into account. This is valid not only for passenger car tires but also for truck tires. Because of their heavy loads and relatively slow driving speeds, the fuel consumption of trucks depends very much on their tires' rolling resistance. The components of passenger and truck tires contribute to different degrees to their rolling resistance [5]. In the case of passenger tires, the main contribution (50%) to rolling resistance is the hysteresis of the tread compound. In the case of truck tires, which are traditionally filled with CB, the other tire parts contribute for up to 70% of the hysteresis loss. Furthermore, another important parameter, especially for truck tires, is retreadability. To reach the goal of excellent retreadability and low rolling resistance, the heat build-up of the tire body has to be reduced.

As can be seen in Figure 7.14, CB and silica have different influences on the rolling resistance (tan δ at 60°C) and the heat buildup (ΔT_{center}) depending on the surface areas [203]. A silica with a BET surface area of 125 m²/g offers a lower tan δ value and less heat buildup, combined with good reinforcing behavior, compared to a low reinforcing CB. This is also the reason to use LSA silica in tire body parts such as the subtread, carcass and belt [204].

TABLE 7.29
Formula and Selected In-Rubber Properties

		I	II	III
Formulation				
SMR 10	Polymer	60	60	60
ESBR 1712	Polymer	55	55	55
N 550	CB	50	25	25
ULTRASIL® 7000 GR	Silica	—	25	—
ULTRASIL® 5000 GR	Silica	—	—	25
X 50-S® (50% Si 69® on CB)		—	3	3
Other chemicals, ZnO₃; stearic acid 1; oil 6; resin 4				
Cure system				
TBBS	Accelerator	1.5	—	—
CBS	Accelerator	—	1.5	1.5
DPG	Accelerator	—	1.5	1.5
Sulfur		2.2	2.2	2.2
In-rubber properties				
ML(1+4), MU		23	35	33
$M_H - M_L$ (165°C), dN m		11.1	11.8	12.7
Tensile strength, MPa		14.5	14.0	14.9
Modulus 100%, MPa		1.4	1.2	1.4
Modulus 300%, MPa		6.9	5.0	5.8
Elongation at break, %		510	560	550
Shore A hardness		52	52	55
tan δ, 60°C (MTS, constant force)		0.094	0.079	0.073

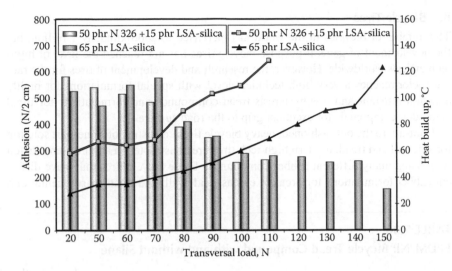

FIGURE 7.32 Heat build-up (lines) and brass adhesion (columns) versus transverse load.

ULTRASIL® VN 2 GR, with a CTAB surface area of 125 m²/g, was tested in a basic carcass compound (III). This LSA silica is compared to ULTRASIL® 7000 GR and CB N 550 (II). In comparison to the compounds with N 550, ULTRASIL® VN 2 GR shows similar reinforcement but advantages in the hysteresis loss, tan δ at 60°C (Table 7.29). The Mooney viscosities of the silica-filled compounds are higher than for the one filled with N 550 because of the higher filler–filler interaction.

The benefit of an LSA silica has been tested in a steel cord adhesion compound (NR/E-SBR). Compared to the compound filled with 50 phr N 326, 15 phr ULTRASIL® VN 2 GR, and 5 phr Cofill 11®, the sample filled with 65 phr of a special HD silica (ULTRASIL® 5000 GR) and 2 phr Si 69® shows a significantly lower heat buildup and a higher lifetime of the vulcanizates, which were tested with the Kainradl flexometer (Figure 7.32).

7. Motorcycle Tires

The development of motorcycles having horsepower beyond 100 hp required high-performance motorcycle tires. Safety (grip and high-speed capability) and endurance, especially under high-severity conditions, are required. The combination of special high-T_g polymers and special CB grades has been used successfully in the past. Presently, special CBs together with the silica/silane system are the benchmark in ultrahigh-performance motorcycle tires. Such a filler system enables the compounder to achieve a compromise in tire performance between high grip under dry and wet conditions, high-speed capability (hysteresis) and durability. The silica/silane filler system is the best approach for achieving the best wet grip and high-speed characteristics. A silica content of approximately 20%–50% of the filler loading is generally used in motorcycle tread compounds to provide the best all-around performance [205].

8. Bicycle Tires

The number of bicycles in the world has increased to approximately 1 billion, but the development of general-purpose bicycle tires has not reached the level of high-tech articles worldwide. However, the research and development of tires for the racing sector reaches a very high technical level with special formulations for highly abrasion-resistant and low-hysteresis tread compounds and formulations for tread (shoulder) compounds for excellent grip to the road surface.

Contrary to the old-fashioned, heavy bicycle tires consisting of a one-part solution for sidewall and tread, modern high-tech tires are constructed in a more complicated way, using many different rubber compounds. These bicycle tires may have different rubber formulations for breaker, carcass, sidewall, and tread (partly center and

---

TABLE 7.30
EPDM/NR Bicycle Tread Compound with and without Silane

		I	II
Formulation			
SMR L, ML(1+4) = 70	Polymer	70	70
EPDM (4.5% diene, 67% ethylene)	Polymer	30	30
ULTRASIL® 7000 GR	Silica	40	40
Si 69®	Silane	0	2
Other chemicals, ZnO 3, stearic acid 2,			
polyethylene glycol 2, wax 1.5, antioxidant 1.5,			
naphthenic oil 5			
Cure system, MBTS 1.2, DPG 0.6, sulfur 1.8	Accelerator system		
In-rubber properties			
ML(1+4), MU		78	68
ODR, 150°C, 1°			
$M_H - M_L$, dN m		27.5	31
$t_{10\%}$, min		3.7	3.2
$t_{90\%}$, min		8.6	10.5
Tensile strength, MPa		20	21.5
Modulus 100%, MPa		1.8	2.5
Modulus 300%, MPa		4	7.3
Elongation at break, %		710	640
Die C tear, N/mm		67	88
Trouser tear, N/mm		32	30
Rebound, %		53	57
Shore A hardness		68	71
DIN abrasion loss, mm^3		164	132
MTS; 16 Hz, 50 N preload, 25 N dynamic load amplitude			
E*, 60°C, MPa		11.5	11.7
tan δ, 60°C		0.129	0.105
E*, 0°C, MPa		28.8	26.3
tan δ, 0°C		0.157	0.148

shoulder treads, both with silica or even with the silica/silane system) [206]. They are made with rather thin and high-strength skinwalls, which results in lighter construction and a more flexible tire [207]. This skinwall construction method was initially used on the new generation of top sport racing tires. The trend in the European bicycle tire industry is to use silica to replace CBs. In addition to the possibility of the introduction of transparent and colored tire parts, another main reason for the use of silica is the reduction of rolling resistance and improvement of grip of the tire to the road surface under wet and dry conditions, without sacrificing the wear properties [208,209].

Two different polymer blends are widely used for bicycle tire treads: NR/SBR for standard-quality tires and EPDM/NR for high-quality tires (colored and black) because of their high ozone resistance (Table 7.30). Because of the requirement of nonstaining properties at the same time, the most effective amine-type antioxidants or antiozonants cannot be used, because they lead to a discoloration of the compounds. Therefore, an EPDM/NR compound is also sometimes used for the tire sidewall. In Table 7.30, the formulations and the in-rubber data for such a compound with and without silane modification are shown. The Mooney viscosity is lower and the moduli, hardness, and resilience are higher for compound II with 2.0 phr Si 69®. The tan δ value at 60°C, correlating with the rolling resistance, and the DIN abrasion are significantly improved by the silane coupling.

The main goal in bicycle tire tread development is a further reduction in rolling resistance, especially for racing bikes and city bikes, whereas for mountain bikes, traction is more important. The abrasion resistance has to be at a good level, which demands the use of silanes in the case of white and colored bicycle tires.

VI. SUMMARY

This chapter has dealt with the use of precipitated silica in combination with a bifunctional organosilane as a unique filler system for the rubber industry. In contrast to the common filler CB, the reinforcement is not established by the physical adsorption of the polymer on the graphite-like surface but by chemical bonding between the silica and the polymer. This bonding occurs during the vulcanization process. The nanoscale of the precipitated silica, in combination with this chemical coupling of the silica to the rubber, results in reinforcement comparable to that of active CBs and is much higher than that for natural fillers like silicates and clays. In addition to this high reinforcement, the hysteresis loss is significantly lower than for CBs with comparable surface areas. Therefore, this filler system is the optimal choice when high reinforcement in combination with reduced hysteresis loss, for example, in tire treads and engine mounts, is demanded. The silica/silane filler system is the best choice also for colored rubber articles such as shoe soles, special sealants, and hoses, which require good abrasion resistance.

To discuss the applications and general benefits of this filler system, first, the chemistry and physics of the precipitated silica and the silanes were described. Besides describing the production process, the focus in these sections was on the analytics of the two products and the understanding of their chemistry. Both the optimal silica morphology—surface area, aggregate structure, and surface activity—and the chemical structure and reactivity of the silane coupling agent are essential for the required performance. Secondly, the reinforcement mechanism of the silica/silane filler system, in contrast to that of CB, was reviewed to give the reader a basic understanding of the advantages of this filler system in various applications and to point out possibilities for further compound optimization. Finally, an overview of the most common applications for this unique filler system was given, including model formulations for several industrial rubber goods as well as car, truck, and bicycle tires.

ACKNOWLEDGMENTS

The permission from Evonik Industries AG to prepare and publish this review is greatly appreciated. We thank all colleagues from Evonik who were involved in preparing this chapter.

REFERENCES

1. Degussa Broschüre. *Stets Geforscht*. Frankfurt, Germany: Degussa AG, 1988.
2. Greenwood NN, Earnshaw A. *Chemistry of the Elements*. Oxford, U.K.: Pergamon Press, 1984.
3. Degussa Brochure. *Harry Kloepfer-Geschichte*. Produkte: Der Weg zur Kieselsäure K 3, 1942.
4. Edelmetalle und Chemie. *Frankfurt am Main*. Frankfurt, Germany: Degussa AG.
5. Uhrlandt S, Blume A. Development of HD-silicas for tires—Process, properties, performance. (a) *ACS Meeting*, April 4–6, 2000, Dallas, TX, Paper 32; (b) *Rubber World* 2002; 226:30; (c) *Kautsch Gummi Kunstst* 2001; 54:520.
6. Iler RK. *The Chemistry of Silica*. New York: Wiley, 1979.
7. Ferch H. Pulverförmige amorphe synthetische Kieseläure-Produkte Herstellung und Charakterisierung. *Chem-Ing-Tech* 1976; 48:922.
8. Iler RK. Colloidal silica. *Surface Colloid Sci* 1973; 6:1.
9. Brunauer S, Emmett PH, Teller E. Adsorption of gases in multimolecular layers. *J Am Chem Soc* 1938; 60:309.
10. Haul R, Duembgen G. Vereinfachte Methode zur Messung von Oberflächengrössen durch Gasadsorption. *Chem-Ing-Tech* 1960; 32:349.
11. Haul R, Duembgen G. Vereinfachte Methode zur Messung von Oberflächengrössen durch Gasadsorption 2. Mitteilung. *Chem-Ing-Tech* 1963; 35:586.
12. Koberstein E, Voll M. Charakterisierung von hochdispersem Siliciumdioxid durch physikalische Adsorption von Gasen. *Z Phys Chem* 1970; 71:275.
13. DIN 66134. Bestimmung der Porengrössenverteilung und der spezifischen Oberfläche mesoporöser Feststoffe durch Stickstoffsorption—Verfahren nach Barrett, Joyner und Halendar, 1998.
14. Saleeb FZ, Kitchener JA. *J. Chem. Soc.* 1965; 9:11.
15. Janzen J. Specific surface area measurements on carbon black. *Rubber Chem Technol* 1971; 44:1287.
16. Voet A, Morawski JC, Donnet JB. Reinforcement of elastomers by silica. *Rubber Chem Technol* 1977; 50:342.
17. Sears GW Jr. Determination of specific surface area of colloidal silica by titration with sodium hydroxide. *Anal Chem* 1956; 28:1981.
18. Heston WA Jr, Iler RK, Sears GW Jr. The adsorption of hydroxyl ions from aqueous solution on the surface of amorphous silica. *J Phys Chem* 1960; 64:147.

19. Kolbstein E, Lakatos E, Voll M. Zur Charakterisierung der Oberflächen von Rußen und hochdispersiblen Kieselsäuren. *Ber Bunsenges Phys Chem* 1971; 75:1105.
20. de Boer JH, Linsen BG, van der Plas Th, Zondervan GJ. Studies on the pore system in catalysts. *J Catal* 1965; 4:649.
21. Ritter HL, Drake LC. Pore-size distribution in porous materials pressure porosimeter and determination of complete macropore-size distribution. *Ind Eng Chem Anal Ed* 1945; 17:782.
22. Drake LC, Ritter HL. Macropore-size distribution in some typical porous substances. *Ind Eng Chem Anal Ed* 1945; 17:787.
23. Calculation method developed by Michelin.
24. Kraus G, Janzen J. Verbesserte physikalische Rußprüfung und Vorhersage des Verhaltens in Vulkanisaten. *Kautsch Gummi Kunstst* 1975; 28:253.
25. Behr J, Schramm G. Über die Bestimmung der Ölzahl von Kautschukfüllstoffen mit dem Brabender-Plastographen. *Gummi Asbest Kunstst* 1966; 19:912.
26. Donnet JB, Voet A. *Physics, Chemistry and Elastomer Reinforcement.* New York: Marcel Dekker, 1976.
27. Blume A. Analytical properties of silica—A key for understanding silica reinforcement. (a) *ACS Meeting*, April 13–16, 1999, Chicago, IL, Paper No. 73; (b) *Kautsch Gummi Kunstst* 2000; 53:338.
28. Zečević J, de Jong KP, de Jongh PE. Progress in electron tomography to assess the 3D nanostructure of catalysts. *Curr Opin Solid State Mater Sci* 2013; 17:115 and literature cited therein
29. Kohjiya S, Kato A, Ikeda Y. Visualization of nanostructure of soft matter by 3D-TEM. *Progr Polymer Sci* 2008, 33, 979.
30. Heimendahl Mv. *Einfuehrung in die Elektronenmikroskopie.* Stuttgart, Germany: Vieweg Verlag, 1979.
31. Wehmeier A. et al. Precipitated silica acids as a reinforcement foller for elastomer mixtures, Prio. 7. April 2008. WO 2009/124829.
32. Stoeber W. Chemische Adsorption von Methylchlorsilanen an kristallinem und amorphem Siliziumdioxyd. *Kolloid Z* 1956; 149:39.
33. Stoeber W, Bauer G, Lieb KT. Chemisorption von Alkoholen an amorphem Siliciumdioxyd. *Ann Chem* 1957; 604:104–110.
34. Boehm HP, Schneider M. Über die Hydroxylgruppen an der Oberflache des amorphen Siliciumdioxyds "Aerosil" und ihre Reaktionen. *Z Anorg Allgom Chem* 1959; 301:326.
35. Bode R, Ferch H, Fratzscher H. Grundlagen und Anwendungen einer durch Flammenhydrolyse gewonnenen Kieselsäure. *Kautsch Gummi Kunstst* 1967; 20:578.
36. Wartmann HJ. Dissertation, Zurich, Switzerland: ETH University Zürich, 1958.
37. Fripiat JJ, Uytterhoeven J. Hydroxyl content in silica gel "Aerosil." *J Phys Chem* 1962; 66:800.
38. Iler RK. (To Du Pont de Nemours). U.S. Patent 2,657,149.
39. Kohlschuetter HW, Best P, Wirzing G. Umsetzung von Trimethylsilicium-monochlorid mit Silicagel. *Z Anorg Allgem Chem* 1956; 258:236.
40. Wirzing G, Kohlschuetter HW. Umsetzung von Trimethylsiliciummonochlorid mit Silicagel. *Z Anal Chem* 1963; 198:270.
41. Boehm HP, Schneider M, Arendt F. Der Wassergehalt "getrockneter" Siliciumdioxyd-Oberflachen. *Z Anorg Allgem Chem* 1963; 320:43.
42. Shapiro I, Weiss HG. Bound water in silica gel. *J Phys Chem* 1953; 57:219.
43. Weiss HG, Knight JA, 4Shapiro I. Change in the ratio of hydroxyl groups attached to silicon and aluminum atoms in silica-alumina catalysts upon activation. *J Am Chem Soc* 1959; 81:1823.
44. Naccache C, Imelik B. Sur la reaction du diborane avec les solides mineraux contenant de l'eau. *CR Acad Sci* 1960; 250:2019.

45. Hesse M, Meier H, Zeeh B. *Spektroskopische Methoden in der organischen Chemie*, 3rd edn. Stuttgart, Germany: Thieme Verlag, 1987.
46. Maciel GE. *Nuclear Magnetic Resonance in Modern Technology*. Dordrecht, the Netherlands: Kluwer Academic Press, 1994, p. 401.
47. Mijatovic J, Binder WH, Gruber H. Characterization of surface modified silica nanoparticles by 29Si solid state NMR spectroscopy. *Mikrochim Acta* 2000; 11:175.
48. Marciel GE, Sindorf DW. Silicon-29 nuclear magnetic resonance study of the surface of silica gel by cross polarization and magic-angle spinning. *J Am Chem Soc* 1980; 102:7606.
49. Evans LR. Non black fillers for the future. Presented at *Intertech Conference Functional Tire Fillers*, Fort Lauderdale, FL, January 29–31, 2001.
50. A. Blume et al. Operando infrared study of the reaction of Thiethoxyptopylsilane with silica. *Kautschuk Gummi Kunsst.* 2008; 8:359.
51. Wood DL, Rubinovich EM, Johnson DW Jr, McChesney JB, Vogel Em. Preparation of high-silica glasses from colloidal gels: III. Infrared spectrophotometric studies. *J Am Ceram Soc* 1983; 66:693.
52. Legrand AP (ed.). The Surface Properties of Silicas. John Willey & Sons, New York 1999.
53. Uhrlandt, S. Internal Report NACF 21950226, Degussa AG, 1995.
54. Hair ML. *IR-Spectroscopy in Surface Chemistry*. New York: Marcel Dekker, 1967.
55. Little LH. *IR-Spectra of Adsorbed Species*. Academic Press: New York, 1960.
56. Pohl ER, Guillet A, Guthier R. Reactivity of silanes with silica. Presented at the *Intertech Conference Functional Tire Fillers*, Fort Lauderdale, FL, January 29–31, 2001.
57. McFarlan AJ, Morrow BA. Infrared evidence for two isolated silanol species on activated silicas. *J Phys Chem* 1991; 95:5388.
58. Morrow BA, McFarlan AJ. Surface vibrational modes of silanol groups on silica. *J Phys Chem* 1992; 96:1395.
59. Burneau A, Barres O, Vidal A, Balard H, Ligner G, Papirer E. Comparative study of the surface hydroxyl groups of fumed and precipitated silicas. Vol. 3. DRIFT characterization of grafted n-hexadecyl chains. *Langmuir* 1990; 6:1389.
60. Wang M-J, Wolff S. Filler elastomer interaction. Part 2. Investigation on the energetic heterogeneity of silica surfaces. *Kautsch Gummi Kunstst* 1992; 45:11.
61. Papirer E, ed. *Adsorption on Silica Surfaces*. Surface Science Series, Vol. 90. New York: Marcel Dekker, 2000.
62. Wehmeier A. et al. Precipitated silica acids as a reinforcement foller for elastomer mixtures, Prio. 7. April 2008. EP 2262730 B1.
63. Blume A, El-Roz M, Thibault-Starzyk F. Infrared study of the silica/silane reaction. *Kautsch Gummi Kunstst*, 10/2013, pp. 63–70; El-Roz M, Thibault-Starzyk F, Blume A. *Infrared Study of the Silica/Silane Reaction* (2nd part), *Kautsch Gummi Kunstst*, 05/2014, pp. 53–57.
64. Hemminger WF, Cammenga HK. *Methoden der Thermischen Analyse*. Berlin, Germany: Springer Verlag, 1989.
65. Wagner MP. Reinforcing silicas and silicates. *Rubber Chem Technol* 1976; 49:703.
66. Kondo S, Muroya M. The dehydration of surface silanol on silica gel. *Bull Chem Jpn* 1970; 43:2657.
67. Tsukruk VV. Screening probe microscopy of polymer surfaces. *Rubber Chem Technol* 1997; 70:430.
68. Van Haeringen DT, Schoenherr H, Vansco GJ, Van der Does L, Noordermeer JWM. Atomic force microscopy of electrons: Morphology, distribution of filler particles, and adhesion using chemically modified TIPS. *Rubber Chem Technol* 1999; 72:862.
69. Galuska AA, Poulter RR, McElrath KO. Force modulation AFM of elastomer blends: morphology, fillers and cross-linking. *Surf Interf Anal* 1997; 25:418.

70. Niedermeier W, Raab H, Stierstorfer J, Kreitmeier S, Goeritz D. The microstructure of carbon black investigated by atomic force microscopy. *Kautsch Gummi Kunstst* 1994; 47:799.
71. Biscoe J, Warren BE. An X-ray study of carbon black. *J Appl Phys* 1942; 13:364.
72. Franklin RE. Crystallite growth in graphitizing and nongraphitizing carbons. *Proc Roy Soc Lond* 1951; A209:196.
73. Martin JE, Hurd AJ. Scattering from fractals. *J Appl Cryst* 1987; 20:61.
74. Freltoft T, Kjems JK, Sinha SK. Power-law correlations and finite-size effects in silica particle aggregates studied by small-angle neutron scattering. *Phys Rev B* 1985; 33:269.
75. Knerr M. Analyse der Struktur von Kieselsaeuren mit Hilfe von Roentgen-und Neutronen-Kleinwinkelstreuung. Dissertation. Regensburg, Osnabrueck: Der Andere Verlag, 2000.
76. Froehlich J, Kreitmeier S, Goeritz D. Surface characterization of carbon blacks. *Kautsch Gummi Kunstst* 1998; 51:370.
77. DIN ISO 787/2, formerly DIN 53198. Allgemeine Prüfverfahren für Pigmente und Füllstoffe Teil 2. Bestimmung der bei 105°C flüchtigen Anteile (ISO 787-2, 1981; EN ISO 787-2, 1995).
78. ISO 3262-19. Füllstoffe für Beschichtungsstoffe—Anforderungen und Prüfverfahren Teil 19, gefällte Kieselsäuren, 2000.
79. Grundke K, Boerner M, Jacobasch H-J. Characterization of fillers and fibres by wetting and electrokinetic measurements. *Colloid Surf* 1991; 58:47.
80. http://www. corporate.evonik.com http://www.corporate.evonik.com.
81. http://www.solvay.com.
82. http://www.ppg.com.
83. http://www.huber.com.
84. Plueddemann EP. *Silane Coupling Agents*, 2nd edn. New York: Plenum Press, 1982.
85. Deschler U, Kleinschmit P, Panster P. 3-Chloropropyltrialkoxysilanes—Key intermediates for the commercial production of organofunctionalized silanes and polysiloxanes. *Angew Chem Int Ed Engl* 1986; 25:236.
86. Moore MJ. Silanes as rubber-to-metal bonding agents. *ACS Meeting*, Cleveland, OH, October 16–19, 2001, Paper 105.
87. Wagner MP. Non-black reinforcers and fillers for rubber. *Rubber World* 1971; 164:46.
88. Wagner MP. The consequences of chemical bonding in filler reinforcement of elastomers. Presented at the *Colloques Internationaux*, September 24–26, 1973. Le Bischenberg-Obernai, France: Editions du Centre National de la Rech Sci, Paris, France, 1975, p. 147.
89. Gent AN, Hsu EC. Coupling reactions of vinylsilanes with silica and poly(ethylene-co-propylene). *Macromolecules* 1974; 7:933.
90. Wolff S. Reinforcing and vulcanization effects of silane Si 69 silica-filled compounds. *Kautsch Gummi Kunstst* 1981; 34:280.
91. Wolff S. Chemical aspects of rubber reinforcement by fillers. *Rubber Chem Technol* 1996; 69:325.
92. Klockmann O, Hasse A. *KGK* 2007; 60(3):82.
93. The next generation of silane coupling agents for silica/silane-reinforced tire tread compounds. *ITEC-Meeting 2002*, Akron, OH or Joshi PG, Cruse RW, Pickwell RJ. Silane coupling agent boosts tire performance. RubberNews published on September 9, 2002.
94. Evonik Industries AG. *Applied Technology Advanced Fillers*. Product Application, PA 701, 2001.
95. Evonik Industries AG. *Applied Technology Aerosols and Silanes*. Product Information Dynasylan® 9498, 2001.

96. Hasse A, Luginsland H-D. Reinforcement of silica-filled EPM compounds using vinyl-silanes. *Kautsch Gummi Kunstst* 2002; 55:294.
97. Evonik Industries AG. Applied Technology Advanced Fillers, Product Appl PA 725, 2001.
98. Bandoypadhyay S, De PP, Tripathy DK, De SK. Interaction between carboxylated nitrile rubber and precipitated silica: role of (3-aminopropyl)-triethoxysilane. *Rubber Chem Technol* 1996; 69:675.
99. Hopkins W, von Hellens W, Koski A, Rausa J. Reinforcement of BIIR with silica. *ACS Meeting*, October 16–19, 2001, Cleveland, OH, Paper 114.99.
100. Klockmann O, Hasse A, Luginsland H-D. Special silanes for special elastomers. Presented at the *5th Kautschuk-Herbst-Kolloquium (DIK)*, October 30–November 1, 2002, Hannover, Germany: *Kautsch Gummi Kunst* 2002; 55:471.
101. Lukevics E, Belyakova ZV, Pomerantseva MG, Voronkov MG. *J Organomet Chem Libr* 1977; 5:1.
102. Meyer-Simon E, Thurn F, Michel R. (To Degussa AG.). German Patent DE 2,141,159, 1973.
103. Childress TE, Schilling CL Jr, Ritscher JS, Tucker OG Jr, Beddow. DG. (To OSI Specialties Inc.). U.S. 5,489,701, 1996.
104. ASTM D 6740. *Standard Test Method for Silanes Used in Rubber Formulations Residue on Ignition*. Conshohocken, PA: American Society for Testing Materials, 2001.
105. Sindorf DW, Marciel GE. Silicon-29 nuclear magnetic resonance study of hydroxyl sites on dehydrated silica gel surfaces, using silylation as a probe. *J Am Chem Soc* 1983; 105:3767.
106. Sindorf DW, Marciel GE. Cross-polarization/magic-angle-spinning silicon-29 nuclear magnetic resonance study of silica gel using trimethylsilane bonding as a probe of surface geometry and reactivity. *J Phys Chem* 1982; 86:5208.
107. Hunsche A, Görl U, Müller A, Knaack M, Gobel T. Investigations concerning the reaction silica/organosilane and organosilane/polymer—Part 1. *Kautsch Gummi Kunstst* 1997; 50:881.
108. Görl U, Munzenberg J. Investigations into the chemistry of the TESPT sulfur chain. (a) *ACS Meeting*, May 6–9, 1997, Anaheim, CA, Paper 38; (b) *Kautsch Gummi Kunstst* 1999; 52:588.
109. (a) Görl U, Hunsche A, Müller A, Koban HG. Investigations into the silica/silane reaction system. *Rubber Chem Technol* 1997; 70:608. (b) Hunsche A, Görl U, Koban HG, Lehmann T. Investigations on the reaction silica/organosilane and organosilane/polymer—Part 2. *Kautsch Gummi Kunstst* 1998; 51:525.
110. Beari F, Brand M, Jenkner P, Lehnert R, Metternich HJ, Monkiewicz J, Schram J. Organofunctional alkoxysilanes in dilute aqueous solution: new accounts on the dynamic structural mutability. *J Organometal Chem* 2001; 625:208.
111. Osterholz FD, Pohl ER. Kinetics of the hydrolysis and condensation of organofunctional alkoxysilanes: A review. In: *Silanes and Other Coupling Agents*. Mittal KL, ed. The Netherlands: VSP, pp. 119–141.
112. Scuati A, Feke DL, Manas-Zloczower I. Influence of powder surface treatment on the dispersion behavior of silica into polymeric material. *ACS Meeting*, October 16–19, 2001, Cleveland, OH, Paper 92.
113. A. Blume, J.-P. Gallas, M. Janik, F. Thibault-Starzyk, A. Vimont, Operando infrared study of the reaction of triethoxypropylsilane with silica. KGK Kautschuk Gummi Kunststoffe 7–8/2008, pp. 359–362.
114. Nieuwenhuizen PJ, Reedijk J, vanDuin M, McGill WJ. Thiuram- and dithiocarbamate-accelerated sulfur vulcanization from the chemist's perspective: methods, materials and mechanisms reviewed. *Rubber Chem Technol* 1997; 70:368.
115. Luginsland H-D. Reactivity of the sulfur function of the disulfane silane TESPD and the tetrasulfane silane TESPT. (a) *ACS Meeting*, April 13–16, 1999, Chicago, IL, Paper 74; (b) *Kautsch Gummi Kunstst* 2000; 53:10.

116. Hasse A, Luginsland H-D. Vulcanization behavior of disulfidic and polysulfidic organic silanes. Presented at the *IRC Rubber Conference*, Helsinki, Finland, June 12–15, 2000.

117. Hasse A, Klockmann O, Wehmeier A, Luginsland H-D. Influence of the amount of TESPT and sulfur on the reinforcement of silica-filled rubber compounds. (a) *ACS Meeting*, October 16–19, 2001, Cleveland, OH, Paper 91; (b) *Kautsch Gummi Kunstst* 2002; 55:236.

118. Thurn F, Wolff S. Neue Organosilane für die Reifenindustrie. *Kautsch Gummi Kunstst* 1975; 28:733.

119. Kraus G. Reinforcement of elastomers by carbon black. *Rubber Chem Technol* 1978; 51:297.

120. Dannenberg EM. Reinforcement of rubber with carbon black and mineral fillers. *Progr Rubber Plast Technol* 1985; 1:13.

121. Donnet J-B, Bansal RC, Wang M-J. *Carbon Black*. New York: Marcel Dekker, 1993.

122. Luginsland H-D, Frohlich J, Wehmeier A. Influence of different silanes on the reinforcement of silica-filled rubber compounds. (a) *ACS Meeting*, April 24–27, 2001, Providence, RI, Paper No. 59; (b) *Rubber Chem Technol* 2002; 67:187.

123. Luginsland H-D. Chemistry and physics of network formation in silica/silane-filled rubber compounds. *ACS Meeting*, April 29–May 1, 2002, Savannah, GA, Paper F; Review on the chemistry and reinforcement of the silica/silane filler system for rubber applications. Aachen, Germany: Shaker Verlag, 2002.

124. Yatsuyanagi F, Suzuki N, Ito M, Kaidou H. Effects of secondary structure on the mechanical properties of silica filled rubber systems. *Polymer* 2001; 42:9523.

125. Payne AR, Whittaker RE. Low strain dynamic properties of filled rubbers. *Rubber Chem Technol* 1971; 44:440.

126. Wang M-J. Effect of polymer–filler and filler–filler interaction on dynamic properties of filled vulcanizates. *Rubber Chem Technol* 1998; 71:520.

127. Wang M-J, Wolff S. Filler–elastomer interactions. Part I. Silica surface energies of and interactions with model compounds. *Rubber Chem Technol* 1991; 64:559.

128. Wang M-J, Wolff S. Filler-elastomer interactions. Part III. Carbon-black surface energies and interactions with elastomer analogs. *Rubber Chem Technol* 1991; 64:714.

129. Wang M-J. Filler–elastomer interactions. Part VI. Characterization of carbon blacks by inverse gas chromatography at finite concentration. *Rubber Chem Technol* 1992; 65:890.

130. Wolff S, Wang M-J. Filler–elastomer interactions. Part IV. The effect of the surface energies of fillers on elastomer reinforcement. *Rubber Chem Technol* 1992; 65:329.

131. Wang M-J, Wolff S. Filler–elastomer interactions. Part V. Investigation of the surface energies of silane-modified silicas. *Rubber Chem Technol* 1992; 65:715.

132. Wolff S, Wang M-J, Tan E-H. Filler–elastomer interactions. Part VII. Study on bound rubber. *Rubber Chem Technol* 1993; 66:163.

133. Tan E-H, Wolff S, Haddeman M, Grewatta HP, Wang M-J. Filler–elastomer interactions. Part IX. Performance of silica in polar elastomers. *Rubber Chem Technol* 1993; 66:594.

134. Wolff S, Wang M-J, Tan E-H. Filler-elastomer interactions. Part X. The effect of filler-elastomer and filler-filler interaction on rubber reinforcement. *Kautsch Gummi Kunstst* 1994; 47:102.

135. Wang M-J, Wolff S, Tan E-H. Filler-elastomer interactions. Part VIII. The role of the distance between filler aggregates in the dynamic properties of filled vulcanizates. *Rubber Chem Technol* 1993; 66:178.

136. Hasse A, Luginsland H-D. Influence of alkylsilanes on the properties of silica-filled rubber compounds. Presented at the *RubberChem'01 Conference*, Brussels, Belgium, April 3–4, 2001.

137. ten Brinke JW, Litvinov VM, van Wijnhoven JEGJ, Noordermeer JWM. Interactions of silicas with natural rubber under influence of coupling agents as studied by *H NMR T2 relaxation. Presented at *IRC 2001*, Birmingham, U.K., June 12–14, 2001.

138. Luginsland H-D. Processing of silica/silane-filled tread compounds. (a) *ACS Meeting*, April 4–6, 2000, Dallas, TX, Paper 34; (b) *Tire Technol* 2000; 3:52.

139. Schaal S, Coran AY. The rheology and processing of tire compounds. *Rubber Chem Technol* 2000; 73:225.

140. CARB/EPA. Legislative impact on fuel system materials and design: LEV II and CAP 2000 Amendments to the California Exhaust and Evaporate Emission Standards and Test Procedures for Passenger Cars, Light-Duty Trucks and Medium Duty Vehicles, and to the Evaporative Emission Requirements of Heavy Duty Vehicles. *Federal Register* 2002; 67:187.

141. Bomal F, Cochet P, Fernandez M, Bomal Y, Advantage of using a highly dispersible silica in terms of mixing and formulation costs. Presented at *ITEC '96 Meeting*, Akron, OH, September 10–12, 1996.

142. Cruse RW, Hofstetter MH, Panzer LM, Pickwell RJ. Effect of polysulfidic silane sulfur content on properties of a low rolling resistance silica-filled tread compound. *ACS Meeting*, Louisville, KY, October 8–11, 1996. Paper 75.

143. ten Brinke JW, van Swaaij PJ, Reuvenkamp LP, Noordermeer JWM. The influence of silane sulfur- and carbon rank on processing of a silica reinforced tyre tread compound. *ACS Meeting*, Cleveland, OH, October 16–19, 2001. Paper 131.

144. Berkemeier D, Haeder W, Rinker M. Mixing of silica compounds from the view of a mixer supplier. *Rubber World* 2001; 224:34.

145. http://www.hf-mixinggroup.com/en/products/mixers/tandem-mixer. HF Mixing group "tandem mixers" Harburg Freudenberger Maschinenbau GmbH, web link, March 24, 2015 accessed.

146. Görl U, Nordsiek KH. Rubber/filler batches in powder form. *Kautsch Gummi Kunstst* 1998; 51:250.

147. Görl U, Schmitt M. Rubber/filler compound systems in powder form: A new raw material generation for simplification of the production processes in the rubber industry. Part 2: Powder rubber based on E-SBR/silica/silane. *Kautsch Gummi Kunstst* 2002; 67:187.

148. Amash A, Bogun M, Schuster R-H, Görl U, Schmitt M. New concepts for continuous mixing of powder rubber. *Plastics, Rubber, Compos* 2001; 30:401.

149. Görl U. Production and application of silica loaded powder rubber based on E-SBR. *ACS Meeting*, Pittsburgh, PA, October 8–11, 2002. Paper 73.

150. Blow CM, Hepburn C. *Rubber Technology and Manufacture*, 2nd edn. London, U.K.: Butterworth Scientific, 1982. Chapter 1.

151. Anon. Über die Wirkung kolloider Kieselsäure in Kautschukmischungen. *Gummi-Ztg* 1925; 39:2102.

152. Scott JR. "Neosyl MH," a new white reinforcing agent. *Trans Inst Rubber Ind* 1941; 17:95.

153. (a) Bode R, Ferch H, Fratzscher H. *Schriftenreihe Pigmente: Aerosil, Fumed Silica*. Frankfurt am Main, Germany. (b) *Kautsch Gummi Kunstst* 1967; 20:578.

154. http://www.sivento.com.

155. Boss AE. Adaptability—A tool for production development. *Chem Eng News* 1949; 27:677.

156. Kraus G. *Reinforcement of Elastomers*. New York: Interscience, 1965.

157. Kerner D, Kleinschmidt P, Meyer J. Precipitated silicas. *Ullmann's Encyclopedia of Industrial Chemistry*, Vol. A23. VCH Weinheim, Germany, 1993, p. 642.

158. Bachmann JH, Sellers JW, Wagner MP, Wolf RF. Fine particle reinforcing silicas and silicates in elastomers. *Rubber Chem Technol* 1959; 32:1286.

159. Wagner MP. Precipitated silicas—A compounding alternative with impending oil shortages. *Elastomerics* 1981; 113:40.
160. Gessler AM, Wiese HK, Rehner J Jr. The reinforcement of butyl and other synthetic rubbers with silica pigments. *Rubber Age* 1955; 78:73.
161. Stearns RS, Johnson BL. Surface treatment of hydrated silica pigments for reinforcement of rubber stocks. *Rubber Chem Technol* 1956; 29:1309.
162. Wolff S, Golombeck P.(To Evonik Industries AG). Ger Patent DE 3,305,373, 1984.
163. Burmester K, Wolff S, Klotzer E, Thurn F. (To Evonik Industries AG). U.S. Patent 3,938,574, 1976.
164. Kern WF. Variation in der Reibungskoeffizienten-Dispersion zur Verbesserung von Reifenlaufflächen auf Eis und winterliche Nässe. Presented at *Int Rubber Conference*, Munich, Germany, 1974.
165. Ranney MW, Solleman KJ, Cameron GM. Applications for silane coupling agents in the automotive industry. *Kautsch Gummi Kunstst* 1975; 28:597.
166. Rauline R. (To Compagnie Generale des Establissements Michelin). Eur Patent 0,501,227; U.S. Patent 5,227,425, 1993.
167. Babbit RO, ed. *The Vanderbilt Rubber Handbook.* Norwalk, CT: RT Vanderbilt Co, 1978, p. 744.
168. Varkey JK, Mathew NM, De PP. Studies on the use of 1,2-polybutadiene in microcellular soles. *Indian J Nat Rubber Res* 1989; 2:13.
169. Evonik Industries AG. Applied Technology Advanced Fillers, Product Info PI 303–306, 2001.
170. Dluzneski PR. The chemistry of peroxide vulcanization; (a) *Rubber World* 2001; 228:34; (b) *Rubber Chem Technol* 2001; 74:451.
171. Meisenheimer H. Ethylene/vinylacetate elastomers (EVM) for environmentally friendly cable applications. *Kautsch Gummi Kunstst* 1995; 48:281.
172. http://www.contitech.de.
173. Wolff S, Panenka R, Haddeman M, Nakahama H. (To Degussa AG). U.S. Patent 6,521,713, 2003.
174. Le Matire U. The rolling resistance. Presented at the *AFICEP/DKG Meeting*, Mulhouse, France, 1992.
175. Grosch KA. A new way to evaluate traction and wear properties of tire tread compounds. *ACS Meeting*, Cleveland, OH, October 21–24, 1997. Paper 119.
176. Grosch KA. A new method to determine the traction and wear properties of tire tread compounds. *Kautsch Gummi Kunstst* 1997; 50:841.
177. Grosch KA, Heinz M. Proposal for a general laboratory test procedure to evaluate abrasion resistance and traction performance of tire tread compounds. Presented at *IRC 2000*, Helsinki, Finland, June 12–15, 2000.
178. Hasse A, Wehmeier A, Luginsland H-D. Aspects concerning the use of the silica-silane reinforcement system in modern tread compounds. Presented at *DKT Meeting*, Nürnberg, Germany, September 4–7, 2000.
179. Wolff S. The influence of fillers on rolling resistance. *Meeting of ACS Rubber Division*, New York, April 8–11, 1986. Paper 66.
180. Agostini G, Berg J, Materne Th. New compound technology. Presented at *Akron Rubber Group Meeting*, Akron, OH, October 27, 1994.
181. Bomal Y, Touzet S, Barruel R, Cochet P, Dejean B. Development in silica usage for decreased tyre rolling resistance. *Kautsch Gummi Kunstst* 1997; 50:434.
182. Wehmeier A. Filler Dispersion Analysis by Topography Measurement. Tech Rep TR 820. Evonik Industries AG, Applied Technology Advanced Fillers, 2002.
183. Wehmeier A. (To Evonik Industries AG). Gen Patent DE 19,917,975, 2000.
184. Geisler H. Bestimmung der Mischgüte. Presented at the *DIK-Workshop*, Hannover, Germany, November 27–28, 1997.

185. Wehmeier A, Wolff W. New silica with improved balance between rolling resistance and abrasion resistance. *IRC 2010*, Mumbai, November 17–19, 2010.
186. Futamura S. Analysis of ice and snow traction of tread material. *Rubber Chem Technol* 1996; 69:648.
187. Ahagon A, Kobayashi T, Misawa M. Friction on ice. *Rubber Chem Technol* 1988; 61:14.
188. Blume A, Luginsland H-D, Uhrlandt S, Wehmeier A. Influence of analytical properties of low surface area silicas on tire performance. Presented at the *Conference Silica 2001*, Mulhouse, France, September 3–6, 2001.
189. Walker LA, Harber JB. Improved durability of OTR mining tires. *Kautsch Gummi Kunstst* 1985; 6:494.
190. Davies KM, Lionnet R. The effect of cure system modification on the performance of silica containing tread compounds. *Rubbercon'81*, Harrogate, England, June 8–12, 1981. Paper G4.
191. Wolff S. Performance of silicas with different surface areas in NR. *ACS Meeting*, New York, April 8–11, 1986. Paper 66.
192. Wolff S. Silica based tread compounds: Background and performance. Presented at *Tyre Tech*, Basel, Switzerland, October 28–29, 1993.
193. Evans LR, Fultz WC. Truck tire tread compounds with highly dispersible silica. *ACS Meeting*, Anaheim, CA, May 6–9, 1997. Paper 5.
194. Tan E-H, Blume A. Silica for tire application. *IRC'99*, Seoul, South Korea, Paper P 2–8.
195. Bomal Y, Cochet Ph, Dejean B, Gelling J, Newell R. Influence of precipitated silica characteristics on the properties of a truck tyre tread II. *Kautsch Gummi Kunstst* 1998; 51:259.
196. Luginsland H-D, Uhrlandt S, Wehmeier A. Silica reinforcement—A key to develop a silica for truck tire treads. Poster Presented at the *Silica 2001 Conference*, Mulhouse, France, August 3–7, 2001.
197. Luginsland D-H, Uhrlandt S, Wehmeier A. Use of a highly dispersible high surface area silica in truck and high performance tire treads. Presented at *ITEC'02*, Akron, OH, September 10–12, 2002.
198. Wolff S, Tan E-H. How crosslinking and reinforcement parameters correlate with relevant natural rubber tread properties. *IRC Conference*, Houston, TX, October 1983. Paper 8.
199. Byers J. *Non-Black Fillers Used in Rubber*. Wadsworth, OH, Byers Rubber Consulting Inc.
200. Okel TA, Waddell WH. Silica properties/rubber performance correlation. Carbon black-filled rubber compounds. *Rubber Chem Technol* 1994; 67:217.
201. http://www.goodyear.com web link, March 24, 2015 accessed.
202. http://www.hyster.co.uk web link, March 24, 2015 accessed.
203. Wolff S. Hochaktive Kieselsäuren als Verstärkerfüllstoffe in der Gummiindustrie. *Kautsch Gummi Kunstst* 1988; 41:674–687.
204. Cochet Ph, Butcher D, Bomal Y. Formula optimization for a steel belt cord insulation compound. *Kautsch Gummi Kunstst* 1995; 48:353.
205. Brasfield E. 101 Sportbike Performance Projects. Motorbooks Workshop. St. Paul, US, 2004; 69.
206. Continental. New Aswers for future mobility. http://www.conti-online.com web link, March 24, 2015 accessed.
207. InfoClip Fahrradreifen ADFC, RADwelt 2 28, 1999.
208. Tufo. tubeless bicylcle tires. http://www.tufo.com web link, 24. March 2015 accessed Flexus Primus 33SG; http://www.tufonorthamerica.com web link, March 24, 2015 accessed.
209. M. Janik, Internal Report 05701, Degussa AG, 2001.

8 General Compounding

*Harry G. Moneypenny, Karl-Hans Menting,
and F. Michael Gragg*

CONTENTS

I.   Introduction 334
     A.   Raw Materials Handling 334
     B.   Mixing 335
     C.   Forming 336
     D.   Vulcanization 336
II. Physical and Chemical Peptizers 338
     A.   Mastication 338
     B.   Processing with Peptizing Agents 341
     C.   Influence of Peptizing Agents on Vulcanizate Properties 342
III. Lubricants 344
     A.   General Discussion 344
     B.   Properties and Mode of Action of Lubricants 346
     C.   Processing with Lubricants 348
     D.   Influence of Lubricants on Vulcanizate Properties 350
IV. Homogenizing Agents 351
     A.   Examples and Function 351
     B.   Processing with Homogenizing Agents 354
V.   Dispersing Agents 355
     A.   Properties of Dispersing Agents 355
     B.   Processing with Dispersing Agents 355
VI. Tackifiers 355
     A.   Definition and Manufacturing Importance 355
     B.   Theories of Autohesion and Tack 356
     C.   Processing with Tackifiers 359
VII. Plasticizers 359
     A.   Functions of Plasticizers 359
     B.   Plasticizer Theory 360
     C.   Compatibility 361
     D.   Selection of Plasticizers 361
     E.   Processing of Plasticizers 363

VIII. Preparations and Blends 363
      A.   General Discussion.. 363
      B.   Sulfur Preparations 364
IX.   Zinc-Free Rubber-Processing Additives.. 366
X.   Process Additives for Silica-Loaded Tread Compounds............................ 366
XI.   Process Oils 369
      A.   General Discussion.. 369
      B.   Manufacturing... 370
      C.   Characterization 371
           1.   Relative Density 372
           2.   Viscosity.. 372
           3.   Aniline Point 373
           4.   Refractive Index 373
           5.   Color.. 373
           6.   Flash Point... 373
           7.   Pour Point.. 373
           8.   Evaporative Loss 374
           9.   Aromatic Content... 374
           10. Viscosity–Gravity Constant 374
           11. Environmental Consideration... 374
      D.   Compatibility 374
      E.   Trends in Aromatic Oils... 375
XII. Summary 376
References.. 377

I. INTRODUCTION

In conjunction with the chemicals used in a rubber formulation to ensure acceptable product characteristics, a number of ingredients may be incorporated to allow or improve processing with the manufacturing equipment available in the plant. The stages of rubber processing may be broken down into raw materials handling, mixing, forming, and vulcanization. Some of the factors that may influence the process economics and product acceptability in these stages are listed in Table 8.1. The function of the processing additives is to minimize or overcome any problems associated with product fabrication while maintaining, or even improving, product performance. Before going into detail, some examples of acceptable performance criteria for processing additives at the four stages in product manufacture are presented briefly.

A. RAW MATERIALS HANDLING

Chemicals are frequently dusty powders that are difficult to handle and to disperse. They can become electrostatically charged, and as a result, incorporation into a product is made more difficult. Also, dusty powders are undesirable for environmental reasons, and this has led to the use of binders and dispersing agents to improve materials handling and weighing. Generally, preparations are coated, nondusting powders, granules, and masterbatches.

TABLE 8.1
Rubber Processing: Performance Factors

Stage	Operation	Performance factors
Raw materials	Storage, handling, weighing, blending, delivery	1. Temperature control 2. Humidity control 3. Handling of dusty and hazardous materials 4. Automatic handling and weighing 5. Weighing accuracy with small quantities 6. Uniformity of blending
Mixing	Internal mixer mill	1. Viscosity reduction 2. Viscosity control 3. Heat generation 4. Filler incorporation 5. Filler dispersion 6. Hydrophobation reaction with silica 7. Homogenization 8. Sticking and release 9. Mix time
Forming	Extrusion, hot/cold feed calendering, sheet/fabric calendering, profile cutting/joining fabric, building	1. Flow 2. Sticking and release 3. Shrinkage and stretching 4. Die swell 5. Dimensional stability 6. Tack 7. Green strength 8. Scorch 9. Surface appearance 10. Bloom 11. Fabric cord penetration
Vulcanization	Compression molding, transfer molding, injection molding, continuous vulcanization	1. Scorch 2. Flow 3. Component state of cure 4. Curative migration and dispersion 5. Mold release, fouling, cleaning 6. Surface appearance

Source: Courtesy of Schill & Seilacher, Hamburg, Germany.

B. MIXING

During mixing in the internal mixer or open mill, the additives should facilitate homogeneous blending of different polymers and enable faster incorporation of fillers and other compounding materials. Mixing should be optimized with respect to time, temperature, and energy. Compound viscosity should be reduced only to that level that allows acceptable processing in the ongoing manufacturing stages.

Uniform distribution and optimum dispersion of all compounding materials should be achieved, and the influence on scorch time has to be minimal and/or controllable. If possible, the tackiness of the compound should be controlled. Both excessive sticking to the machines and bagging on the mill due to a lack of stickiness must be avoided.

C. FORMING

Downline processing, that is, shaping of semiproducts, requires compounds with good flow properties. Profile compounds should calender and extrude easily, fast, and uniformly. The profiles should exhibit dimensional stability, smooth surface appearance, and exact edge definition. Temperature and die swell or shrinkage should be controllable and acceptable. For sheet calendering, a smooth surface, uniform shrinkage, and freedom from blisters are required. For metal wire or textile calendering, cutting, and joining, good flow properties and acceptable tack are required. Last but not least, bloom should be avoided.

D. VULCANIZATION

In the vulcanization process, good flow properties are needed in order to

1. Obtain adequate compound–compound adhesion
2. Obtain compound–metal and/or compound–textile adhesion
3. Fill the mold quickly, uniformly, and free of blisters or trapped air, particularly with transfer and injection molding equipment

---

TABLE 8.2
Processing Additives: Chemical Structure

Group	Examples
Mainly hydrocarbons	Mineral oils
	Paraffin waxes
	Petroleum resins
Fatty acid derivatives	Fatty acids
	Fatty acid esters
	Fatty alcohols
	Metal soaps
	Fatty acid amides
Synthetic resins	Phenolic resins
Low-molecular-weight polymers	Polyethylenes
	Polybutenes
Organothio compounds	Peptizers

Source: Courtesy of Schill & Seilacher, Hamburg, Germany.

Finally, the vulcanizates should demold easily without tear and not produce mold-fouling residues.

Processing additives may be subdivided according to their chemical structures (Table 8.2) or according to their application (Table 8.3). Several classes of substances can have more than one application. For example, fatty acid esters act as lubricants and dispersing agents. Mineral oils act as physical lubricants in rubber compounds, reducing viscosity, and also help in the filler dispersion process. In this chapter, we discuss the following compounding ingredients with respect to

TABLE 8.3
Processing Additives: Applications

Processing Aid	Application	Examples
Chemical peptizer	Reduces polymer viscosity by chain scission	2,2'-Dibenzamidodiphenyl disulfide
		Pentachlorothiophenol
Physical peptizer	Reduces polymer viscosity by internal lubrication	Zinc soaps
Dispersing agent	Improves filler dispersion	Mineral oils
	Reduces mixing time	Fatty acid esters
	Reduces mixing energy	Metal soaps
		Fatty alcohols
Lubrication agent	Improves compound flow and release	Mineral oils
		Metal soaps
		Fatty acid esters
		Fatty acid amides
		Fatty acids
Homogenizing agent	Improves polymer blend compatibility	Resin blends
	Improves compound uniformity	
Tackifier	Improves green tack	Hydrocarbon resins
		Phenolic resins
Plasticizer	Improves product performance at low and high temperatures	Aromatic di- and triesters
		Aliphatic diesters
		Alkyl and alkylether monoesters
Stiffening agent	Increases hardness	High-styrene resin–rubber
		Masterbatches
		Phenolic resins
		Trans-polyoctenamer
Softening agent	Lowers hardness	Mineral oils
Mold release agent	Eases product release from mold	Organosilicones
		Fatty acid esters
	Decreases mold contamination	Metal soaps
		Fatty acid amides

Source: Courtesy of Schill & Seilacher, Hamburg, Germany.

their influence on processing behavior and their relevant compound vulcanizate properties:

Physical and chemical peptizers
Lubricants
Homogenizing agents
Dispersing agents
Tackifiers
Plasticizers
Masterbatches—that is, sulfur and accelerator
Mineral oils

Considerable effort has been expended in recent years to develop more environmentally friendly rubber compounds. Consequently, we will mention some processing aids for various applications obtained from renewable sources and not petroleum based.

II. PHYSICAL AND CHEMICAL PEPTIZERS

A. MASTICATION

Mastication is the process whereby the average molecular weight of a polymer is reduced by mechanical work. The resulting lower viscosity of the polymer facilitates the incorporation of fillers and other compounding ingredients and can improve their dispersion. Because it is often difficult to homogeneously blend rubbers with very different viscosities, mastication of the higher viscosity rubber will enable improved blending with other, lower viscosity elastomers. Improved compound flow leads to easier downline processing such as calendering and extrusion. Shorter processing time and lower power consumption are generally obtained.

Because most of today's synthetic rubbers are supplied with easy-to-process viscosity levels, the mastication process is mainly restricted to natural rubber.

Although the natural rubber mastication process may be accomplished on an open mill, it is generally carried out in an internal mixer. During mechanical breakdown, the long-chain rubber molecules are broken under the influence of high shear from the mixing equipment. Chain fragments with terminal free radicals are formed, which recombine to form long-chain molecules if they are not stabilized (Figure 8.1). Through atmospheric oxygen, the radicals are saturated and stabilized. The chains are shorter, the molecular weight is reduced, and the viscosity drops. The course of the chain breakdown of natural rubber is shown in Figures 8.2 and 8.3.

Temperature is an important factor in the mastication of natural rubber. When the breakdown of natural rubber is plotted versus temperature (Figure 8.4), it can be seen that the effect is lowest in the range of 100°C–130°C. Chain cleavage by the mechanical process is more efficient at low temperatures (below 90°C) because owing to the viscoelastic nature of elastomers, the shear is higher the lower the temperature. With increasing temperature, the mobility of the polymer chains increases; they slide

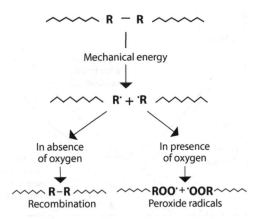

FIGURE 8.1 Physical peptization of rubber. (Courtesy of Schill & Seilacher, Hamburg, Germany.)

$$ROO^{\cdot} + RH \longrightarrow ROOH + R^{\cdot}$$

$$R^{\cdot} + O_2 \longrightarrow ROO^{\cdot}$$

$$2ROOH \longrightarrow RO^{\cdot} + ROO^{\cdot} + H_2O$$

$$RO^{\cdot} + RH \longrightarrow ROH + R^{\cdot}$$

FIGURE 8.2 Physical peptization of rubber—reaction sequence. (Courtesy of Schill & Seilacher, Hamburg, Germany.)

$$\begin{array}{c}
\qquad\quad CH_3 \qquad\qquad\qquad\qquad\quad CH_3 \\
\qquad\quad | \qquad\qquad\qquad\qquad\qquad\quad | \\
-CH_2-C=CH-CH_2-CH_2-C=CH-CH_2- \longrightarrow \\
\\
\qquad\quad CH_3 \qquad\qquad\qquad\qquad\quad CH_3 \\
\qquad\quad | \qquad\qquad\qquad\qquad\qquad\quad | \\
-CH_2-C=CH-{}^{\cdot}CH_2 + {}^{\cdot}CH_2-C=CH-CH_2-
\end{array}$$

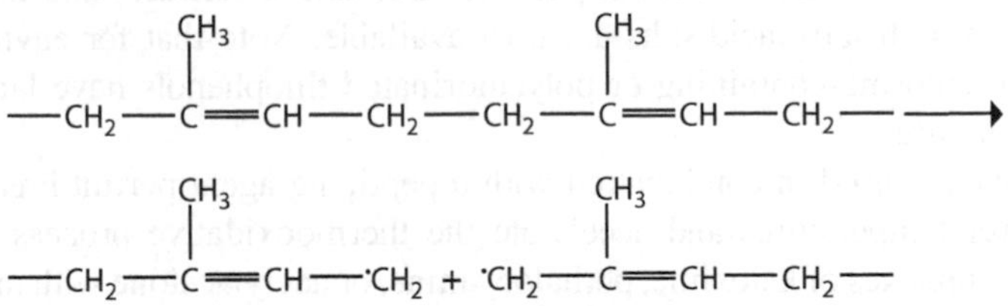

FIGURE 8.3 Physical peptization of polyisoprene. (Courtesy of Schill & Seilacher, Hamburg, Germany.)

over one another, and the energy input and generated shear force drop. However, although the mechanical breakdown process is minimal around 120°C, above this temperature, another breakdown process with a different mechanism, thermooxidative scission of the polymer chains, takes over and becomes more severe as temperature increases. An envelope curve is formed by the curves of the thermomechanical mastication and thermooxidative breakdown at elevated temperatures. In practice, the two reaction modes superimpose. Whereas the mechanical breakdown at low

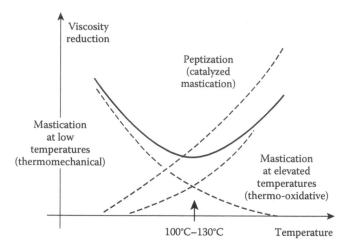

FIGURE 8.4 Peptization of NR. Viscosity reduction versus temperature. (Courtesy of Schill & Seilacher, Hamburg, Germany.)

temperatures largely depends on the mixing parameters, the thermooxidative breakdown is accelerated by temperature and catalysts, that is, peptizing agents.

Free radicals are generated when the molecular chains of the rubber are broken by mechanical or thermooxidative means. These radicals may recombine, and consequently, no reduction in molecular weight and viscosity will be observed. Moreover, branching is likely to occur. The peptizing agents can act as radical acceptors, thus preventing recombination of the generated chain-end free radicals.

All peptizing agents shift the start of thermooxidative breakdown to lower temperatures. Of the peptizing agents used in former times (Figure 8.5), only combinations of specific activators with thiophenols, aromatic disulfides, and mixtures of the activators with fatty acid salts are now available. Note that for environmental reasons, the chlorine-containing or polychlorinated thiophenols have largely been removed from use.

The activators used in combination with a peptizing agent permit breakdown to start at lower temperatures and accelerate the thermooxidative process. They are chelates—complexes of ketoxime, phthalocyanine, or acetylacetone with metals such as iron, cobalt, nickel, or copper, but nowadays almost exclusively iron complexes. These chelates facilitate the oxygen transfer by formation of unstable coordination complexes between the metal atom and the oxygen molecule. This loosens the O–O bond, and the oxygen becomes more reactive. Because of the high effectiveness of the activators or boosters, they are used only in small proportions in the peptizing agents.

During recent times, physical peptizers have gained major importance. They act as internal lubricants and reduce viscosity without breaking the polymer chains. Generally, zinc soaps have proved to be very effective in this role. Mechanical and chemical breakdown of the elastomer results in chain scission, lower molecular weight, broader molecular weight distribution, and an increased number of free chain ends. Normally, this leads to an increase in compound heat buildup and a

Dibenzamido
diphenyldisulfide (DBD)

+Activator

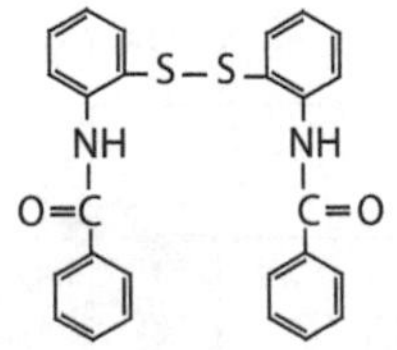

Pentachlorothiophenol
(PCTP)

+Activator

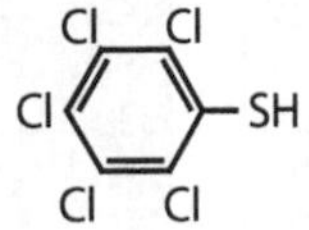

Zinc pentachlorothiophenol

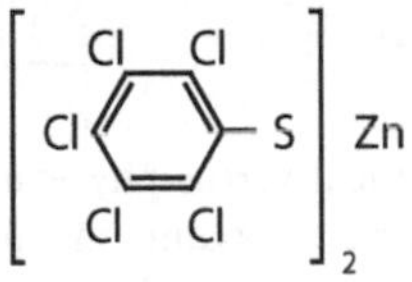

FIGURE 8.5 Common peptizing agents. (Courtesy of Schill & Seilacher, Hamburg, Germany.)

decrease in abrasion resistance. Lubricants do not change the molecular chains, that is, the chains are not broken.

As mentioned previously, synthetic rubbers are normally supplied with easy-to-process viscosity levels. If viscosity reduction is needed, mechanical mastication in an internal mixer has virtually no effect. In comparison to natural rubber, viscosity reduction of synthetic rubbers is more difficult owing to the (1) lower number of double bonds (SBR, NBR); (2) electron-attracting groups in the chain, which stabilize the double bonds; (3) vinyl side groups, which foster cyclization at high temperatures (NBR, SBR, CR); and (4) lower green strength due to the absence of strain-induced crystallization (NBR, SBR).

Synthetic rubbers can be broken down by means of peptizing agents. However, they require higher dosage levels and temperatures than natural rubber. For this reason, they are nowadays mostly physically peptized with salts of unsaturated fatty acids.

B. PROCESSING WITH PEPTIZING AGENTS

At one time, it was common practice to have a separate mastication stage whereby the peptizer was added to the NR and the mixing cycle was controlled to obtain an acceptable viscosity reduction. Nowadays normally, only one stage is used, with the filler addition being delayed in order to allow the peptizing agent to be incorporated in the rubber. The early addition of the filler, while enhancing shearing and breakdown, also has a positive effect on dispersion. However, as the activators used in combination with the peptizing agent may be adsorbed by the filler, it is normal to slightly increase the loading.

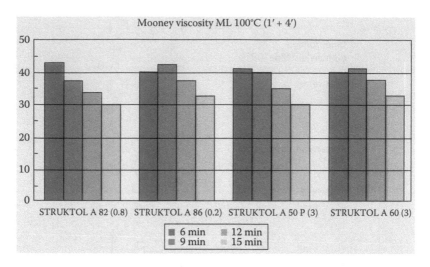

FIGURE 8.6 Chemical versus physical peptizers in NR. STRUKTOL® A 82 is a chemical peptizer containing an organic metal complex booster. STRUKTOL® A 86 combines a chemical peptizer and a booster. Its composition is similar to that of STRUKTOL A 82 but with a higher concentration of active substance. STRUKTOL® A 50 P contains zinc soaps of high-molecular-weight fatty acids. STRUKTOL® A 60 is similar to STRUKTOL A 50 P but has a lower melting range, allowing open mill mixing. (Courtesy of Schill & Seilacher, Hamburg, Germany.)

When natural rubber is blended with synthetic rubber that has a lower viscosity, it is useful to peptize the natural rubber before the synthetic rubber is added.

Because antioxidants inhibit the oxidative breakdown of rubber, they should be added late in the mixing cycle during the processing of natural rubber. With synthetic rubbers, an early antioxidant addition can avoid cyclization.

Figure 8.6 shows the influence of a number of chemical and physical peptizing agents on the breakdown, as measured by Mooney viscosity, of natural rubber (RSS1) in a 1 L laboratory internal mixer at 65 and 49 rpm and a start temperature of 90°C. Samples for Mooney viscosity testing were taken after 6, 9, 12, and 15 min.

Comparable results are obtained when physical peptizers are used at higher dosage levels than the chemical peptizers. The raw RSS1 had a Mooney viscosity of 104.

C. INFLUENCE OF PEPTIZING AGENTS ON VULCANIZATE PROPERTIES

The effects of a chemical peptizer (STRUKTOL* A 86, an aromatic disulfide in combination with a metal organic activator), a physical peptizer (STRUKTOL A 60, based on unsaturated fatty acid salts of zinc), and mechanical mastication on viscosity reduction and the tensile properties of NR (SIR 5 L) have been investigated. Apart from the usual evaluation of viscosity at low shear rates (i.e., Mooney viscosity, ML 1 + 4, 100°C), viscosity at higher shear rates, using a rubber-processing analyzer, was measured. The data are shown in Table 8.4.

---

* STRUKTOL is a registered trademark of Schill & Seilacher "Struktol" AG, Hamburg, Germany.

TABLE 8.4
Influence of Chemical and Physical Peptizers on Viscosity Reduction

	Mechanical Mastication	Chemical Peptizer	Physical Peptizer
ML 1+4, 100°C	94	65	83
Shear stress (s⁻¹)		Viscosity (Pa s)	
55.3	5.017	5.536	4.325
99.7	3.125	3.125	2.626
299.1	1.344	1.329	1.113

Source: Courtesy of Schill & Seilacher, Hamburg, Germany.

TABLE 8.5
Influence of Chemical and Physical Peptizers on Tensile Properties

	Mechanical Mastication		Chemical Peptizer		Physical Peptizer	
Cure at 150°C (min)	40	120	40	120	40	120
300% Modulus (MPa)	13.1	12.1	14.4	13.1	14.6	14.1
Tensile strength (MPa)	26.1	24.4	25.8	22.1	26.8	25.6

Source: Courtesy of Schill & Seilacher, Hamburg, Germany.

Under low shear conditions, the chemical peptizer is by far the more effective method for viscosity reduction. However, under higher shear, which is a better simulation of factory conditions, the physical peptizer performs better. This is of special importance because there are sometimes concerns with the use of chemical peptizers regarding their effect on long-term physical properties of compounds. The changes in modulus and tensile strength (Table 8.5) show a greater degradation of natural rubber with the chemical peptizer system, whereas the physical peptizer highlights better retention of physical properties.

Chemical peptizers give the greatest fall in rubber viscosity, but they cause an increase in the amount of very low-molecular-weight polymer. They also adversely affect dynamic heat buildup, increasing tan δ in comparison with mechanical mastication, especially on overcure and aging. Similar rubber viscosities can be achieved by masticating the rubber in the presence of fatty acid soaps without a significant change in tan δ [1].

In summary, the benefits of peptizing agents are as follows. They

Accelerate viscosity reduction, decreasing mixing time
Reduce power consumption
Promote batch-to-batch uniformity
Facilitate blending of elastomers
Reduce mixing costs
Improve dispersion

III. LUBRICANTS

A. GENERAL DISCUSSION

Lubricants are processing additives that are used to improve compound flow and release. In the early days of rubber processing, it was recognized that stearic acid, zinc stearate, and wool grease were effective in improving the flowability of rubber compounds. Barium, calcium, and lead stearate were used but were withdrawn some time ago for environmental reasons. The essential raw materials for this class of products are fatty acids, fatty acid salts, fatty acid esters, fatty acid amides, and fatty alcohols. In addition, hydrocarbons such as paraffin wax are of importance. More recently, low-molecular-weight polyethylene and polypropylene have been used because of their waxlike character.

Modern lubricants on the market are normally composed of the earlier listed basic materials. Among the fatty acids, stearic acid still finds widespread application as a material that improves both the processability of compounds and their curing characteristics. Because of their low melting point (M.P.) and waxlike character, fatty acids enhance both mixing and downline processing. They reduce the stickiness of compounds. The fatty acids produced from vegetable oils and animal fats are predominantly mixtures of C_{16}–C_{18} fatty acids. Even though they have a higher volatility, fatty acids having a shorter chain length, such as lauric acid (C_{12}), are occasionally used.

The limited compatibility of stearic acid with synthetic rubbers and the need for specialty products to solve complex processing problems have been the driving force for the development of more modern lubricants. Raw materials for most lubricants are mixtures of glycerides such as vegetable oils and animal fats. Typical examples are listed in Table 8.6. Through saponification of the glycerides, mixtures of fatty acids are obtained that vary in carbon chain length distribution and in their degree of unsaturation.

The most important fatty acids are listed in Table 8.7. Separation and purification processes lead to specified technical grade fatty acids that are the basis for tailor-made lubricants in rubber processing.

TABLE 8.6
Important Raw Materials for Fatty Acids

Castor oil	Cotton oil
Coconut oil	Groundnut oil
Herring oil	Linseed oil
Olive oil	Palm oil
Palm kernel oil	Rapeseed oil
Soybean oil	Sunflower oil
Tallow	

Source: Courtesy of Schill & Seilacher, Hamburg, Germany.

TABLE 8.7
Important Fatty Acids

Fatty Acid	Chain Length[a]	Double Bonds
Palmitic	16	0
Stearic	18	0
Oleic	18	1
Erucic	22	1
Ricinoleic[b]	18	1
Linoleic	18	2
Linolenic	18	3

Source: Courtesy of Schill & Seilacher, Hamburg, Germany.
[a] Number of C atoms.
[b] 12-Hydroxyoleic acid.

The fatty acids tend to be incompatible and therefore insoluble in the rubber hydrocarbon, and consequently, they can migrate to the surface of the uncured rubber to form a bloom. This will be detrimental to the tack-building ability of the component and may lead to downline assembly problems. This has led to the development of fatty acid esters, fatty acid amides, and metal soaps that are soluble in rubber and minimize bloom formation.

Fatty acid esters are produced through reaction of fatty acids with various alcohols. Apart from good lubricating effects, they promote the wetting and dispersion of compounding materials. The carbon chain lengths of the acid and alcohol components vary between C_{16} and C_{34}.

Metal soaps are produced through the reaction of water-soluble fatty acid salts with metal salts (e.g., $ZnCl_2$) in an aqueous solution (precipitation process). Metal soaps are also obtained via a direct reaction of fatty acid with metal oxide, hydroxide, or carbonate.

The most important metal soaps are zinc and calcium soaps, with the zinc soaps having the largest market share. Because calcium soaps have less influence on the crosslinking reaction and scorch time, in most cases, they are used in compounds based on halogen-containing elastomers such as CR or halobutyl. The metal soaps are mostly based on C_{16}–C_{18} fatty acids. Because of better solubility in the rubber and lower melting points, modern lubricants frequently contain the salts of unsaturated fatty acids.

When 2–5 phr of a metal soap is present in a compound, the stearic acid level should be reduced to 1 phr to minimize bloom.

The most well-known soap, zinc stearate, is also used as a dusting agent for uncured slabs based on nonpolar rubbers. Owing to its high crystallinity, the compatibility of zinc stearate is often limited. Bloom can occur, which may lead to ply separation in assembled articles.

In general, metal soaps are also good wetting agents. Under the influence of higher shear rates, they promote compound flow, but without shear, the viscosity remains high (green strength).

As discussed in the previous section, soaps of unsaturated fatty acids are also used as physical peptizers because of their lubricating effect; they exhibit high compatibility with rubber.

Mixtures of zinc salts based on aliphatic and aromatic carboxylic acids are cure activators, strongly delaying the reversion of NR compounds. The effect is most pronounced in semi-EV systems.

Fatty alcohols are obtained through reduction of fatty acids. Straight fatty alcohols are rarely used as processing additives for rubber compounds because of their very limited solubility. They act as internal lubricants and reduce the viscosity.

Fatty acid amides are reaction products of fatty acids or their esters with ammonia or amines. All products of this group reduce scorch safety, which needs to be allowed for in compound development.

Organosilicones are relatively new in the range of lubricants. They are produced through condensation of fatty acid derivatives with silicones and combine good compatibility through the organic component with the excellent lubricating and release properties of the silicones. Depending on their structure, they can be adapted to standard or specialty elastomers. They have high thermal stability. Because of their high compatibility, the organosilicones are not prone to reduced adhesion, delamination, or general contamination, which are generally associated with the presence of silicones in a rubber factory. They significantly improve calendering and demolding and reduce mold fouling in critical polymers such as ethylene oxide epichlorohydrin copolymer (ECO) or fluoropolymers such as FKM.

Polyethylene and polypropylene waxes of low molecular weight are easily dispersed in natural rubber and synthetic rubbers. They act as lubricants and release agents. They improve the extrusion and calendering of dry compounds in particular and reduce the stickiness of low-viscosity compounds. Their compatibility with polar rubbers such as polychloroprene or acrylonitrile–butadiene copolymer (NBR) is limited. This can lead to adhesion and knitting problems when higher dosage levels are used.

B. PROPERTIES AND MODE OF ACTION OF LUBRICANTS

The major positive effects that can be achieved in various processing stages by using lubricants are listed in Table 8.8.

A strict classification of the products into internal and external lubricants is difficult, because practically all lubricants for rubber compounds combine internal and external lubricating effects. This depends not only on their chemical structure but also on the specific polymer in which they are used. In general, the solubility in the elastomer is the determining factor.

A processing additive predominantly acting as an internal lubricant will serve mainly as a bulk viscosity modifier and improve filler dispersion. Slip performance is influenced only to a minor extent. A lubricant with predominantly external action will greatly improve slip and reduce friction between the elastomer and the metal surfaces of the processing equipment. Its influence on compound viscosity is marginal. Filler dispersion can be improved through accumulation at the interface between

---

TABLE 8.8

Lubricants: Processing Benefits

Process Step	Benefit
Mixing	Faster filler incorporation
	Better dispersion
	Lower dump temperature
	Reduced viscosity
	Improved release
Processing	Faster and easier calendering and extrusion
	Improved release
	Less energy consumption
Molding	Faster cavity fill at lower operating pressure
	Reduced stress in molded parts through easy cavity fill
	Shorter cycle times
	Improved release

Source: Courtesy of Schill & Seilacher, Hamburg, Germany.

---

elastomer and filler. Higher dosage levels, however, can lead to "overlubrication" (overconcentration) and subsequent blooming.

Lubrication is achieved through a reduction of friction. In the initial phase of addition, the lubricant is coating the elastomer and possibly other compounding materials, and friction against the metal parts of the processing equipment is reduced. With increasing temperature, the lubricant begins to melt and is worked into the matrix by the shearing action of the mixer. The rate and extent of incorporation of the lubricant into an elastomer is determined by its melting point, melt viscosity, and solubility. These factors depend on its chemical structure and polarity.

The chemical criteria for the efficiency of organic lubricants are the length of the hydrocarbon chain, the degree of branching, the unsaturation, and the structure and polarity of the terminal groups. The action of fatty acid–based lubricants has been explained by means of the micelle theory of surfactant chemistry [2].

Rubber may be considered to be mostly nonpolar and as such is similar to a mineral oil but with far higher molecular weight. When dispersed in this medium, metal soaps that have a sufficiently long hydrocarbon chain can form spherical or lamellar micelles. The nonpolar hydrocarbon chain of the soaps is soluble in the rubber, whereas the polar terminal group remains insoluble. Because of their limited solubility, the micelles can aggregate in stacks (Figure 8.7). Under the influence of the high shear rates that occur during rubber processing, these layered aggregates can be shifted against one another, and the rubber compound flows more easily (Figure 8.8).

Relatively strong cohesion of the aggregates formed by zinc stearate can be noted through a slight increase in the green strength of NR compounds that contain this metal soap at higher concentrations.

The structure-related effect of fatty acid–based lubricants is shown in Tables 8.9 and 8.10.

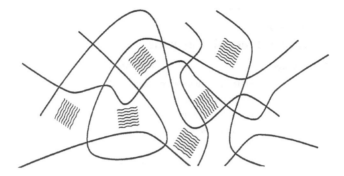

FIGURE 8.7 Metal soaps as surfactants in a polymer matrix. (Courtesy of Schill & Seilacher, Hamburg, Germany.)

Zero shear stress   Flow

FIGURE 8.8 Metal soaps as rheological additives. (Courtesy of Schill & Seilacher, Hamburg, Germany.)

The polar groups of certain fatty acids and their derivatives exhibit a high affinity to metal surfaces and are adsorbed easily. A film is formed at the metal surface. The film is extremely thin, is quite stable, and withstands relatively high shear. The formation of a film should in theory facilitate demolding, and the high thermal stability of the lubricant should reduce mold contamination. This is not, however, always the case in practice. Because limited compatibility is the essential and determining factor for the effectiveness of external lubricants, an overdosage has to be avoided; otherwise, undesirable bloom will occur. The lubricant concentration required, under practical conditions, depends on the processing procedures used and in particular on other compounding materials included in the formulation and their individual dosage levels. Therefore, it is necessary to check the compatibility of the lubricant chosen for a specific formulation. Additives are easily adsorbed by fillers. Therefore, higher dosages are required when highly active fillers or high filler loadings are used. Certain plasticizers can reduce compatibility and make the additives bloom.

Many commercial zinc soaps are indeterminate blends resulting from the "cut" of natural fatty acids used in the manufacture.

C. PROCESSING WITH LUBRICANTS

Most lubricants are easily incorporated. In some cases, they are added at the beginning of the mixing cycle, along with the fillers, to make use of their dispersing effects. Many of them can also be added at the end of the cycle. Because of their

TABLE 8.9

Structure–Property Relationships of Zinc Soaps

Structure	Property
Carbon chain length	
Below C_{10}	Unable to form effective micelles
Above C_{10}	Acts as surfactant
Carbon chain length distribution (blend)	
Narrow	Highly crystalline
	Higher M.P.
	Poor dispersibility
	Can bloom easily
Broad	Amorphous
	Lower M.P.
	Disperses readily
	Reduced bloom tendency
	Solubility is increased
Polarity	
High (functional groups, metal salts)	Increased affinity to metal surfaces
	More surface active
Low	Acts internally
	Less blooming
Branching	Disrupts crystallinity
	Totally soluble
	Nonblooming

Source: Courtesy of Schill & Seilacher, Hamburg, Germany.

TABLE 8.10

Structure–Property Considerations of Zinc Soaps

Most zinc soaps are rubber soluble and therefore act as intermolecular lubricants.
Increased hydrocarbon chain length improves surfactant action.
Presence of unsaturation improves dispersibility.

Source: Courtesy of Schill & Seilacher, Hamburg, Germany.

relatively low melting points, the products will soften early and give a good and uniform dispersion.

When the lubricating effect is of major importance, the processing additives should be incorporated in the final stage. The effects of selected lubricants on spiral mold cavity fill, when they are added in the first pass or final stage, are shown in Figure 8.9.

Depending on requirements and compatibility, the dosage varies between 1 and 5 phr. Usually, the minimum dosage is 2 phr. For an exceptionally high lubricating

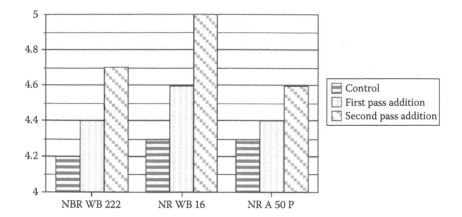

FIGURE 8.9 Spiral mold cavity fill with lubricant added in the first or final stage. Struktol WB 222 is an ester of saturated fatty acids. It is a lubricant and release agent predominantly used for polar elastomers. STRUKTOL® WB 16 is a mixture of calcium soaps and amides used as a lubricant for nonpolar polymers. STRUKTOL A 50 P is a zinc soap of unsaturated fatty acids. It is used as a physical peptizer in NR compounds. (Courtesy of Schill & Seilacher, Hamburg, Germany.)

effect in tacky compounds or where high extrusion rates and easy demolding are critical, even higher dosages might be useful. This applies also to compounds with high filler loadings.

D. INFLUENCE OF LUBRICANTS ON VULCANIZATE PROPERTIES

The effects of the lubricants STRUKTOL WB 16 and STRUKTOL A 50 P on the physical properties of a natural rubber compound are shown in Table 8.11. The lubricants lead to a decrease in 300% modulus in conjunction with a small drop in tensile strength and increase in elongation to break. No difference is observed in Shore hardness, but compression set increases slightly [3].

TABLE 8.11
Influence of Lubricants on NR Physical Properties

Property	Control	STRUKTOL WB 16	STRUKTOL A 50 P
Cure at 150°C (min)	10	9	12
300% Modulus (MPa)	13.6	10.7	10.3
Tensile strength (MPa)	19.1	18.3	18.4
Elongation at break (MPa)	430	480	480
Shore A hardness	60	60	60
Compression set, 22 h at 70°C (%)	21	24	24

Source: Courtesy of Schill & Seilacher, Hamburg, Germany.

IV. HOMOGENIZING AGENTS

A. EXAMPLES AND FUNCTION

Homogenizing agents are used to improve the homogeneity of difficult-to-blend elastomers. They assist in the incorporation of other compounding materials, and intrabatch and batch-to-batch viscosity variation are reduced by their use.

They are resin-based mixtures that exhibit good compatibility with various elastomers and facilitate blending through early softening and wetting of the polymer interfaces. Because the softening resins exhibit a certain tackiness, polymers that tend to crumble and polymer blends will coalesce more easily; energy input is maintained at a high level, that is, mixing is more effective; and mixing times can often be reduced.

Fillers are incorporated at a faster rate and are more evenly distributed owing to the wetting properties of the homogenizing agents. Filler agglomeration is minimized.

Apart from their compacting effects, the homogenizers lead to increased green strength when used as a partial replacement for processing oil, and compound flow is facilitated through improved homogeneity and a certain softening effect. They increase the green tack of many compounds and boost the efficiency of tackifying agents.

In summary, homogenizing agents promote the (1) blending of elastomers, (2) batch-to-batch uniformity, (3) filler incorporation and dispersion, (4) shortening of mixing cycles, (5) energy savings, and (6) building of tack.

The greater the difference in the solubility parameter and/or viscosity of each component elastomer in a blend, the more difficult it is to produce a homogeneous mix (Table 8.12). Blends of plasticizers that are each compatible with different elastomers can in theory be effective at improving blend homogeneity, provided that they have a viscosity sufficiently high to maintain high shear on mixing. Plasticizers have the disadvantage of being prone to migration and bloom. Therefore, mixtures of high-molecular-weight products such as resins are more often used.

The homogenizing resins are themselves complex blends and contain parts that are compatible with aliphatic and aromatic structures in a blend. Potential raw materials for use as homogenizing resins can be divided into three groups:

1. Hydrocarbon resins including coumarone–indene resins, petroleum resins, terpene resins, bitumens, tar, and copolymers, for example, high-styrene reinforcement polymers
2. Rosins and their salts, esters, and other derivatives
3. Phenolic resins of various kinds, such as alkylphenol–formaldehyde resins, alkylphenol and acetylene condensation products, and lignin and modifications thereof

Coumarone resins, produced from coal tar, were the first synthetic resins used as processing additives because of their ability to act as dispersing agents to improve filler incorporation and as tackifiers. They are typical aromatic polymers, consisting mainly of polyindene. The structural elements of these copolymers are methylindene,

TABLE 8.12

Solubility Parameters of Elastomers and Plasticizers

Solubility Parameter	Elastomer	Plasticizer
	AU, EU	
11		
	NBR (high nitrile)	Polar ethers
		Highly polar esters
	NBR (medium nitrile)	
	NBR (low nitrile)	
		Low polar esters
	CR	
10		Aromatic
	SBR	
	NR	
	BR	Naphthenic
	IIR	
	EPDM	Paraffinic
9	EPM	

Source: Courtesy of Schill & Seilacher, Hamburg, Germany.

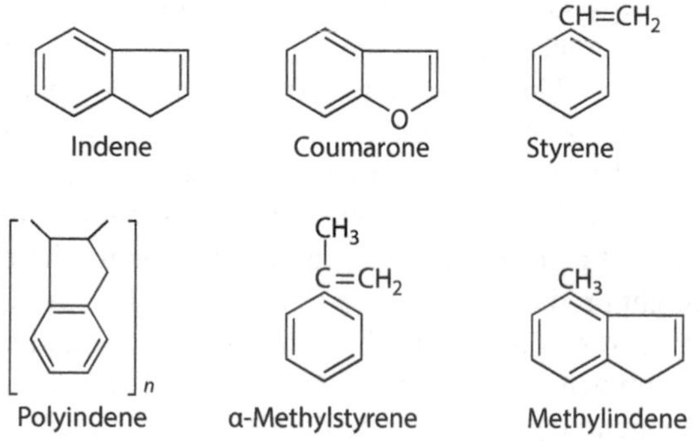

FIGURE 8.10 Coumarone resins—structural members. (Courtesy of Schill & Seilacher, Hamburg, Germany.)

coumarone, methylcoumarone, styrene, and methylstyrene (Figure 8.10). The melting range of these products is between 35°C and 170°C.

Petroleum resins are relatively inexpensive products that are often used at fairly high dosages, up to 10 phr or more. They are polymers produced from the C_5 cut of highly cracked mineral oils. The petroleum resins are relatively saturated and are also available with a high content of aromatic structures. Grades with a lower content of aromatic compounds have a stronger plasticizing effect. The highly saturated grades are used by the paint industry. Apart from cyclopentadiene, dicyclopentadiene and its methyl derivatives, styrene, methylstyrene, indene, methylindene, and higher homologs of isoprene and piperylene are found in these resins. This may explain their high compatibility with different elastomers.

Copolymers such as high-styrene resin masterbatches are used for high-hardness compounds. Whereas straight polystyrene can hardly be processed in rubber compounds, copolymers of styrene and butadiene with higher styrene contents have proven their worth.

Terpene resins are very compatible with rubber and give high tackiness. However, they are used mainly for adhesives. The polymers are based on α- and β-pinene. The cyclobutane ring is opened during polymerization and polyalkylated compounds are formed (Figure 8.11). Terpene resins improve aging performance and resistance against oxidation of rubbers.

Asphalt and bitumen are products that have been used since the very beginning of rubber processing. Their tackifying effect is not very distinct. They are relatively inexpensive products. Whereas asphalt is a naturally occurring product, bitumen is produced from the residues of mineral oil production. Blown bitumen, oxidized to achieve higher solidification points, is also known as mineral rubber and is a good processing additive, for example, in difficult-to-process compounds that have a high percentage of polybutadiene. Mineral rubber is also successfully used to improve the collapse resistance of extrusions.

Rosins are natural products obtained from pine trees. They are mixtures of organic substances, for the most part doubly unsaturated acids, such as abietic acid, pimaric acid, and their derivatives (Figure 8.12). To reduce their sensitivity to oxidation, resins are partially hydrogenated or disproportionated. Because of their acidity, they have a slight retarding effect on vulcanization. Abrasion resistance, in particular that of SBR, is said to be improved. Rosin acid is widely used (as a salt) in the production of synthetic rubbers (SBR) because of its emulsifying properties.

Phenolic resins (Figure 8.13) are used mainly as tackifiers, reinforcing resins, curing resins, and adhesives. Their use is determined by the degree of parasubstitution and the presence of methylol groups.

Lignin has a complex structure and is based on various substituted phenols that are in part linked via aliphatic hydrocarbon units. As a by-product of the cellulose industry and especially the paper industry, it is available in large quantities and is

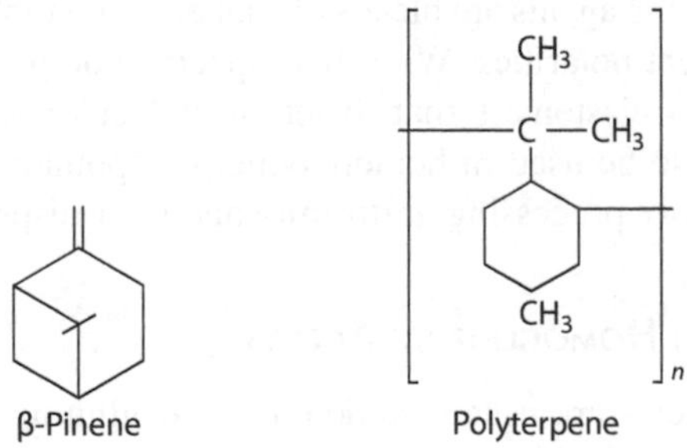

β-Pinene   Polyterpene

FIGURE 8.11 Terpene resins—main constituents. (Courtesy of Schill & Seilacher, Hamburg, Germany.)

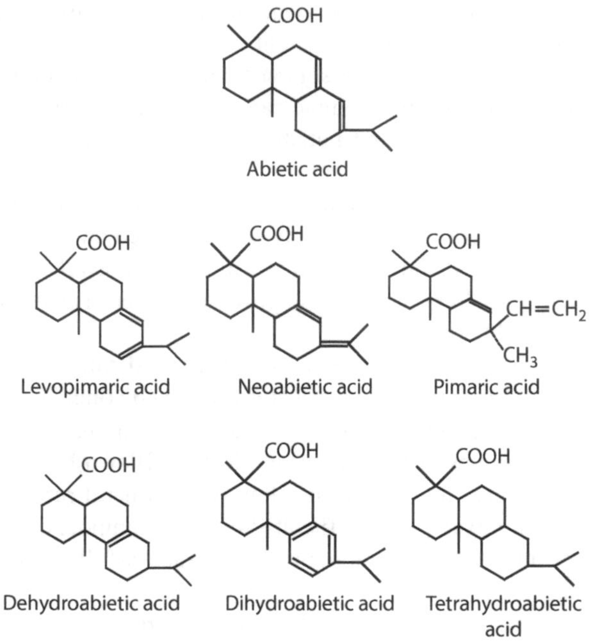

FIGURE 8.12 Rosin acids. (Courtesy of Schill & Seilacher, Hamburg, Germany.)

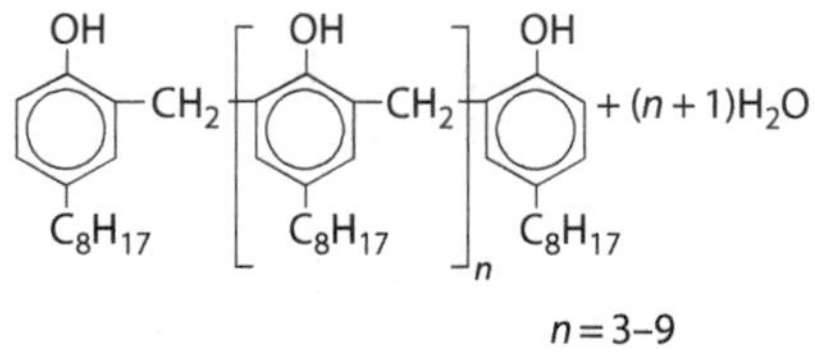

FIGURE 8.13 Alkylphenol resins. (Courtesy of Schill & Seilacher, Hamburg, Germany.)

quite cheap. It is often used in shoe soles, where it improves the incorporation and dispersion of high mineral filler loadings.

Modern homogenizing agents are blends of rubber-compatible nonhardening synthetic resins of different polarities. With their specific compositions, they promote the homogenization of elastomers that differ in molecular weight, viscosity, and polarity. They may also be used in homopolymer compounds where, among other effects, they can improve processing uniformity and filler dispersion.

B. PROCESSING WITH HOMOGENIZING AGENTS

The homogenizing agents are usually added at the beginning of the mixing cycle, particularly when elastomer blends are used. They exhibit optimum effectiveness at around their softening temperature. The recommended dosage is between 4 and 5 phr. Difficult-to-blend elastomers will require an addition of 7–10 phr.

As an example, the processing of butyl compounds may be improved though the use of a mixture of aromatic hydrocarbon resins, such as STRUKTOL® 40 MS Flakes, as a homogenizing agent. Filler dispersion, splice adhesion, physical properties, and impermeability are significantly improved through the use of this resin blend.

V. DISPERSING AGENTS

A. PROPERTIES OF DISPERSING AGENTS

Because dispersing agents are mostly fatty acid derivatives, they can be looked at as a subgroup of lubricants. The central property, however, is dispersibility. In particular, they improve dispersion of solid compounding materials. They reduce mixing time and have a positive influence on subsequent processing stages.

Dispersing agents have distinct wetting properties. They are often less polar fatty acid esters. Because often a combination of dispersing properties and good lubrication is desirable, the dispersing agents available on the market are occasionally mixtures of higher-molecular-weight fatty acids and metal soaps. Most products on the market are offered as "dispersing agents and lubricants" and are not listed separately in the product ranges. Their mode of action has already been described in Section III.

B. PROCESSING WITH DISPERSING AGENTS

Dispersing agents are usually added together with the fillers. Their product form and low melting point facilitate easy incorporation. When fillers are added in two steps, the dispersing agents should be added at the beginning. The dosage of these parts is between 1 and 5 phr. Because of their high effectiveness, however, low dosages are often sufficient. Very high filler loadings may require higher dosages.

A typical product is STRUKTOL® W 33 Flakes, a mixture of fatty acid esters and metal soaps that allows fillers to be rapidly incorporated and dispersed, particularly when high loadings have to be processed. Agglomerations are avoided and batch-to-batch uniformity is significantly improved. Their lubricant action leads to shorter mixing cycles, less power consumption, and lower mixing temperature. Downline processing is facilitated, and release performance is improved.

VI. TACKIFIERS

A. DEFINITION AND MANUFACTURING IMPORTANCE

The use of tackifiers in the tire industry has been reviewed by Lechtenboehmer et al. [4]. Tack is considered the ability of two uncured rubber compound surfaces to adhere together or resist separation after being in contact under moderate pressure for a short period of time. Two types of tack may be defined: autohesive tack, in which both materials are of the same chemical composition, and heterohesive tack, in which the materials have different compositions. A factor inherent in tack is compound green strength, the resistance to deformation and fracture of a rubber stock in the uncured state.

The tack, or autohesion, and green strength of the unvulcanized rubber compound components are of considerable importance in tire manufacture. Tack properties must be optimized; too high a tack value will cause difficulties in positioning the components during the building operation and may lead to trapped air between tire parts, giving after-cure defects. Simultaneously, sufficient tack must be present in order that the components of the green tire will hold together until the curing process. In addition, to prevent creep with resultant component distortion, or tear occurring during molding in the curing press, good green strength is required.

B. Theories of Autohesion and Tack

The principal theories that have been proposed to explain the mechanism of autohesion have been reviewed by Wake [5,6] and Allen [7]. These can be considered to have four fundamental modes: absorption theory, diffusion theory, electronic theory, and mechanical interlocking theory. The adsorption theory depends on the formation of an autohesive bond due to the van der Waals attraction between molecules and hence between surfaces. If the adsorption theory is correct, a correlation would be expected between the energy of adsorption and the autohesive bond strength. Although there is a tendency for this to be observed, it is not sufficient to form the basis for a precise and quantitative theory [7]. In addition, the adsorption theory should predict that autohesion increases with increasing polarity of polymeric materials.

The diffusion theory associated mainly with Voyutskii [8] and Vasenin [9] states that autohesive bonding takes place as a result of self-diffusion of the polymer molecules across the interface between two similar polymer surfaces. The strength of the autohesive bond is controlled by the self-diffusion, owing to the ability of the polymeric chains to undergo micro-Brownian motion of the surface polymer molecules across the interface. Skewis [10] measured the self-diffusion coefficients for a series of elastomers including natural, styrene–butadiene, and butyl rubbers. Rates were determined by application of a layer of radioactive-labeled polymer to the top of an unlabeled polymer base film and following the decay of radioactivity at the surface of the system due to self-adsorption. Results indicated that when two pieces of unvulcanized rubber are brought into contact, diffusion of polymer chains across the interface can occur with a subsequent increase in adhesion between the samples.

Anand [11,12] proposed and, in conjunction with coworkers [13–15], developed the contact theory, which states that the bonding between two similar polymer surfaces consists in, first, the development of molecular contact, followed by instantaneous physical adsorption brought about by van der Waals forces.

Wake [6] suggested that the mechanism of autohesion has not been fully confirmed and may be different for different rubbers. This, in conjunction with Vasenin's [16] comment that the individual theories cannot explain all the facts of the adhesion phenomena, led to the proposal [5,7] that some combination of the common mechanisms is required to represent the real-life situation. Rhee and Andries [17], when investigating the factors influencing the autohesion of natural and styrene–butadiene rubbers, considered that a combined diffusion–adsorption mechanism was operative.

Wool [18], in treating strength development at a polymer/polymer interface in terms of the dynamics and statistics of random coil chains, regarded the interdiffusion of chain segments across the interface to be the controlling parameter in determining tack and green strength of uncured linear elastomers. However, the theory included the concept of time-dependent molecular contact (wetting) development occurring first, followed by interdiffusion.

Wake [6], in considering the conditions that must be fulfilled by any comprehensive theory for the tack of elastomeric materials, and Hamed [19], in listing the conditions that must be met for a rubber to exhibit high tack, make the following points:

1. The two rubber surfaces must come into intimate contact in order for an appreciable common interface to be formed.
2. The degree of contact will be a function of the pressure applied during the contact time, rubber viscosity, surface microroughness, surface impurities, adsorbed gases, and, if a yield point exists for viscous flow, whether or not the applied pressure exceeds the yield pressure for the area concerned.
3. For increased contact area, the rubber must undergo viscous flow and displace pockets of trapped gases. In addition, the viscous flow should allow sufficient dissipation of local elastic stresses at the interface during the contact period to maintain adhesion.
4. After achieving molecular contact, polymer chains from each surface must diffuse across the initial interface. If interdiffusion between the two surfaces is sufficient or complete, the interface will disappear and the strength of the tack bond will equal the cohesive strength of the material.
5. After formation, the tack bond strength must be sufficient to resist high stress before rupture.

Because most synthetic rubbers are less tacky than natural rubber, it is often necessary to add tackifying substances. These should lead to improved uncured building tack on assembly and improved knitting of contact joints. They are also used in highly filled "dry" natural rubber compounds. They should give rubber compounds a high degree of tack that is maintained on storage and facilitate processing through viscosity reduction. On the other hand, the compounds must not stick to the processing equipment or lead to sticky vulcanizates. Physical properties and aging performance should not be adversely affected. Tackiness should not be reduced by compounding materials like waxes.

Tackifiers are products that may occasionally act as homogenizing agents (which have been discussed previously). They comprise rosin, coumarone–indene resins, alkylphenol–acetylene resins, and alkylphenol–aldehyde resins. Other hydrocarbon resins such as petroleum resins, terpene resins, asphalt, and bitumen can also be included, although their effectiveness is generally not very high.

The most extensively used tackifying resins are of the phenol–formaldehyde type, generally "novolaks" (prepared under acidic conditions with a P/F ratio >1) having the general structure [20] shown in Figure 8.14. R is typically a tertiary alkyl group. Wolney and Lamb [21] studied, in a blend of oil-extended SBR and NR, the effect on tack of novolaks prepared using *o-sec*-butylphenol, *p-sec*-butylphenol,

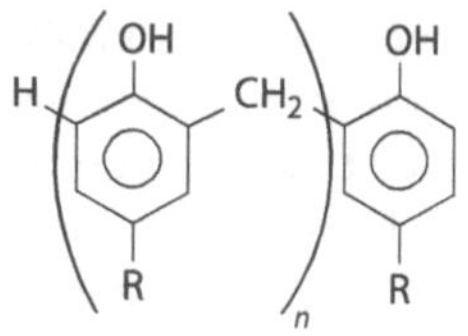

FIGURE 8.14   General structure of phenol–formaldehyde novolak resin.

p-tert-butylphenol, *p-tert*-amylphenol, and *p-tert*-octylphenol. All polymers had approximately the same molecular weight, free monomer level, and melting point.

It was found that novolaks based upon the alkylphenol with the largest alkyl group had the best initial tack. The best retention of tack values was found with a novolak based on *p-tert*-butylphenol. It was noted that to remain effective as a tackifier, a resin had to display limited compatibility, and because increasing the length of the alkyl groups increases compatibility, the novolak based upon *p-tert*-butylphenol was recommended as the best compromise. The effects of novolak molecular weight and free monomer level on autohesion were also studied. Tack and tack retention decreased with increasing free monomer level. The optimum number-average molecular weights for tack with resins based on *p-tert*-butylphenol and *p-tert*-octylphenol were 850 and 1350, respectively. Rhee and Andries [17] studied the effect of molecular weight and loading of tackifying resin on autohesion of NR and SBR rubbers. It was concluded that the optimum number-average molecular weight of *tert*-octylphenol–formaldehyde resin was 2095, corresponding to approximately 10 alkylphenol units, for maximum autohesion. The results also suggested that a critical level of phenolic resin is required to provide sufficient improvement of tack retention and that an optimum resin level may exist for maximum tack retention that is a function of the compound formulation.

Belerossova et al. [22] proposed a theory to explain the effectiveness of phenolic resins as tackifiers and the dependence of autohesion on molecular weight. They suggested that the phenolic resin molecules diffuse from one surface layer to another and form a hydrogen bonding network across the interface. The molecules should also have a minimum length so that the resin molecules can be attached in both surface layers, and at the other extreme, a maximum molecular weight should not be exceeded in the phenolic resin because the solubility of the resin in the rubber will fall, consequently decreasing autohesion. Therefore, an optimum resin molecular weight is considered to exist between the two extremes in order to optimize autohesion.

Two other tackifying resins are in use:

1. Koresin® (BASF AG, Ludwigshafen, Germany), a polymeric addition product from *p-tert*-butylphenol and acetylene. Its effectiveness is only marginally influenced by heat, humidity, and atmospheric oxygen. It has an exceptionally high melting point, approximately 130°C.
2. Xylene–formaldehyde resins are highly effective tackifiers with good plasticizing properties that improve knitting, for example, on injection

molding. Their tack improvement properties have been known for a long time but because of their high viscosity and stickiness, they are not very popular.

Hamed [19] noted that the function of tackifying agents in rubber mixes may be mainly to prevent tack degradation upon aging. Schlademan [23] proposed that the phenolic resin tackifiers act as antioxidants to prevent surface cross-linking and showed that if aging is carried out in nitrogen instead of air, even formulations with no tackifier will exhibit no tack loss with aging. Forbes and McLeod [24] showed that surface oxidation is detrimental to tack.

C. PROCESSING WITH TACKIFIERS

In general, the melting range of the tackifying resins is between 80°C and 110°C. Resins that have a high melting point should be added early in the mixing cycle in order to guarantee melting and sufficient dispersion. Soft resins can be added together with fillers to make use of their wetting and dispersing properties. A relatively late addition can be useful for maximum building tack. High-viscosity resins are occasionally prewarmed for easier handling. Normal dosage levels can vary between 3 and 15 phr.

An example of an aliphatic–aromatic soft resin that is an effective tackifier and exhibits a good plasticizing effect is STRUKTOL® TS 30. It significantly enhances building tack of compounds based on synthetic rubber, such as SBR, BR, NBR, and CR, provides improved filler incorporation and dispersion, and has relatively good resistance against extraction by aliphatic hydrocarbons and mineral oils.

VII. PLASTICIZERS

A. FUNCTIONS OF PLASTICIZERS

Although plasticizers represent a separate group of compounding materials, they can also be considered processing additives. They not only modify the physical properties of the compound and the vulcanizate but can also improve processing. The principal functions of plasticizers as modifiers of compound physical properties or processes are listed in Table 8.13. A plasticizer is defined as "a substance or material incorporated in a plastic or an elastomer to increase its flexibility, workability, or distensibility" [25].

As a property modifier in rubber compounds, a plasticizer can reduce the second-order transition temperature (glass transition temperature) and the elastic modulus. As a result, cold flexibility is improved. The static modulus and tensile strength are lowered in most cases, and correspondingly, a higher elongation at break results. Specialty plasticizers improve flame retardance, antistatic properties, building tack, and permanence.

The softening effect of plasticizers leads mostly to improved processing through easier filler incorporation and dispersion, lower processing temperatures, and better flow properties.

TABLE 8.13
Influence of Plasticizers on Physical Properties and Processing

Influence on physical properties

Lowers hardness

Increases elongation

Improves flex life

Improves low-temperature performance

Modifies swelling tendency

Imparts flame resistance

Improves antistatic performance

Influence on processing

Lowers viscosity

Speeds up filler incorporation

Eases dispersion

Lowers power demand and decreases heat generation during processing

Improves flow

Improves release

Enhances building tack

Source: Courtesy of Schill & Seilacher, Hamburg, Germany.

B. PLASTICIZER THEORY

The four main theories [26–29] that describe the effects produced by plasticizers have been summarized by O'Rourke [30]:

> The Lubricity Theory basically states that the plasticiser acts as a lubricant between the large polymer molecules. As the polymer flexes, it is believed that the polymer molecules glide back and forth with the plasticiser lubricating the guide planes. The theory assumes that the polymer macromolecules have, at the most very weak bonds and/or plasticizer–polymer bonding.
>
> The Gel Theory of plasticization starts with a model of the polymer in a 3-D honeycomb structure. The stiffness of the polymer results from this structure and the gel is well-formed by weak attachments which occur at intervals along the polymer chains. The points of attachment are close together and so provide little movement. The elasticity of the polymer is low. Plasticiser selectively solvates these points of attachment along the polymer chain therefore, the rigidity of the gel structure is reduced. Free plasticiser that is not solvating the polymer attachments can also swell the polymer providing further flexibility.
>
> The Free Volume Theory is based on the difference in volume observed at absolute zero temperature, –273°C, and the volume measured for the polymer at a given temperature. When adding plasticiser to a polymer, the free volume of the polymer increases. With rising temperatures, the free volume increases, thus allowing more movement of the polymer chains. The most important application of the theory to plasticisation has been to clarify the lowering of the glass transition temperature, T_g, by a plasticiser. Plasticisers have a smaller molecular size compared to polymers, which

provides greater free volume, thus allowing more mobility of the polymer. The lower T_g of plasticisers has the effect of lowering the T_g of the polymer.

The Mechanistic Theory of Plasticisation (also referred to as solvation–desolvation equilibrium), supplements the other three theories previously mentioned. This theory closely resembles the Gel Theory in which a plasticiser selectively solvates the points of attachment along the polymer chains. The essential difference is that in the Gel Theory, the plasticiser stays attached to the polymer chain, whereas the Mechanistic Theory states that the plasticiser can be exchanged by other plasticizer molecules along the polymer matrix. This exchange results in a dynamic equilibrium between solvation and desolvation of the polymer.

C. Compatibility

It is common practice to divide plasticizers into mineral oils and synthetic plasticizers. Mineral oils, by-products of the lubricating oil industry, have the largest market share as relatively inexpensive plasticizers that are used on a large scale in tire compounds and general rubber goods to reduce costs. At high dosage levels, they allow for higher filler loadings. The mineral oils are split into paraffinic, naphthenics, and aromatic types. They all exhibit a high compatibility with the weakly polar or nonpolar diene rubbers.

Compatibility of plasticizers with the elastomer is of major importance for their optimum effectiveness. It is largely determined by the relative polarity of both polymer and plasticizer. A homogeneous and stable mixture of plasticizer and elastomer is achieved if their polarities are nearly the same. In any case, sufficient compatibility is required to achieve the processability and physical properties intended without separation problems, which are observed as an exudation or bloom during processing.

Table 8.14 lists a number of elastomers and plasticizers according to their polarity and facilitates the selection of suitable plasticizers. Mineral oils are not included. Among them, the high aromatic products have a higher polarity, whereas the paraffinic ones are practically nonpolar.

D. Selection of Plasticizers

Plasticizers act on elastomers through their solvent or swelling power. They can be split into two groups: primary or true plasticizers, which have a solvating effect, and secondary plasticizers or extenders, which are nonsolvating and act as a diluent.

Liquid elastomers are plasticizers that can be viewed as processing additives. They co-cross-link during vulcanization and cannot be extracted. The vulcanizate properties are insignificantly changed, but hysteresis tends to be slightly higher.

Among the synthetic plasticizers, esters are the most widely used type. For compatibility reasons, they are used mainly in polar polymers. Their main function is to modify properties rather than to improve processing. In many cases, they enhance low-temperature flexibility and the elasticity of the vulcanizates. They are preferably used in NBR, CR, and chlorosulfonyl polyethylene (CSM).

The ester plasticizers can be split up into general-purpose plasticizers and specialty plasticizers, with the latter used mainly to modify such properties as (1) cold

TABLE 8.14

Polarity of Elastomers and Plasticizers

Elastomer	Plasticizer
High	
NBR, very high nitrile	
	Phosphates
AU, EU	
NBR, high nitrile	Dialkylether aromatic esters
	Dialkylether diesters
NBR, medium nitrile	
ACM, AEM	Tricarboxylic esters
	Polymeric plasticizers
CO, ECO	Polyglycol diesters
CSM	Alkyl alkylether diesters
	Aromatic diesters
CR	Aromatic trimesters
NBR, low nitrile	
CM	
HNBR	
SBR	Aliphatic diesters
BR	Epoxidized esters
NR	
Halo-IIR	Alkylether monoesters
EPDM	
EPM	Alkyl monoesters
IIR	
FKM	
Q	
Low	

Source: Courtesy of Schill & Seilacher, Hamburg, Germany.

flexibility, (2) heat resistance, (3) resistance to extraction, (4) flame retardance, and (5) antistatic behavior.

Of the monomeric ester plasticizers, the phthalic acid esters represent the largest group because they are relatively inexpensive. The carbon chain length of the alcohol components ranges from C_4 to C_{11}, and often mixed alcohols are used in the esterification process. The number of C atoms on the chains and the degree of branching determine the properties of the esters. A greater number of C atoms reduces compatibility, volatility, and solubility in water. It degrades processability and enhances oil solubility, viscosity, and cold flexibility. A higher degree of branching leads to poor low–temperature performance, higher volatility, easier oxidation, and higher resistivity. However, for environmental reasons, the phthalic acid esters have generally been replaced, mainly with sebazates and adipates.

Plasticizers that improve low-temperature performance and elasticity of the vulcanizates are aliphatic diesters of glutaric, adipic, azelaic, and sebacic acid. They are mostly esterified with alcohols having branched chains, such as 2-ethylhexanol or isodecanol. Oleates and thioesters are often used in polychloroprene. Esters based on triethylene glycol and tetraethylene glycol or glycol ethers of adipic and sebacic acid and thioethers are used as low-temperature plasticizers in nitrile and polychloroprene.

A wide variety of low-temperature plasticizers are available, although differences in effectiveness are often marginal. The choice is fully determined by properties such as volatility or compatibility.

Heat-resistant vulcanizates require plasticizers that have low volatility. It should be noted that it is not the volatility of the pure product that is decisive but the volatility of the vulcanizate, which depends on compatibility and migration.

Particularly, suitable plasticizers for polar elastomers are, for example, trimellitates or pentaerythritol esters, polymeric esters, and aromatic polyethers, which also act as tackifiers. In comparison with common ester plasticizers, the processability of these plasticizers is more difficult. The polymeric esters in particular exhibit a remarkable resistance to extraction by oils and aliphatic solvents. This group of plasticizers has proven to be of use in heat-resistant vulcanizates based on thermally stable elastomers such as hydrogenated acrylonitrile–butadiene copolymer (HNBR), ethyl acrylate polymers (ACM), and CSM.

Flame retardant ester plasticizers play a relatively important role, because halogen-containing products, such as the chlorinated paraffins, are not generally permitted in use. Phosphate esters are often used. Several types are commercially available, permitting a proper choice regarding heat resistance or low-temperature performance. They are alkyl, aryl, and mixed esters.

Antistatic plasticizers are another important group. Having limited compatibility, they accumulate at the vulcanizate surface and reduce the surface resistance by absorption of moisture from the atmosphere. The best known representatives of this group are polyglycol esters and ethers.

E. PROCESSING OF PLASTICIZERS

The incorporation of plasticizers, at moderate dosage levels, on two-roll mills or in internal mixers is relatively easy. They act to increase dispersion during filler incorporation, and at the same time, the compound viscosity and consequently the processing temperatures are reduced. Plasticizer-containing compounds generally have enhanced building tack and better extrusion performance.

In general, the synthetic plasticizers have very little influence on the shelf life or scorch safety of the compounds in which they are incorporated.

VIII. PREPARATIONS AND BLENDS

A. GENERAL DISCUSSION

Some compounding ingredients may be difficult to incorporate and disperse during mixing; for example, high melting point or agglomeration of the ingredient may cause problems. Other ingredients are highly active and are added at only very low loadings. In these cases, a dispersing system can be used to produce a preparation or blend with significantly better processing performance.

Some rubber chemicals such as a few accelerators exhibit limited storage stability; others are sensitive to humidity or oxidation. Suitable binders or coatings can protect these materials.

Often, chemicals are dusty powders that are difficult to handle and to disperse. They can become electrostatically charged and as a result be difficult to incorporate. Dusty powders are undesirable for toxicological and ecological reasons, and this led to the relatively early use of binders and dispersing agents by the chemical industry. Generally, preparations are coated, nondusting powders, granules, and master-batches; a few are pastes.

Powders that are easy to process are mostly mixtures of fine particle size chemicals with oil and/or dispersing agents. The very homogeneous mixtures are non-dusting, are easy to handle and weigh, and can be easily and evenly dispersed in the compound. Oil and dispersing agent can also have a protective function for the chemical.

Granular chemicals are widely used because they are easy to handle. The simplest forms are granules obtained through fusion of pure low-melting chemicals. Granules are often mixtures of chemicals and various binders. Waxes, oils, latex, fatty acid derivatives, and elastomers are used as binders. The forms of granules are micro-beads, macrobeads, pastilles, cylinders, spheres, cubes, and compressed granules.

In most granules, the chemicals are very finely dispersed so that outstanding dispersion in the compound is guaranteed. Additional advantages of granules are freedom from dust; ease of weighing, in particular automatic weighing; good stability; and rapid dispersion, which can reduce mixing time and heat generation. An example of the use of an ingredient preparation follows.

B.   Sulfur Preparations

Sulfur is known to cause dispersion problems in rubber compounds. However, it is important to distinguish between soluble, insoluble, and colloidal sulfur, all of which may be used.

Colloidal sulfur, produced through grinding in colloid mills or precipitation of sulfur from colloidal solutions, is a material of very fine particle size that is very suitable for latex compounds. It scarcely settles and can be very well dispersed.

In solid rubber compounds, natural, soluble, ground sulfur of high purity (≥99.5%) is mostly used. A medium particle size, which can be easily dispersed, is preferable.

In most cases, rubber compounds contain more sulfur than is soluble in the respective elastomer at room temperature. Usually, however, complete dissolution is achieved during mixing because the mixing temperature is high enough to melt the sulfur. On cooling, a supersaturated solution is formed in the compound that is a source of sulfur crystals visible at the surface after migration.

Crystallization occurs once the solubility limit is reached. The migration rate depends on the filler content and the elastomer. Highly loaded compounds exhibit a lower migration rate. Significantly more sulfur is soluble in NR and SBR than in

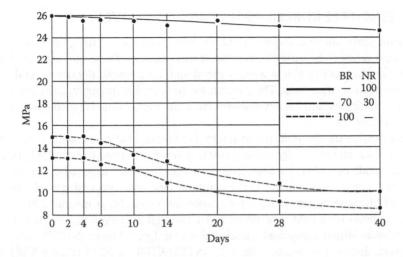

FIGURE 8.15 Tensile strength versus time. (Courtesy of Schill & Seilacher, Hamburg, Germany.)

NBR, EPDM, or IIR. This explains the long mixing time required for sulfur in IIR. Differences in solubility and migration rate can give rise to problems when elastomer blends are stored for too long a period of time. NR/BR or SBR/BR blends can show a reduction in tensile strength and elongation at break when vulcanization is performed after prolonged storage (Figure 8.15).

Because sulfur is less soluble in BR and its diffusion rate is higher than in NR or SBR, relatively large rhombic sulfur crystals can be formed in the BR phase. Therefore, it is advisable to rework such blends intensively after prolonged storage before shaping is performed, and the vulcanization should take place as early as possible. To counteract these problems effectively, insoluble sulfur is used in place of ground sulfur when the dosage level is above the solubility limit of sulfur. The benefit of insoluble sulfur is that it is insoluble in rubber, does not migrate, and does not bloom.

Insoluble sulfur is produced by melting soluble sulfur and instantly cooling the hot sulfur to room temperature. Polymeric sulfur is formed that is insoluble in organic solvents and elastomers. Like an inert filler, on mixing it is present in a rubber compound as a particulate suspension.

During processing, the stability of insoluble sulfur has to be taken into account. Being a metastable modification, it can rapidly revert to rhombic sulfur, particularly at elevated temperatures and under the influence of alkaline substances. Therefore, the processing temperature should not exceed 100°C for extended times.

For a good distribution of insoluble sulfur in a compound, a particularly fine particle size is required. This, however, makes its dispersion in the elastomer more difficult. Furthermore, insoluble sulfur is strongly prone to developing an electrostatic charge.

The preceding problems have led to the development of sulfur preparations that are easy to incorporate and disperse and thus require only a short mixing time at relatively low temperature. The sulfur is normally treated with special dispersing agents and surfactants.

IX.   ZINC-FREE RUBBER-PROCESSING ADDITIVES

In recent years, the zinc content of rubber, especially that of tires, has come under increased scrutiny because of environmental concerns. These concerns are based mainly on the fact that zinc is a heavy metal and has a potentially detrimental influence on aquatic organisms. The concern for tires is that an appreciable amount of the zinc dissipates into the environment from the rubber dust abraded from the tire in normal service.

The trend in the tire industry, therefore, is to reduce the zinc content of the products. To date, no technically viable possibility has been found to completely replace the zinc oxide in rubber, but considerable success has been achieved in the development of a zinc-free alternative to the common zinc soap processing additives [31]. This is illustrated by comparing the extrusion performance, as measured by extruding compound on a cold feed lab extruder through a Garvey die, of an NR/SBR carbon black–filled compound containing a zinc-free additive (STRUKTOL® HT 207) with that of a traditional zinc soap (STRUKTOL A 50 P) (Figure 8.16). The new zinc-free additive gives the best extrusion performance with respect to extrusion speed and rate while maintaining the same level of surface quality (Garvey rating 9A). No significant differences are observed in tensile, hardness, tear, and hysteretic properties (Table 8.15).

X.   PROCESS ADDITIVES FOR SILICA-LOADED TREAD COMPOUNDS

Tread compounds for low-rolling-resistance tires, based on blends of solution-polymerized SBR and BR polymers reinforced predominantly with silica together with a silane coupling agent, were first patented [32] in 1991. The claim was that they

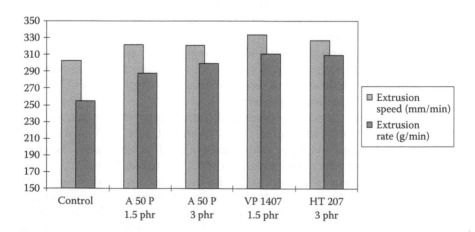

FIGURE 8.16 Extrusion speed and extrusion rate of NR/SBR compounds. (Courtesy of Schill & Seilacher, Hamburg, Germany.)

TABLE 8.15

Physical Properties of NR/SBR Compounds

Property	Control	STRUKTOL A 50 P		STRUKTOL HT 207	
		1.5	3	1.5	3
Unaged					
Shore A hardness (Sh.U.)	63	63	63	64	65
Rebound (%)	31	30	30	30	31
Elongation at break (%)	539	517	553	539	544
Tensile strength (MPa)	23.4	21.3	21.2	22.0	21.2
Modulus 100% (MPa)	2.2	2.1	2.0	2.2	2.1
Modulus 300% (MPa)	10.4	9.9	8.9	10.0	9.4
Tear resistance (Graves) (N/mm)	24,1	27,7	25,1	23.7	26.8
After aging (1 week at 100°C in air)					
Shore A hardness (Sh.U.)	75	75	75	75	73
Elongation at break (%)	215	202	218	201	248
Tensile strength (MPa)	13.7	13.4	12.8	12.3	13.0

Source: Courtesy of Schill & Scilacher, Hamburg, Germany.

had equal wet grip properties and superior ice grip properties compared to carbon black–filled tread compounds. This type of tread compound is now well established in the industry, but there are still a number of problems associated with them:

- Energy-intensive, multiple-stage mixing cycles are required to achieve processable compound. In many cases, the processing properties are still very poor.
- Close time and temperature control are required during mixing to achieve silica-to-silane coupling and to avoid silane degradation [33,34].
- The compounds tend to have high viscosities that become even higher with storage time, leading to more difficult processing.
- The compounds tend to have short scorch times.

The silica-to-silane coupling reaction is believed to be hindered by a number of normal compound additives [35], some of which might be used to improve mixing or processability. Consequently, considerable effort has been expended on the development of processing additives, two of which are illustrated in the following, to overcome some of the problems associated with silica-containing compounds.

An ester process additive, STRUKTOL® WB 217, is added at 2 phr in the first-stage mix with the first silica addition. A zinc–potassium soap process additive, STRUKTOL® EF 44, is added at 3 phr in the second mix stage in place of stearic acid. This product cannot be added during the first mixing stage because its polarity gives it a strong affinity with the silica surface that causes it to interfere with the silane coupling. These two process additives improve the mixing and processing

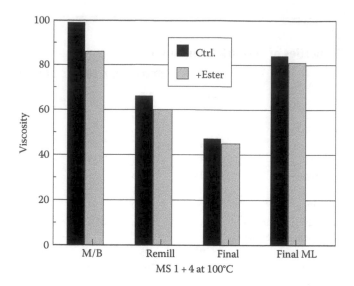

FIGURE 8.17 Improvement in mixing and processing behavior of a silica-loaded compound containing an ester process additive. (Courtesy of Schill & Seilacher, Hamburg, Germany.)

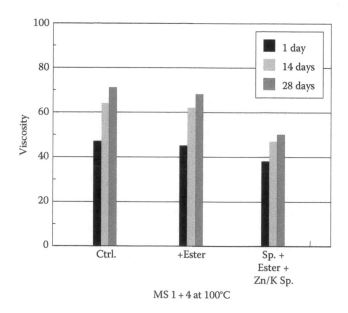

FIGURE 8.18 Improvement in mixing and processing behavior of a silica-loaded compound containing a zinc–potassium soap process additive. (Courtesy of Schill & Seilacher, Hamburg, Germany.)

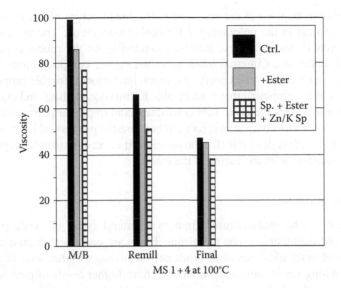

FIGURE 8.19 Influence on viscosity of a silica-loaded compound containing a zinc–potassium soap process additive. (Sp. = special mixing.) (Courtesy of Schill & Seilacher, Hamburg, Germany.)

characteristics of silica-loaded compounds [36] (Figures 8.17 and 8.18). In particular, when the masterbatch is mixed with only polymer, silica, silane, oil, and the ester additive present, very good silane coupling and a low degree of silane degradation are observed. This results in favorable dynamic properties, especially a low tan δ at 70°C, indicating a low rolling resistance, and a higher tan δ at 0°C, indicating a better wet grip.

The addition of the zinc–potassium soap also gives a large reduction in viscosity and virtually stops the increase in viscosity with storage time (Figure 8.19).

A zinc-free processing aid especially designed for use with silica-filled elastomers, STRUKTOL HT 207, is available. It improves extrusion rate, reduces hysteresis, and reduces mill roll sticking. It has a slight activation effect in sulfur-cured compounds.

XI. PROCESS OILS

A. GENERAL DISCUSSION

Mineral oils have been added to rubber compounds for over 150 years. Mineral oils are made from crude petroleum rather than animal fats or vegetable oils. Mineral oils serve three major functions:

1. Improving processing during mixing, milling, and extruding
2. Modifying the physical properties of the rubber
3. Reducing the cost of the rubber compound

Mineral oils can be used as extender oils in the manufacture of the polymer or as process oils to aid in the processing of the rubber compound. The same oil or different oils may be used for these purposes, depending on the rubber compound. A process oil serves as an internal lubricant in the rubber compound and allows the use of higher-molecular-weight polymers, which have more desirable properties and still give a rubber compound that is acceptable for mixing, milling, and extruding. A process oil also lowers the cured rubber hardness and improves pigment dispersion. The cost of the rubber compound is reduced because the process oil is less expensive than the polymer. It is desirable that the process oil be as cure neutral as possible so it does not interfere with the curing of the rubber.

B. MANUFACTURING

Process oils can be manufactured from two general types of crude petroleum: paraffinic and naphthenic crude petroleum. These are complicated mixtures of the same types of molecules. Paraffinic crude petroleum has a higher level of paraffinic or saturated long-chain molecules. It tends to have higher levels of petroleum wax, which has straight-chained paraffinic molecules. Naphthenic crude petroleum has higher levels of saturated ring compounds and tends to be low in wax content.

Process oils are largely manufactured by either an extraction–hydrotreating–solvent dewaxing process as shown in Figure 8.20 or by a newer hydrocracking–isodewaxing process as shown in Figure 8.21. In the process shown in Figure 8.20, crude petroleum is first distilled into streams according to boiling point, which roughly relates to molecular weight and hydrocarbon type. The heaviest oil stream is first deasphalted to remove asphaltenes from the oil. The oil streams are next extracted with a solvent such as phenol or furfural to remove the highly aromatic molecules (three or more rings). These highly aromatic oils are used as a process oil in SBR compounds. The oil streams are hydrotreated to improve color and oxidation

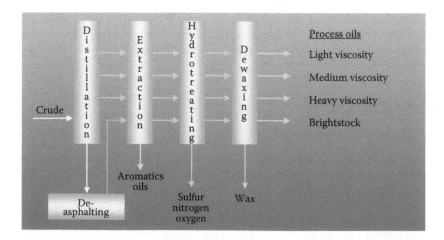

FIGURE 8.20 Manufacture of process oils for elastomer use.

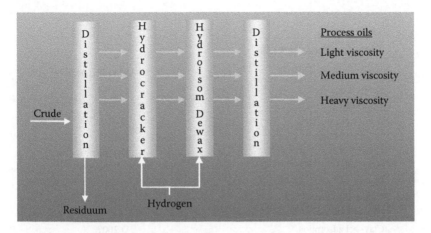

FIGURE 8.21   Alternative process oil manufacture.

stability and are then dewaxed to improve the low-temperature handling properties and improve compatibility with the rubber.

In the hydrocracking–isodewaxing process shown in Figure 8.21, the crude petroleum is distilled and then goes into a hydrocracker that breaks up the larger molecules into smaller molecules, opening ring compounds and saturating double bonds with hydrogen. This process converts the aromatic molecules rather than removing them. The oil streams then go to a hydroisomerization processing step, which branches the normal paraffins, making them no longer wax-type molecules. This process has a higher yield of process oil than conventional processing, because the aromatic molecules and wax molecules are converted to process oil rather than being removed. This process does not produce the highly aromatic oils used in SBR, wax, or the heaviest process oil.

C.   CHARACTERIZATION

The processes just described produce three types of process oils:

1. *Paraffinic*: High levels of isoparaffinic molecules; lower odor and more oxidative stability than naphthenic and aromatic oils; levels of monoaromatics similar to those of the aromatic oils, but much lower levels of multiring aromatics than aromatic process oils
2. *Naphthenic*: Higher level of saturated rings than aromatic process oils and paraffinic process oils; similar odor to paraffinic process oils
3. *Aromatic*: High levels of unsaturated single- and multiple-ring compounds, higher odor, lower oxidation stability, and higher reactivity than paraffinic oils

Process oils can be characterized by the analytical tests listed in Table 8.16. The basic selected properties of the three classes of process oils are outlined in Table 8.17. The terms defining these oils are described in turn in the following sections.

TABLE 8.16
Tests for Process Oils

Inspection	ASTM Test Method
Relative density	D 1298
Viscosity	D 445
Aniline point	D 611
Refractive index	D 1218
Color	D 1500 or D 156
Flash point	D 92 or D 93
Pour point	D 97
Evaporative loss	D 972
Composition	
Clay–gel adsorption	D 2007
Carbon type	D 2140
VGC	D 2501

TABLE 8.17
Properties of Oils Used in the Rubber Industry

Physical property	ASTM	Paraffinic	Naphthenic	Aromatic
Specific gravity	D 1250	0.85–0.89	091–0.94	0.95–1.00
Pour point (°C)	D 97	−18 to −9	−40 to −18	+0 to +32
Refractive index	D 1747	1.48	1.51	1.55
Aniline point (°C)	D 611	95–127	65–105	35–65
Molecular weight	D 2502	320–650	300–460	300–700
Aromatic content (%)	D 2007	19–30	20–40	65–85

1. Relative Density

Relative density, also known as specific gravity, is a measure of the density of the oil relative to the density of water. Relative density increases with the aromatic and naphthenic content of the oil and with molecular weight. Relative density is measured at 15.6°C. Relative density is used to convert from a volume basis to a weight basis. Relative density can be converted to density at 15.6°C by dividing by 0.99904, the density of water at that temperature.

2. Viscosity

There are two different types of viscosity. Dynamic viscosity is a measure of a liquid's resistance to movement and it is measured in centipoise (cP). Kinematic viscosity is a measure of the velocity of a liquid and is obtained by measuring the time taken for a certain quantity of liquid to pass through a capillary tube. It is measured in centistokes, where 1 cSt = 1 mm²/s. The relationship between the two viscosities can be described as

$$\text{Kinematic viscosity } (T) = \frac{\text{Dynamic viscosity } (T)}{\text{Density } (T)}$$

where T is the temperature at which the viscosity and density are determined.

3. Aniline Point

The aniline point is measured by ASTM D 611 and is based on measurement of the temperature at which aniline dissolves in the oil. The aniline point is a measure of the solvency of the oil. The lower the solvency of the oil, the higher the aniline point. Aniline point can be used to help determine the compatibility of an oil with a particular polymer. The aniline point depends on the molecular weight of the oil. Oils with higher molecular weights have less solubility for aniline and thus higher aniline points.

4. Refractive Index

Refractive index is measured by ASTM D 1218 and is a measure of the ratio of the velocity of light in air to the velocity of light in the substance being tested. It can be used to measure batch-to-batch consistency. It is also used to calculate the refractive intercept used in the carbon-type composition calculation.

5. Color

The color of an oil is affected principally by the presence of heterocyclic polar compounds, generally aromatic groups that include sulfur, nitrogen, or oxygen. Color can be measured by comparing the color of the oil with a preset color chart. Color of the process oil can be important when the rubber compound is light in color. Color is generally measured by either ASTM D 156 (Saybolt Color) or D 1500.

6. Flash Point

The flash point of an oil is specified for safety reasons and is indicative of the oil's volatility. The lightest few percent of the oil determines the flash point. A correlation exists between the 5% point in the boiling range and the flash point. The lighter the products, the lower the flash point. Thus, two oils with the same viscosity (50% point) may have different flash points, depending on the amount of light products in the oil.

The flash point of an oil is the temperature at which enough flammable vapors exist above the oil that they will ignite or flash when presented with an open flame. The Pensky–Marten (PM) closed-cup method, ASTM D 93, may better represent the flash point of the oil in tankage or closed container and gives the best repeatability. Another method, the Cleveland Open Cup (COC), ASTM D 92, may better represent the flash point of the oil in open mixing devices or vessels and gives approximately 5°C–10°C higher flash point values than the PM closed-cup method.

7. Pour Point

Pour point is the temperature at which the oil will no longer flow and is measured by ASTM D 97. Paraffinic and aromatic oils tend to have wax pours, in which the oil will not flow owing to the formation of wax crystals. Naphthenic oils, because they generally contain very little wax, tend to have viscosity pours, where they stop flowing because of high viscosity at low temperature. Pour point is important in determining the handling characteristics of the oil at low temperature. It is also related to the wax content of paraffinic and aromatic oils.

8. Evaporative Loss

Evaporative loss is measured by ASTM D 972 and is the measure of loss of volatile materials under controlled conditions. This can be important in selecting process oils, especially if the rubber will be subjected to high temperatures. Evaporative loss from the rubber compound will be influenced by the compatibility of the process oil with the rubber polymer.

9. Aromatic Content

Two methods can be used to measure the aromatic content of an oil. Clay/silica gel testing, ASTM D 2007, gives the percentage of aromatic, saturated, polar, and asphaltenic molecules in the oils. ASTM D 2140 calculates the weight percent of carbon atoms involved in each type of bond—aromatic, naphthenic, and paraffinic— from the viscosity–gravity constant (VGC), refractive intercept, and density.

10. Viscosity–Gravity Constant

The VGC is a dimensionless constant that is based on a mathematical processing of the viscosity and density values and is measured by ASTM D 2501. The VGC increases as the hydrocarbon distribution changes from paraffinic to naphthenic to aromatic. As a general rule, paraffinic oils have a VGC ranging from 0.79 to 0.85, naphthenic oils from 0.85 to 0.90, and aromatic oils above 0.90.

11. Environmental Consideration

The environmental considerations regarding a process oil are related to its polyaromatic content (PAC). There are a number of ways to measure the PAC of an oil: IP 346 (an analytical method essentially measuring the level of certain polyaromatic compounds through selective extraction with a solvent), high-pressure liquid chromatography, and gas chromatography. The results of the various methods will differ significantly because they measure different things. The IP 346 method is used for deciding which oils have to be labeled under European Community (EU) legislation. It measures the content of substances that are soluble in dimethyl sulfoxide (DMSO). DMSO dissolves all polyaromatics and a number of single aromatics and naphthenes, especially if they contain a heteroatom. It has been shown, using skin painting on mice, that there may be a correlation between IP 346 test results and possible physiological effects. Oils with a value of 3% (by weight) and above have to be labeled in Europe. Values obtained by IP 346 are significantly higher than the true PAC of interest, and this is especially true for naphthenic oils.

D. COMPATIBILITY

It is important that the process oil selected for a rubber compound be compatible with the polymer. Poor polymer compatibility can cause bleeding of the oil, lack of adhesion, poor distribution of pigment, and poor physical properties. Good polymer compatibility results in more efficient mixing, better cure development, and improved physical properties. In order to have good polymer compatibility, the oil and the polymer need to have similar molecular units as well as optimized viscosity and molecular weight levels.

The oil molecular units in order of increasing polarity are paraffins, naphthenes, aromatics, and polars. The polymer molecular units along the backbone and bonded to the backbone are phenyl, halogen, nitrile, and vinyl. Similar molecules are more compatible. Aromatic oils are very compatible with SBR because the aromatic molecules and the phenyl group on the SBR backbone are both polar. Paraffinic oils, which are the least polar, are more compatible with polymers such as EPDM, which has a nonpolar backbone. As a general guide, naphthenic oils tend to be used in EPDM, CR, SBR, BR, and butyl-based compounds. Paraffinic oils can be used in natural rubber, butyl rubbers, CR, and SBR. Aromatic oils are used with natural rubber, SBR, and BR compounds.

E. Trends in Aromatic Oils

Highly aromatic oils, distillate aromatic extract (DAE) were used in SBR for tire tread. It was the single largest use of oil in rubber. As discussed previously, these highly aromatic oils are very suitable for use in the very polar SBR. However, they are carcinogenic. These oils have now been replaced by "nonlabeled" oils. The main replacement oils are mild extraction solvate (MES), treated distillate aromatic extract (TDAE), and residual aromatic extract (RAE). The manufacture of these oils is diagramed in Figure 8.22; the properties are shown in Table 8.18. The TDAE is further processed to remove the multiringed aromatic molecules in DAE oil that are of concern. TDAE is the closest to DAE in properties, making substitution easier. It does require investment for the additional processing step at the petroleum refinery, and the yields tend to be low.

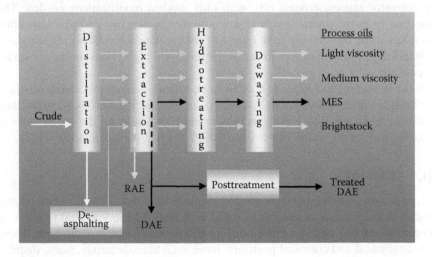

FIGURE 8.22 Aromatic oil replacement.

TABLE 8.18

Properties of DAE Replacements: RAE, TDAE, MES, and Naphthenic Oils

Property	DAE	RAE	TDAE	MES	Naphthenic
Category	LA	NLA	NLA	NLA	Naphthenic
IP 346% PAH	18	5.7	2.8	0.4	<3
Mutagenicity index	3.2	n/a	0	0.2	n/a
Kinematic viscosity (mm^2/s)	30	51	20	15	13
VGC	0.945	0.953	0.900	0.850	0.854
T_g (°C)	−40	−53	−50	−60	−60
Aniline point (°C)	45	83	70	90	92
Carbon type (aromatic, mass%)	40	27	25	15	12

LA, labeled aromatic; NLA, nonlabeled aromatic.

RAE is produced from the extract of the heaviest stream. This stream is rendered environmentally acceptable by restricting the oil molecules to the larger polyaromatic molecules, which are then not bioavailable because of their size. RAE has a viscosity three to four times as high as that of DAE, and there may not be enough supply to meet the marketplace demand.

MES is manufactured similarly to a heavy process oil. MES is more lightly extracted than a heavy process oil. The oil extraction conditions are set to remove only selected fractions of multiringed molecules that are not desirable in the final product. MES has a much higher aromatic content than a similar heavy process oil. The higher paraffinic content in MES relative to that of DAE makes substitution of DAE with MES more difficult. Formulations will need to be modified. MES does have the advantage that it can be made in a number of plants around the world that are currently making process oils, with only modest modifications needed. The transition from the aromatic oils to a more environmentally desirable substitute has taken some time owing to the need to reformulate the rubber compounds to maintain their performance in tires and the relatively high prices for the substitute process oil.

Work is ongoing to replace part or all of petroleum-derived processing oils such as TDAE in tire compounds with bioplasticizers. Four such materials, palm oil, flaxseed oil, cashew nut shell liquid, and low-saturated soybean oil showed promise in replacing part, or all, of the TDAE in an SBR/NR silica-loaded compound [37].

XII. SUMMARY

In conjunction with the chemicals used in a rubber formulation to ensure acceptable product characteristics, a number of ingredients may be incorporated to allow or improve processing with the manufacturing equipment available in the plant. These include physical and chemical peptizers, lubricants, homogenizing agents, dispersing agents, tackifiers, plasticizers, masterbatches such as sulfur and accelerator, and mineral oils. In this chapter, these compounding ingredients have been discussed with respect to their influence on processing behavior and their relevant compound vulcanizate properties.

With the advent of automatic tire building equipment such as the Pirelli MIRS system, the use of process additives to ensure component uniformity in terms of compounds and dimensions has become increasingly more relevant.

REFERENCES

1. Stone C, Hensel M, Menting K. Peptizers, mastication and internal lubricants for natural rubber. In: *IRC97*, Kuala Lumpur, Malaysia, 1997.
2. Umland H. Schill + Seilacher "Struktol" Aktiengesellschaft, Hamburg, Germany, Private communication.
3. Krambeer M. Schill + Seilacher "Struktol" Aktiengesellschaft, Hamburg, Germany, Tech Bull 1677.
4. Lechtenboehmer A, Mersh F, Moneypenny H. Polymer interfaces in tire technology. *Br Poly J* 1990;22:291.
5. Wake WC. *Adhesion and the Formulation of Adhesives*, 2nd edn. London, U.K.: Applied Science Publishers, 1982.
6. Wake WC. In: Hollwink R, Salomon G, eds., *Adhesion and Adhesives*, Vol. II. New York: Elsevier, 1965.
7. Allen KW. In: Alner DJ, ed., *Aspects of Adhesion*. Cleveland, OH: CRC Press, 1969.
8. Voyutskii SS. *Autohesion and Adhesion of High Polymers*. New York: Wiley Interscience, 1963.
9. Vasenin RM. *Adhesion: Fundamentals and Practice*. London, U.K.: Mclaren, 1969.
10. Skewis JD. Self-diffusion coefficients and tack of some rubbery polymers. *Rubber Chem Technol* 1966;39:217.
11. Anand JN. *J Adhes* 1969;1(1):31.
12. Anand JN. *J Adhes* 1970;2(1):23.
13. Anand JN, Balwinski RZ. *J Adhes* 1969;1(1):24.
14. Anand JN, Dipzinski L. *J Adhes* 1970;21(1):16.
15. Anand JN, Karam HJ. *J Adhes* 1969;1(1):16.
16. Vasenin RM. *Adhes Age* 1965;8(5):21; 1965;8(6):30.
17. Rhee CK, Andries JC. Factors which influence auto-adhesion of elastomers. *Rubber Chem Technol* 1981;54:101.
18. Wool RP. *Rubber Chem Technol* 1984;57:307.
19. Hamed GR. Tack and green strength of elastomeric materials. *Rubber Chem Technol* 1981;54:576.
20. Hooser ER, Diem HE, Rhee CK. Analytical characterization of tackifying resins. *Rubber Chem Technol* 1982;55:442.
21. Wolney FF, Lamb JJ. In: *Conference Paper, ACS Rubber Division Meeting*, Houston, TX, October 1983.
22. Belerossova AG, Farberov MI, Epshtein VG. *Colloid J* 1956;18:139.
23. Schlademan JA. In: *Conference Paper, ACS Rubber Division Meeting*, Cleveland, OH, October 1977.
24. Forbes WG, McLeod LA. *Inst Rubber Ind Trans* 1958;34:154.
25. ASTM D 883. Plastics nomenclature. American Society for Testing and Materials, Philadelphia, PA.
26. Bernol JD. General introduction. In: *Swelling and Shrinking: A General Discussion held at the Royal Institution, London, U.K.* London, U.K.: Faraday Society, 1946, pp. 1–5.

27. Doolittle AK. Mechanism of plasticization. In: Bruins PF, ed., *Plasticizer Technology*. New York: Reinhold, 1965, Chapter 1.
28. Doolittle AK. *The Technology of Solvents and Plasticizers*. New York: Wiley, 1954, p. 1056.
29. Kurtz SS, Sweeley JS, Stout WJ. Plasticizers for rubber and related polymers. In: Bruins PF, ed., *Plasticizer Technology*. New York: Reinhold, 1965, Chapter 2.
30. O'Rourke SE. The function and selection of ester plasticisers. *Rubber Technol Int* 1996;60.
31. Galle-Gutbrecht R, Hensel M, Menting KH, Mergenhagen T, Umland H. Zinc-free rubber processing additives for the tyre industry. Presented at *the Tire Technology Expo 2002*, Hamburg, Germany.
32. Rauline R. Patent Appl EP 0501 227 (to Michelin), February 25, 1991.
33. Degussa Information for the Rubber Industry. Compounding of Si69-Stocks. Degussa AG.
34. Gorl U, Hunsche A. Advanced investigation into the silica/silane reaction system. In: *150th Meeting of the Rubber Division*, ACS, Louisville, KY, Paper 76.
35. Wolff S. Optimization of silane-silica OTR compounds, Part 1. Variations of mixing temperature and time during the modification of silica with bis-(3-triethoxysilylpropyl)-tetrasulfide. *Rubber Chem Technol* 1982;55:967.
36. Stone CR, Menting KH, Hensel M. Improving the silica "green tyre" tread compound by the use of special process additives. Presented at *the ACS Rubber Division*, Orlando, FL, September 21–24, 1999.
37. Flanigan C, Beyer L, Klekamp D, Rohweder D, Haakenson D. Using bio-based plasticizers, alternative rubber plasticizers. *Rubber & Plastics New*, February 11, 2013.

9 Resins

James E. Duddey

CONTENTS

I.   Introduction 379
II.   Naturally Derived Resins.. 382
       A.   Rosin and Terpene-Based Resins.. 382
III.   Hydrocarbon Resins.. 384
       A.   Coumarone–Indene Resins 384
       B.   Petroleum-Based Hydrocarbon Resins 384
       C.   Properties of Hydrocarbon Resins 385
            1.   Glass Transition Temperature/Softening Point 385
            2.   Viscosity... 386
            3.   Molecular Weight and Molecular Weight Distribution................ 387
            4.   Solubility and Compatibility 387
            5.   Residual Unsaturation 387
            6.   Color... 388
            7.   Acidity and Basicity 388
            8.   Performance of Hydrocarbon Resin... 388
            9.   Processing at High Temperature 388
            10. Performance of Compatible Resins... 389
            11. Performance of Incompatible Resins 391
III.   Phenolic Resins... 393
       A.   In Situ Reinforcing Resin Networks 393
       B.   Phenolic Resins 396
            1.   Novolak Resins... 397
            2.   Resole Phenolic Resins... 399
            3.   Polymerized Cashew Nutshell Oil 400
IV.   Resin Bonding Systems for Rubber Reinforcements....................................401
V.   New Resin Technologies for Wire Adhesion...407
VI.   New Trends in Resin Utilization 412
VII.   Identification of Resins in Cured Rubber 415
VIII. Summary 416
References... 416

I. INTRODUCTION

Resins or in situ resins are considered to be polymeric materials with weight-average molecular weights (M_w) in the approximate range of 800–4000. They are amorphous, thermoplastic materials that can be either liquids (low softening points) or solids (high softening points). They serve many functions in rubber compounds. In uncured rubbers, they function as process aids, softeners, tackifiers, pigment dispersing aids, and homogenizing agents. In cured rubbers, they function as plasticizers, extenders, and reinforcing agents.

There are no definitive lines between process aids, plasticizers, softeners, and tackifiers. Barlow [1] defined process aids as materials included in the rubber formulation to reduce the time and energy required to break down the polymer, for example, peptizers or other materials that reduce the viscosity of the rubber mix. At the same time, they can help improve the dispersion of dry materials, produce smoother stocks, improve the dispersion rate of fillers, and, in some instances, increase the homogeneity of rubber blends [1]. The ASTM (American Society for Testing Materials) definition for a softener is "a compound material used in a small proportion to soften a vulcanizate or facilitate processing or incorporation of fillers [uncured compound]." Tackifiers are specialized types of softeners that are used to increase the ability of uncured rubber formulations to form bonds with themselves or other surfaces under pressure by increasing the wettability or contact at the interface while maintaining or even increasing the green strength of the rubber formulation [2]. The ATSM definition of a plasticizer is "a compounding material used to enhance the deformability of a polymeric compound."

At processing temperatures, softeners and plasticizers have the same function, reducing viscosity and improving processing. Both materials are compatible with most rubbers. In the cured rubber, plasticizers are materials that lower the glass transition temperature of the rubber and improve low-temperature performance, for example, oils or phosphate esters, whereas softeners are generally high softening point materials that increase the glass temperature of the formulated rubber but reduce the modulus of the cured compound. At normal operating temperatures, softeners generally have little influence on the physical properties of cured rubber but can influence dynamic performance. Recently, this effect on viscoelastic properties has increasingly gained interest due to the regulatory banning of labeled aromatic oils in tire rubber compounds and the ever-increasing demand for improving specifically wet grip performance.

Several theories have been proposed to explain the performance of resins as softeners and plasticizers. Solvating and fluid effects on polymers were reviewed by Files [3]. Stephens [4] reviewed a number of theories and recognized three as prominent: the lubricity theory, the gel theory, and the free volume theory. Barlow [5] provided simplified summaries of each of these theories. A more fundamental approach to understanding the performance of resins is that of compatibility or miscibility, which is grounded in the thermodynamics of mixing [6–8]. The oldest of these approaches is that of the "solubility parameter" developed by Hildebrand, and it is the approach that in one form or another is most frequently utilized by the resin technologist. The solubility parameter of a liquid (δ) is defined as the square root of the cohesive energy density. The cohesive energy density in turn is defined as the ratio of energy of vaporization to the molar volume ($\Delta E_v/V$) where ΔE_v is the energy of vaporization at a given temperature and V is the corresponding molar volume at the same temperature. For low-molecular-weight liquids, the solubility parameter at a given temperature can be calculated from the heat of vaporization (ΔH_v) using the equation

$$\delta = \left(\frac{\Delta E_v}{V}\right)^{1/2} = \left(\frac{\Delta H_v}{V} - \frac{RT}{V}\right)^{1/2}$$

Although properly classified as liquids, amorphous, high-molecular-weight polymers have vapor pressures that are too low to detect. Hence, indirect methods must be used to estimate their solubility parameter. Two of the more widely used methods are the determination of equilibrium swelling of a cross-linked polymer in a variety of solvents that have a range of δ values (the extent of swelling will be maximum when the δ value of the solvent matches that of the polymer) [9] and the measuring of the intrinsic viscosity of an un-cross-linked polymer in a series of solvents (the δ value for the polymer is taken to be the same as that of the solvent in which the polymer has the greatest viscosity) [10,11]. More satisfactory results can be obtained by using a 3D solubility parameter approach developed by Hansen in which the solubility parameter is separated into dispersion, polar, and hydrogen bonding contributions [12].

One of the simplest indirect methods of determining δ values is based on the assumption that atomic and group increments exist that can be summed over the known structure of the substance (liquids as well as high-molecular-weight polymers) to provide estimates for δ. Fedors [13] reviewed a number of approaches and proposed a simplified approach for estimating both the solubility parameters and the molar volumes of liquids and high-molecular-weight amorphous polymers that requires only a knowledge of the chemical structure. Finally, it must be kept in mind that, although the terms compatible and miscible tend to be used interchangeably, a compatible blend may have useful technological properties but may or may not exhibit true thermodynamic miscibility. Above the softening point, resins become viscous liquids, and the viscosity drops rapidly as the temperature increases above the softening point. At process and cure temperatures, the resins are low viscosity fluids. They function as process aids by lowering the viscosity of the formulation, by improving filler dispersion, and by improving the kneading of the batch during mixing. At higher temperatures, the increased compatibility of resins with a number of polymers can improve stock homogeneity of a two-polymer blend by bridging differences in the compatibility of the two polymers. Processing of the formulation is improved by the reduced viscosity and the improved batch uniformity. The performance of the resin in the rubber formulation at room temperature is determined by the compatibility of the resin with rubbers in the formulation and the other compounding ingredients. An incompatible resin will phase separate and return to a hard, brittle, glass state. The presence of the hard resin increases compound hardness, modulus, and green strength. A compatible resin will increase the rubber glass transition temperature, soften the compound by lowering the viscosity, and function as an extender of the rubber phase. In the cured compound, an incompatible resin will be phase separated and function as additional filler up to the softening point of the resin. A compatible resin will shift the glass transition temperature of cured rubber formulations but have little effect on cured properties unless there is substantial unsaturation in the resin. The residual unsaturation will reduce the cure state of the formulation by tying up a portion of the sulfur in the cure system. This is especially important in butyl and ethylene propylene diene rubber (EPDM) formulations in which there is a limited amount of curative available. Any residual acidity or basicity present in the resin will interfere with the curing of the rubber formulation, particularly the cure rate. Finally, the color of the resin must be considered, particularly in the preparation of light-colored compounds.

Historically, the first resins available to the rubber industry were the crude materials obtained from the distillation of coal tar, pine gum, and petroleum stocks. These products contained many impurities and their quality varied. Over time, suppliers learned to isolate monomer streams and to polymerize them to yield resins that were light in color and consistent in performance and still performed the functions of the crude predecessors. By combining monomer streams from different sources, resins could be produced to meet the requirements of specific applications.

Three types of synthetic hydrocarbon resins are available today: (1) polyterpene resins produced from monomers isolated from pine gum, crude sulfate turpentine (CST), or extracts from orange peel, (2) coumarone–indene resins produced from the coumarone–indene fractions isolated from the distillation of coal tars, and (3) hydrocarbon resins produced from monomers obtained from the steam cracking of heavy petroleum streams. A second group of synthetic resins, the phenolic resins, is available for use in rubber compounds. They are produced from the reaction of phenol, resorcinol, and other aromatic compounds containing hydroxyl groups with formaldehyde and/or other higher aldehydes.

II. NATURALLY DERIVED RESINS

A. ROSIN AND TERPENE-BASED RESINS

There are various sources of raw materials for rosin and terpene-based resins [51]. Pine gum is captured by the tapping of pine trees similar to the tapping of *Hevea brasiliensis* for natural rubber. Distillation of this pine gum results in oleo gum terpenes and rosins. Steam distillation of aged pine tree stumps generates the so-called wood rosin. The availability of these two raw materials derived from pine gum has been in a long-term decline in countries like Portugal and Greece because of the labor-intensive process of collecting the pine gum, but China, Brazil, and Indonesia are still producing annual amounts of a million tons per year and more. The cost of collecting and processing pine tree stumps for steam distillation however caused the closure of corresponding operations. Another source of these two naturally occurring raw materials is the side stream of the Kraft paper pulping process of major paper manufacturers in southern United States, Scandinavia, Russia, South America, and South Africa. The volatile fraction generated during wood refining is called crude sulfate turpentine CST and contains the terpene fraction that is separated by fractionated distillation. The condensed water phase collected from the pulping refiners is called black liquor and contains, besides lignins and hemicellulose, the sodium salts of resin and fatty acids as a cake that is skimmed off. Upon acidulation, this cake is converted into crude tall oil that is separated by fractionated distillation into tall oil fatty acid and tall oil rosin. The third source or terpenes is occurring during the processing of oranges to produce frozen orange juice. D-Limonene is either separated directly from the juice or extracted from orange peel by steam distillation.

Fractional distillation of the crude fractions of pine gum yields terpenes and residual gum rosin, while the crude fraction of CST results in terpenes only. The most abundant terpene is α-pinene, followed by D-limonene, β-pinene, and Δ-3-carene.

Rosin, particularly after hydrogenation, is an efficient softener and tackifier, especially for natural rubber. It assists in the dispersion of carbon blacks and imparts some degree of age protection to rubber compounds. However, due to its acidity and unsaturation, it will retard the cure of a rubber compound. It is very dark in color and cannot be used in white or light-colored compounds. Rosin soaps are used as emulsifiers for the manufacture of emulsion SBR, providing the benefit of building tack in synthetic rubber. With the introduction of disproportionated rosin soaps, the sensitivity of synthetic rubber to oxidation was significantly reduced. Disproportionation is a catalyzed isomerization of resin acids with conjugated double bonds in rosin into aromatic ring–containing and single unsaturated resin acid molecules.

Rosin esters form another product class with excellent tackification properties and less impact on the vulcanization behavior due to their reduced acidity.

Polyterpene resins are produced by the acid catalyzed polymerization of the purer fractions of the terpene monomers α-pinene, β-pinene, and D-limonene. A typical manufacturing process is outlined in Figure 9.1. Terpene resins vary from viscous liquids to hard, brittle solids and range in color, with some products being nearly water white. They are thermoplastic, very tacky when soft, and stable to heat and UV radiation. Molecular weights are relatively low (range of 550–2200), and the molecular weight distribution is broad. They are compatible with a wide range of rubbers, such as natural rubber (NR), styrene butadiene rubber (SBR), isobutylene isoprene rubber (IIR) and chloroprene rubber (CR) and numerous other compounding ingredients. The polyterpene resins are excellent tackifiers and can be used in compounds that come in contact with foods and skin.

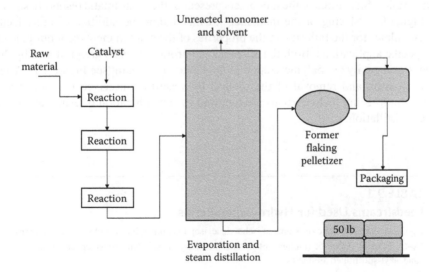

FIGURE 9.1   Manufacture of hydrocarbon resin.

III. HYDROCARBON RESINS

A. COUMARONE–INDENE RESINS

The coumarone–indene resins are produced by the acid catalyzed polymerization of the coumarone–indene fractions obtained from coal tar solvent fractions in a process similar to that used for the production of the terpene resins (see Figure 9.1). These coal tar fractions are by-products of coke manufacturing. As is the case with terpene resins, production of the coumarone–indene resins is limited by the scarcity of feedstock resulting from the reduced use of coke in modern steel manufacturing and the decline in steel production in the United States. Synthetic coumarone–indene resins are light in color, have good color stability, and function as process aids, improving the dispersion of fillers without having a negative influence on cured properties. They are particularly useful for improving the dispersion of mineral fillers in high strength, white or light-colored SBR compounds.

B. PETROLEUM-BASED HYDROCARBON RESINS

Because of the limited availability of feedstocks for the production of the terpene and coumarone–indene resins, they have largely been replaced by petroleum-based hydrocarbon resins or simply hydrocarbon resins. Monomer feedstocks for these hydrocarbon resins are obtained from the steam cracking of heavy petroleum stocks. The feedstocks are a complex mixture of resin-forming monomers and non-resin-forming hydrocarbons. The feedstocks are classified by the carbon content of the monomers (see Table 9.1).

Resins produced from the C_4–C_6 fraction are identified as C_5 or aliphatic resins. Structures of the monomers present in the C_5 aliphatic resins are shown in Figure 9.2. The resins produced from the C_8–C_{10} fraction are identified as C_9 aromatic resins. Structures of the monomers present in the C_9 aromatic resins are shown in Figure 9.3. Mixing of the monomer streams and/or the addition of other feedstocks allows for the tailoring of the properties of the resin to meet the requirements of specific applications. Both the aliphatic and aromatic resins are produced by the Lewis acid catalyzed polymerization of the monomer stream (see Figure 9.1). After neutralization and removal of the catalyst by washing, unreacted monomers and non-resin-forming hydrocarbons are removed by a combination of evaporation and steam distillation.

TABLE 9.1

Feedstreams Used for Hydrocarbon Resins

C_4–C_5 fraction: Aliphatic (mixtures of isobutylene, isoprene, piperylenes, and saturated paraffins)

C_8–C_{10} fraction: Aromatic (mixtures of styrene, methylstyrenes, indene, dicyclopentadiene, and alkyl-substituted aromatics)

Pure monomers: Aromatics (styrene, vinyl toluene, and α-methylstyrene)

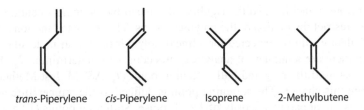

trans-Piperylene   cis-Piperylene   Isoprene   2-Methylbutene

FIGURE 9.2   Monomers used in preparation of C$_5$ aliphatic resins.

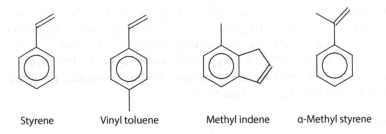

Styrene   Vinyl toluene   Methyl indene   α-Methyl styrene

FIGURE 9.3   Monomers used in preparation of C$_9$ aromatic resins.

Cyclopentadiene is removed from the C$_4$–C$_5$ stream prior to polymerization of the C$_4$–C$_5$ stream and is dimerized to form dicyclopentadiene (DCPD). Resins obtained from DCPD are produced by thermal polymerization. They are inherently reactive, have a distinct odor, and are very dark in color. Hydrogenation is used to improve the color and stability of these resins. The DCPD resins have some aromatic characteristics and exhibit solubility characteristics intermediate between those of the aliphatic and aromatic resins. Resins produced from the pure monomer streams exhibit high softening points, are used where an increase in hardness and modulus is required, and are classified as reinforcing resins. They are generally water white and are used where color and stability are important and cost is not a major concern.

C.   PROPERTIES OF HYDROCARBON RESINS

The properties of the hydrocarbon resins, glass transition temperature (T_g), softening point (T_s), viscosity, molecular weight (MW), molecular weight distribution (M_n/M_w), and compatibility or miscibility (as discussed in Section I) are determined by the mix of monomers in the feedstream and the polymerization conditions. In turn, these properties determine how the resin performs in the rubber compound. Other properties such as residual unsaturation, color, and acidity/basicity determine the secondary influences that the resins have on the cure and performance of compounds [14].

1.   Glass Transition Temperature/Softening Point

The glass transition temperature (T_g) is the temperature at which a material changes from a rigid glass to a highly viscous liquid. The T_g of a resin is a function of the monomers, the molecular weight (MW), and the degree of branching. The bulkiness resulting from branching and the rigidity introduced by monomers containing ring structures restrict the motion of the polymer chains. The restricted motion leads to a higher T_g than would be expected for a linear polymer of similar molecular weight. Although there are a number of analytical procedures for determining T_g, the resin industry has used the ring and ball softening point (T_s, ASTM Test Method E 28) to measure this change. The softening point is defined as the temperature at which the resin has an apparent viscosity of 1×10^6 poise. Softening points (T_s) are related to T_g, but they are also determined by the increase in temperature above T_g required for the viscosity of the resin to drop to 1×10^6 poise. The differential between T_g and T_s is small for low-molecular-weight materials but increases as the molecular weight increases, because it requires higher temperatures to reduce the viscosity of the high MW resin to the 1×10^6 poise level. Hydrocarbon resins range from liquids with softening points below room temperature to hard, brittle materials with softening points up to 180°C–190°C. Resins with low softening points less than 110°C are used as process aids, softeners, and tackifiers. The high softening point resins are used as reinforcements for the rubber matrix.

2. Viscosity

At the glass transition point, the viscosity of the resin is extremely high. As the temperature of the resin is raised above the T_g, there is a marked drop in viscosity. The viscosity continues to drop sharply as the temperature is increased to the ranges encountered in the mixing and processing of rubber compounds. The drop of melt viscosities with increased temperature is summarized in Table 9.2 for several types of hydrocarbon resins.

The dynamic modulus of resins also drops dramatically as the temperature of the resin passes through the glass transition point and continues to drop as mixing and processing temperatures are reached (see Figure 9.4). In comparison, there is an initial large drop in the dynamic modulus of rubber at the glass transition point, but the decrease in dynamic modulus with further increases in temperature is less severe.

TABLE 9.2
Resin Melt Viscosity versus Temperature

	Temperature (°C)	
Viscosity (Poise)	Aliphatic Resin (T_s = 100)	Aromatic Resin (T_s = 100)
1,000,000[a]	100	100
1,000	118	118
100	136	133
10	169	157

Source: Napolitano, M.J., The use of hydrocarbon resins in rubber compounding, in: *ACS Rubber Division Meeting*, Pittsburg, PA, September 1994. With permission.

[a] At softening point.

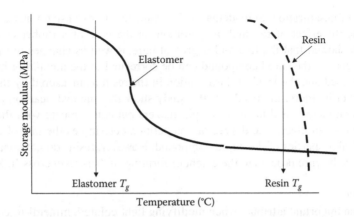

FIGURE 9.4 Schematic of dynamic mechanical spectra.

3. Molecular Weight and Molecular Weight Distribution

Molecular weight and molecular weight distribution are important because they determine the T_g/T_s ratio, the melt viscosity, and the solubility or compatibility of the resin. M_m and M_w are measured by gel permeation chromatography (GPC) and compared to those of a polystyrene standard. Resins generally have weight-average molecular weights (M_w) ranging from 800 to 4000 and broad molecular weight distributions M_m/M_w 72.0. The highly branched structures leading to the broad molecular weight distribution limit the solubility and compatibility of these materials. Low molecular weights and narrow molecular weight distributions favor solubility and compatibility.

4. Solubility and Compatibility

The performance of resins in rubber is directly related to their solubility or compatibility in the rubber. Although, as mentioned previously, there are a number of theoretical approaches for determining compatibility [5–8], the rubber and resin industries have relied on the old adage "like dissolves like," which is a simplified approach to Hildebrand's "solubility parameter" approach. Resin compatibility is estimated by use of cloud point measurements in suitable solvents. As a hot solution of the resin is cooled, the temperature at which the resin comes out of solution is defined as the cloud point. Lower temperatures are indicative of better solubility. By testing in solvents of known characteristics, the solubility can be correlated with compatibility in rubber. Odorless mineral spirits and a mixture of methylcyclohexane and aniline are used in the industry to gauge the aromatic character of hydrocarbon resin. By combining the information obtained from cloud point measurements with the known characteristics of rubbers, it is possible to estimate of the compatibility of the resin with the rubber.

5. Residual Unsaturation

Any residual unsaturation in the resin has a potential to interfere with the cure of the rubber compound. ASTM D 1959 (iodine number) is the test used to determine the level of unsaturation. Schlademan [15] conducted an extensive kinetic study to determine the effect of residual unsaturation on the cure of a rubber compound. His study showed that the rate and degree of cure as well as changes in the physical properties of the cured compound can be correlated to the amount of hydrocarbon resin used and the level of unsaturation in the resin as measured by the iodine number. Furthermore, the results of the study show that the hydrocarbon resins do not function merely as diluents or as plasticizers but can compete with the rubber for the sulfur curatives. The degree of competition determines the rate of cure and the physical properties of the cured compound. However, resins do not appear to be involved in the cure process to the extent of forming rubber–resin cross-links.

6. Color

Color is an important attribute when modifying light-colored, mineral-filled stocks. Any color in the resin will influence the color of the rubber compound. Two test procedures have been developed to measure color. ASTM D 1544 is used for resins ranging from moderately light to very dark in color. ASTM D 1209 is used to measure the color differences of water-white resins (usually the pure monomer-based resins).

7. Acidity and Basicity

Sulfur curing of rubber is known to be retarded by acids and accelerated by bases. Any residual acidity or basicity in the resin can change the cure of a resin-modified compound relative to the control. It is possible to compensate for this change with cure system adjustments if the levels of acidity or basicity are constant, but it is impossible to make changes if the levels are constantly changing. In order to control cure rates, levels of acidity or basicity are specified. The level of acidity is controlled by measuring acid number, whereas the level of basicity is controlled by measuring the base number or saponification number.

8. Performance of Hydrocarbon Resin

Napolitano [14] pointed out that hydrocarbon resins with the combination of high T_g and low molecular weights make these resins unique among ingredients available to rubber compounders (see Figure 9.5). Rubber, on the other hand, is a low T_g, high-molecular-weight material exhibiting a high viscosity over a wide temperature range above the glass transition point, whereas oils are low T_g materials with low viscosities. Through an understanding of the unique properties of hydrocarbon resins, we can begin to understand the contributions of these materials for improving processing performance and for altering the properties of cured compounds.

9. Processing at High Temperature

At mixing, processing, and cure temperatures, most of the resins used in rubber compounding are low viscosity liquids. During mixing, the addition of the resin reduces the viscosity of the rubber mix and improves filler dispersion by acting as a wetting agent and binder. Compound homogeneity in compounds containing rubber blends is improved by the wetting and by the potential to bridge the gap between marginally compatible polymers. The lower compound viscosity and improved plasticity

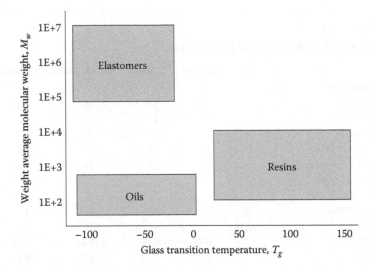

FIGURE 9.5 Chemistry of resins.

of compounds containing resin improve performance during processing operations such as injection molding, extrusion, and calendering.

10. Performance of Compatible Resins

The improvements obtained by the introduction of a compatible resin to a rubber formulation were illustrated by Napolitano [14] using an SBR formulation modified with 7 phr of three different hydrocarbon resins. The results of his study are summarized in Table 9.3. In the uncured compound, the addition of resin lowered the Mooney viscosity, suggesting better processability, improved scorch time (10-point rise), and improved tack. The improved processability was achieved with little impact on hardness (Shore A). The increase in tensile strength and elongation and the decrease in modulus are a reflection of the reduced state of cure resulting from the interference of the resin with curatives.

Hilner [16] reported similar findings in a study of the addition of 10 phr of three different hydrocarbon resins to an SBR tread compound in place of 10 phr free aromatic process oil (note that there is still 37.5 phr bond aromatic oil in the formulation introduced with the SBR rubber). The base formulations are summarized in Table 9.4. He reports that (1) cure times were extended from 12.3 to 13 min by the introduction of resins, but cross-link densities of the modified compounds were equal to those of the control; (2) physical properties such as rebound, Shore A hardness, compression set, abrasion wear, and tensile strength were not affected; and (3) de Mattia crack propagation was improved up to fivefold with the C_9 resin (Figure 9.6). He also reports that the C_9 resin and the aliphatic-modified C_9 resin were particularly effective in improving wet grip and high-speed performance without harming other tire performance properties such as rolling resistance (see Table 9.5 for a summary of results).

TABLE 9.3

Styrene Butadiene Tread Compound

	Control	C₅ Resin	DCPD Resin	C₉ Resin
Component				
SBR 1502	100	100	100	100
Stearic acid	1	1	1	1
Zinc oxide	3	3	3	3
N234 carbon black	52	52	52	52
Sundex 790 process oil	10	10	10	10
Piccopale® 100	—	7	—	—
Piccodiene® 2215	—	—	7	—
Picco® 6100	—	—	—	7
TMQ	1	1	1	1
TMTD	0.18	0.18	0.18	0.18
TBBS	0.75	0.75	0.75	0.75
Sulfur	1.8	1.8	1.8	1.8
Processability				
ML(1 + 4) 212°F	71	59	61	58
MS, 250°F, Δ10 (min)	45.5	48.8	68.3	63.2
Physical, Cured T_{90} at 320°F				
300% modulus (psi)	1863	1460	1178	1360
Tensile strength psi)	3368	3546	3286	3499
Elongation (%)	469	573	610	594
Hardness, Shore A	70	69	69	69
Tack (psi)	21	26	28	29

Source: Napolitano, M.J., The use of hydrocarbon resins in rubber compounding, in: *ACS Rubber Division Meeting*, Pittsburg, PA, September 1994. With permission.

Aromatic process oils are known to contain a number of polycyclic aromatic compounds such as benzo[a]pyrene and may be subject to future regulatory attention. Because of possible health risks, the use of this group of process oils has been restricted in Europe, and the restriction is likely to spread to the rest of the world. Simple substitution of naphthenic oils for the aromatic oils is not an acceptable solution, because the modified compounds do not meet all the requirements of the original compound. Hilner reported that a combination of 20 phr hydrocarbon resin with 10 phr of naphthenic oil can be used to replace the aromatic process oil. Most mechanical and dynamic properties showed improvement over those of compounds containing only aromatic oil. The C₉ aromatic resin was particularly effective (see Table 9.6). Tear propagation performance of formulations modified with a phenol-modified C₉ aromatic resin or a coumarone–indene resin exceeded that of a formulation containing only aromatic oil. The tear propagation performance of the formulation containing a nonmodified C₉ resin was better than that of a formulation

TABLE 9.4
Composition of Tread Compound

Ingredient	Base Formulation (phr)	Resin Based
Styrene butadiene rubber (Buna EM1712)	137.5	137.5
Carbon black N339	80.0	80.0
Zinc oxide	3.0	3.0
Stearic acid	2.5	2.5
Sulfur	1.8	1.8
IPPD	1.5	1.5
CBS	1.2	1.2
DPG	0.4	0.4
Aromatic process oil	10.0	—
Hydrocarbon resin		10.0

Source: Hilner, K., *Tire Technol. Int.*, 106, 1994. With permission.

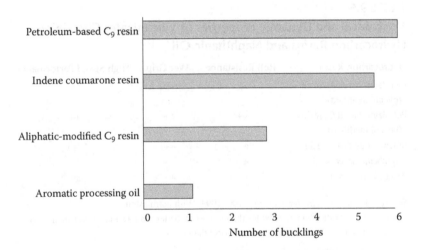

FIGURE 9.6 Crack propagation in vulcanizates containing hydrocarbon resins with a softening point of 100°C. (From Hilner, K., *Tire Technol. Int.*, 106, 1994. With permission.)

with all naphthenic process oil but not as good as that of the formulation with all aromatic oil (Figure 9.7).

11. Performance of Incompatible Resins

Even if a resin is incompatible with rubber at room temperature, compatibility will improve with increasing temperature. At typical mixing and processing temperatures, the resin will exist in the liquid state and have a higher level of compatibility with rubber. In this state, it should function similarly to a fully compatible resin.

TABLE 9.5
Mechanical Dynamic Properties of Vulcanizates Containing Hydrocarbon Resins with 100°C Softening Point

Hydrocarbon Resin	Property		
	Roll Resistance	Wet Grip	High-Speed Performance
Petroleum-based C_9 resin	++	++	+(+)
Aliphatic-modified C_9 resin	+(+)	+(+)	+(+)
Indene–coumarone resin	+	–	+
Aromatic process oil	++	– –	a

Source: Hilner, K., *Tire Technol. Int.*, 106, 1994. With permission.

a, average; +, better than a; +(+), better than +; ++, better than +(+); –, worse than a; –(–), worse than –; – –, worse than –(–).

TABLE 9.6
Mechanical and Dynamic Properties of Vulcanizates Containing Hydrocarbon Resins and Naphthenic Oil

Hydrocarbon Resin	Roll Resistance	Wet Grip	High-Speed Performance
Petroleum-based C_9 resin (phenol modified)	–	– –	(+)
Petroleum-based C_9 resin (phenol modified)	++	++	++
Indene–coumarone resin	a	++	(+)
Naphthenic process oil	+	– –	+
Aromatic process oil	–	+	a

Source: Hilner, K., *Tire Technol. Int.*, 106, 1994. With permission.

a, average; (+), better than a; +, better than (+); +(+), better than +; ++ = better than +(+); –, worse than a; –(–), worse than –; – –, worse than –(–).

But when the compound is cooled to room temperature, it will phase separate. As long as the compound is below the softening point of the resin, the hard resin phase will act as additional filler in the uncured compound and increase stiffness. This increase in stiffness can interfere with the handling of the compound in the plant. In the cured compound, the incompatible resin will also exist in the phase-separated state. Hard resins will function as a reinforcement up to the softening point of the resins; hence they are classified as reinforcing resins. Complex modulus (E^*) measured on the dynamic mechanical analyzer can be used to determine the stiffness of a rubber compound under dynamic conditions. The complex modulus (E^*) is defined by the equation

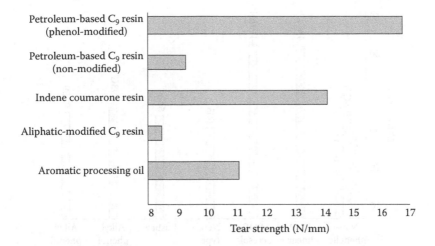

FIGURE 9.7 Tear propagation resistance of vulcanizates made with liquid hydrocarbon resins. (From Hilner, K., *Tire Technol. Int.*, 106, 1994. With permission.)

$$E^* = (E'^2 + E''^2)^{1/2}$$

where
E^* is the complex modulus
E' is the elastic modulus
E'' is the viscous modulus

Stuck and Souchet [17] showed that a compound containing a high-styrene reinforcing resin exhibits high static stiffness as measured by Shore A durometer testing but has poor dynamic hardness E^*, no higher than that of the nonmodified control (see Figure 9.8). Compounds produced with additional carbon black or with phenolic reinforcing resin to a similar level of static hardness maintained a high level of dynamic stiffness under dynamic testing.

III. PHENOLIC RESINS

A. IN SITU REINFORCING RESIN NETWORKS

Phenol, alkylphenols, and resorcinol can be reacted in bulk or in a polymeric formulation with methylene donors. Typical donors are 2-nitro-2-methyl propanol (NMP), hexamethylenetetramine (HMTA), and hexamethoxymethylmelamine (HMMM) to produce a thermoset resin network in the rubber compound. NMP has a melting point (m.p.) of 89°C–90°C and a boiling point (b.p.) of 94°C–95°C. HMTA is a solid with a m.p. of 289.9°C. It does not have a b.p. because it sublimes. HMMM consists of a mixture of monomeric and condensed products. It is a liquid at room temperature but it has no definitive b.p.

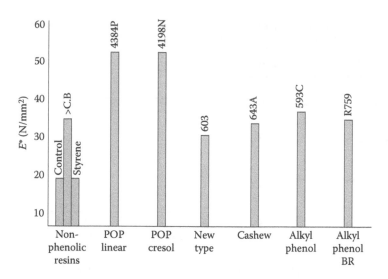

FIGURE 9.8 Complex modulus, E^*, of rubber formulations modified with resin. (From Stuck, B.L. and Souchet, J.-C., *Tire Technol. Int.*, 131, 1997.)

Phenol (m.p. 43°C; b.p. 181°C–187°C) and alkylphenols have not found direct use in rubber compounding because of their volatility and environmental issues that would necessitate special handling. Resorcinol (m.p. 111°C; b.p. 181.7°C), even though it is as volatile as phenol, has been used directly in rubber compounds because up to now, the vapors have not been of significant concern. However, resorcinol is hydroscopic and must be protected from moisture during handling and storage. Resorcinol is generally added to the rubber formulation in one of the earlier Banbury mixing stages. Although rubbers are generally mixed, processed, and cured at temperatures (340°F/170°C) below the b.p. of resorcinol, there is a substantial amount of vaporization (fuming) of resorcinol in the Banbury mixer, during the calendering and extrusion of the uncured compound and during the cure of the rubber compounds. Condensation of these fumes on the process equipment and the factory walls and ceiling can cause significant manufacturing problems. More recently, concerns have been raised about the safety of resorcinol, so it may be banned from rubber plants in the future. HMTA and HMMM are the preferred methylene donors used in rubber formulations. Both of these curatives are added in the lower temperature, final mixing stage. HMTA must be isolated from the other rubber curatives during storage and batch preparation because its basicity can cause premature decomposition of the rubber cure accelerators and can accelerate the conversion of insoluble sulfur to the soluble form.

The structure of HMTA and the reaction with resorcinol are illustrated in Figure 9.9. The cure begins with the formation of dihydroxybenzoazine (not shown) and progresses through a tris(dihydroxybenzyl)amine intermediate onto a bis(dihydroxyphenyl)methane and eventually onto a branched thermoset resin structure when three methylene bridges are attached to the same ring [18]. Classical chemical studies indicate that as much as 75% of the nitrogen remains chemically bonded

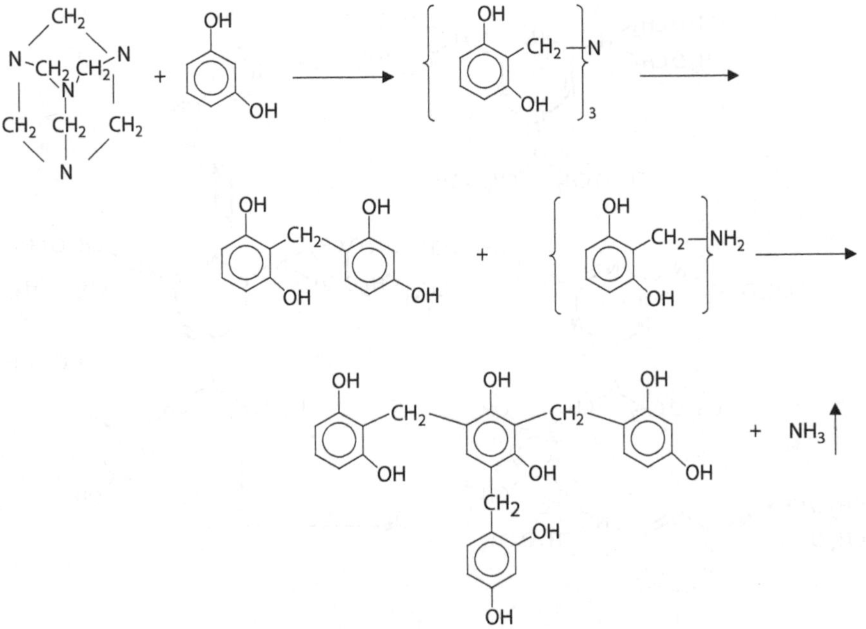

FIGURE 9.9 Reaction of HMTA with resorcinol. (From Kopf, P.W., Phenolic/resins, in: Kraschwitz, J.I., ed., *Kirk-Othmer Encyclopedia of Chemical Technology*, 4th edn., Vol. 18, Wiley, New York, 1996, pp. 603–644.)

to the rubber (for more details of the mechanism, see pp. 616–617 of Ref. [18]). However, some free ammonia is released during the cure of the resin and the rubber. This free ammonia and the alkylamine can have detrimental effects on rubber composites reinforced with synthetic fibers or steel cords. Similar reactions account for polymerization of phenols and the cure of phenol- and resorcinol-based novolak resins.

The base structure of HMMM and the reaction with resorcinol are given in Figure 9.10. Although it is conceivable that a methylene group might be transferred from the HMMM to the resorcinol as is the case with HMTA, it is generally accepted that the entire melamine structure is joined to the resorcinol molecule through a methylene bridge. Formation of linear and branched structures occurs through one of two paths: the addition of a second and third melamine methylene bridge to the same resorcinol or the reaction of additional resorcinol units on to the remaining di(methoxymethylene) amine sites of the original melamine unit. Steric factors and chemical reactivities will determine the extent to which each pathway is followed in the formation of the branched resin structure. As the case with HMTA, the same reactions can account for the polymerization of phenols and the cure of phenolic and resorcinol resins with HMMM. Because HMMM is acidic, the cure rate of compounds containing this material will be retarded in comparison to the unmodified control. Hence, adjustments in cure systems might be needed.

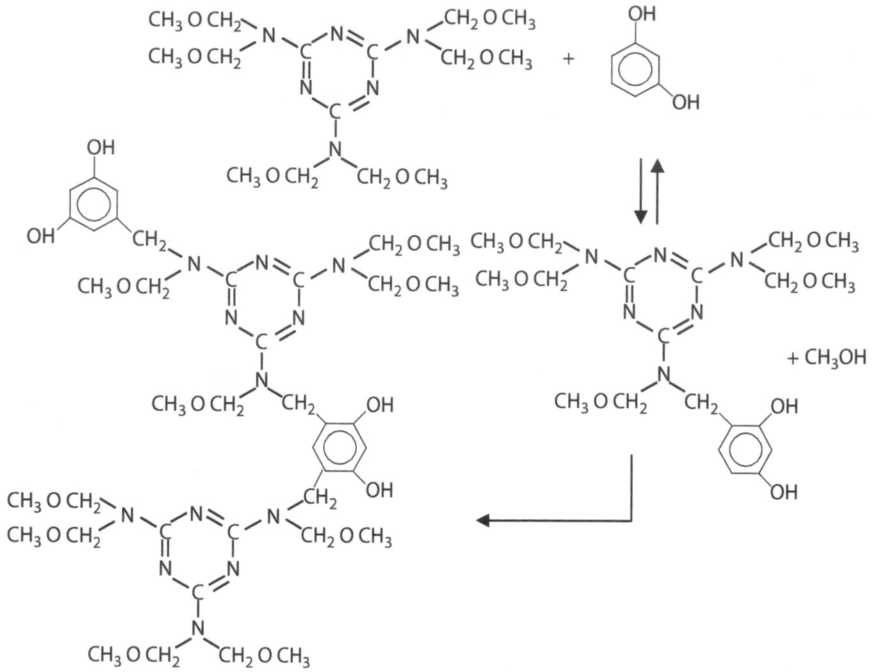

FIGURE 9.10   Reaction of HMMM with resorcinol.

Before cure, the free resorcinol and methylene donors have little effect on mixing and processing of the rubber formulation. The structure of the resin network developed when the resorcinol reacts with HMTA is significantly different from that obtained with HMMM. In both systems, the cure of the resin proceeds in parallel with the rubber cure. With either of the two methylene donors, resorcinol is cured to insoluble, infusible thermoset networks dispersed in the cured rubber. After cure, these resin networks appear to be incompatible with the rubber. The cured resin reinforces the rubber, increasing the hardness and stiffness of the compound. Compounds containing these resin networks maintain the high level of stiffness at elevated temperatures. Normally, compounds with high levels of hardness and stiffness would require high loadings of carbon black, which would make the compound difficult to mix and process.

B.   PHENOLIC RESINS

Phenol, *p*-substituted phenols, and resorcinol can be reacted with formaldehyde or paraformaldehyde to generate phenol–formaldehyde or resorcinol–formaldehyde resins. The reactions can be run in bulk or in aqueous solution. The resins are prepared in closed systems to eliminate exposure to the toxic phenols and formaldehyde. By controlling the monomer ratios, the catalyst selection, and the reaction conditions, a wide variety of resins can be produced. These resins can be linear or branched, have low to moderate molecular weights, and have different chemical functionalities and chemical reactivities. Resins prepared with an acid catalyst are known as novolak resins, and those prepared with a base catalyst are known as resole resins. A third type of phenolic resin is prepared by the polymerization of monomers isolated from cashew nutshell oil.

1. Novolak Resins

The novolak resins are prepared with strong, protonic acid catalysts such as sulfuric acid, sulfonic acid, oxalic acid, and occasionally phosphoric acid. Occasionally, weak or Lewis acids, such as divalent metal (Zn, Mg, Mn, Cd, Co, Pb, Cu, and Ni) carboxylates, are used for specialty resins. A molar ratio of formaldehyde to phenol of between 0.5:1 and 0.8:1 is used in these preparations, and the reactions are run to their energetic completion. A general mechanism for the reaction is outlined in Figure 9.11 (for more details, see pp. 606–609 of Ref. [18]). If protonic acids are used as the catalyst, the resins produced contain 50%–75% of the 2,4'-methylene-linked bisphenol. If divalent metals are used as catalyst, equal amounts (45%) of 2,2'- and 2,4'-methylene-linked bisphenols are produced. Smaller amounts of 4,4'-methylene-linked bisphenol are produced with both types of acid catalysts. Typical properties of the two different types of acid-catalyzed phenolic resins are summarized in Table 9.7.

The straight phenolic novolak resins are thermoplastic with molecular weights of 500–5000 and glass transition temperatures (T_g) of 45°C–70°C. They are essentially linear at molecular weights up to 1000 because of the lower reaction rate for the addition of a third methylene bridge to the doubly substituted phenol ring. Incorporation of some p-substituted phenol in the preparation of the resin broadens the range of potential performance. If all of the phenol is replaced with a p-substituted phenol,

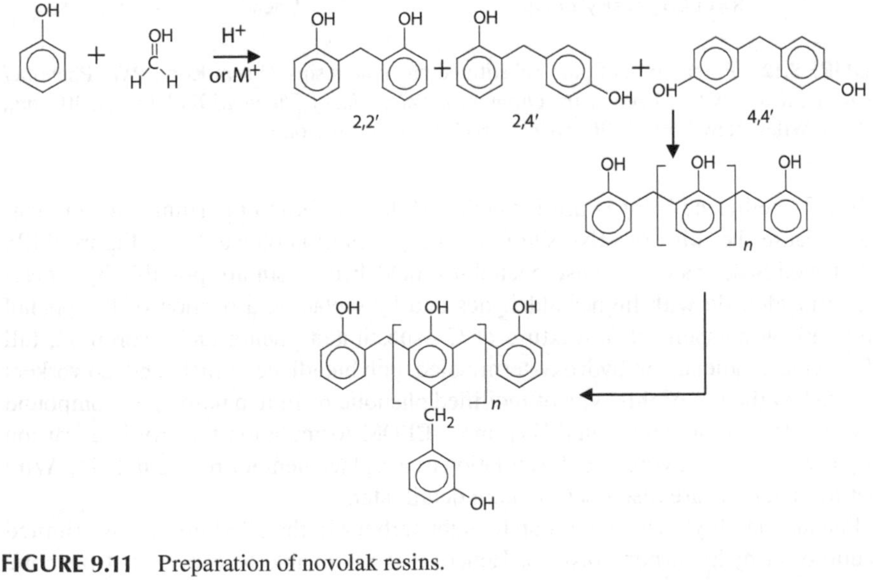

FIGURE 9.11 Preparation of novolak resins.

TABLE 9.7

Effect of Catalyst on Properties of Novolak Resins

Property	Catalyst	
	Acid	Zn Acetate
Formaldehyde/phenol mole ratio NMR analysis%	0.75	0.60
2,2'	6	45
2,4'	73	45
4,4'	21	10
GPCA analysis		
Phenol%	4	7
M_n	900	550
M_w	7300	1800
Water%	1.1	1.9
T_g (°C)	65	48
Gel time (s)	75	25

Source: Kopf, P.W., Phenolic/resins, in: Kraschwitz, J.I., ed., *Kirk-Othmer Encyclopedia of Chemical Technology*, 4th edn., Vol. 18, Wiley, New York, 1996, pp. 603–644. With permission.

R = Cl, CH₃, or alkyl group

Linear

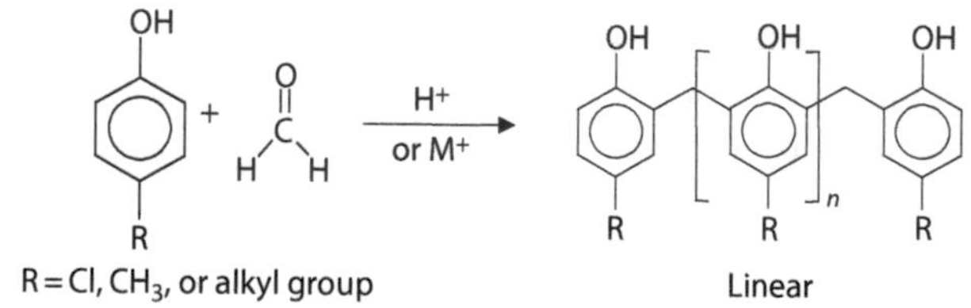

FIGURE 9.12 Preparation of para-substituted phenolic resins. (From Kopf, P.W., Phenolic/resins, in: Kraschwitz, J.I., ed., *Kirk-Othmer Encyclopedia of Chemical Technology*, 4th edn., Vol. 18, Wiley, New York, 1996, pp. 603–644. With permission.)

only 2,2'-methylene bridges are formed, and there is little opportunity for branching, because the third reactive site (the para position) is blocked (see Figure 9.12). Other modifications of the base phenol–formaldehyde resin are possible by replacing formaldehyde with higher aldehydes and by replacing a portion of the phenol with cashew nutshell oil (a mixture of C_{15}-substituted phenol and resorcinol), tall oil, cyclopentadiene, or hydroxy-terminated polybutadiene. Stuck and coworkers reported on the use of this type of modified phenolic resin in a hard apex compound [19], in a farm tread compound [17], in an EPDM formulation for profile extrusion [20], and in a typical wire coat formulation as a replacement for resorcinol [21]. Wire coat formulations are discussed in more detail later.

Phenol and alkylphenols with up to eight carbons in the alkyl group have limited solubility in hydrocarbon solvents. Hence, it is reasonable to expect that phenolic resins would have limited solubility or compatibility with hydrocarbon rubbers. Higher levels of the alkylphenols and the other modifications of phenolic resins are directed to improving the compatibility with rubber. There are commercial phenolic resins available that are recommended for use as tackifiers in rubber compounds. These materials are likely low molecular weight, linear resins that have been modified to produce a softening point below room temperature and to enhance solubility. Higher softening point phenolic resins are used in place of resorcinol as precursors to the formation of reinforcing resin networks in rubber compounds. Phenolic resins usually contain small amounts of unreacted phenol. They are added to the rubber formulation in the early stages of the Banbury mixing process. At mixing and processing temperatures, these resins are low viscosity liquids that do not interfere with mixing and processing. At room temperature, the resins have limited compatibility with rubber, as evidenced by the stiffening of the uncured rubber (frequently referred to as boardiness). These thermoplastic resins must be cured with HMTA or HMMM to form a thermoset resin network that is not compatible with the rubber. The hard, intractable resin network provides increased stiffness and hardness in the rubber without interfering with processing.

In the 1940s and 1950s, the first resorcinol novolak resin was developed as a way to reduce the fuming associated with resorcinol when it is mixed into a rubber compound. However, it still contained 18% free resorcinol. In the 1960s, INDSPEC Chemical, the world's largest producer of resorcinol, discovered that adding a resorcinol homopolymer to the resorcinol novolak helped to reduce the free resorcinol content of the resin [22]. In the 1980s, introduction of styrene to the resorcinol–formaldehyde resin helped to reduce the free resorcinol content of the resin even further and made the resin less hydroscopic. Continued development of the resorcinol–formaldehyde–styrene system has led to resins with less than 1% free resorcinol [23]. More recently, resorcinol has been reacted with a bisphenol A epoxy resin to produce unique resins that are no longer hydroscopic and contains less than 1% free resorcinol [24].

2. Resole Phenolic Resins

Resole-type phenolic resins are produced with a molar ratio of formaldehyde to phenol in the range of 1.2:1 to 3.0:1. If substituted phenols are used, the molar ratio is usually 1.2:1 to 1.8:1. The reactions are catalyzed by strong bases such as NaOH, $Ca(OH)_2$, and $Ba(OH)_2$. Resole resins cover a wider spectrum of structures and properties than the novolak resins. They can be solids or liquids, water-soluble or water-insoluble, alkaline or neutral, slowly curing or highly reactive. Details of the preparation of resole resins can be found in Ref. [18], pp. 609–612, but the general path is outlined in Figure 9.13. The initial reaction of phenol with formaldehyde produces a hydroxyl-substituted benzyl alcohol. On heating, the benzyl alcohol condenses to form a dibenzyl ether as an intermediate. Further heating converts the dibenzyl ether to a bis(hydroxyphenyl)methane and eliminates a molecule of formaldehyde. Free formaldehyde continues to add to the phenolic rings until all of the formaldehyde has been converted to methylene bridges and pendant hydroxymethylene units. Generally, the formation of the resole resin is stopped short of a completely cross-linked structure. Finally, cross-linking

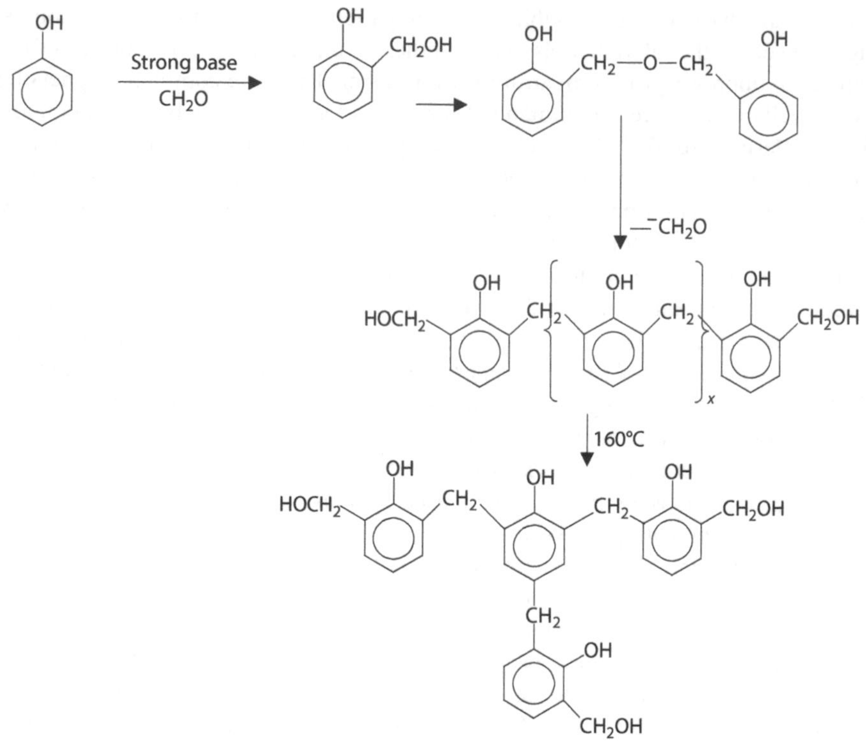

FIGURE 9.13 Preparation of resole phenolic resin. (From Kopf, P.W., Phenolic/resins, in: Kraschwitz, J.I., ed., *Kirk-Othmer Encyclopedia of Chemical Technology*, 4th edn., Vol. 18, Wiley, New York, 1996, pp. 603–644. With permission.)

is achieved by heating the hydroxymethylene-terminated resin to form a highly branched thermoset resin.

Because the resole phenolic resins are heat reactive, they have not been used extensively in rubber compounding. An exception is the use of resole resins prepared with alkylphenols to cure butyl rubber bladders. The pendant hydroxymethyl groups can react with the residual double bonds of the isoprene units in the polymer backbone to produce methylene bridges between the resin and the rubber. Lewis acids are used to catalyze the reaction of the resole resin with the double bonds. Typical cure systems for butyl [25] and halobutylrubbers [26] are presented in Table 9.8. If metal oxides (Lewis acids) are used as a catalyst in the cure formulation for halobutyl rubber, additional carbon–carbon cross-links are formed by the reaction of the pendant hydroxymethylene groups with the halogen on the backbone of the rubber. Butyl rubbers cured with resole resins have outstanding resistance to heat and steam aging because of the stability of the carbon–carbon cross-links (Figure 9.14).

3. Polymerized Cashew Nutshell Oil

Another approach to overcoming the fuming of compounds containing resorcinol is the use of naturally occurring cashew nutshell oil. This oil is a mixture of

TABLE 9.8
Resole Resin Cures for Butyl Rubbers

	Butyl Rubber	Halobutyl Rubber
Cure package		
ZnO	5	3
Steric acid	1	
SP-1045 resole resin[a]		5
SP-1055 resole resin[a]	12	
Cure conditions		
Temperature (°C)	180	160
Time (min)	80	15

Source: Kresge, E. and Wang, H.C., Butyl rubber, in: Kroschwitz, J.I., ed., *Kirk-Othmer Encyclopedia of Chemical Technology*, 4th edn., Vol. 8., Wiley, New York, 1993, pp. 934–966. With permission.

[a] Products of Schenectady Chemical, Inc.

anacardic acid, cardol, and anacardol (see Figure 9.15). Each of these monomers contains a 15-carbon alkyl chain with one or two double bonds. The anacardic acid can be decarboxylated to produce a simple mixture of cardol and anacardol. The cashew nutshell oil can be used to modify phenol–formaldehyde resins but cannot be used directly in rubber compounding because the cardol in the mixture can create handling difficulties in the factory. However, the oil can be converted to a liquid resin by acid catalyzed polymerization of the double bonds in the alkyl side chains to minimize the effect of cardol. A commercial product, Cardolite NC 360®, is available and is recommended as a softener or tackifier to improve the physical strength and adhesion of rubber [27]. It has been evaluated as a resorcinol replacement for the preparation of hard apex compounds [28]. Data (summarized in Table 9.9) extracted from the patent demonstrate that the resin does function as softener and plasticizer. When it is added to a formulation without a methylene donor (sample 1), the 300% elongation of the compound is increased and the 150% modulus is decreased (compare sample 1 and the control). When the compound with the Cardolite resin is cured with HMTA (sample 2), the 300% elongation is lower than that of the control, and the 150% modulus is greater. These changes indicate that the resin can be cured to form a hard, intractable resin network that reinforces and stiffens the compound.

IV. RESIN BONDING SYSTEMS FOR RUBBER REINFORCEMENTS

The first synthetic fiber, viscose rayon, began to replace cotton fabric as the reinforcement for rubber products in the 1930s. But the physical adhesion obtained with the coarse fibrous surface of cotton was absent with the smooth surface of the continuous filament synthetic fiber. In addition, the level of adhesion developed through polar interactions was much lower because of the low polarity of the synthetic rayon.

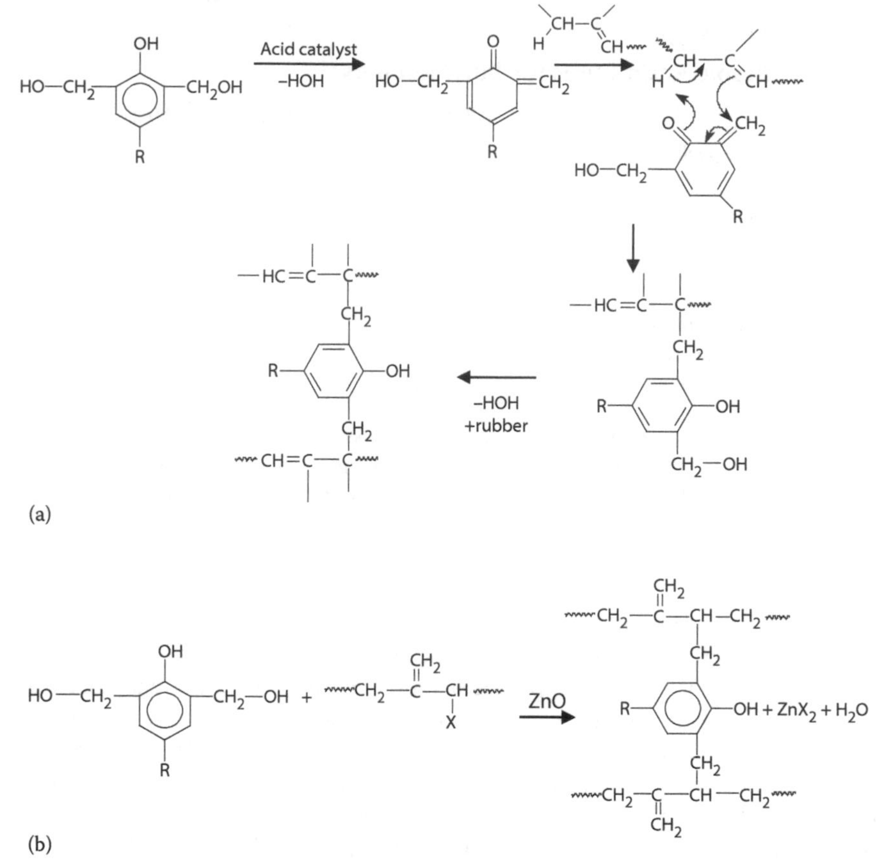

(a)

(b)

FIGURE 9.14 (a) Phenolic resin cures for butyl rubbers. (From Coran, A.Y., Vulcanization, in: Mark, J.E., Erman, B., Eirich, F.R., eds., *Science and Technology of Rubber*, Academic Press, New York, 1994, Chapter 7. With permission.) (b) Schematic of phenolic resin cures for halobutyl rubbers. (From Kresge, E. and Wang, H.C., Butyl rubber, in: Kroschwitz, J.I., ed., *Kirk-Othmer Encyclopedia of Chemical Technology*, 4th edn., Vol. 8, Wiley, New York, 1993, pp. 934–966. With permission.)

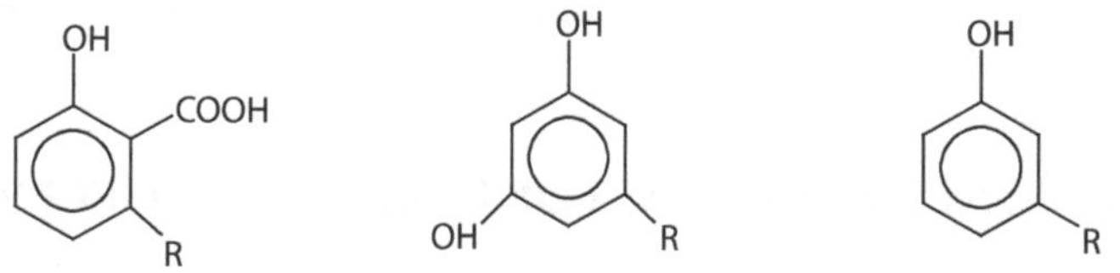

Anacardic acid   Cardol   Amacardol

$$R = C_{15}H_{27} \text{ or } C_{15}H_{29}$$

FIGURE 9.15   Components of cashew nutshell oil.

TABLE 9.9

Preparation and Performance of Polymerized Cashew Nutshell Oil Modified Rubber

	Control	1	2
First stage mix			
Polybutadiene	90	90	90
Natural rubber	10	10	10
Carbon black	90	90	90
Process oil	7.25	7.25	7.25
Softener	2.0	2.0	2.0
Fatty acid	1.0	1.0	1.0
ZnO	3.0	3.0	3.0
Polymerized cashew nutshell oil	—	6.0	6.0
Final stage mix			
Retarder	0.3	0.3	0.3
Accelerator	1.6	1.6	1.6
ZnO	2.0	2.0	2.0
Sulfur	5.0	5.0	5.0
HMTA	—	—	0.9
Cured properties			
Tensile strength (MPa)	21.3	19.87	18.0
Elongation at break (%)	260.9	335.6	191.3
150% modulus (MPa)	12.43	9.51	14.65
Shore D at 23°C	NA	31	35.0

Source: Duddey, J.E., U.S. Patent 6,467,520 Tire with apex rubber containing in-situ resin (to Goodyear Tire & Rubber Co.), October 22, 2002.

As other high modulus synthetic fibers such as nylon, polyester, glass fiber, steel, and Aramid were introduced, the problem became more acute. In order to use the full potential of these synthetic fibers, it was necessary to design and develop new adhesion systems for bonding rubber to them.

The primary function of the adhesive is to transfer the load from the matrix (rubber) to the reinforcement. Ideally, the adhesive system should have a modulus intermediate between those of the rubber and the fiber. It should have a low viscosity for maximum wetting and contact with the fiber surface, and then it should be capable of drying and/or curing into a tough, flexible, fatigue-resistant coating.

Several modes of adhesion have been identified for bonding cord reinforcements to rubber: (1) mechanical–physical fastening or anchoring, (2) adsorption–surface wetting or polar interactions, (3) diffusion–migration of the adhesive into the adherent, and (4) chemical–primary covalent bond formation [29]. Ideally, one adhesive formulation would be preferred for a wide combination of fibers and rubber compounds.

But each fiber and rubber compound has its own unique surface chemistry and morphology. Hence, specialized formulations have been developed to optimize performance.

The first adhesive treatment was developed in the 1930s [30] for use on rayon. It was based on a resorcinol–formaldehyde resin and a rubber-based resorcinol-formaldehyde latex (RFL dip). Current RFL dips can be complex, but they essentially consist of either an in situ resorcinol–formaldehyde resin or preformed resorcinol–novolak resin dispersed in a synthetic styrene–butadiene–vinylpyridine terpolymer latex (VP latex). Representative starting formulations [31] are listed in Table 9.10. The in situ resins are low-molecular-weight oligomers prepared with a base catalyst and a formaldehyde–resorcinol molar ratio of <1. The chemistry of the in situ resin is similar to that previously presented for the resole phenolic resins. During the drying and heat treatment of the dipped fabrics, the hydroxymethyl groups of the resin can condense to form reinforcing networks that increase the modulus of the VP rubber phase, interact with any hydroxyl or amine functions available on the fibers, form hydrogen bonds with the polar sites on the fibers, and form methylene bridges with the VP polymer. The VP rubber forms a tough, flexible coating on the fiber, increases diffusion into the rubber and promotes interaction with the polar groups present in the rubber layer, and provides reactive sites for conventional sulfur cross-linking with the rubber. Aqueous preformed resorcinol–novolak resins are used to reduce the aging time required with the in situ resin systems and to improve the quality and consistency of the final RFL dip. In the presence of base and some free form-aldehyde, the novolak resins are converted to reactive resole-type structures. A less reactive RFL system is generated if ammonia is used to replace part of the sodium

TABLE 9.10
Basic RFL Formulations

	Formulation	
Resin Solution	In Situ Resin RFL[a]	Preformed Resin RFL
Soft water	235.8	192.6
Sodium hydroxide (10%)	3.0	4.5
Resorcinol	11.0	—
Aqueous RF novolak resin (75%)	—	16.8
Formaldehyde (37%)	11.3	12.2
RFL dip[b]		
VP latex (41% solids)	244.0	244.0
Soft water	60.0	72.2
Ammonium hydroxide (28%)	11.3	12.0
Total	581.3	534.3

Source: Peterson, A., Adhesion promoters for bonding rubber to metals and textiles, in: *ACS Rubber Division Meeting*, Pittsburgh, September 1994. With permission.

[a] Age the in situ resin solution 1–4 h to allow for resin formation.
[b] Age the RFL dip 18–24 h before fabric treatment.

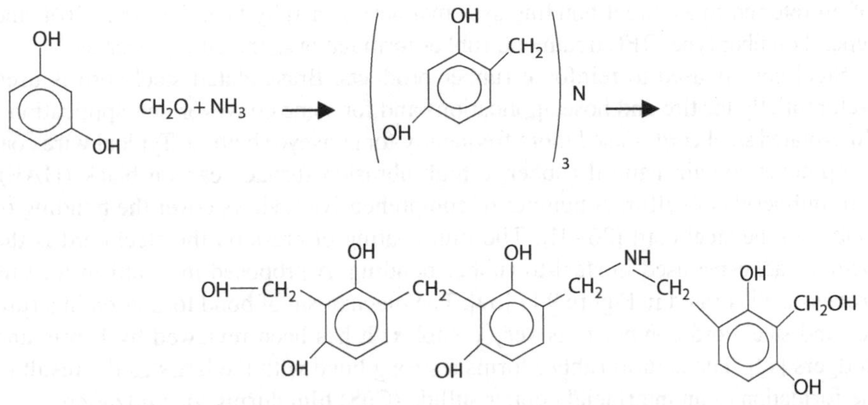

FIGURE 9.16 Ammonia-modified RFL dip. (From Rijpkema, B. and Weening, W.E., in: *ACS Rubber Division Meeting*, Orlando, October 1993, Paper 142. With permission.)

hydroxide (Figure 9.16). This RFL formulation produces a lower modulus, more flexible, and more fatigue-resistant coating on the cord [32]. Resorcinol, formaldehyde, and ammonia react to form bis- and tris(dihydroxybenzyl)amines. If ammonia is the only base employed, the alkylated amines precipitate from the remaining mixture of resorcinol and resorcinol oligomers. Precipitation of the amines is avoided by the use of a small amount of sodium hydroxide or by delaying the addition of formaldehyde until all other ingredients have been added [33].

Application of the adhesive is achieved by passing the reinforcement fabric through the RFL dip and then passing the treated fabric through drying and curing ovens at optimum temperatures, time, and tension to develop the best balance of strength, fatigue resistance, and adhesive properties. The number of dipping and heat treatments depends on the fiber type, adhesive requirements, and targeted mechanical properties. Adequate adhesion can be obtained on rayon, the various nylons, and polyvinyl alcohol fabric with a single dip. However, nylon 66 requires a higher level of vinyl pyridine in the RFL formulation. Glass fiber must be sized with special organofunctional silanes prior to a single application of the RFL treatment. The RFL solids and the dip pickup must be higher for glass fiber than for other fibers because the fabric must be thoroughly coated to protect it from breakage and to increase cord strength and fatigue resistance. Polyester (polyethylene terephthalate) and aramid fabrics require surface activation prior to the application of the RFL adhesive because of the low polarity and low concentration of functional groups on the surfaces of these fibers. Treatments based on blends of resorcinol, chlorophenol, epoxy resin, and a blocked diisocyanate are added to the RFL dip or are applied directly to the cord surface prior to application of the RFL dip [34]. For additional details and references, the review article by Peterson [31] is recommended.

In some cases, the RFL-treated reinforcements do not produce sufficient levels of adhesion. Modification of the rubber coat stock with resorcinol or a resorcinol–novolak resin, a methylene donor such as HMTA or HMMM, and precipitated silica has been used to overcome this deficiency [35]. Improvements obtained with this approach (often referred to as direct bonding systems) have generally been less than 30% and depend on fiber type, RFL treatment, rubber formulations, and cure properties.

Steel cord is used to reinforce rubber products. Brass-plated steel cord is used preferentially for tire and hose applications and for some conveyor belt applications. Zinc-plated steel cord is used more frequently for conveyor belting. Typical wire coat compounds contain natural rubber, a high abrasion furnace carbon black (HAF), and high levels of sulfur. A number of comprehensive reviews cover the bonding of rubber to the steel cord [36–41]. The thin coating of brass on the steel cord is the primary adhesive used in steel-to-rubber bonding. A proposed mechanism for this process is illustrated in Figure 9.17 [39]. The mechanism of bond formation in a rubber and steel cord composite is very complex. It has been reviewed by Kovac and Rodgers [42]. The natural rubber forms a strong bond with the brass as the result of the formation of an interfacial copper sulfide (CuS) film during vulcanization.

Copper sulfide domains are created on the surface of the brass film during the vulcanization reaction. These domains have a high specific surface area and grow within the wire coat compound before the viscous polymer phase is cross-linked into an elastomeric network. Important factors governing this bonding are the formation of copper sulfide, cohesive strength, adhesion to the brass substrate, and the rate of secondary corrosion reactions underneath the copper sulfide film. A primary requirement is the formation of the copper sulfide domains before the initiation of cross-linking. Mechanical stability of the copper sulfide domains is essential to retain long-term durability of the rubber-to-wire adhesion. However, breakdown of the wire–rubber adhesive bond is catalyzed by Zn^{2+} ions, which diffuse through the interfacial CuS layer. This will eventually result in an excess of either ZnS or $ZnO/Zn(OH)_2$. Under dry conditions, this process is slow. Nevertheless, Zn^{2+} will migrate to the surface, with a consequent drop in the mechanical interlocking of the CuS domains and rubber followed by loss of adhesion. Migration of Zn^{2+} ions

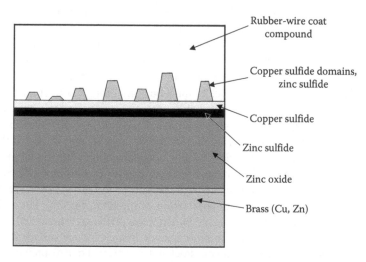

FIGURE 9.17 Model for interfacial copper sulfide film in rubber–brass bonding. (From van Ooij, W.J., *Rubber Chem. Technol.*, 57, 421, 1984. With permission.)

is a function of the electrical conductivity of the brass coating. Addition of Co^{2+} or Ni^{2+} ions will reduce this conductivity. Both di- and trivalent cobalt ions reduce the electrical conductivity of the ZnO lattice, thereby reducing the diffusion of Zn^{2+} ions through the semiconducting film. Thus if a cobalt salt is used, formation of copper sulfide at the cord surface will be accelerated, whereas ZnS generation will be hindered.

In addition, it has been found that incorporation of an resorcinol-formaldehyde (RF) resin–HMMM system and silica into the rubber formulation enhances adhesion and performance of a steel–rubber composite. An optimum rubber formulation likely contains

1. Relatively high levels of insoluble sulfur (to form sufficient copper sulfide).
2. A slow-acting cure accelerator (to allow time for the sulfur to react with copper).
3. A cobalt salt (to activate Cu_xS formation). (Note: The cobalt ions also activate the oxidation of rubber, which can lead to premature failure in the resin–rubber layer.)
4. A resin–rubber layer (the RF resin–HMMM mixture migrates to the high energy wire surface and forms a resin-rich rubber layer on the wire surface).
5. Silica.

The resin–rubber layer protects the steel cord from attack by moisture and oxygen (rust), reduces the depletion of zinc from the brass, stabilizes the Cu_xS layer, and improves both the initial and humidity-aged adhesion. Lawrence and de Almeida [43] showed that high levels of wire adhesion can be maintained when a styrenated resorcinol–formaldehyde resin is used in place of resorcinol in a wire coat compound. The use of the resin eliminates the problems associated with the handling of the hydroscopic resorcinol and with the fuming of compounds containing resorcinol.

Zinc-plated steel wire cord also requires adhesion promoters in the rubber compound. The Zn/ZnO surface of the zinc-plated wire has inherently low adhesion to cured rubber compounds [31]. The addition of organic cobalt compounds at higher levels than those normally required for brass wire helps increase zinc wire adhesion by activating the formation of zinc sulfide while suppressing dezincification [37]. As with brass wire bonding, the addition of an RF resin adhesion promoter makes it possible to reduce cobalt levels and increase both the initial and aged wire adhesion.

V.   NEW RESIN TECHNOLOGIES FOR WIRE ADHESION

In the past several years, two new resin technologies have been proposed to promote the bonding of rubber to steel cord without the use of resorcinol. The most advanced is the family of "one-component" melamine resins [44] introduced by Cytec Industries as Cyrez CRA132L®, CRA132S®, CRA138L®, and CRA138S® (where the L refers to liquid products and S refers to the liquid products adsorbed on a high surface area silica carrier). The preparation of HMMM involves two reaction steps, the methylation of melamine with formaldehyde and the alkylation of the intermediate hydroxymethylmelamine with alcohols (see Figure 9.18). By adjusting the ratio of reactants and the

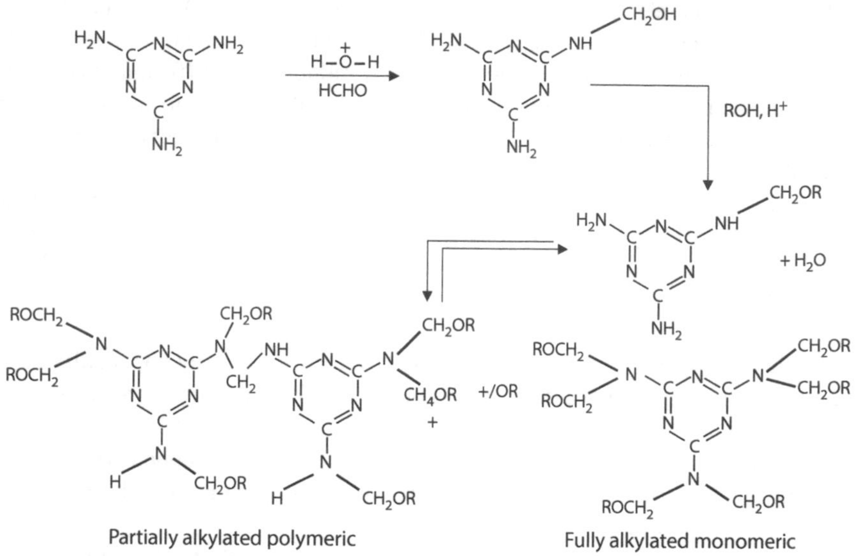

FIGURE 9.18 Formation of alkylated melamine–formaldehyde resin. (From Hoff, C.M. Jr., Wire adhesion—A review of present day technology and a look to the future, in: *ACS Rubber Division Meeting*, Cleveland, OH, October 1997. With permission.)

reaction conditions, partially alkylated melamine intermediates are formed. On heating, the hydroxymethylene units condense to form dimers containing bismethylene ether and/or methylene bridges between the pendant nitrogen groups of two melamine molecules. The dimers contain alkoxymethylene (NCH_2OR), hydroxymethylene (NCH_2OH), and imino (NHR) sites. On heating, these sites can condense to form resins (see Figure 9.19). These partially alkylated melamine dimers are the basis of the one-component resins. The two series, which differ from each other by the alcohol used in the alkylation reaction, exhibit different cure characteristics (see Figure 9.20). These resins were evaluated by Cytec in a typical wire coat formulation containing cobalt and compared with formulations containing a resorcinol–novolak resin cured with either liquid HMMM (Cyrez® 963L), HMMM on silica carrier (Cyrez® CRA100), or HMTA. A rubber formulation containing cobalt without resin was also included for comparison (see Table 9.11 for a summary of formulations). Formulations containing the self-condensing resins exhibit similar tensile properties, higher elongation, lower moduli, and less hardness (Table 9.12). Original and aged wire adhesion testing did not readily differentiate between samples (Table 9.13), with the exception of a 96 h salt aging test. This test showed that the "one-component" resins outperformed the resorcinol–novolak resin control cured with HMMM. Dynamic testing showed no major differences in performance, but de Mattia flex testing showed that the "one-component" melamine resins were superior to the formulations containing the resorcinol–novolak resin cured with HMMM (Table 9.14).

The second new technology has not been commercialized but offers some unique opportunities. van Ooij and Jayaseelan [45] showed that a mixture of

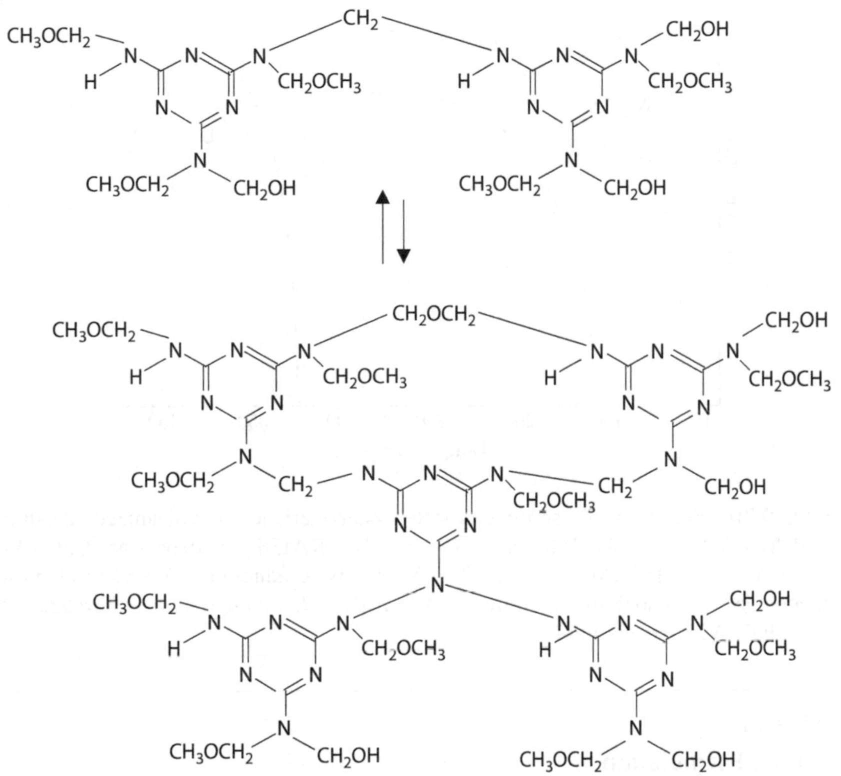

FIGURE 9.19 Self-condensation reaction of imino melamine dimers. (From Hoff, C.M. Jr., Wire adhesion—A review of present day technology and a look to the future, in: *ACS Rubber Division Meeting*, Cleveland, OH, October 1997. With permission.)

bis(trimethoxysilylpropyl)amine and bis(triethoxysilylpropyl)tetrasulfide in a ratio of 1:3 by volume promotes the bonding of rubber to different metal substrates [46]. The mixture of silanes can be applied as a neat liquid or as a solution in alcohol to control film thickness. The silane film is cured at 160°C for 40 min. The aminosilane is the primary dry-film former leading to polysiloxane linkages (Si–O–Si) and metallosiloxane linkages (metal–O–Si) at the metal surface. Bis(triethoxysilylpropyl)tetrasulfide is not a film former by itself but is incorporated in the polysiloxane (Si–O–Si) network in the presence of the aminosilane. The tetrasulfide function promotes bonding to rubber. A mechanism has been proposed for the polymerization of the trialkoxylsilanes. A model for the bonding of rubber to a metal surface has been suggested (see Figure 9.21). The same mixture of silanes works (1) for high or low sulfur compounds, (2) in the presence or absence of cobalt salts, and (3) on various metal substrates such as brass, zinc, and steel. Additional studies by van Ooij and coworkers [46,47] demonstrated the ability of the polysiloxane coating to inhibit corrosion of metals. On the basis of these findings, they suggested that it might be

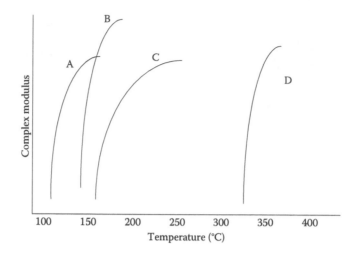

FIGURE 9.20 Schematic illustrating expected characteristics of vulcanized melamine-formaldehyde MF resins. A, HMMM + resorcinol; B, CRA138 precursor resin; C, CRA132 precursor resin; D, HMMM. (From Hoff, C.M. Jr., Wire adhesion—A review of present day technology and a look to the future, in: *ACS Rubber Division Meeting*, Cleveland, OH, October 1997. With permission.)

TABLE 9.11
Wire Coat Formulation

				Master Batch				
			SIR-20	100.00				
			N330	60.00				
			ZnO	8.00				
			ST acid	1.25				
			Co naphthalene	1.50				
			Naphthalene oil	1.50				

	963L	CRA100	HMTA	Cobalt	CRA132L	CRA132S	CRA138L	CRA138S
B19S	3.5	3.5	3.5	—	—	—	—	—
Crystx	5.0	5.0	5.0	8.0	5.0	5.0	5.0	5.0
DCBS	0.75	0.75	0.75	0.75	0.75	0.75	0.75	0.75
963L	3.00							
CRA100		4.50						
HMTA			1.50					
CRA132L					3.80			
CRA132S						5.00		
CRA138L							4.25	5.00
Total	184.50	186.00	183.00	181.00	181.8	183.00	182.25	183.00

Source: Hoff, C.M. Jr., Wire adhesion—A review of present day technology and a look to the future, in: *ACS Rubber Division Meeting*, Cleveland, OH, October 1997. With permission.

TABLE 9.12
Cured Properties

Sample	Hardness	Tensile (MPa)	Elong (%)	100% Mod	300% Mod
963L	81	19.1	309	5.8	18.5
CRA100	83	18.4	287	6.6	19.1
HMTA	81	20.2	272	7.0	Broke
Cobalt	77	22.6	341	6.7	20.6
CRA132L	78	22.4	389	5.0	17.1
CRA132S	80	22.1	372	5.4	18.0
CRA138L	79	20.6	358	5.7	18.1
CRA138S	78	20.1	348	5.5	17.5

Source: Hoff, C.M. Jr., Wire adhesion—A review of present day technology and a look to the future, in: *ACS Rubber Division Meeting*, Cleveland, OH, October 1997. With permission.

TABLE 9.13
Adhesion Results—ASTM D 2229-93a

Sample	Original	Heat Aged[a]	Humidity Aged[b]	Salt Aged[c] 48 h	96 h
963L	843 (95)	515 (95)	812 (95)	541 (95)	488 (100)
CRA100	888 (100)	550 (100)	674 (100)	754 (95)	630 (100)
HMTA	950 (100)	728 (100)	692 (100)	821 (95)	657 (*)
Cobalt	843 (100)	768 (100)	812 (100)	660 (100)	408 (*)
CRA132L	985 (100)	510 (100)	674 (95)	843 (95)	692 (*)
CRA132S	883 (100)	684 (100)	760 (100)	901 (100)	785 (90)
CRA138L	852 (100)	510 (100)	652 (100)	812 (100)	737 (100)
CRA138S	821 (100)	515 (100)	621 (100)	812 (100)	830 (100)

Source: Hoff, C.M. Jr., Wire adhesion—A review of present day technology and a look to the future, in: *ACS Rubber Division Meeting*, Cleveland, OH, October 1997. With permission.

[a] Heat aged = 120 h at 90°C in forced air oven. Newtons (appearance rating).
[b] Humidity aged = 7 days at 98% rel. humidity, 80°C.
[c] Salt aged = 48 and 96 h at 70°C in 10% NaCl solution.
(*) = wire breakage due to corrosion.

possible to eliminate the brass plating of steel cord and replace it with zinc-plated cord and that in some applications it might be possible to use untreated steel cord. They also suggested that the sulfur level in the rubber compound could be reduced and the cobalt salt could be eliminated. These improvements would allow the rubber wire coat compound to be formulated for maximum performance.

TABLE 9.14
ASTM de Mattia Crack Growth

	Cycles to Failure
963L/RF resin/cobalt	3000
CRA100/RF resin/cobalt	2500
HMTA/RF resin/cobalt	1300
Cobalt	2000
CRA132L/cobalt	6500
CRA132S/cobalt	5200
CRA138L/cobalt	4500
CRA138S/cobalt	5400

Source: Hoff, C.M. Jr., Wire adhesion—A review of present day technology and a look to the future, in: *ACS Rubber Division Meeting*, Cleveland, OH, October 1997. With permission.

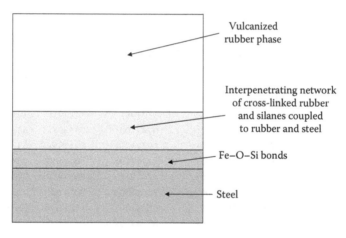

FIGURE 9.21 Simplified schematic for bonding rubber to steel with silane adhesion promoter. (From van Ooij, W.J. and Jayaseelan, S.K., Bonding rubber to metals by silanes, in: *ACS Rubber Division Meeting*, April 1999. With permission.)

VI. NEW TRENDS IN RESIN UTILIZATION

W. Pille-Wolf

As already pointed out earlier, resins do influence the dynamic properties of a cured rubber compound when added to a compound formulation and even more so when resins replace partially or fully mineral process oil. With the introduction of EU legislation on clean oil in 2010, highly aromatic oil was banned for use in tire rubber compounds. Null [52] pointed out that a replacement of distilled aromatic extract by treated distillate aromatic extract (TDAE) is accompanied by a drop in wet traction performance. There was a need to compensate this performance drop.

Already in 1985, a patent application [53] was filed in which resins were specifically used to improve the wet traction of tread compounds. Since then, the number of patent applications that cover the design and use of resins for the modification of dynamic properties have strongly increased, and a continuation of this trend can be observed (e.g., [54–62]). As described earlier, the impact of resinous oligomers on the dynamic properties is driven by their high glass transition temperature T_g. A high resin T_g however does not necessarily translate in a raised compound T_g. This is dictated by the resin compatibility in the elastomer. Thielen has studied the influence of molecular weight and polarity of styrenic oligomers with various chain lengths in typical tire-related elastomers SBR, BR, and NR [63]. Phase diagrams revealed that in NR and in BR, a nonlinear relationship between resin dosage levels and T_g [64] occurs, which means that the Fox equation [72] is not followed. The T_g transition signal of a filled rubber compound is weak and therefore not easy to identify. It has become common practice to determine the influence of resin on the viscoelastic properties by DMA temperature sweeps. The tan δ maximum position and the tan δ values at specific temperature bands are used as predictors for potential tire performances for wet traction in the area of 0°C–10°C and for rolling resistance between 50°C and 70°C [65]. In Figure 9.22, the influence of two resins with different T_g on the hysteresis of silica-filled tread rubber compounds is shown. Even with dosages levels of 4 phr, an effect on hysteresis is noticeable. Although both resins have different glass transition temperatures (34°C and 60°C, respectively), the corresponding tan δ maxima are positioned at the same temperature. This phenomenon can be attributed to differences in resin compatibility, the pure aromatic resin being more compatible than the terpene phenol resin. Due to the

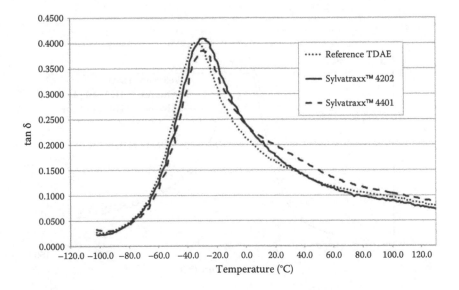

FIGURE 9.22 Temperature sweep DMA of silica PCR-tread compounds (formulations in Table 9.15) containing a pure vinyl-aromatic-based resin Sylvatraxx 4401 (T_g 34°C) and a terpene phenol resin Sylvatraxx 4202 (T_g 60°C). Sylvatraxx is a trademark of Arizona Chemical.

plasticizing effect of resins, their addition to rubber compounds usually lowers the rubbery plateau that can be correlated to increased rolling resistance [69]. However, the terpene phenol resin in Figure 9.22 does not follow this phenomenon in having lower hysteresis in the temperature range of 50°C–70°C than the reference compound without resin addition that indicates a potential for lower rolling resistance. There are an increasing number of recent publications on the mechanism of resin influence on dynamic rubber compound properties for tires. Class studied polymer resin blends used in related application of pressure sensitive adhesives PSA. For both, their viscoelastic characteristics determine the practical use, and the corresponding elastomers and resins are very similar. He pointed out that highly compatible resins shift the tan δ, and incompatible resins are not affecting the rubbery plateau [69–71]. Shekleton compared differential scanning calorimetry (DSC) and dynamical mechanical analysis (DMA) measurements with solubility calculations of hydrocarbon resin blended with a variety of elastomers, but found only in specific cases a good correlation [65]. Vleugels tested naturally derived resins such as a polyterpene and a terpene phenol as well as petrochemically derived pure aromatic resin and found good correlation between the Payne effect measured by rubber process analyzer (RPA) and DMA temperature sweeps indicating improvements of the wet traction indicator without compromising the rolling resistance indicator [66]. Carvagno tested a variety of petrochemically derived resins [67], and Robertson published a corresponding rheological model [68].

Unlike the use of resins for improving building tack, polymer homogenization, or filler dispersion, the dosage to improve wet traction are far beyond the 2–4 phr level and are reported to be as high as 20 phr up to the point that additional process oil is completely replaced by resin. Figure 9.23 shows the effect of increasing

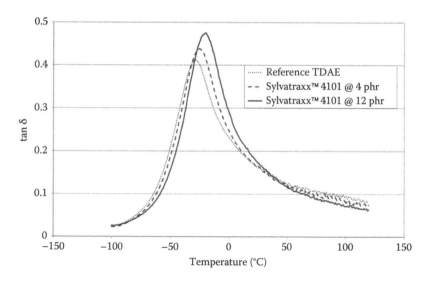

FIGURE 9.23 Temperature sweep DMA of silica PCR-tread compounds with increasing amount of polyterpene resin Sylvatraxx 4101 (formulations in Table 9.15). Sylvatraxx is a trademark of Arizona Chemical.

TABLE 9.15

PCR Silica Compound Formulations as Referenced in Figures 9.22 and 9.23

Raw Materials in phr	Reference				
S-SBR Buna® VSL 5025–0 HM	70	70	70	70	70
BR Buna® CB 24	30	30	30	30	30
Silica Ultrasil® 7000 GR	80	80	80	80	80
Carbon Black Statex® N 234	10	10	10	10	10
Silan Si 69®	8	8	8	8	8
Mineral oil TDAE Viva Tec® 500	20	16	16	16	8
IPPD Vulkanox® 4010	1	1	1	1	1
6PPD Vulkanox® 4020	2	2	2	2	2
TMQ Vulkanox® HS	0.5	0.5	0.5	0.5	0.5
Ozon wax Antilux® 654	1	1	1	1	1
ZnO Zinkoxid Rotsiegel®	3	3	3	3	3
Stearic acid	1	1	1	1	1
Sylvatraxx™ 4401	—	4	—	—	—
Sylvatraxx™ 4202	—	—	4	—	—
Sylvatraxx™ 4101	—	—	—	4	12
Sulfur	1.5	1.5	1.5	1.5	1.5
CBS Vulkacit® CZ	1.5	1.5	1.5	1.5	1.5
DPG Vulkacit® D	2	2	2	2	2

dosage of a polyterpene resin to a silica-filled low rolling resistance tread compound (Table 9.15). The tan δ maximum shifts to higher temperature as expected, but also is higher. Despite the high resin dosage, the hysteresis in the temperature band from 50°C to 70°C is actually lower.

The published results refer to single elastomer resin systems only, while in tire applications, binary or ternary elastomer blends are very common. The distribution of resins in such systems is unknown so far as well as the interaction between resin and process oil and reinforcing filler. Future research on theses aspects is to be expected.

VII. IDENTIFICATION OF RESINS IN CURED RUBBER

Compound analysis of unknown cured rubber is a complex task for the analytical chemist. Commercial rubber formulations always contain a number of ingredients. Many kinds of separation and identification methods have been developed for rubber component analysis using thermal analysis, spectroscopy, chromatography, and mass spectroscopy (MS). Identification of resins in cured rubbers is more difficult than analysis of other ingredients because of the high molecular weights, low volatilities, and structure complexity. Kim et al. [48] identified several combinations of analytical procedures such as thermogravimetric analysis/Fourier transform infrared and pyrolysis gas chromatography/MS that can be used to identify a number of tackifying and reinforcing resins that are present in cured rubber.

VIII. SUMMARY

Resins, unlike many other rubber compounding ingredients, do not have any precise definition. The class of materials known as resins covers many low-molecular-weight materials such as natural products, petroleum-based products, and a variety of other synthetic materials produced to fulfill specific functions. This review is necessarily brief, and the reader is encouraged to view other more in-depth works such as that by Mildenberg et al. [49] or Gardziella et al. [50]. Resins play a critical role in rubber compounding in that they have a major influence on the processing characteristics, compound vulcanization, final compound mechanical properties, performance properties such as component-to-component adhesion, and product durability. This is of particular importance in tire compounds. Future work will most likely center on these parameters with the view to further improving the environmental impact encountered in resin manufacturing, final end product performance characteristics such as durability, and other factors such as tire rolling resistance. The role of resin systems in tire and other rubber product compounding formulas will therefore continue to be of considerable importance and present many opportunities in the future.

REFERENCES

1. Barlow F. *Rubber Compounding: Principles, Materials, and Techniques*, 2nd edn. New York: Marcel Dekker, 1993, p. 159, Chapter 12.
2. Barlow F. *Rubber Compounding: Principles, Materials, and Techniques*, 2nd edn. New York: Marcel Dekker, 1993, p. 167, Chapter 13.
3. Files E. A basic look at fluid effects on elastomers. *Rubber Plastic News*, June 10, 1991, pp. 39–41.
4. Stephens HL. Plasticizer theory—An overview based on the works of Doolittle, Bueche and Ferry. In: *Education Symposium No. 10, ACS Rubber Division*, Toronto, Ontario, Canada, May 1983.
5. Barlow F. *Rubber Compounding: Principles, Materials, and Techniques*, 2nd edn. New York: Marcel Dekker, 1993, p. 168, Chapter 13.
6. Flory PJ. *Principles of Polymer Chemistry*. Ithaca, NY: Cornell University Press, 1953.
7. Doty PM, Zable HS. Determination of polymer-liquid interaction by swelling measurements. *J Polym Sci* 1946;1:90.
8. Hildebrand JH, Scott RL. *Solubility of Non-Electrolytes*. New York: Reinhold, 1950.
9. Gee G. *Trans Rubber Ind* 1943;18:266.
10. Gardon JL. *Cohesive-energy density*. In: Birkales NM, ed., *Encyclopedia of Polymer Science and Technology*, Vol. 3. New York: Wiley, 1965, pp. 833–862.
11. Sheehan CJ, Bisio AL. *Rubber Chem Technol* 1966;39:149.
12. Hanson CM. *Ind Eng Chem Prod Res Dev* 1968;8:2.
13. Fedors RF. *Polym Eng Sci* 1974;14:147, 472.
14. Napolitano MJ. The use of hydrocarbon resins in rubber compounding. In: *ACS Rubber Division Meeting*, Pittsburg, PA, September 1994.
15. Schlademan JA. Hydrocarbon resins and their effect on the sulfur vulcanization of SBR. In: *ACS Rubber Division Meeting*, Cleveland, OH, October 24, 1979.
16. Hilner K. The action of hydrocarbon resins in rubber formulations for the tire industry. *Tire Technol Int* 1994;106.
17. Stuck BL, Souchet J-C. Phenol formaldehyde reinforcing resins in tire compounds. *Tire Technol Int* 1997;131.

18. Kopf PW. Phenolic/resins. In: Kraschwitz JI, ed., *Kirk-Othmer Encyclopedia of Chemical Technology*, 4th edn., Vol. 18. New York: Wiley, 1996, pp. 603–644.

19. Stuck BL, Souchet J-C, Morel-Fourbier C. Chemical variations on novolak phenol formaldehyde resins in a hard bead apex compound. In: *ITEC Conference*, Akron, OH, September 1994.

20. Stuck BL, Morel-Fourrier C, Souchet J-C. Reinforcing resins for EPDM profiles. *Rubber Plastics News*, December 18, 1995, pp. 15–20.

21. Souchet J-C, Morel-Fourrier C. *CECA Novolak Resins as Adhesion Promoters for Steel Cord/Rubber Bonding*. Paris, France: CECA, October 23, 1996.

22. Lawrence MA. *Current Resorcinol Formaldehyde Resins and Their Use in Optimized Steel Cord Adhesion*. Pittsburg, PA: INDSPEC Chem Corp Literature.

23. Hood RT, Lamars RM. U.S. Patent 5,049,641 Rubber compounding resin (to INDSPEC Chem Corp), September 17,1991; Durairaj RB, Peterson A Jr. U.S. Patent 5,936,056, Non-volatile resorcinolic resins and methods of making and using the same. August 10, 1999; U.S. 5,945,500, August 31, 1999 (to INDSPEC Chem Corp).

24. Dressler H, Peterson A Jr. U.S. Patent 5,244,725 Hydroxyalkyl aryl ethers of di- and polyhydric phenols (to INDSPEC Corp), September 14, 1993; Durairaj RB, Peterson A Jr. U.S. Patent 5,936,056 Non-volatile resorcinolic resins and methods of making and using the same, August 10, 1999; U.S. 5,945,500, August 31, 1999 (to INDSPEC Chem Corp).

25. Coran AY. Vulcanization. In: Mark JE, Erman B, Eirich FR, eds., *Science and Technology of Rubber*. New York: Academic Press, 1994, Chapter 7.

26. Kresge E, Wang HC. Butyl rubber. In: Kroschwitz JI, ed., *Kirk-Othmer Encyclopedia of Chemical Technology*, 4th edn., Vol. 8. New York: Wiley, 1993, pp. 934–966.

27. Cardolite NC-369 product literature. Newark, NJ: Cardolite Corp, December 11, 1995.

28. Duddey JE. U.S. Patent 6,467,520 Tire with apex rubber containing in-situ resin (to Goodyear Tire & Rubber Co.), October 22, 2002.

29. Shanahan M.E.R., Adhesion and wetting: Similarities and differences. *Rubber World* 1991;205:29.

30. Charch WH. U.S. Patent 2,128,635 Laminated structure and method for preparing sam (to DuPont), 1938.

31. Peterson A. Adhesion promoters for bonding rubber to metals and textiles. In: *ACS Rubber Division Meeting*, Pittsburgh, September, 1994.

32. Rijpkema B, Weening WE. In: *ACS Rubber Division Meeting*, Orlando, October 1993, Paper 142.

33. van Gils GE. *J Appl Polym Sci* 1969;13:835.

34. Solomon TS. In: *ACS Rubber Division Meeting*, Orlando, October 1993, Paper 53.

35. Iyengar R. *Rubber World* 1977;197:24.

36. Buchan R. *Rubber to Metal Bonding*. London, U.K.: Crosby Lockwood & Sons, 1959, pp. 119–152.

37. van Ooij WJ. *Rubber Chem Technol* 1979;52:605.

38. Haemers G. *Rubber World* 1980;182:26.

39. van Ooij WJ. *Rubber Chem Technol* 1984;57:421.

40. Ishikawa Y. *Rubber Chem Technol* 1984;57:855.

41. van Ooij WJ. *Kautsch Gummi Kunstst* 1991;44:348.

42. Kovac FJ, Rodgers MB. Tire engineering. In: Mark JE, Erman B, Eirich FR, eds., *Engineering Science and Technology of Rubber*. New York: Academic Press, 1994, Chapter 14.

43. Lawrence MA, de Almeida J. Maximize steel cord adhesion. *Tire Technol Int* 2001 pp. 58–61.

44. Hoff CM Jr. Wire adhesion—A review of present day technology and a look to the future. In: *ACS Rubber Division Meeting*, Cleveland, OH, October 1997.

45. van Ooij WJ, Jayaseelan SK. Bonding rubber to metals by silanes. In: *ACS Rubber Division Meeting*, Chicago, April 1999.

46. Jayaseelan SK, van Ooij WJ. Rubber to metal bonding by silanes. In: *Adhesion'99, Conf Proc* Cambridge, U.K., September 1999, pp. 43-48.

47. van Ooij WJ, Zuh D, Prasad G, Jayaseelan SK, Fu Y, Teredesai N. *Surf Eng* 2000;16:386.

48. Kim SW, Lee GH, Heo GS. Identification of tackifying resins and reinforcing resins in cured rubber. *Rubber Chem Technol* 1999;72:181.

49. Mildenberg R, Zander M, Collin G. *Hydrocarbon Resins*. Weinheim: VCH Verlagsgesellschaft mbH/VCH Publishers. A Wiley Company, 1977.

50. Gardziella A, Pilato LA, Knop A. *Phenolic Resins: Chemistry, Applications, Standardization, Safety, and Ecology*, 2nd edn. Berlin, Germany: Springer-Verlag, 2000.

51. Zinkel DF, Russel JR, eds. *Naval Stores: Production, Chemistry, Utilization*. New York: Pulp Chemicals Association, 1989, Chapter 8: Chemistry of turepentine, p. 225; Chapter 9: Chemistry of rosin, p. 261.

52. Null V. Safe process oils for tires with low environmental impact. *KGK* 1999;52(12):799.

53. Bridgestone. Vinyl aromatic resin. U.S. 4487892, 1984.

54. Goodyear. Vinyl aromatic resin. U.S. 5877249, 1999.

55. Pirelli. Vinyl aromatic resin. U.S. 646910, 2002.

56. Michelin. Polyterpene resin. U.S. 7084228, 2005.

57. The Yokohama Rubber Corp. Rubber composition for tire tread. U.S. 7989550, 2007.

59. Arizona Chemical. Tire and tread formed from phenol-aromatic-terpene resin. U.S. 8637606, 2007.

60. Continental AG. Pneumatic Vehicle Tyre. EP 1526002, 2005.

61. Exxonmobile Chemical, Elastomeric compositions comprising hydrocarbon polymer additives, US20090186965, 2009.

62. Sumitomo Rubber Ind., Rubber compositions for tread and pneumatic tire. EP 2338698, 2010.

63. Thielen, G. Thesis: Molekulare Zusammenhaenge zu Verhalten und Wirkung unpolarer Modell-Verarbeitungshilfsmittel in Kautschuken, University of Hannover, Hannover, Germany, 1991.

64. Schuster RH, Thielen G, Hallersleben ML. Loeslichkeit und Verteilung von Kohlenwasserstoff-Modellharzen in Kautschukverschnitten. *KGK* 1991;44:232.

65. Shekleton LE, Henning S.K. Measuring the compatibility of petroleum-based hydrocarbon resins in elastomers. *Rubber World* 2013;247(6):38–44.

66. Vleugels N, Pille-Wolf W, Dierkes WK, Noordermeer JWM. Influence of oligomeric resins on traction and rolling resistance of silica-reinforced tire treads. In: *Fall 184th Technical Meeting of the Rubber Division*, American Chemical Society, Cleveland, OH, October 8–10, 2013, Vol. 1, pp. 656–679.

67. Carvagno T. Performance resins in tire compounding. In: *184th Technical Meeting, ACS Rubber Division*, Cleveland, 2013.

68. Robertson C. Viscoelastic properties of elastomer-resin blends and relevancy to tire tread performance. In: *185th Technical Meeting, ACS Rubber Division*, Louisville 2014.

69. Class JB, Chu BS. The viscoelastic properties of rubber-resin blends, I: The effect of resin structure. *J Appl Polym Sci* 1985;30:805–814.

70. Class JB, Chu BS. The viscoelastic properties of rubber-resin blends, II: The effect of resin molecular weight. *J Appl Polym Sci* 1985;30:815–824.

71. Class JB, Chu BS. The viscoelastic properties of rubber-resin blends, II: The effect of resin concentration. *J Appl Polym Sci* 1985;30:825–842.

72. Fox TG. *Bull Am Phys Soc (Series 2)* 1956;1:123.

10 Antioxidants and Other Protectant Systems

Sung W. Hong

CONTENTS

I.   Introduction 420
II.   Antioxidants...421
      A.   Oxidative Degradation 421
      B.   Mechanisms of Antioxidants 423
           1.   Phenolic Antioxidants 423
           2.   Aromatic Amine Antioxidants...425
           3.   Hydroperoxide-Decomposing Antioxidants................................427
III. Experiments...428
      A.   Phenolic Antioxidants 428
           1.   White Sidewall Compound 429
           2.   Stabilization System for SBR 429
      B.   Aromatic Amine Antioxidants...431
           1.   Tire Casings and Other Internal Components..............................431
           2.   Bead Filler...436
           3.   Hydroperoxide-Decomposing Antioxidants................................437
IV.   Antiozonants 441
      A.   Chemical Reactivity...441
      B.   Migrations of Various Antiozonants in NR/BR and SBR Compounds.....443
           1.   Static Migration..444
           2.   Dynamic Migration..446
           3.   Measurements..447
      C.   Solubility..448
           1.   Solubilities of Various Antiozonants in SBR, NR/BR,
                and NR Compounds 448
           2.   Solubility in Water and Acid Rain 448
      D.   Conclusion...450
V.   Petroleum Waxes 451
VI.   Compounding Evaluation of Waxes 451
      A.   Evaluation of Static Ozone Resistance..451
           1.   Preparation of Compounds...452
           2.   Results and Discussion..452
           3.   Conclusion 453

VII. Vulcanization System 454
    A. Sulfur Cure System...454
    B. Peroxide Cure System 454
    C. Compounding Evaluation..455
        1. Various Cure System in Natural Rubber..455
        2. Results and Discussion 455
        3. Conclusion 457
VIII. Summary: Antioxidants and Other Protective Systems............................458
Abbreviations...458
References..459

I. INTRODUCTION

To extend the service life of vulcanized rubber goods, it is very important to protect them from oxygen, ozone, light, heat, and flex fatigue. Most natural and synthetic rubbers containing unsaturated backbones—natural rubber (NR), styrene butadiene rubber (SBR), polybutadiene rubber (BR), and nitrile rubber (NBR), for example— must be protected against oxygen and ozone. Usually, internal components that are not exposed to atmosphere require only antioxidants. However, external components that are exposed to the environment require both antiozonants and antioxidants.

Antioxidants react with oxygen to prevent oxidation of vulcanized rubber and react with free radicals that degrade vulcanized rubber.

There are four principal theories on the mechanisms of antiozonant protection of vulcanized rubber. The first is the scavenger theory, which postulates that the antiozonant competes with the rubber for ozone. The second theory is that the ozonized antiozonant forms a protective film on the surface of the vulcanized rubber, preventing further attack. The third mechanism postulated is that the antiozonants react with elastomer ozonide fragments, relinking them and essentially restoring the polymer chain. The fourth theorized mechanism suggests that Criegee zwitterions are formed from the ozonide produced.

Paraffinic and microcrystalline waxes are often added to the rubber as protective agents. Usually, waxes have poor solubility in rubbers, so they migrate to the surface of the vulcanized rubber and form a protective film or barrier that prevents ozone attack. During dynamic flexing, these barriers can be broken, exposing the rubber to ozone attack. Therefore, waxes can protect against ozone only under static conditions. Antiozonants alone do not prevent ozone cracking at the initial stage, because their migration rate is much slower than that of waxes owing to their better solubility. One theory supporting the combination of antiozonant with wax is that the wax would accelerate migration of the antiozonant to the surface for protection against ozone. Therefore, most exterior rubber goods contain both antiozonant and wax for both static and dynamic protection from ozone cracking.

Commercially available antioxidants usually protect vulcanized rubber at temperature below 120°C. Above this temperature, special polymers or types of cross-link systems provide better protection. For example, a peroxide cure system would provide carbon–carbon cross-links whose bonding energy is the strongest. Also, a monosulfuric cross-link has much higher energy than a polysulfuric bond. The lower the bonding energy, the more easily cross-links break, which would deteriorate the vulcanized rubber's physical properties after heat aging. Cure systems with a higher level of accelerators and lower sulfur levels are known as semiefficient cure systems (semi-EV cure). Such a system provides better heat aging properties than the conventional cure system, which has a smaller amount of accelerator and a higher level of sulfur. In this chapter, the protection of mechanical rubber goods through selection of antioxidants, antiozonants, waxes, and the design of the vulcanization system will be discussed.

II. ANTIOXIDANTS

A. OXIDATIVE DEGRADATION

Generally, the greater the amount of unsaturation in the polymer, the more susceptible it is to degradation. Highly unsaturated polymers can be attacked by oxygen, especially when energy is applied. The application of energy may come from heat, shear, and ultraviolet (UV) light, which promote faster oxidation.

Oxidative degradation of the polymers is a free radical process. This oxidation process, known as autoxidation, consists of three steps: initiation, propagation, and termination, as depicted in Figure 10.1.

Free radicals are formed during initiation reactions. Energy from heat, mechanical shearing, or high-energy radiation can dissociate the chemical bonds in the polymers (RH) resulting in the formation of free radicals (R˙) (reaction 10.1). In autoxidation mechanisms of polymers, the molecular reaction of oxygen with the

Initiation

$$RH \xrightarrow{\text{Energy}} R˙ \tag{10.1}$$

$$2RH + O_2 \longrightarrow [RH\cdots O_2] + RH \longrightarrow 2R˙ + H_2O_2 \tag{10.2}$$

$$ROOH \longrightarrow RO˙ + ˙OH \tag{10.3}$$

$$2\,ROOH \longrightarrow RO˙ + ROO˙ + H_2O \tag{10.4}$$

Propagation

$$R˙ + O_2 \longrightarrow ROO˙ \tag{10.5}$$

$$ROO˙ + RH \longrightarrow ROOH + R˙ \tag{10.6}$$

Termination

$$2R˙ \longrightarrow \text{Nonradical products} \tag{10.7}$$

$$R˙ + ROO˙ \longrightarrow \text{Nonradical products} \tag{10.8}$$

$$2ROO˙ \longrightarrow \text{Nonradical products} \tag{10.9}$$

FIGURE 10.1 Mechanisms of polymer degradation by oxidation under thermal energy.

polymers by thermal energy has been suggested for the initiation of the first free radicals in the polymer [1–4] (reaction 10.2). Hydroperoxide concentration builds up as the autoxidation proceeds, and consequently, decomposition of the hydroperoxide eventually becomes the dominant initiation process (reactions 10.3 and 10.4). This is usually preceded by a short period of induction.

The alkyl (R·) and alkylperoxyl (ROO·) radicals resulting from the initiation reactions are the chain-propagating species. The alkyl radicals react rapidly with atmospheric oxygen to form alkylperoxyl radicals (reaction 10.5). The alkylperoxyl radicals abstract labile hydrogens on the polymer, regenerate new alkyl radicals, and yield hydroperoxides (ROOH) as the primary oxidation product (reaction 10.6). Reactions (10.5) and (10.6) form a cycle in the propagation step. As the hydroperoxide concentration builds up, more alkoxyl and alkylperoxyl radicals are formed via decomposition of the hydroperoxide to start new cycles.

Termination occurs when two free radicals, alkyl and/or alkylperoxyl radicals, react to form the stable nonradical products. The termination reaction in the solid polymers, where oxygen is limited, usually involves two alkyl radicals (R·), which undergo recombination to form R–R or disproportionate to form saturated and unsaturated products. On the other hand, in the presence of sufficient oxygen such as in liquid hydrocarbons, the hydroperoxyl radical (ROO·) concentration is much greater than the alkyl radical concentration (R·). Chain termination occurs predominantly by reaction between two hydroperoxyl radicals.

Antioxidants are used to stabilize organic polymers. Antioxidants inhibit autoxidation by reducing the rate of autoxidation during processing, storage, and service. There are two major groups of antioxidants, commonly known as primary and secondary antioxidants. Primary antioxidants act as chain terminators, and secondary antioxidants act as hydroperoxide decomposers. The primary antioxidants remove the chain-carrying species (R· and ROO·), while the secondary antioxidants convert the hydroperoxides to nonradical species.

A schematic chain-termination mechanism is shown in Figure 10.2. In reaction (10.10), the alkylperoxyl radical abstracts the reactive hydrogen from the antioxidant (AH). The resulting antioxidant radicals (A·) are stabilized via electron delocalization. Consequently, the antioxidant radicals (A·) do not readily continue the radical chain either via hydrogen abstracting from the substrate (reaction 10.11a) or via reaction with oxygen (reaction 10.11b). The resonance structures for various antioxidant radicals derived from typical antioxidants will be illustrated in the next section. Transformation products derived from typical antioxidants will also

$$ ROO^{\cdot} \;+\; A{-}H \;\xrightarrow{\;k_1\;}\; ROOH \;+\; A^{\cdot} \qquad\qquad (10.10) $$

$$ A^{\cdot} \;+\; RH \;\xrightarrow{\;k_{2a}\;}\!\!\!\!/\;\; AH \;+\; R^{\cdot} \;\xrightarrow{\;O_2\;}\; ROO^{\cdot} \qquad (10.11a) $$

$$ A^{\cdot} \;+\; O_2 \;\xrightarrow{\;k_{2b}\;}\!\!\!\!/\;\; AOO^{\cdot} \;\xrightarrow{\;RH\;}\; AOOH \;+\; R^{\cdot} \qquad (10.11b) $$

FIGURE 10.2 Mechanisms of reaction of peroxyl radical with antioxidant and explanation of the stabilization of antioxidant radical by electron delocation, without continuing formation of radicals.

be discussed to elucidate the antioxidant mechanism. Hindered phenolics and secondary aromatic amines are the two most commonly used primary antioxidants.

Examples of secondary antioxidants that act as hydroperoxide decomposers include phosphite esters such as **I** and sulfur-containing compounds such as thioester **II**. As the hydroperoxides are removed from the organic substrates, fewer free radicals are produced via the decomposition of the hydroperoxides. Consequently, the rate of the autoxidation is reduced. The mechanism of converting hydroperoxides to non-radical species will be discussed in Section II.B.3.

B. MECHANISMS OF ANTIOXIDANTS

Commercially available antioxidants can be divided into three categories: phenolic antioxidants, aromatic amine antioxidants, and hydroperoxide-decomposing antioxidants. Each antioxidant performs in a specific way to protect polymers and rubber compounds from oxidation. The performance of each antioxidant is related to its chemical reactivity, its rate of staining or migration to the surface of vulcanized rubber or polymers, and its volatility. Therefore, it is very important to understand the mechanisms of various antioxidants before applying them for experiments.

1. Phenolic Antioxidants

Hindered phenolics, which act as chain terminators, are excellent antioxidants. Phenolic antioxidants are in general nonstaining and nondiscoloring. Many of them are approved for use in food packaging.

A simplified mechanism commonly used to show the process of chain termination by a hindered phenolic is shown in Figure 10.3. The alkylperoxyl radicals (ROO$^\bullet$) abstract the reactive hydrogen from the phenolics. The resulting phenoxy radical **1** is stabilized through electron delocalization as indicated by the resonance

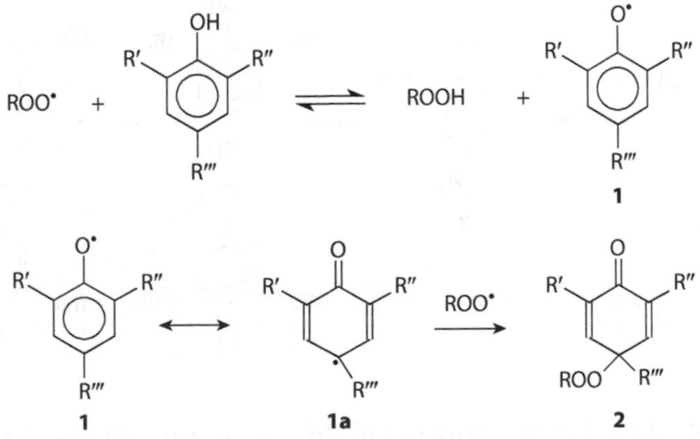

FIGURE 10.3 Chain termination by hindered phenolics.

structures **1** and **1a**. Reaction of the alkylperoxyl radical with **1a** produces the nonradical product **2**.

More detailed information about the antioxidant mechanism was made possible with the identification of the transformation products. The mechanism depicted in Figure 10.4 illustrates the formation of transformation products using 2,6-di-*t*-butylhydroxytoluene (BHT) as an example. The alkylperoxyl radical attacks the reactive hydrogen from the BHT, yielding the phenoxy radical **3**, which is stabilized via delocalization to the carbon-centered radical **3a**. Bimolecular reaction between **3** and **3a** would produce **4** and **5**. Reaction of **3a** with the alkylperoxyl radical forms **6**, which thermally decomposes to the *p*-quinone **7**. The phenoxy radical **3a** would also be the precursor for the formation of compounds **9** and **10** [6].

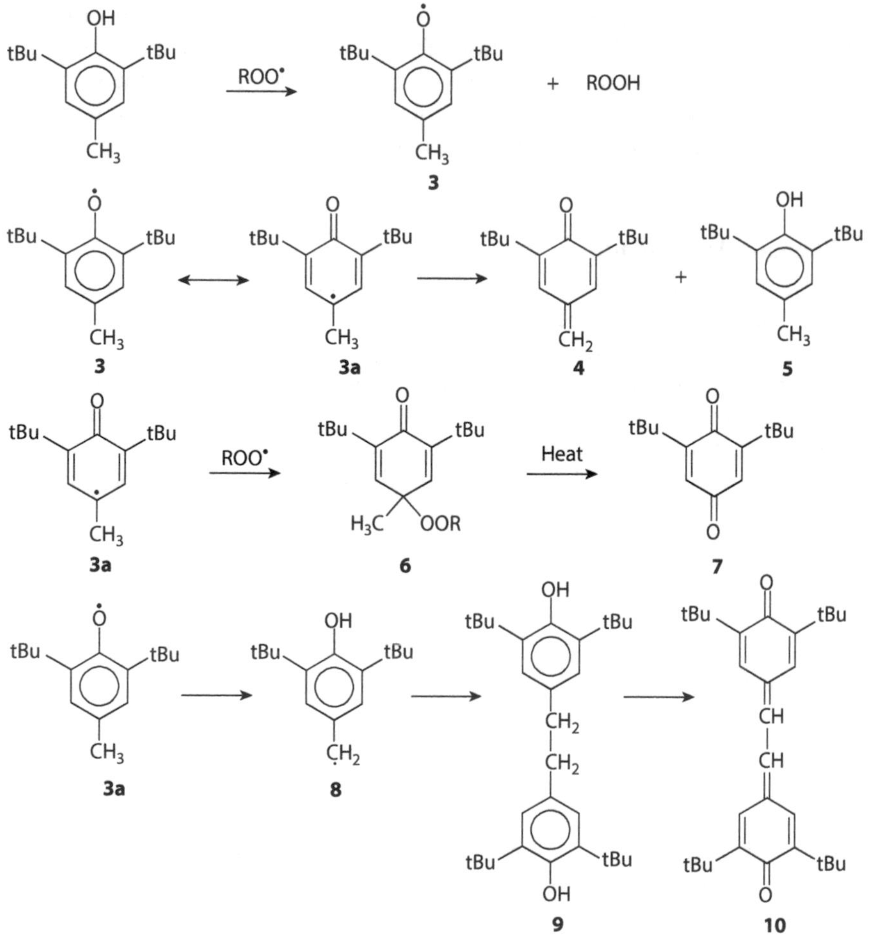

FIGURE 10.4 Antioxidation mechanism of BHT (illustrated with some transformation products).

2. Aromatic Amine Antioxidants

Amine antioxidants in general are better antioxidants than phenolic antioxidants. However, most amine antioxidants are discoloring and staining and have limited approval for food contact use. The mechanism of chain termination by secondary aromatic amines is shown in Figure 10.5, with N,N'-dialkyldiphenylamines as an example [5,6]. The alkylperoxyl radicals abstract the reactive hydrogen (N–H) from the N,N'-dialkyldiphenylamines. The resulting aminyl radical **11** is stabilized through electron delocalization as indicated by the resonance structures **11**, **11a**, and **11b**. Reaction of the alkylperoxyl radical with **11b** produces a nonradical product **12**. Reaction of **11** with the primary alkylperoxyl radicals leads to a stable nitroxyl radical **13** that is capable of trapping the free alkyl radical and producing the stable product, alkoxyamine **14**. Reactions of the aminyl radical **11** with secondary and tertiary alkylperoxyl radicals produce the hydroxylamine **15** and the nitroxyl radical **13**, respectively [17,18].

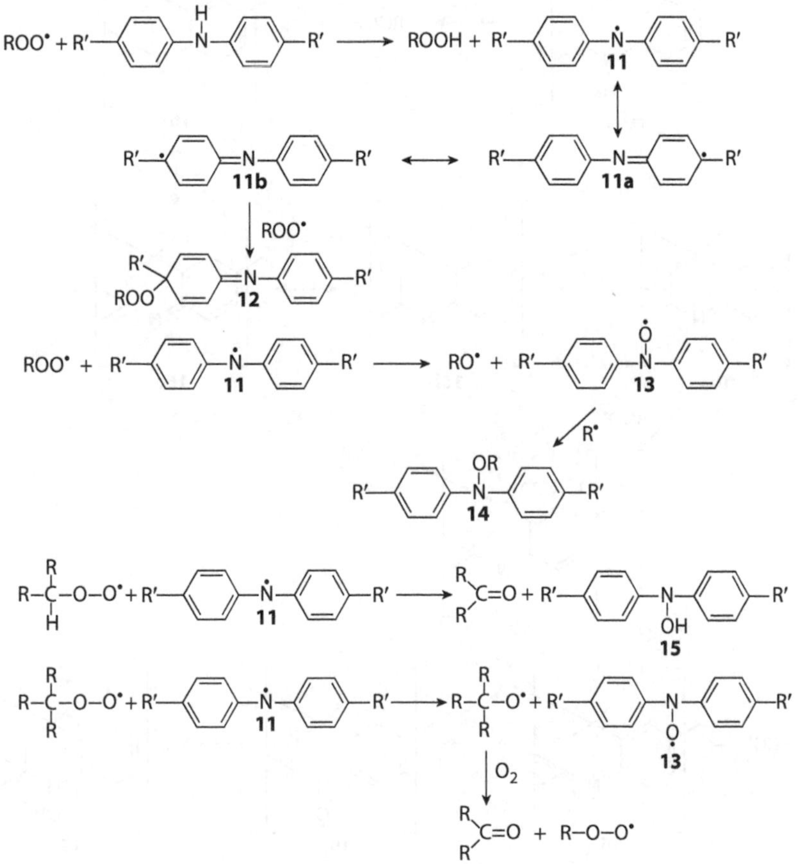

FIGURE 10.5 The chain-termination mechanism by N,N'-dialkyldiphenylamines.

Another commonly used secondary amine is polymerized 1,2-dihydro-2,2,4-trimethylquinoline (TMQ). A mechanistic illustration with the resonance structures and the transformation products is shown in Figure 10.6. The alkylperoxyl radicals abstract the reactive hydrogen (N–H) from TMQ. The resulting aminyl radical **16** is stabilized through electron delocalization as indicated by the resonance structures **16a**, **16b**, and **16c**. However, the transformed radical **16a** would be capable of trapping alkylperoxyl radicals, leading to the nonradical product **17**. The aminyl radical **16** would also trap alkylperoxyl radicals to form the nitroxyl radical **18**, which in turn traps alkyl radicals to form the alkoxyamine **19**.

N,N'-Dialkylated p-phenylenediamines are excellent antioxidants. A mechanism that illustrates chain termination by the N,N'-diphenyl-p-phenylenediamine (DAPD) **20** is shown in Figure 10.7 [6,7]. Two alkylperoxyl radicals attack the two reactive hydrogens from N-phenyl-N'-alkyl-p-phenylenediamine, yielding the

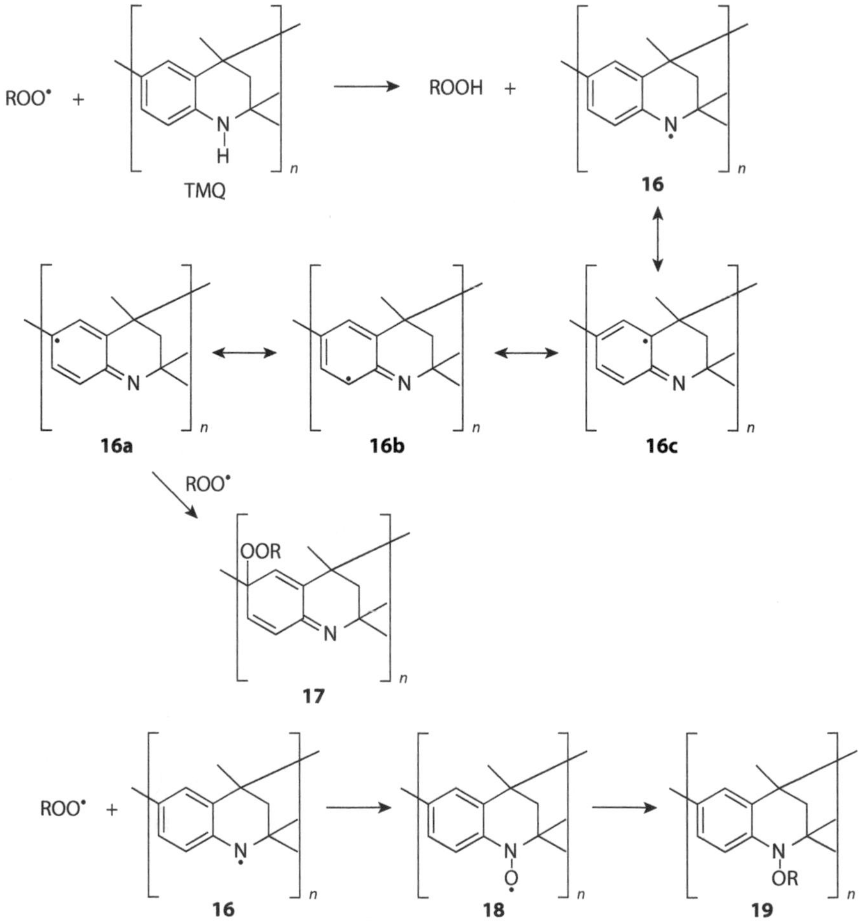

FIGURE 10.6 Antioxidation mechanism illustrated with the resonance structures and the transformation products from TMQ.

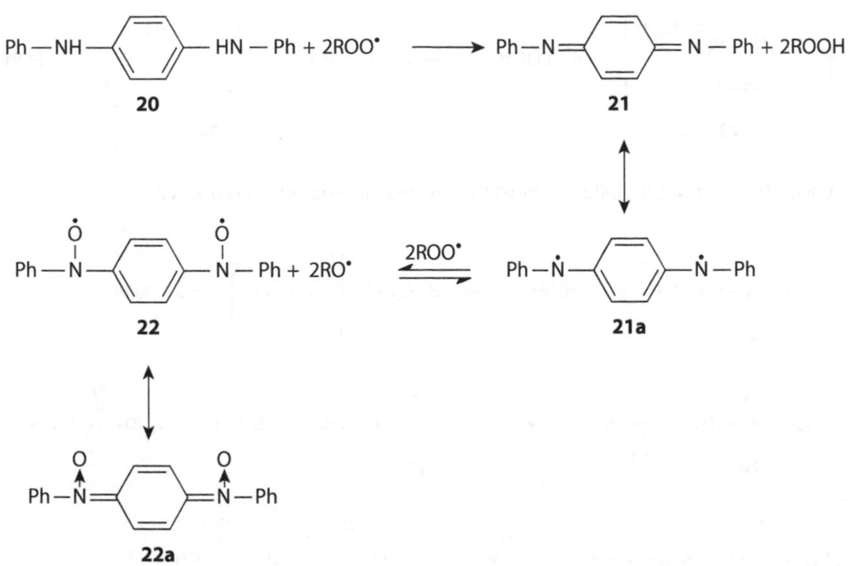

FIGURE 10.7 Chain termination by DAPD.

quinonediiamine **21**, which is stabilized through electron delocalization as shown by the resonance structure **21a**. Further reaction of the aminyl diradical **21a** with two alkylperoxyl radicals produces the dinitroxyl radical **22**, which is converted via electron delocalization to the dinitrone **22a** [12,15].

3. Hydroperoxide-Decomposing Antioxidants

Hydroperoxide-decomposing antioxidants reduce the rate of chain initiation by converting hydroperoxide, ROOH, into nonradical products. Two major classes of the hydroperoxide-decomposing antioxidants are organic phosphite esters and sulfides. During the reaction with hydroperoxide, the hydroperoxide is reduced to alcohol (ROH) and the phosphite and sulfide antioxidants are oxidized to phosphates and sulfoxides, respectively (Figure 10.8). Further transformation of the sulfoxides has been reported [8]. In general, phosphites decompose hydroperoxides at substantially lower temperatures than the sulfides. The sulfide antioxidants are active at temperatures exceeding 100°C but are not active at ambient temperatures [9].

FIGURE 10.8 Oxidation of phosphites and sulfides by hydroperoxide.

FIGURE 10.9   Tris(nonylphenol)phosphite as hydroperoxide decomposer.

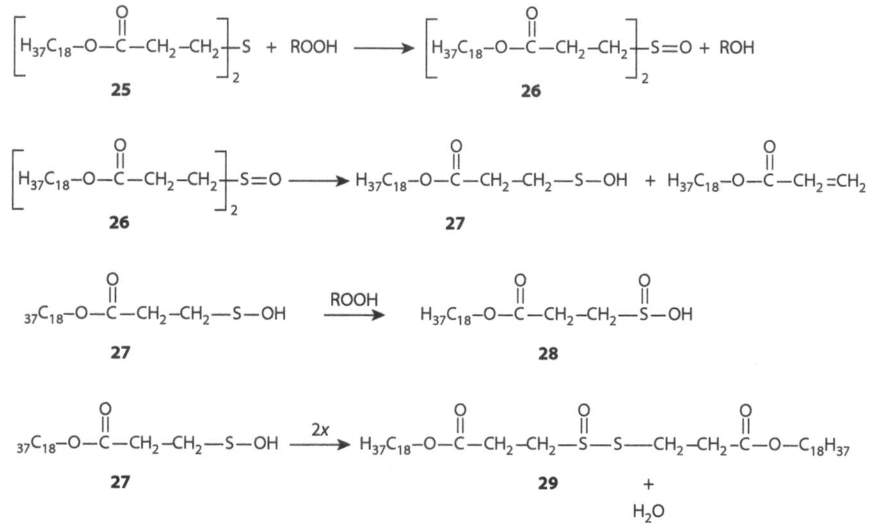

FIGURE 10.10   Distearyl thiopropionate as a hydrogen peroxide decomposer shown with certain transformation products.

An example of commonly used phosphite antioxidants is tris(nonylphenol) phosphite **23**. Its hydroperoxide-decomposing mechanism is depicted in Figure 10.9. A stable trialkyl phosphoric acid ester **24** and an alkyl alcohol are formed.

Thioesters are often used in combination with phenolics and are more widely used in thermoplastics, where sulfur will not interfere in the vulcanization process [10]. A mechanistic explanation of a thioester as a hydroperoxide decomposer is shown in Figure 10.10 with distearyl thiopropionate **25** as an example. A stable sulfoxide **26** of the thioester and an alkyl alcohol are formed. Thermal decomposition of the sulfoxide **26** produces sulfenic acid **27** and stearyl acrylate. Oxidation of sulfenic acid **27** by hydroperoxide yields the sulfinic acid **28**. The sulfenic acid **27** could also undergo a bimolecular reaction to produce thiosulfinate ester **29** [11].

III.   EXPERIMENTS

A.   PHENOLIC ANTIOXIDANTS

Usually, phenolic antioxidants are used in white sidewall compounds due to their nonstaining and nondiscoloring properties. They are also used as the stabilization

TABLE 10.1

Recipe for White Sidewall Compounds

Component[a]	MB	A-1-1	A-1-2
NR	35		
EPDM	15		
CIIR	50		
Titanium dioxide	40		
Hydrated aluminum silicate	20		
Stearic acid	1.0		
SP-1068	3.0		
Ultramarine blue	0.2		
Blended wax	3.0		
Hindered bisphenol			1.5
Zinc oxide		15	15
TBBS		1.5	1.5
Alkyl phenol disulfide (APD)		1.0	1.0
80% Insoluble sulfur		1.0	1.0

[a] EPDM, ethylene propylene diene terpolymer; CIIR, chlorobutyl rubber; SP-1068, alkyl phenol formaldehyde resin; TBBS, N-t-butyl-2-benzothiazole sulfenamide.

system for nonstaining SBR or other polymers along with phosphite antioxidants, which are hydroperoxide-decomposing antioxidants. Two examples are introduced in this section.

1. White Sidewall Compound

One master batch for a white sidewall without an antioxidant and curatives (accelerators and sulfur) was prepared. This master batch was finalized by adding only curatives (A-1-1) and a hindered bisphenol antioxidant with curatives (A-1-2), respectively, which are presented in Table 10.1. Their Mooney viscosity at 100°C, Mooney scorch at 132°C, unaged and aged physical properties, DeMattia flexing, and weatherometer tests were measured. The results indicated that slight improvements in flex fatigue, color retention, and retention of physical properties were achieved with the addition of 1.5 phr (parts per hundred rubber) of a hindered bisphenol antioxidant (Table 10.2). Therefore, tire manufacturers use 1–2.0 phr of hindered bisphenol in white sidewall and cover strip compounds to ensure improved performance.

2. Stabilization System for SBR

SBR latex (type 1502) was prepared for evaluating phosphite in comparison with a blend of phosphite and hindered bisphenol. Antioxidants were added as an emulsion and coagulated by addition of latex to $Al_2(SO_4)_3/H_2SO_4$, whose pH was controlled to pH 3 for 30 min. The crumb was washed twice at 50°C and dewatered on a two-roll mill. The milled sheet was dried at 40°C–50°C. The sample was molded to form

TABLE 10.2
Physical Properties of White Sidewall Compounds A-1-1 and A-1-2[a]

Component	A-1-1	A-1-2
Phenolic antioxidant		1.5
ML 1 + 4 at 100°C	32	31
MS at 132°C 3 pt rise time, min	12.6	13.1
Cured 15 min at 160°C		
Tensile strength at RT, MPa	17.4	16.0
% Elongation	610	650
300% Modulus	6.1	5.5
Shore A hardness	52	51
Aged 1 month at 70°C, % retention		
Tensile strength	80.5	87.57
Elongation	90.0	95.1
300% Modulus	121.0	120.0
Increase in hardness	+3	+4
DeMattia flexing, kilocycles to failure	452.4	499.4
Weatherometer for 1 week, color	Slightly yellow	Very slightly yellow

[a] See Table 10.1.

TABLE 10.3
Aging Test: Delta Mooney Viscosity at 100°C

| Time (Weeks) | Additive[a] | | |
	None	0.7 phr TNPP	0.7 phr TNPP/AO
0	0	0	0
1	8	3	4
2	17	8	5
3	25	10	7
4	35	18	14

[a] TNPP, tris(mono- and dinonylphenyl)phosphite; AO, 2,2′-methylenebis(4-methyl-6-nonylphenol).

a 1 cm thick plaque and oven aged at 70°C. The Mooney viscosity at 100°C was measured over a period of 4 weeks of aging, with the results as shown in Table 10.3. These results indicated a slight improvement with the blend system. However, longer aging is necessary to differentiate these two antioxidant systems. The blend system used phosphite and hindered bisphenol antioxidants in a 5:1 ratio. The phosphite antioxidant was tris(mononnonylphenol)phosphite (TNPP), and the hindered bisphenol antioxidant was 2,2-methylenebis(4-methyl-6-nonylphenol).

B. AROMATIC AMINE ANTIOXIDANTS*

Secondary aromatic amines such as TMQ, a high-temperature reaction product of diphenylamine and acetone (BLE), and 4,4'-bis(α,α'dimethylbenzyl)diphenylamine (AO 445), produced by Crompton Corporation, are commonly used as antioxidants in tire compounds. These three antioxidants nonetheless provide different antifatigue efficiencies for rubber compounds. This section reports the evaluation of these three antioxidants in carcass compounds to explain the differences in performance.

1. Tire Casings and Other Internal Components

Tire body or carcass rubber compounds must form strong and durable bonds to the coated fabric with resorcinol formaldehyde latex dipped solution. Usually, polyester, rayon, or nylon fibers are used to make cords for tire casings, except for steel monoply truck tires. Tire casing strength and durability should be sufficient to insulate the tire cords and hold them in place. A tire casing compound must, however, be soft enough to permit a slight change in cord angles when the tire is flexed. The body rubber serves as insulation between the fabric plies. Outstanding fatigue resistance and heat aging resistance are required of the carcass compounds in order to withstand cyclic deformation.

Currently, most tire companies use TMQ for protecting carcass compounds. In this experiment, BLE, TMQ, AO 445, and a BLE/TMQ blend were evaluated in an NR/BR/SBR carcass compound, which is presented in Table 10.4.

To minimize experimental variations, one master batch without antioxidants and curatives was prepared in a 10 L internal mixer. Using the same master batch, antioxidants and curatives were added in a 1 L Banbury mixer. Mooney viscosity at 100°C, Mooney scorch, and curometer at 177°C of all five compounds were measured (Table 10.5). No significant differences were obtained. The results indicated that BLE, the BLE/TMQ blend, and AO 445 would provide better flex fatigue properties, and TMQ and AO 445 the best heat protection. The blend of TMQ and BLE is not only better in heat aging than BLE alone but also improves flex fatigue over TMQ itself. We would like to explore the difference in the reaction mechanisms for the antifatigue and antioxidation. A rationale is also proposed to explain why these three amines perform differently under various conditions. The chemical structures of TMQ, BLE, and AO 445 are shown in Figures 10.11 through 10.13.

The fatigue phenomenon of elastomers at room temperature is a degradation process caused by the shear of repeated mechanical stress under limited access to oxygen. The mechanical shear generates macroalkyl radicals (R'). A small fraction of the macroalkyl radicals react with oxygen to form alkylperoxyl radicals, with a high concentration of macroalkyl radicals remaining. Consequently, removal of the macroalkyl radicals in a catalytic process is the prevailing antifatigue process [1].

On the other hand, the macroalkyl radicals are rapidly converted to alkylperoxyl radicals by aging in air at oven temperatures. The autoxidation propagated by the

---

* Based on Ref. [12].

TABLE 10.4
Tire Ply Compound

Master Batch 1[a] (MB-1)

NR	60
BR1205	20
SBR 1707	27
N-660	50
Zinc oxide	4
Naphthenic oil	7.5
Stearic acid	1.5
Octylphenol formaldehyde	2.0
R-6	2.0
Total	174.0

Blend

Component[b]	B-1-1	B-1-2	B-1-3	B-1-4	B-1-5
MB-1	174	174	174	174	174
TMQ		1.0			0.5
BLE			1.0		0.5
AO 445				1.0	
M3P	1.0	1.0	1.0	1.0	1.0
MBTS	1.2	1.2	1.2	1.2	1.2
DPG	0.25	0.25	0.25	0.25	0.25
80% Insoluble sulfur	3.00	3.00	3.00	3.00	3.00

Source: Hong, S.W. and Lin, C.Y., Improved flex fatigue and dynamic ozone crack resistance through the use of antidegradants in tire compounds, Presented at the *156th Meeting of ACS Rubber Division*, Orlando, FL, September 21–24, 1999.

[a] R-6, resorcinol resin.

[b] TMQ, polymerized 1,2-dihydro-2,2,4-trimethylquinoline; BLE, high-temperature reaction product of diphenylamine and acetone; AO 445, 4,4'-bis(α,α'-dimethylbenzyl)diphenylamine; M3P, 1-aza-5-methylol-3,7-dioxabicyclo[3,3,0]octane; MBTS, benzothiazyl disulfide; DPG, diphenylguanidine.

---

alkylperoxyl radicals thus dominates the degradation process. Therefore, removal of the alkylperoxyl radicals becomes the primary function of an antioxidant.

It has been shown that the diarylamines (Figure 10.14, **II**) are good antifatigue agents and that diarylamine nitroxyl radicals (Figure 10.14, **I**) are even more effective than the parent amines [13]. The antifatigue mechanism of the amine antidegradants shown in Figure 10.15 has been proposed, with the formation of the intermediate nitroxyl radicals playing an active role [13]. Generation of the nitroxyl radicals **I** from the free amines **II** is depicted in Figure 10.14. In the fatiguing process (Figure 10.15), macroalkyl reaction (10.1), removal of the macroalkyl radicals by the nitroxyl radicals is shown in reactions (10.2) and (10.3). The resulting hydroxylamine **III** can be reoxidized by alkylperoxyl to regenerate the nitroxyl radicals in an autoxidation chain-breaking process (reaction 10.4). The nitroxyl radicals **I** can be partially

TABLE 10.5
Physical Properties of Five Blends[a]

	B-1-1	B-1-2	B-1-3	B-1-4	B-1-5
Mooney viscosity at 100°C	49	49	48	49	48
Mooney scorch at 132°C					
3 Pt rise time (min)	11.9	11.6	11.6	11.8	11.4
Curometer at 177°C					
ML, N m	0.40	0.40	0.41	0.41	0.40
MH, N m	3.45	3.43	3.49	3.48	3.45
t1, min	0.93	0.94	0.95	0.97	0.93
tc50, min	1.71	1.72	1.74	1.74	1.73
tc90, min	2.96	3.00	3.05	3.04	3.03
Cured 10 min at 177°C					
Tensile at RT, MPa	19.5	18.8	19.1	16.7	19.0
% Elongation	450	460	470	430	460
300% Modulus, MPa	10.50	10.2	10.1	10.5	10.3
Shore A hardness	57	56	55	54	55
Tear strength (die C) kN/m	40.3	43.8	40.8	45.5	43.8
Aged 2 weeks at 70°C					
Tensile, % retention	63	79	74.4	81	77.7
Elongation, % retention	53	65	62.7	60	66
Change in shore A hardness	+5	+4	+6	+6	+5
Tear strength, % retention	61	68	71	70	69
DeMattia flex					
Kilocycles to full cracking unaged	133	361	603	460	538
Aged 70 h at 100°C	25	80.5	102	90.5	105
Monsanto flex fatigue					
Kilocycles to failure unaged	Not tested	68.2	110.0	79.0	98.0
Aged 70 h at 100°C	Not tested	11.8	12.0	16.0	15.5

Source: Hong, S.W. and Lin, C.Y., Improved flex fatigue and dynamic ozone crack resistance through the use of antidegradants in tire compounds, Presented at the *156th Meeting of ACS Rubber Division*, Orlando, FL, September 21–24, 1999.

[a] Blend recipes given in Table 10.4. ML, minimum torque; MH, maximum torque.

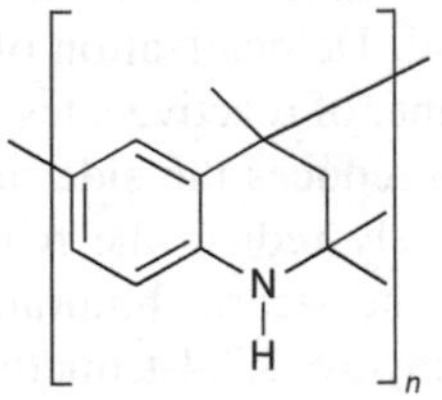

FIGURE 10.11   Structure of TMQ (polymerized 1,2-dihydro-2,2,4-trimethylquinoline).

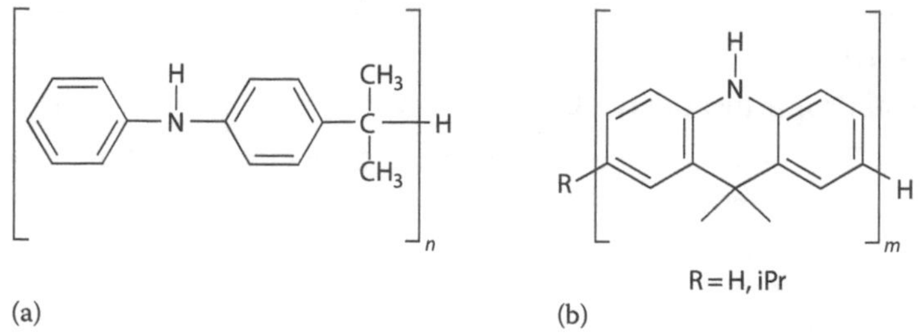

(a)   (b)

R = H, iPr

FIGURE 10.12 Structures for major components of BLE (a high-temperature reaction product of (a) acetone and (b) diphenylamine).

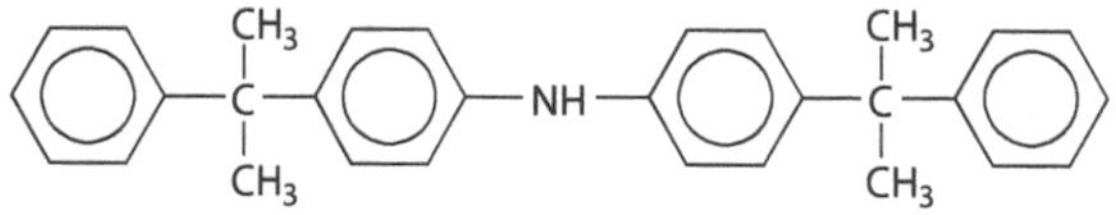

FIGURE 10.13 Structure of AO 445 (4,4′-bis(α,α′-dimethylbenzyl)diphenylamine).

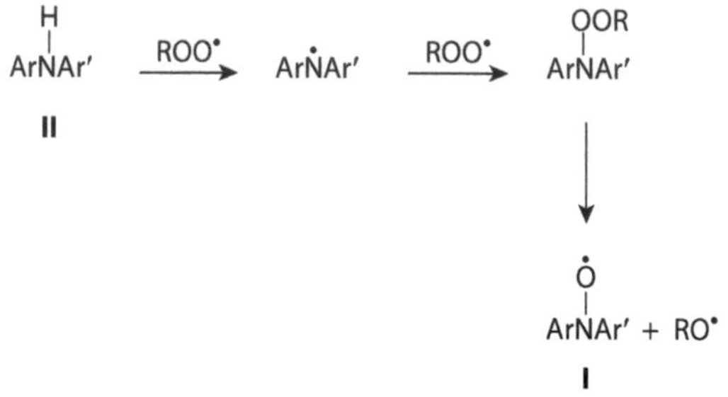

FIGURE 10.14 Formation of nitroxyl radical.

converted back to the free diarylamine **II** during vulcanization through the reductive action of thiyl radicals of thiols (reaction 10.5). Radicals are generated. The free diarylamine **II** thus regenerated would repeat the steps shown in Figure 10.14 to form more nitroxyl radicals **I**.

Reactivity of the nitroxyl radicals is affected by delocalization, stearic hindrance, and substitutions [14]. Delocalization of the unpaired electron into the aromatic ring increases the number of reactive sites and decreases its stability. Ring substitution at the para position reduces the side reaction of the nitroxyl radicals. Stearic hindrance would obviously reduce the reactivity of the nitroxyl radicals as an antifatigue agent. Thus, the stearic hindrance in the TMQ-nitroxyl radical (TMQ, polymerized 1,2-dihydro-2,2,4-trimethylquinone) (**IV** Figure 10.16) reduces its reactivity in intercepting the macroalkyl radicals (e.g., Equation 10.3); consequently, the TMQ-nitroxyl radical **IV** is less efficient as an antifatigue agent

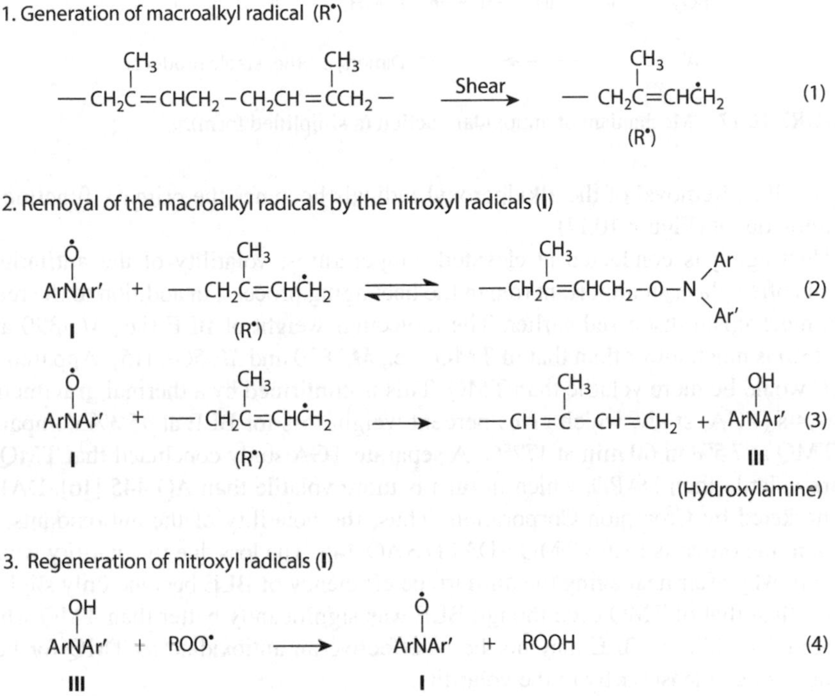

1. Generation of macroalkyl radical (R•)

$$-CH_2C(CH_3)=CHCH_2-CH_2CH=C(CH_3)CH_2- \xrightarrow{\text{Shear}} -CH_2C(CH_3)=CH\overset{\bullet}{C}H_2 \quad (1)$$

(R•)

2. Removal of the macroalkyl radicals by the nitroxyl radicals (I)

$$Ar\overset{\overset{\bullet}{O}}{N}Ar' + -CH_2C(CH_3)=CH\overset{\bullet}{C}H_2 \rightleftharpoons -CH_2C(CH_3)=CHCH_2-O-N\overset{Ar}{\underset{Ar'}{}} \quad (2)$$

I   (R•)

$$Ar\overset{\overset{\bullet}{O}}{N}Ar' + -CH_2C(CH_3)=CH\overset{\bullet}{C}H_2 \longrightarrow -CH=C(CH_3)-CH=CH_2 + ArNAr'(OH) \quad (3)$$

I   (R•)   III

(Hydroxylamine)

3. Regeneration of nitroxyl radicals (I)

$$Ar\overset{OH}{N}Ar' + ROO^{\bullet} \longrightarrow Ar\overset{\overset{\bullet}{O}}{N}Ar' + ROOH \quad (4)$$

III   I

FIGURE 10.15 Antifatigue mechanism.

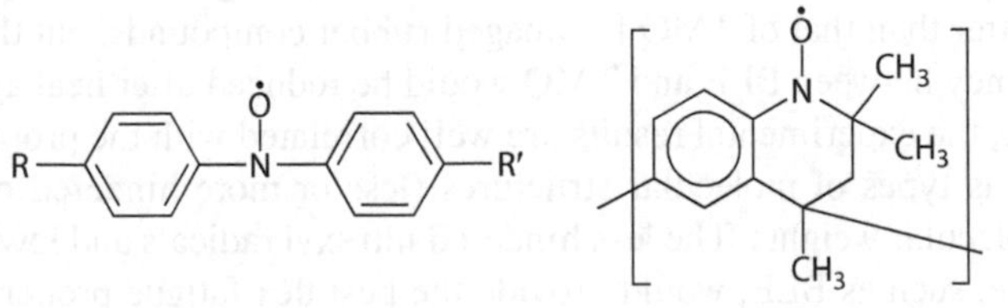

FIGURE 10.16 Nitroxyl radicals. The less sterically hindered nitroxyl radical **I** is more reactive than the more sterically hindered **IV**.

than the less hindered diarylamine nitroxyl radical **I**. Thus, the free amine precursors of the nitroxyl radical **I** would be more efficient antifatigue agents than the free amine precursors of the nitroxyl radical **IV** shown in Figure 10.16. Therefore, the diarylamines in general, such as BLE and AO 445, are more efficient antifatigue agents than TMQ. This argument is in agreement with the data given in a previous report [13].

Generation of the alkylperoxyl radicals is rapid during air–oven heat aging of rubbers, in contrast to the fatiguing process. The alkylperoxyl radicals propagate an autoxidation degradation process. The autoxidation mechanism is depicted in

$$RO_{2'} \quad + \quad AH \quad \longrightarrow \quad ROOH \quad + \quad A'$$

$$A' \quad \longrightarrow \quad \text{Dimers, ... other stable products}$$

FIGURE 10.17 Mechanism of antioxidant action (a simplified form).

Figure 10.1. Removal of the alkylperoxyl radicals becomes the primary function of an antioxidant (Figure 10.17).

Heat aging is conducted at elevated temperatures; volatility of the antifatigue agents often plays an important role in the heat aging process, in addition to the reaction mechanism discussed earlier. The molecular weight of BLE (i.e., M_w 390 and M_n 230) is much lower than that of TMQ (i.e., M_w 820 and M_n 560) [15]. Apparently, BLE would be more volatile than TMQ. This is confirmed by a thermal gravimetric analysis (TGA) study, which gives percent weight loss for BLE at 45.3% compared to TMQ at 7.5% in 60 min at 177°C. A separate TGA study concluded that TMQ is more volatile than DAPD, which in turn is more volatile than AO 445 [16]. DAPD is marketed by Crompton Corporation. Thus, the volatility of the antioxidants, in descending order, is BLE > TMQ > DAPD > AO 445. The loss due to volatility would explain why after heat aging the antifatigue efficiency of BLE became only slightly better than that of TMQ even though BLE was significantly better than TMQ when unaged. In addition, BLE may not be as effective an antioxidant as TMQ for heat aging owing at least partly to the volatility.

Based on the antifatigue mechanism and the volatility, we came to the following conclusion. For fatigue protection of unaged rubber compounds, AO 445 would be better than TMQ. After heat aging, the efficiency of the AO 445 would become much better than that of TMQ. On the other hand, the antifatigue efficiency of the BLE would be better than that of TMQ for unaged rubber compounds, but the difference in the efficiency between BLE and TMQ would be reduced after heat aging.

Therefore, the experimental results are well correlated with the proposed mechanisms, such as types of molecular structures (less or more hindered nitroxyl radicals) and molecular weights. The less hindered nitroxyl radicals and lower molecular weight amine, such as BLE, would provide the best flex fatigue property, while the higher molecular weight amines, such as AO 445 and TMQ, would provide better heat aging property.

2. Bead Filler

To confirm our proposed mechanisms for DAPD and *N*-1,3-dimethylbutyl-*N'*-phenyl-*p*-phenylenediamine (6PPD), they were evaluated in bead filler compound along with TMQ and BLE.

In both passenger vehicle and truck radial ply tires, a stiff lower sidewall construction is very important for handling performance. The stiffness controls the tire's movement at elevated speeds to provide improved handling and cornering. The tire manufacturers continue to develop higher hardness bead filler compounds for radial tires. The current bead filler compounds are highly filled with carbon black with increased cross-link density. The highly filled bead filler compound will cause problems in mixing and extrusion because of its high Mooney viscosity. Several resin

TABLE 10.6
Bead Filer Compound[a]

NR	100
SP-6700	10
N-351 black	55
Aromatic oil	5
Zinc oxide	10
Stearic acid	2
SP-1068	2
Antidegradant	2
M3P	2
TBBS	0.6
TBzTD	0.25
CPT	0.25
80% Insoluble sulfur	5.0

Source: Hong, S.W. and Lin, C.Y., Improved flex fatigue and dynamic ozone crack resistance through the use of antidegradants in tire compounds, Presented at the *156th Meeting of ACS Rubber Division*, Orlando, FL, September 21–24, 1999.

[a] SP-6700, oil-modified phenol formaldehyde two-step resin; SP-1068, alkylphenol formaldehyde resin; M3P, 1-aza-5-methylol-3, 7-dioxabicyclo[3.3.0] octane; TBBS, *N-t*-butyl-2-benzothiazole sulfenamide; TBzTD, *N,N,N',N'*-tetrabenzylthiuram disulfide; CPT, *N*-(cyclohexylthio)phthalimide.

manufacturers have developed oil-modified phenol formaldehyde two-step resin (SP-6700) to meet the tire manufacturers' requirements such as lower Mooney viscosity and higher hardness. The typical bead filler compound consists of 100% NR, which requires protection against heat and flexing with antioxidants.

Five batches were prepared using TMQ, BLE, DAPD, and 6PPD along with a blank compound. Ten parts per hundred rubber of SP-6700 resin was added to 100% NR compound as listed in Table 10.6. Mooney viscosity at 100°C and Mooney scorch value at 132°C were determined. Curometer at 177°C and unaged and aged physical properties were determined, and unaged and aged DeMattia testing was run. The compounds were cured 10 min at 177°C, which simulated vulcanization of passenger car radial tires. There were no significant differences in Mooney viscosity, Mooney scorch, and unaged physical properties (Table 10.7). However, significant improvement of unaged DeMattia flex was obtained by the addition of BLE, DAPD, or 6PPD (Table 10.7), and heat aging properties were improved with less volatile antidegradants such as DAPD, 6PPD, or TMQ. These results also agree with our proposed mechanisms of antioxidants.

3. Hydroperoxide-Decomposing Antioxidants

Phosphite antioxidants are usually used for nonstaining polymer stabilization systems along with phenolic antioxidants and sulfides because these antioxidants are reduced from ROOH to RH or ROH, which is more stable. Ciba Geigy developed

TABLE 10.7
Physical Properties

	B-2-1	B-2-2	B-2-3	B-2-4	B-2-5
TMQ		2			
BLE			2		
DAPD				2	
6PPD					2
Mooney viscosity at 100°C	51	49	49	52	50
Mooney scorch at 132°C					
3 pt rise time	16.0	16.4	15.9	15.1	14.9
Curometer at 177°C					
t1, min	1.05	1.04	1.06	0.99	0.95
tc50, min	2.50	2.51	2.55	2.47	2.40
tc90, min	5.10	5.09	5.13	4.93	4.88
ML	0.34	0.34	0.35	0.38	0.37
MH	3.40	3.31	3.11	3.21	3.15
Cured 10 min at 177°C					
Tensile at RT, MPa	16.80	16.20	16.00	17.70	17.50
% Elongation	320	350	320	420	430
300% Modulus, MPa	15.20	13.8	14.60	13.50	13.40
Shore A hardness	85	85	84	87	86
Aged 2 days at 100°C					
Tensile, % retention	43	68	59	74	73
Elongation, % retention	25	34	31	33	35
Shore A hardness change	+5	+4	+3	+1	+2
Aged 2 weeks at 70°C					
Tensile, % retention	60	77	73	79	78
Elongation, % retention	31	41	37	42	43
Change in shore A hardness	+6	+4	+3	+2	+3
DeMattia flex test					
Kilocycles to failure unaged	5.8	41.1	83.5	85.5	86.0
Aged 2 weeks at 70°C	0.5	11.7	11.7	21.5	22.5

Source: Hong, S.W. and Lin, C.Y., Improved flex fatigue and dynamic ozone crack resistance through the use of antidegradants in tire compounds, Presented at the *156th Meeting of ACS Rubber Division*, Orlando, FL, September 21–24, 1999.

2,4-bis[(octylthio)methyl]-*o*-cresol (CG 1520) for a polymer stabilization system [17]. In this section, phosphite, a phosphite-phenolic blend, and CG 1520 are evaluated in *cis*-BR and SBR, respectively.

    Stabilization system for cis-BR: Butadiene cement was diluted 1:1 by volume in *n*-hexane and coagulated with water at 80°C by adding emulsified stabilizers. It was dewatered and dried at 40°C. The dried crumb was then compression molded at

85°C for 10 min. Then the pressed samples were aged at 80°C until the formation of gel reached 2% or higher.

The results indicated that the blends of phosphite and phenolic antioxidants or CG 1520, which has both phenolic and sulfide antioxidant functions, are much better than phenolic antioxidants (shown in Figure 10.18).

Stabilization system for SBR: The same procedure as described in Section III.A.2 was adopted for evaluating phosphite, blends of phosphate–phenol antioxidants, and CG 1520. The results are shown in Table 10.8. The results indicated that CG 1520 is superior to both tris(mono- and dinonylphenyl) phosphite (TNPP) and its blend with 2,6-di-*t*-butyl-4-methylphenol (BHT).

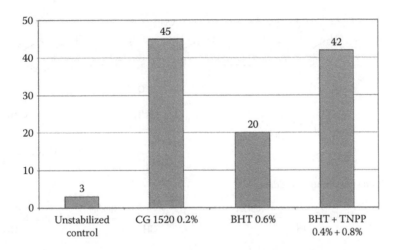

FIGURE 10.18 Stabilization of low-*cis* polybutadiene and the influence of antioxidants on oven aging performance. CG 1520, 2,4-bis[(octylthio)methyl]-*o*-cresol; BHT, di-*t*-butyl-4-methylphenol; TNPP, tris(mono- and dinonylphenyl)phosphite. Scale at left is for days aging at 80°C until a 2% gel is formed. (From Bruck, D. et al., *Kautsch Gummi Kunstst*, 44(11), 1014, 1991.)

TABLE 10.8
Aging Test: Change in Mooney Viscosity at 100°C

	Additive			
Time (Weeks)	None	0.7 phr TNPP	0.7 phr TNPP/AO	0.5 phr CG 1520
0	0	0	0	0
1	9	3	4	2
2	17	8	5	4
3	25	10	7	6
4	35	18	14	9

Source: Bruck, D. et al., *Kautsch Gummi Kunstst*, 44(11), 1014, 1991.

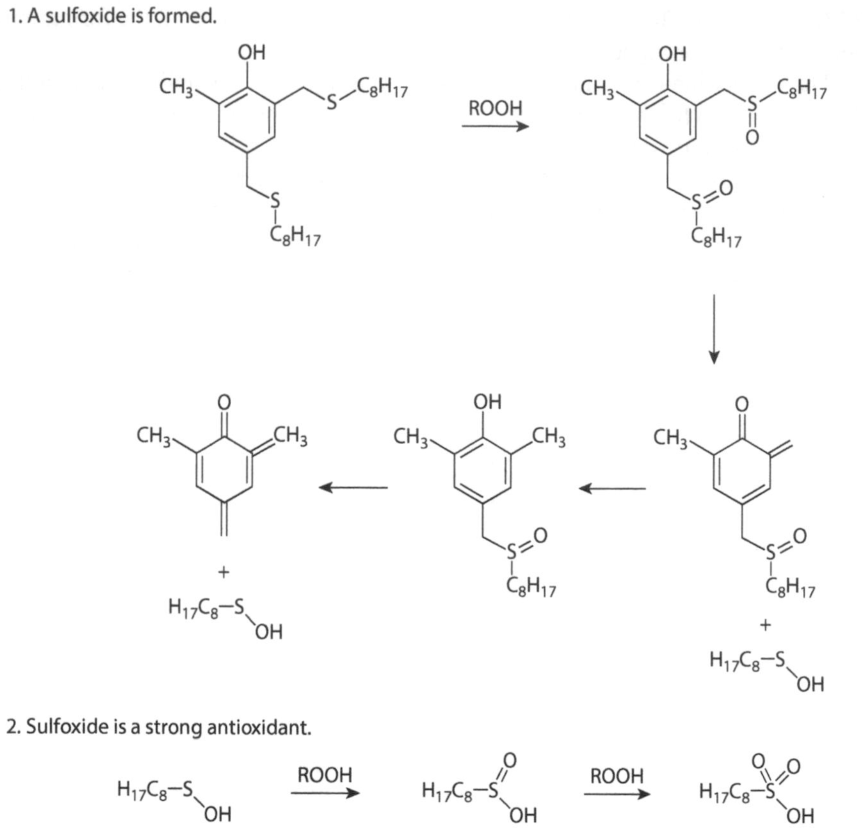

1. A sulfoxide is formed.

2. Sulfoxide is a strong antioxidant.

FIGURE 10.19   Mechanism of hydroperoxide decomposition by CG 1520.

The mechanism for both phenolic and sulfidic activity is shown in Figure 10.19. First, a sulfoxide is formed by reaction with ROOH to reduce to ROH, which is more stable. Sulfoxide is a strong antioxidant that reacts with ROOH again to reduce to a more stable form of $R(SO_2)OH$. Therefore, one molecule is twice or more as effective as either phosphite or phenolic antioxidant alone.

Thioesters are used in thermoplastics.

Stabilization system for SBR. In another experiment, SBR latex (type 1502) was prepared for evaluating phosphite in comparison with a phosphate–phenolic blend. Antioxidants were added as an emulsion and coagulated by addition of latex to $Al_2(SO_4)_3/H_2SO_4$ with the pH controlled to pH 3 for 30 min. The crumb was washed at 50°C twice and dewatered on a two-roll mill. The milled sheet was dried at 40°C–50°C. The sample was molded to a 1 cm thick plaque and oven aged at 70°C. The Mooney viscosities were measured at 100°C over 4 weeks of aging at 70°C, with the results shown in Table 10.9.

The results indicated a slight improvement with the blend system. However, longer term aging is necessary to differentiate these antioxidant systems. The blend

TABLE 10.9
Aging Test: Change in Mooney Viscosity

	Additive		
Time (Weeks)	None	0.7 phr TNPP	0.7 phr TNPP/AO[a]
0	0	0	0
1	8	3	4
2	17	8	5
3	25	10	7
4	35	18	14

[a] AO, 2,2'-methylenebis(4-methyl-6-nonylphenol).

system used a 5:1 phosphite/phenol ratio. TNPP was tris(mononnonylphenol)phosphite, and the hindered bisphenol antioxidant was 2,2-methylenebis(4-methyl-6-nonylphenol).

IV. ANTIOZONANTS

The commercially available antiozonants are dialkylparaphenylenediamine, alkyl-arylparaphenylenediamine, and diarylparaphenylenediamine. Their functions and mechanisms are different. Therefore, the selection of antiozonant is very important to obtain a long-term service life without compound cracking. Chemical antiozonants are required to protect all of the external vulcanized rubber components from ozone-induced cracking. In this section, various antiozonant systems are illustrated for chemical reactivity with ozone, solubility in various polymers, solubility in solutions with various pH values, and static and dynamic migration in rubber compounds. These are related to the functions and mechanisms in the actual service life of various antiozonants.

Antiozonants that were employed in this study are N-(1,3-dimethylbutyl)-N'-phenyl-p-phenylenediamine (6PPD), N-isopropyl-N'-phenyl-p-phenylenediamine (IPPD), N,N'-bis-(1,4-dimethylpentyl)-p-phenylenediamine (77PD), N,N'-diaryl-p-phenylenediamine (DAPD), and 2,4,6-tris(N-1,4-dimethylpentyl-p-phenylenediamine)-1,3,5-triazine (TAPDT), whose structures are shown in Figure 10.20.

A. CHEMICAL REACTIVITY

In a PPD molecule, the arylalkyl-substituted NH group is more reactive than the bisaryl-substituted NH group. It is known that amines are attacked by ozone at the free electron pair of the N atom. The charge density on the nitrogen atom, qN, is therefore believed to be important for the antiozonant effect. The calculated charge density qN on the N atom for arylalkyl- and bisaryl-substituted NH groups are shown in Figure 10.21 [18]. The arylalkyl-substituted NH group possesses a higher charge density than the bisaryl-substituted NH group (i.e., −0.208 vs. −0.188) and would thus be more reactive than the bisaryl-substituted NH group.

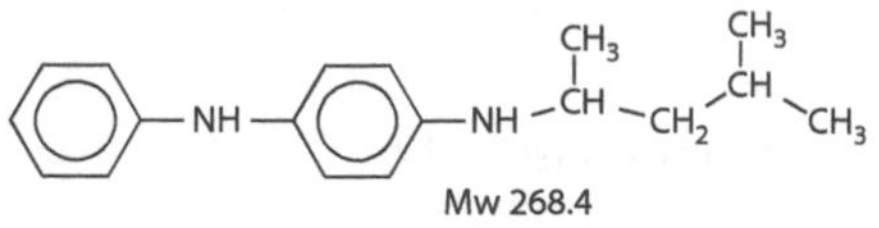

Mw 268.4

N-(1,3-Dimethylbutyl)-N'-phenyl-p-phenylenediamene (6PPD)

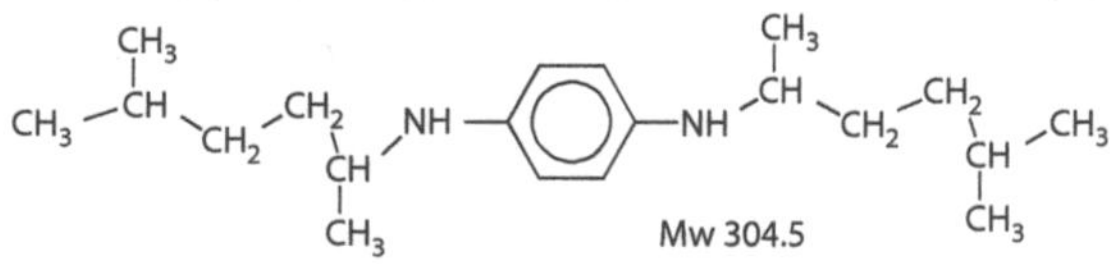

Mw 304.5

N,N'-Bis-(1,4-dimethylpentyl)-p-phenylenediamine (77PD)

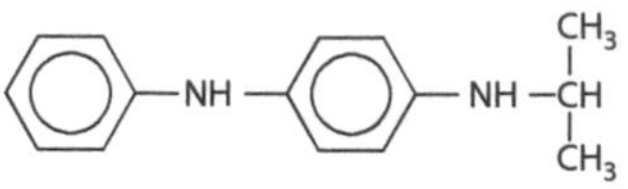

N-Isopropyl-N'-phenyl-p-phenylenediamine (IPPD)

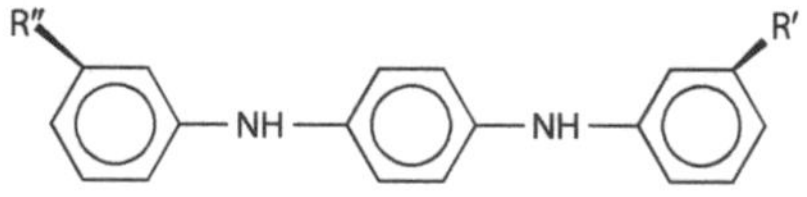

N,N'-Phenyl-p-phenylenediamine (DAPD)

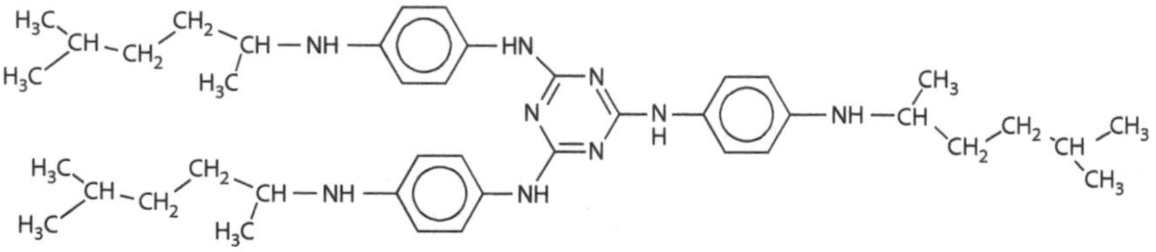

2,4,6-Tris-(N-1,4-dimethylpentyl-p-phenylenediamine)-1,3,5-triazine (TAPDT)

FIGURE 10.20 Various commercially available antiozonants.

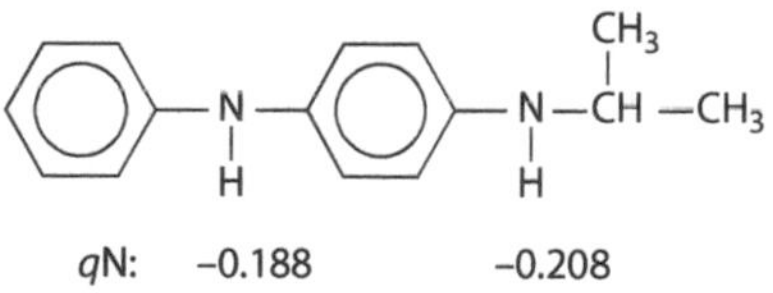

qN: −0.188   −0.208

FIGURE 10.21 Charge density on the nitrogen atom, qN.

The characterization of the ozonation products led to the discovery that among other products, 6PPD (an arylalkyl-PPD) produced nitrone and the 77PD (a bisalkyl-PPD) produced dinitrone instead [16,19,20]. This difference is exemplified by the simplified reaction mechanism for arylalkyl- and bisalkyl-PPD, depicted in Figures 10.22 and 10.23 using 6PPD and 77PD as examples. Note that further reactions of the nitrone with the ozone (step 4) are very slow in the case of 6PPD reaction

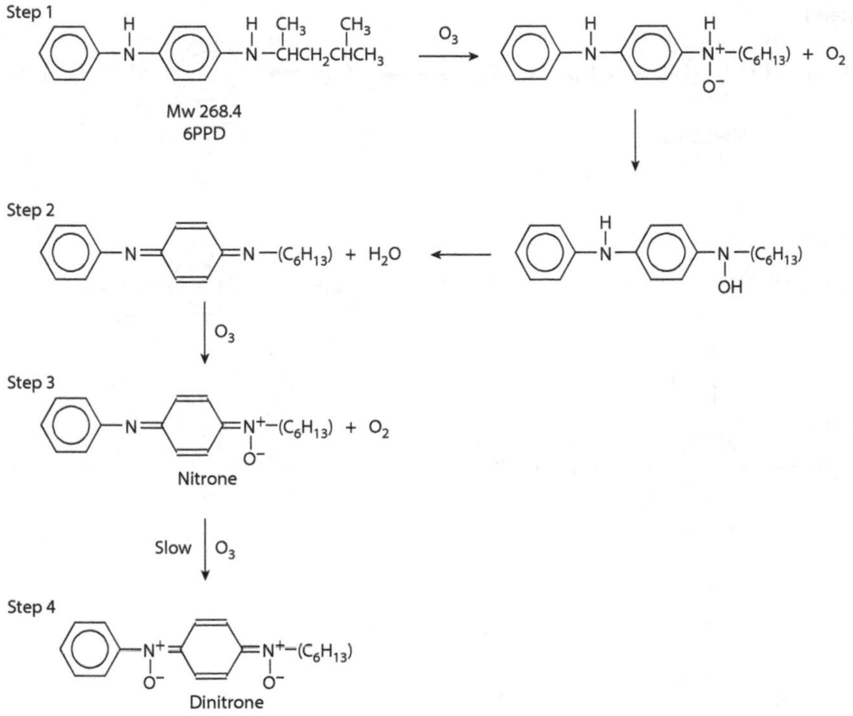

Step 1

Mw 268.4
6PPD

Step 2

Step 3

Nitrone

Step 4

Dinitrone

FIGURE 10.22 The ozonation mechanism for the arylalkyl-PPDs shown in simplified form.

(Figure 10.22) but fast for the 77PD mechanism (Figure 10.23). Apparently, the stabilizing effect of the *N*-aryl group on the nitrone (Figure 10.22) inhibits further reaction of the nitrone with ozone. Consequently, the bisalkyl-PPD such as 77PD (Mw 304) would scavenge more ozone molecules than the arylalkyl-PPD such as 6PPD (Mw 268) on an equal basis. TAPDT resembles an arylalkyl-PPD. The chemical reaction mechanism of TAPDT (Mw 694) with ozone would be similar to that of 6PPD. However, TAPDT possesses three arylalkyl-substituted NH groups, which would scavenge more ozone molecules than the 6PPD on an equal basis (Figure 10.24).

In summary, under conditions such that the antiozonant diffusion is limited, the key protective mechanism would be the scavenging role played by the antiozonant. This reasoning is supported by the data we observed with 77PD and TAPDT, which provide better static ozone protection than 6PPD [5,12,15].

B. Migrations of Various Antiozonants in NR/BR and SBR Compounds

Two separate experiments were conducted in an effort to quantify the rates of migration of commonly used rubber antiozonants and a high-molecular-weight nonstaining TAPDT antiozonant. One evaluation involves the study of antiozonant static migration, and the other addresses migration rates under dynamic conditions.

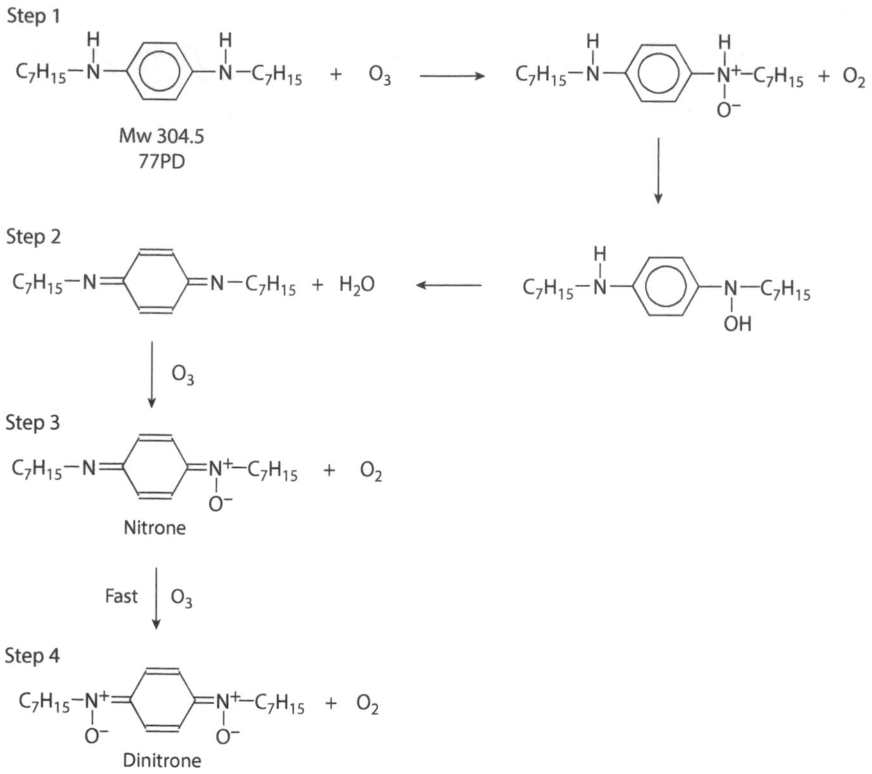

Step 1

C₇H₁₅—N(H)—[ring]—N(H)—C₇H₁₅ + O₃ → C₇H₁₅—N(H)—[ring]—N⁺(H)(O⁻)—C₇H₁₅ + O₂

$C_7H_{15}-\overset{H}{\underset{}{N}}--\overset{H}{\underset{}{N}}-C_7H_{15} + O_3 \longrightarrow C_7H_{15}-\overset{H}{\underset{}{N}}--\overset{H}{\underset{O^-}{N^+}}-C_7H_{15} + O_2$

Mw 304.5
77PD

Step 2

$C_7H_{15}-N{=}{=}N-C_7H_{15} + H_2O \longleftarrow C_7H_{15}-\overset{H}{\underset{}{N}}--\underset{OH}{N}-C_7H_{15}$

$\downarrow O_3$

Step 3

$C_7H_{15}-N{=}{=}\overset{}{\underset{O^-}{N^+}}-C_7H_{15} + O_2$

Nitrone

Fast $\downarrow O_3$

Step 4

$C_7H_{15}-\overset{}{\underset{O^-}{N^+}}{=}{=}\overset{}{\underset{O^-}{N^+}}-C_7H_{15} + O_2$

Dinitrone

FIGURE 10.23 The ozonation mechanism for the dialkyl-PPDs shown in simplified form.

The static migration study involves two compounds (NR/BR and SBR). NR/BR is for sidewall and SBR for tread application. Each of these formulations was compounded without wax and evaluated in static and dynamic ozone tests. The dynamic migration study involves compounds based on NR/BR only.

1. Static Migration

The relative rates of migration of antiozonants in cured rubber compounds were measured by submerging 3.6 cm diameter × 3.8 cm high cylinders of cured sample in methanol and measuring the antiozonant concentration of the methanol at various times. This technique simulates the material migration of the antiozonant to the surface of the compound if the following assumptions are made: (1) Methanol does not appreciably penetrate the polymer and therefore extraction of the antiozonant does not occur and (2) the methanol acts only as an antiozonant acceptor and little if any antiozonant is transferred back into or onto the polymer once it has migrated out. Before being submerged in the methanol, the samples, which had been aged for 1 month, were flash washed with acetone for 10 s. This was done to remove any antiozonant bloom from the surface prior to starting static migration in order to eliminate antiozonant solubility in NR/BR and SBR compounds.

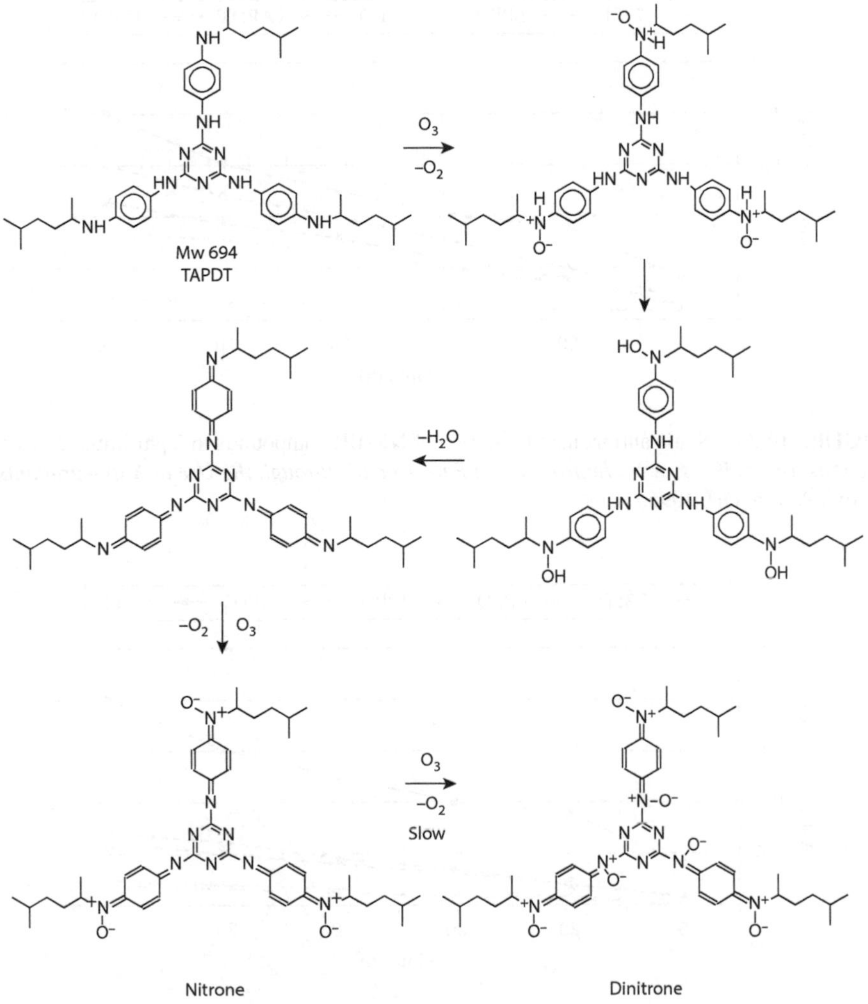

FIGURE 10.24 The ozonation mechanism for TAPDT shown in simplified form.

Following the acetone wash, the buttons were immediately immersed in 35 mL of methanol, which completely covered the sample, and allowed to stand at room temperature (25°C) for 4, 22, 28, 46, 72, and 96 h time periods. At the end of each period, the methanol was decanted and fresh methanol added to the vessel; the next interval would then begin at 0 h. Figures 10.25 and 10.26 show the rates of antiozonant migration from NR/BR and SBR rubber samples, respectively. The rates vary depending upon the type of rubber compound and type of antiozonant. The fastest migration is observed with IPPD, followed by 6PPD, DAPD, 77PD, and TAPDT, which migrates at a significantly slower rate. Usually, the highest-molecular-weight antiozonants such as TAPDT typically had the slowest rates of migration.

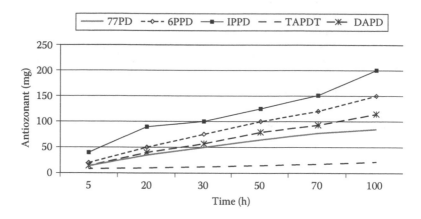

FIGURE 10.25 Static antiozonant migration of NR/BR compounds in 2 phr initial concentrations. (From Hong, S.W., *Improved Tire Performance through the Use of Antidegradants*, ITEC, Akron, OH, 1996.)

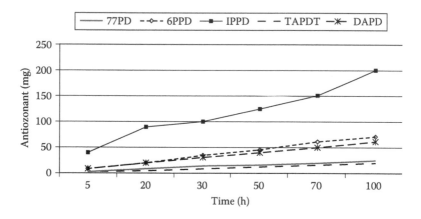

FIGURE 10.26 Static antiozonant migration of SBR compounds in 2 phr initial concentrations. (From Hong, S.W., *Improved Tire Performance through the Use of Antidegradants*, ITEC, Akron, OH, 1996.)

2. Dynamic Migration

The effects of flexing on the relative rates of the antiozonant migration were investigated by analyzing 1.3 cm × 10.2 cm × 0.20 cm cured rubber strips for antiozonant remaining after various durations of flexing. Strips that were flexed for 2, 4, 8, 16, 32, and 48 days were diced into approximately 2 mm cubes and Soxhlet extracted for 16 h with an ethanol–toluene azeotropic (ETA) mixture. The antiozonants were compounded at 2.0 phr, and the migration rates are graphically represented in Figure 10.27. The lower levels of antiozonant in NR/BR were evaluated because of the lower solubility of DAPD and TAPDT in NR/BR compound.

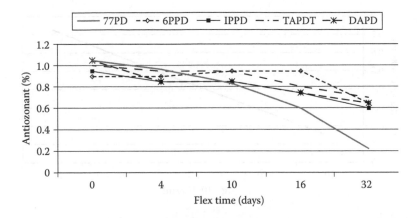

FIGURE 10.27 Antiozonant retention of NR/BR compounds in 2 phr initial concentrations. (From Hong, S.W., *Improved Tire Performance through the Use of Antidegradants*, ITEC, Akron, OH, 1996.)

The results of the dynamic migration are somewhat different than those observed in the static migration study. TAPDT shows the slowest migration, and DAPD and 6PPD exhibit good retention in the rubber compound following flexing, whereas 77PD is depleted at a much more rapid pace. It is postulated that this accelerated loss is due primarily to chemical reactivity, which would lead to rapid oxidation of the 77PD molecule both in and on the rubber during flexing. Due to the high reactivity of 77PD with oxygen and ozone, it was quickly depleted in this experiment. However, the static methanol migration study did not permit exposure to oxygen or ozone and thus truly measured migration as opposed to antiozonant depletion. If interest is only in migration rate without chemical reaction with ozone or oxygen, dynamic flex testing in a nitrogen atmosphere would undoubtedly provide a more accurate evaluation of dynamic migration rates of these antiozonants.

3. Measurements

For both the static and dynamic tests, the antiozonant content of the resultant solutions was measured colorimetrically. After the initial dilution, all subsequent dilutions necessary for the colorimetric analysis were made with *ETA* (ethylene toluene azeotropic solution). The 77PD samples were diluted with methanol, whereas a cupric acetate oxidizing reagent was used for the 6PPD and IPPD samples. The 6PPD, IPPD, and 77PD solutions were oxidized with cupric acetate, and the absorbance of each was measured at the specified wavelength of each antiozonant in a 1 cm cell. TAPDT and DAPD solutions were oxidized with benzoyl peroxide, and the absorbance of each was measured at 400 and 450 nm, respectively, in a 1 cm cell. Background absorbances were subtracted for each solution, and external standard calibration curves were generated for each antiozonant. Figure 10.28 shows the calibration curve for TAPDT as an example.

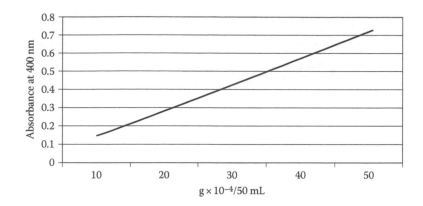

FIGURE 10.28 External calibration curve for TAPDT. 1 cm cell in *ETA*. (From Bruck, D. et al., *Kautsch Gummi Kunstst*, 44(11), 1014, 1991.)

C. SOLUBILITY

1. Solubilities of Various Antiozonants in SBR, NR/BR, and NR Compounds

The solubility of an antiozonant is very important in determining the maximum amount of antiozonant that can be added to the compound effectively. The solubility of each antiozonant is different from one rubber to another. Limited solubility allows for very fast diffusion to the rubber surface. An excessive amount of bloom may cause lower tack or separation of the interface due to surface bloom. This would prevent the formation of cross-links between interfaces.

Solubility studies were carried out by preparing cylindrical buttons that were then aged for 1 month. The aged buttons were flash washed with acetone for 10 s to remove any antiozonant bloom to the surface for analysis. The amount of bloomed antiozonant is shown in Figure 10.29. It was clearly indicated that DAPD has poor solubility in NR and NR/BR compounds at the 2.0 phr level. The solubility of TAPDT varies depending on the base polymer or polymer blend. However, in general, TAPDT is more soluble than the diaryl PPD. This slight insolubility property and better chemical reactivity with three alkyl arms would provide static ozone crack resistance without using wax. Antiozonants such as 6PPD and 77PD have higher solubility in SBR, NR, and NR/BR compounds [21].

2. Solubility in Water and Acid Rain

One environmental condition that must be accounted for in today's society is acid rain. The pH levels and the amount of rain will have a significant effect on the level of chemical antiozonant on both tread and sidewall surfaces. The Environmental Protection Agency has numerous sites throughout the United States that sample rainwater. Figure 10.30 shows the yearly average pH for rainfall in the Northeast, Southeast, Rocky Mountains, and West Coast regions of the United States. These data show that the pH of rain varies between 4.2 and 5.2 depending on the area in which the rain is collected. Therefore, it is important to study the water and

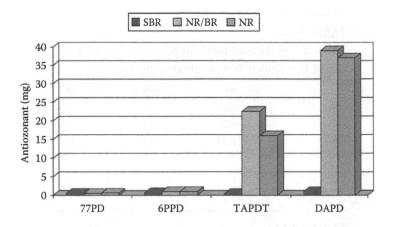

FIGURE 10.29 Solubility of various antiozonants in rubber compounds. Acetone wash from cylindrical button samples aged 1 month. (From Bruck, D. et al., *Kautsch Gummi Kunstst*, 44(11), 1014, 1991.)

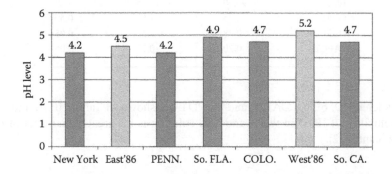

FIGURE 10.30 pH levels for acid rain within various areas of the United States showing changes over a 10-year span. Average over 1-year period (1996), reported by the EPA.

acid rain solubility of each antiozonant that would be inversely correlated with the performance of ozone cracking resistance.

Each of the antiozonants was mixed at a rate of 2.0 phr into NR/BR compounds. The completely mixed compounds were vulcanized, and the vulcanized samples were aged in each pH solution for 24 and 72 h. The samples were removed from solution and dried, and ozone cracking resistance testing was initiated. The individual solutions were tested for absorbance of the antiozonant. A Beckman DU-70 UV/visible spectrophotometer was used for the analysis. The range scanned was from 500 to 200 nm, with a 1 cm cell used for all measurements. Each sample was read undiluted to determine the level of antiozonant. If needed, a dilution was made to establish concentration within the range of Beer's law [22]. The appropriate multiplier should be used to calculate original concentration. The absorbance

TABLE 10.10
Solubility of Antiozonants at pH 3, pH 5, and pH 7
for 24 h at Room Temperature (Absorbance)

	pH 3	pH 5	pH 7
IPPD	76.80	8.18	2.54
6PPD	4.25	1.1	0.25
DAPD	0.24	0.12	0.05
TAPDT	0.20	0.09	0.05

TABLE 10.11
Solubility of Antiozonants at pH 3, pH 5, and pH 7
for 72 h at Room Temperature (Absorbance)

	pH 3	pH 5	pH 7
IPPD	90.00	12.95	5.20
6PPD	4.75	1.65	0.50
DAPD	0.48	0.31	0.05
TAPDT	0.20	0.12	0.06

results for IPPD, 6PPD, DAPD, and TAPDT are shown in Tables 10.10 and 10.11. The solubility of IPPD in water and acid solution is far greater than that of other antiozonants [22].

D.   CONCLUSION

The solubility of various antiozonants decreases in the order

IPPD > 6PPD > DAPD > TAPDT

Ozone crack resistance under static conditions is in the order

TAPDT > 6PPD > DAPD > IPPD

And for migration and chemical reactivity determined by dynamic testing,

6PPD > IPPD > TAPDT > DAPD

Solubility under acid rain conditions is

TAPDT > DAPD > 6PPD > IPPD

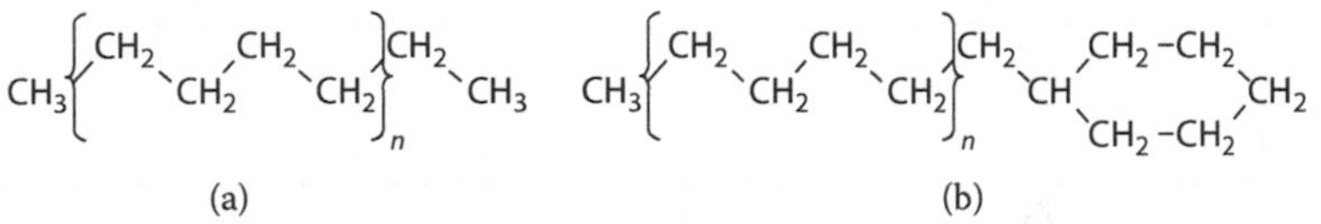

FIGURE 10.31 Structures of (a) paraffin wax and (b) microcrystalline wax.

V. PETROLEUM WAXES

Waxes are divided into two major types—paraffin and microcrystalline—whose structures are shown in Figure 10.31. Usually, waxes are insoluble in polymers such as NR and synthetic rubbers. Due to their poor solubility, the waxes migrate out to the surface, forming a film or barrier that would prevent the vulcanized rubber surface from attack by ozone. Even though the solubility of waxes in rubber compounds is inferior, the migration rates vary depending upon the molecular weight and molecular structures of the wax. Usually, carbon number distribution curves indicate the distribution of molecular weight that is proportional to the melting point of the waxes. Therefore, some producers of waxes do not provide the carbon number distribution of each wax and indicate only their melting points. However, the performance of a wax can be explained from the carbon number distribution, but not with the melting point. Paraffin wax, having a lower carbon number distribution than microcrystalline wax, would migrate faster, whereas microcrystalline wax would diffuse out to the surface much more slowly. Also, the migration rate is dependent upon temperature. Therefore, at a colder temperature, a faster migration wax such as paraffinic wax is needed, and at warmer temperatures, a microcrystalline wax would be required. The migration rate should be adjusted to form a barrier of the proper thickness to protect the rubber product from ozone attack.

The carbon number distribution curves are measured by gas chromatography with increases of temperature that meet the boiling point of each carbon number wax; this is shown in Figure 10.32. A major portion of wax 75 is paraffinic wax, whereas wax 96 is mostly microcrystalline wax. Wax 76 is a 50/50 blend of wax 75 with wax 96.

VI. COMPOUNDING EVALUATION OF WAXES

A. EVALUATION OF STATIC OZONE RESISTANCE

It was already mentioned that paraffinic waxes generally protect better at low temperatures, whereas microcrystalline waxes are more effective at elevated temperatures. This is based on carbon number distribution, which offers insight into the molecular weight distribution shown in Figure 10.32. Due to the slow migration of microcrystalline wax, it provides longer term static protection from ozone cracking. Blends offer a wider temperature range of protection. Waxes are used for only static ozone protection. During dynamic flexing, the barrier of wax is exposed owing to breaking of the wax film. Therefore, reactive antiozonants are required for dynamic ozone crack resistance.

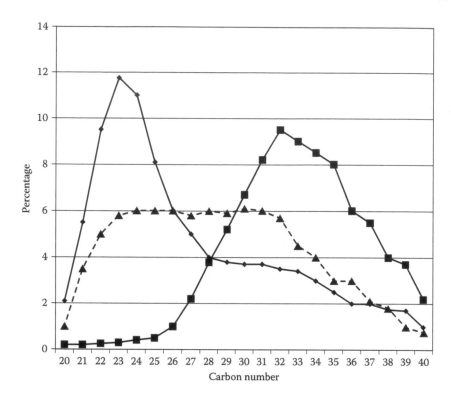

FIGURE 10.32 Wax carbon number profiles for (♦) wax 75, a paraffinic wax; (■) wax 96, a microcrystalline wax; and (▲) wax 76, a 50/50 blend of wax 75 with wax 96.

1. Preparation of Compounds

In order to verify the aforementioned hypothesis based on carbon number distribution, three waxes (paraffinic wax, microcrystalline wax, and a 50/50 blend) were used in an NR/BR black sidewall compound with 2.0 phr of 6PPD antiozonant. The NR/BR master batch without any wax, antiozonant, or curative was mixed in a 10 L internal mixer at the first stage. Four mixes were made in a 1 L Banbury, using the same master batch, which may eliminate or reduce experimental error. There was one control (A) that did not contain wax, and the other three compounds had 2.0 phr of either the paraffinic, microcrystalline, or blended wax. The curatives were also added to each master batch in a 1 L Banbury as listed in Table 10.12.

2. Results and Discussion

The completely mixed compounds were cured for 10 min at 160°C. Usually, a 20% or 40% extension test is carried out for static ozone testing, but in this case, testing was carried out at 75% extension and 30 pphm ozone by varying temperatures from

TABLE 10.12

Recipes for Mixes A–D

	A	B	C	D
SMR CV 60	55.00	55.00	55.00	55.00
BR 1203	45.00	45.00	45.00	45.00
N660	50.00	50.00	50.00	50.00
Naphthenic oil	7.00	7.00	7.00	7.00
Zinc oxide	3.00	3.00	3.00	3.00
Stearic acid	1.00	1.00	1.00	1.00
Total	161.00	161.00	161.00	161.00
MB-1	161.00	161.00	161.00	161.00
Wax 75	—	2.00	—	—
Wax 96	—	—	2.00	—
Wax 76	—	—	—	2.00
6PPD	2.00	2.00	2.00	2.00
Total	163.00	165.00	165.00	165.00
MB-2	163.00	165.00	165.00	165.00
TBBS	1.00	1.00	1.00	1.00
Crystex 80%	2.00	2.00	2.00	2.00
Total	166.5	168.50	168.50	168.50

0°C to 40°C to try to obtain accelerated results. The ratings were set from 1 to 5, where a higher number indicates more severe cracking. The results clearly indicated that the lower carbon number distribution wax performed well at lower temperature, whereas a higher carbon number wax protects from ozone cracks at elevated temperatures. The blended wax performed well between 20°C and 30°C but did not perform well at the temperature extremes.

3. Conclusion

The static ozone crack resistance ratings are as follows:

Temperature (°C)	Rating Comparison[a]
0	B>D>C>A
10	B>D>C>A
20	B=D>C>A
30	C=D>B=A
40	C>D>B>A

[a] A, no wax; B, low carbon number wax (paraffinic); C, high carbon number wax (microcrystalline); D, broad carbon number wax (blend).

VII. VULCANIZATION SYSTEM

Commonly used vulcanizing agents are sulfur and peroxides, because metal oxides easily promote oxidation of elastomers, which results in failure. Because of differences in bonding energy, the thermal stability of sulfur cross-links varies, as shown in Table 10.13. In this section, conventional, semiefficient, and efficient cure systems will be discussed along with a peroxide vulcanization system that would provide carbon-to-carbon linkage.

A. SULFUR CURE SYSTEM

Sulfur was the original vulcanizing agent used by Charles Goodyear in 1839 and today is the most common vulcanizing agent used in the rubber industry. Sulfur in the presence of heat reacts with adjoining olefinic bonds in the polymeric backbone chains or pendent chains of two elastomeric molecules to form cross-links between the molecular chains. However, the sulfur may combine in many ways to form the cross-link network of vulcanized rubber. Sulfur may be present as monosulfide, disulfide, and polysulfide linkages. It may also be present as pendent sulfide and pendent cyclic mono- and polysulfides.

The conventional cure systems feature a high sulfur level and low accelerator concentration, which forms a higher level of polysulfide cross-links whose bonding energy is the lowest among cross-links. Therefore, conventional cure systems show poor heat and oxidation resistance, and a longer chain of polysulfide cross-links would provide better flex fatigue properties with lower modulus.

The efficient or semiefficient cure systems feature a higher level of accelerators and sulfur donors and a lower level of free sulfur. These systems form mono- or disulfide cross-links whose bonding energies are much higher than that of polysulfide cross-links. They show good heat stability and oxidation resistance. However, they are inferior in flex fatigue because they have a shorter chain of cross-links, which would have a higher modulus. A higher modulus compound generates higher heat energy during flex fatigue test, which would lead to break cross-link for an early failure. This relationship is shown in Figure 10.33.

B. PEROXIDE CURE SYSTEM

Organic peroxides are used to vulcanize elastomers that have both a saturated and an unsaturated backbone. The cross-linking occurs through a carbon-to-carbon bond by reacting with hydrogen from polymers with peroxides. This reaction is normally

---

TABLE 10.13
Bonding Energy (kcal/mol)

Polysulfide	$-S_x-$	34
Disulfide	$-S-S-$	54
Monosulfide	$-S-$	74
Carbon to carbon	$-C-C-$	80

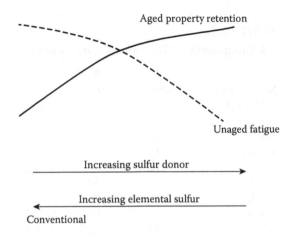

FIGURE 10.33 Fatigue/heat aging compromise.

initiated at the tertiary carbon atoms located along the molecular chain. Various peroxides may be employed, but two of the most common are dicumyl peroxide and benzoyl peroxide. Peroxides, upon heating, cleave to form two or more oxy radicals, which react by abstracting hydrogen atoms, usually tertiary ones, from the polymer chain, forming polymeric radicals. Two of these polymeric radicals then combine to form a cross-link that is a more stable bond (higher bonding energy) with superior heat aging and oxidation resistance.

C. COMPOUNDING EVALUATION

1. Various Cure System in Natural Rubber

A model master batch with 100% NR was prepared in a 10 L internal mixer. The same master batch was used to finalize the compounds by the addition of curatives in a 1 L Banbury (Table 10.14). Mooney scorch values were measured at 132°C and vulcanization kinetics at 150°C and 170°C. The unaged, overcured, and aged physical properties and unaged/aged flex fatigue were measured.

2. Results and Discussion

The results clearly indicated that shorter term scorch safety was obtained with either a higher level of accelerators or the addition of peroxides. However, a faster cure was achieved with only a higher level of accelerators (Table 10.15). The slower cure was measured by peroxides. As expected, a higher modulus and less elongation were measured with a peroxide cure system, even though the maximum torque in the rheometer was much lower than in either a conventional or semi-EV cure system. Usually, the maximum torque is correlated with cross-link density. This means that lower cross-link density with peroxide cure system has the highest modulus and the lowest elongation among three compounds. By overcuring compounds for 30 min at 170°C, the peroxide cure system increased modulus, tensile strength, and elongation, which other cure systems slightly reduced. Heat-aging properties at both 70°C and

TABLE 10.14
NR Compounds with Various Cure Systems

	A	B	C
Natural rubber	100	100	100
N-330 black	50	50	50
Aromatic oil	7	7	7
Zinc oxide	4	4	4
Stearic acid	1.5	1.5	1.5
TBBS	1.0		1.2
DCP 60		3.0	
TMTD			0.5
MBTS		0.5	0.5
Sulfur	2.0	0.5	0.5

Notes: TBBS, *N-t*-butyl-2-benzothiazole sulfenamide; DCP 60, dicumyl peroxide; TMTD, tetramethylthiuram disulfide; MBTS, benzothiazyl disulfide.

TABLE 10.15
Processing Data

	A	B	C
TBBS	1.0	—	1.2
DCP 60	—	3.0	—
TMTD	—	—	0.5
MBTS	—	0.5	0.5
Sulfur	2.0	0.5	0.5
Mooney scorch at 132°C			
t_3, min	25	10	14
Rheometer at 150°C			
ts_1, min	7.3	3.1	4.8
t_{90}; min	13.5	22.4	8.9
MH, lb in.	37.1	24.6	27.4
Rheometer at 170°C			
ts_1, min	2.3	1.1	1.9
t_{90}, min	4.6	6.9	3.5
MH, lb in.	34.0	27.3	30.8

100°C indicated that the best retention was achieved by peroxides, than by the semi-EV and conventional cure systems. The conventional cure system was inferior in heat aging because of polysulfide cross-links. The Monsanto flex fatigue test was run to measure flex fatigue resistance. The conventional cure systems provided the best resistance to flex fatigue. All the physical property results are shown in Table 10.16.

TABLE 10.16
Physical Properties

	A	B	C
TBBS	1.0	—	1.2
DCP 60	—	3.0	—
TMTD	—	—	0.5
MBTS	—	0.5	0.5
Sulfur	2.0	0.5	0.5
Unaged physical properties			
Cured 10 min at 170°C			
Tensile at RT, MPa	22.68	23.37	26.06
% Elongation	560	470	550
300% Modulus, MPa	9.03	11.44	11.10
Tear, die C, kN/m	85.8	47.28	71.80
Cured 30 min at 170°C			
Tensile, % retained	87	108	95
Elongation, % retained	100	104	98
300% Modulus, % retained	90	107	93
Tear, die C, % retained	39	74	63
Aged 2 weeks at 70°C			
Tensile, % retained	82	95	90
Elongation, % retained	80	96	96
300% Modulus, % retained	121	107	97
Tear, die C, % retained	57	68	91
Aged 48 h at 100°C			
Tensile, % retained	78	92	85
Elongation, % retained	77	89	91
300% Modulus, % retained	126	90	101
Tear, die C, % retained	42	74	56
Flex fatigue, unaged			
Kilocycles to failure	82	54	62
Flex fatigue, aged 48 h at 100°C			
Kilocycles to failure	28	22	24
% Retained	34	41	39

3. Conclusion

Property	Rating[a]
Scorch safety	CV > SEV > PV
Faster cure	SEV > CV > PV
Heat-aging property	PV > SEV > CV
Flex fatigue	CV > SEV > PV

[a] CV, conventional vulcanization system; SEV, semiefficient vulcanization system; PV, peroxide vulcanization system.

VIII. SUMMARY: ANTIOXIDANTS AND OTHER PROTECTIVE SYSTEMS

In this chapter, functions and mechanisms for antioxidants, antiozonants, waxes, and vulcanization systems to protect vulcanized rubber goods against oxygen, ozone, heat, light, and shear by flexing have been discussed, and experimental data have been provided.

Usually, antioxidants have been used for internal components to improve heat aging and flex fatigue properties. Another important factor in heat aging is the volatility of the antioxidant, whereas flex fatigue is related to the type of hindered nitroxyl radical, which depends on the antioxidant structure. For nonstaining and nondiscoloring rubber goods, phenolic antioxidants have been used along with secondary antioxidants such as phosphates. However, they are not as effective as secondary amine-type antioxidants, which are usually staining and discoloring.

Antiozonants protect rubber goods from ozone cracking, and they are usually used for external components of vulcanized rubber goods. The performance of antiozonants is directly related to the charge density of the nitrogen atoms attached to either aryl or alkyl groups, both static and dynamic migration rates, and their solubility in water and acid rain. The migration rates are dependent on the molecular weight and structure of the antiozonant; but solubility in water is dependent on structure only.

There are two types of waxes: paraffin and microcrystalline. The performance of a wax is dependent upon its migration rate, which is directly related to the environmental temperature. Usually, the migration rate of a paraffinic wax is much faster than that of a microcrystalline wax. In order to form a proper barrier on the surface at different temperatures, selection of the wax is very important to protect from ozone cracking in static conditions. At elevated temperatures, microcrystalline wax performs well, whereas at lower temperatures, paraffin wax can migrate out to the surface for protection.

The vulcanization system is very important to protect from heat and shear energy. There are polysulfide, disulfide, monosulfide, and carbon-to-carbon cross-links for vulcanization. A stronger bonding energy provides better heat resistance, and the flex property can be achieved with a longer chain (polysulfide) cross-link whose bonding energy is the weakest. The selection of the proper vulcanization system would compromise not only better heat aging but also better flex fatigue properties.

ABBREVIATIONS

Primary accelerators
MBTS	Benzothiazyl disulfide
TBBS	N-t-Butyl-2-benzothiazole sulfenamide

Secondary accelerators
APD	Alkylphenol disulfide
DPG	Diphenylguanidine
TBzTD	Tetrabenzylthiuram disulfide
TMTD	Tetramethylthiuram disulfide

Vulcanizing agent
   DCP-60   Dicumyl peroxide, 60% active
Primary antioxidants
Phenolic antioxidants
   BHT   2,6-Di-*t*-butyl-4-methylphenol
   Naugawhite 2,2'-Methylenebis(4-methyl-6-nonylphenol)
   CG-1520   2,4-Bis[(octylthio)methyl]-o-cresol
Secondary amine antioxidants
   AO 445   4,4'-Bis(α,α-dimethyl benzyl) diphenylamine
   BLE   High-temperature reaction product of diphenylamine and acetone
   TMQ   Polymerized 1,2-dihydro-2,2,4-trimethylquinoline
Secondary antioxidant
   TNPP   Tris(mono- and dinonylphenyl)phosphite
Antiozonants
   DAPD   *N*,*N*'-Diphenyl-*p*-phenylenediamine
   IPPD   *N*-Isopropyl-*N*'-phenyl-*p*-phenylenediamine
   6PPD   *N*-(1,3-Dimethylbutyl)-*N*'-phenyl-*p*-phenylenediamine
   77PD   *N*,*N*'-Bis-(1,4-dimethylpentyl)-*p*-phenylenediamine
   TAPDT   2,4,6-Tris-(*N*-1,4-dimethylpentyl-*p*-phenylenediamine)-1,3,5-
   triazine
Resins
   SP-1068   Alkylphenol formaldehyde resin
   SP-6700   Oil-modified phenol formaldehyde two-step resin
Bonding agents
   M3P   1-Aza-5-methylol-3,7-dioxabicyclo[3,3,0]octane
   R-6   Resorcinol resin, an acetaldehyde condensation product, produced
              by Crompton
Retarder
   CPT   *N*-(Cyclohexylthio)phthalimide
Waxes
   Wax 75   Paraffinic wax
   Wax 76   50/50 Blend of paraffinic and microcrystalline waxes
   Wax 96   Microcrystalline wax
Polymers
   BR   Polybutadiene rubber
   CIIR   Chlorobutyl rubber
   EPDM   Ethylene propylene diene terpolymer
   NBR   Nitrile rubber
   NR   Natural rubber
   SBR   Styrene butadiene rubber

REFERENCES

1. Scott G, ed. *Atmospheric Oxidation and Antioxidants*, Vol. 1. London, U.K.: Elsevier, 1993, p. 48.
2. Denisov ET. *Russ J Phys Chem* 1964; 38:1.

3. Carlsson DJ, Robb JC. Liquid-phase oxidation of hydrocarbons. Part 4.—Indene and tetralin: Occurrence and mechanism of the thermal initiation reaction with oxygen, *Trans Faraday Soc* 1966; 62:3403.

4. Dulog L. *Makromol Chem* 1964; 77:206.

5. Mazzeo RA, Boisseau NA, Hong SW. Presented at the *145th Rubber Division, ACS Meeting*, Chicago, IL, April 1994.

6. Pospisil JP. In: Scott G, ed. *Developments in Polymer Stabilization*, Vol. 1. London, U.K.: *Appl Sci*, 1979, pp. 19–20.

7. Schulz M, Wegwart H, Stampehl G, Reidiger W. New aspects in the activity of nitrogen containing antioxidants, *J Polym Sci Polym Symp* 1976; 57:329.

8. Shelton JR. The role of certain organic sulfur compounds as antioxidants. *Rubber Chem Technol* 1974; 47:949.

9. Pospisil J. In: Pospisil J, Klemchuk PP, eds. *Oxidation Inhibition in Organic Materials*, Vol. 1. Boca Raton, FL: CRC Press, 1989, p. 39.

10. Pospisil J, Klemchuk PP, eds. *Oxidation Inhibition in Organic Materials*, Vol. 1. Boca Ráton FL: CRC Press, 1989, p. 21.

11. Hong SW, Lin C-Y. Functions and mechanisms of various antioxidants in tire polymers and compounds for improved tire performance. Presented and published at *ITEC ASIA 2001*, Busan, Korea, September 18–20, 2001.

12. Hong SW, Lin CY. Improved flex fatigue and dynamic ozone crack resistance through the use of antidegradants in tire compounds. Presented at the *156th Meeting of ACS Rubber Division*, Orlando, FL, September 21–24, 1999.

13. Dweik HS, Scott G. Mechanisms of antioxidant action: Aromatic nitroxyl radicals and their derived hydroxylamines as antifatigue agents for natural rubber, *Rubber Chem Technol* 1984; 57:735.

14. Forrester FA, Hay JM, Thompson RH. *Organic Chemistry of Stable Free Radicals*. London, U.K.: Academic Press, 1968, p. 180.

15. Hong SW. *Improved Tire Performance through the Use of Antidegradants*. Akron, OH: ITEC, September 10–12, 1996.

16. Lattimer RP, Hooser ER, Layer RW, Rhee CK. Mechanisms of ozonation of N-(1,3-dimethylbutyl)-N′-phenyl-p-phenylenediamine, *Rubber Chem Technol* 1983; 56:431.

17. Knohloch G, Raue D, Patel A. *Kautsch Gummi Kunstst* 1991; 41 (8): 623–628.

18. Bruck D, Dormagen, Engles WH. *Kautsch Gummi Kunstst* 1991; 44(11):1014–1018.

19. Lattimer RP, Hooser ER, Diem HE, Layer RW, Rhee CK. *Rubber Chem Technol* 1980; 53:1170.

20. Scott G. Mechanisms of ozonation of N,N′-di-(1-methylheptyl)-p-phenylenediamine, *Rubber Chem Technol* 1985; 58:269.

21. Birdsall DA, Hong SW, Hajdasz DJ. A Review of recent developments in the mechanisms of antifatigue agents, *The TYRETECH'91 Conference*, Berlin, Germany.

22. Greene PK, Hong SW, Gallaway JK, Landry ES. Solubility of various antiozonants in low pH solutions and resultant effect on compound performance. Presented at *ACS Rubber Division Meeting*, Cleveland, OH, October 21–24, 1997.

11 Vulcanization

Frederick Ignatz-Hoover and Byron H. To

CONTENTS

I.   Background Fundamentals 462
    A.   Vulcanization 463
    B.   Scorch 463
    C.   Scorch Safety 463
    D.   Rate of Cure 464
    E.   Cure Time 464
    F.   State of Cure 464
    G.   Reversion 464
    H.   Network Maturation 464
    I.   Vulcanizing Agents 465
    J.   Accelerator 465
    K.   Activators 465
    L.   Retarders 465
        1.   Ml or Rmin 466
        2.   Mh or Rmax 466
        3.   ts2 466
        4. t_{25} 466
        5. t_{90} 466
II. Vulcanizing Agents 466
    A.   Sulfur and Sulfur Donors 466
        1.   Tetramethylthiuram (TMTD) 468
III. Activators 471
IV. Accelerators 473
    A.   Accelerator Classes 474
    B.   Mechanism of Zinc-Mediated Accelerated Sulfur Vulcanization 476
        1.   Historical and General Aspects Related to the Mechanism
            of Sulfur Vulcanization 476
        2.   Historical QSAR Studies 478
        3.   Early QSAR Studies of Sulfur Vulcanization 479
        4.   Recent QSAR Studies 480
        5.   Molecular Explanations of Various Accelerator Activities 483
        6.   Molecular Effects on the Activation Energy for Vulcanization 484

C. Practical Comparison of Primary Accelerators 485
   1. Natural Rubber 485
   2. SBR 485
   3. Performance Comparison 485
D. Comparison of Secondary Accelerators 491
   1. Natural Rubber 491
   2. SBR 492
   3. Nitrile 495
V. Retarders 497
VI. Cure Systems for Specialty Elastomers 503
  A. EPDM 503
  B. Nitrile Rubber 510
  C. Neoprene 511
  D. Butyl Rubber 512
VII. Effective Curative Dispersion for Uniform Mechanical Properties 513
  A. Accelerator Dispersion: A Review 514
VIII. Conclusions 519
Abbreviations 519
References 520

About 150 years ago, Goodyear[1] in the United States and Hancock[2] in England discovered that Indian rubber could be changed by heating it with sulfur so that it was not greatly affected by heat, cold, and solvents. This process was termed vulcanization by Brockedon, who derived the word from the association of heat and sulfur with the Roman god of fire (including the fire of volcanoes) Vulcan.

Since that time, many other chemicals have been examined as possible vulcanizing agents with some degree of success. Sulfur vulcanizates provide an outstanding balance of cost and performance, providing excellent strength and durability for relatively low cost. No other cure system has, on its own, successfully competed on scale with sulfur as a general purpose vulcanizing agent. One limitation imposed upon the use of sulfur as a vulcanizing agent is that the elastomer must contain some chemically activated group such as chemical unsaturation. In saturated elastomers, other chemicals, particularly organic peroxides, have been found quite useful.

Vulcanization is fundamentally a set of chemical reactions whereby a mass of polymer chains (plastic in character) are chemically linked into a three-dimensional network having tough elastic properties. This transformation from a soft, plastic material to a tough, elastic material is the basis by which the engineering properties of vulcanized rubber are based. Vulcanized rubber is the foundation upon which the $138 billion (2013 estimated) tire industry has grown.

I. BACKGROUND FUNDAMENTALS

The following is a short list of terms and concepts commonly used within rubber industry discussions of vulcanization, where indicated reference is made to specific test *methodologies*.

A. VULCANIZATION*

Vulcanization is the process of treating an elastomer with a chemical to decrease its plasticity, tackiness, and sensitivity to heat and cold, and to give it useful properties such as elasticity, strength, and stability. Ultimately, this process chemically converts thermoplastic elastomers into three-dimensional elastic networks.

This process converts a viscous entanglement of long-chain molecules into a three-dimensional elastic network by chemically joining (crosslinking) these molecules at various points along the chain. The process of vulcanization is depicted graphically in Figure 11.1. In this diagram, the polymer chains are represented by the lines and the crosslinks by the heavy dots.

B. SCORCH

Scorch refers to the initial formation of an extensive three-dimensional network rendering the compound elastic. The compound is thus no longer plastic or permanently deformable and thus cannot be shaped or further processed.

C. SCORCH SAFETY

Scorch safety refers to the time the compound can be maintained at an elevated temperature during which the compound remains plastic. This time marks the point at which the plastic material begins the chemical conversion to the elastic network. The consequence of scorch is that the compound develops enough elasticity that it can no longer be formed or otherwise shaped or mechanically processed. Thus, if the compound scorches before it is formed into the desirable shape or composite structure, it can no longer be used. Time to scorch is important for factory processing considerations since it indicates the amount of time (heat history) the compound may be exposed to during shaping and forming operations before it becomes an intractable mass.

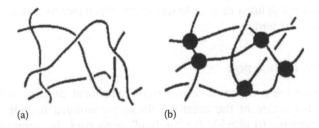

(a) (b)

FIGURE 11.1 Crosslinks. (a) Raw rubber. (b) Vulcanized rubber.

---

* While based upon ASTM D-1566-80b, these definitions have been modified to fit this discussion.

D. Rate of Cure

Rate of cure or cure rate describes the "rate" at which crosslinks form. After the point of scorch, the chemical crosslinking continues providing more crosslinks and thus greater elasticity or stiffness (modulus). The rate of cure determines how long a compound must be cured in order to reach "optimum" properties. The maximum rate of cure (M×R) is often estimated by the maximum slope observed in the rheometer cure.

E. Cure Time

Cure time is the time required to reach a desired state of cure. Most common lab studies will utilize the t_{90} *cure time* or the time required reaching 90% of the maximum cure.

F. State of Cure

State of cure refers to the degree of crosslinking (or crosslink density) of the compound. State of cure is commonly expressed as a percentage of the maximum attainable cure (or crosslink density) for a given cure system. The elastic force of retraction, elasticity, is directly proportional to the crosslink density or number of crosslinks formed in the network.

G. Reversion

Reversion refers to the loss of crosslink density as a result of nonoxidative thermal aging. *Reversion* is primarily observed in isoprene-containing polymers to the extent that such network contains polysulfidic crosslinks. The nature of *reversion* is to convert a polysulfidic network into a network rich in monosulfidic and disulfidic crosslinks and, most importantly, having a lower crosslink density than the original polysulfidic network. *Reversion* is not apparent or hardly occurs in isoprene polymers cured with vulcanization systems designed to produce networks rich in monosulfidic and disulfidic crosslinks. *Reversion* is commonly characterized by the time required for a defined drop in torque in the rheometer as measured from the maximum observed torque.

H. Network Maturation

Network maturation is a term used to describe chemical changes to the network imparted by the action of the curatives through continued heating beyond the "cure time" required to provide for "optimal" properties. In isoprene polymers, the effect is commonly referred to as reversion. However, in butadiene-containing polymers, the effect is to reduce polysulfidic networks to networks rich in monosulfidic and disulfidic crosslinks having *greater crosslink density than the original network*. This slow increase in modulus with time is often called a "marching modulus."

I. Vulcanizing Agents

Vulcanizing agents are chemicals that will react with active sites in the polymer to form the chemical connections or crosslinks between chains.

J. Accelerator

An *accelerator* is a chemical used in small amounts with a vulcanizing agent to reduce the time of (accelerate) the vulcanization process. In sulfur vulcanization today, accelerators are used to control the onset, speed, and extent of reaction between sulfur and elastomer.

K. Activators

Activators are materials components of an accelerated vulcanization system to improve acceleration and to permit the system to realize its full potential of crosslinks.

L. Retarders

Retarders are chemicals used to reduce the tendency of a rubber compound to vulcanize prematurely by increasing scorch delay (time from beginning of the heat cycle to the onset of vulcanization). Ideally, a *retarder* would have no effect on the rate of vulcanization. Such an ideal *retarder* has been called a pre-vulcanization inhibitor (PVI).

The kinetics of vulcanization are studied using curemeters or rheometers that measure the development of torque as a function of time at a given temperature. An idealized cure curve is given in Figure 11.2. Several important values derived from the curemeter characterize the rate and extent of vulcanization of a compound. Critical values include the following.

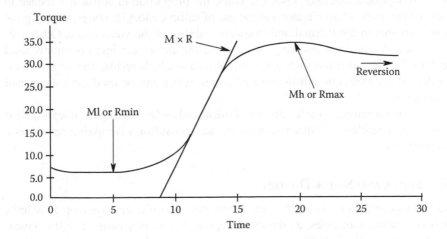

FIGURE 11.2 Rheometer curve.

1. Ml or Rmin

Ml is the minimum torque in the rheometer. This parameter often correlates well with the Mooney viscosity of a compound.

2. Mh or Rmax

Mh or Rmax represents the maximum torque achieved during the cure time.

3. ts2

ts2 is the time required for the state of cure to increase to two torque units above the minimum at the given cure temperature. This parameter often correlates well with the Mooney scorch time.

4. t_{25}

t_{25} is the time required for the state of cure to reach 25% of the full cure defined as (Mh–Ml). Generally, a state of cure of about 25%–35% is necessary to prevent the development of porosity from occurring when a large rubber article is removed from a curing press. This level of cure also often provides enough strength to prevent the article from tearing as it is removed from a curing mold.

5. t_{90}

t_{90} is the time required to reach 90% of full cure defined as (Mh–Ml). t_{90} is generally the state of cure at which most physical properties reach optimal results.

II. VULCANIZING AGENTS

Sulfur is the oldest and most widely used vulcanizing or crosslinking agent and will be the vulcanizing agent of interest in most of this discussion. For the general purpose elastomers, the majority of cure systems in use today involve the generation of sulfur-containing crosslinks, usually with elemental sulfur in combination with an organic accelerator. Over the years, the proportion of sulfur has tended to fall and the levels of accelerator and the use of sulfur donors increased to give great improvements in the thermal and oxidative stability of the vulcanizate. Other vulcanization systems not involving sulfur or sulfur donors are less commonly used and involve various resins such as resorcinol-formaldehyde resins, urethanes, or peroxides. Metal oxides or sulfur-activated metal oxides can be used for halogenated elastomers.

We will, therefore, consider elemental sulfur and sulfur-bearing chemicals (sulfur donors) as one class of vulcanizing agents and nonsulfur vulcanizing agents as a second class.

A. Sulfur and Sulfur Donors

Sulfur vulcanization occurs by the formation of sulfur linkages or crosslinks between rubber molecules, as shown in Figure 11.3. In conventional sulfur vulcanization (generally formulated as higher sulfur to accelerator ratio), the resultant network is rich in polysulfidic sulfur linkages. Sulfur chain linkages can contain

Elastomer + S$_8$

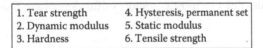

Heat

FIGURE 11.3   Sulfur vulcanization.

four-to-six or more sulfur atoms. Lower sulfur to accelerator ratios produce networks that are characterized by a greater number of sulfur linkages containing fewer sulfur atoms. Thus, the so-called efficient vulcanization systems produce higher crosslink densities for the same loading of sulfur. At very low sulfur to accelerator ratios, networks can be produced that are composed predominately of monosulfidic and disulfidic crosslinks.

Unaccelerated sulfur vulcanization is a slow, inefficient process. Over a century of research efforts have been directed toward the development of materials to improve the efficiency of this process. The activators, accelerators, and retarders to be discussed in later sections have resulted from these endeavors.

Figure 11.4 depicts the general changes in vulcanizate physical properties as the vulcanization state of the rubber changes. As the crosslink density of the vulcanizate increases (or the molecular weight between crosslinks decreases), elastic properties such as tensile and dynamic modulus, tear and tensile strength, resilience and hardness increase while viscous loss properties such as hysteresis decrease. Further increases in crosslink density will produce vulcanizates that tend toward brittle behavior. Thus, at higher crosslink densities, such properties as hardness and tear

1. Tear strength	4. Hysteresis, permanent set
2. Dynamic modulus	5. Static modulus
3. Hardness	6. Tensile strength

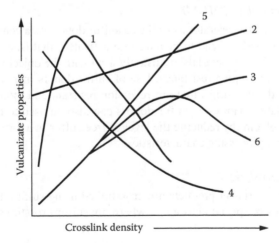

FIGURE 11.4   Effects of vulcanization.

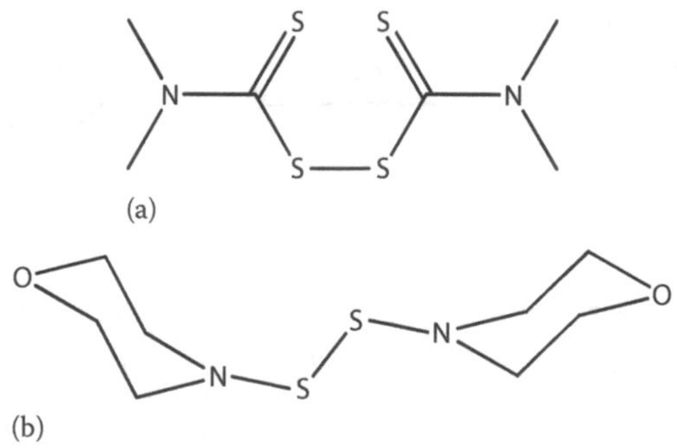

FIGURE 11.5   Sulfur donors: (a) Tetramethyl thiuram disulfide and (b) dithiodimorpholine.

and tensile strength plateau or begin to decrease. As a consequence, proper compounding must be done to provide the best balance in properties for the specified application.

Unaccelerated sulfur vulcanization is a slow, inefficient process. Over a century of research efforts have been directed toward the development of materials to improve the efficiency of this process. The activators, accelerators, and retarders to be discussed in later sections have resulted from these endeavors. Another class of chemicals, known as sulfur donors, has been developed to improve the efficiency of sulfur vulcanization. These materials are used to replace part or all of the elemental sulfur normally used in order to produce vulcanized products containing fewer sulfur atoms per crosslink. In other words, these materials make more efficient use of the available sulfur. The two most common sulfur donors are the disulfides shown in Figure 11.5*:

1.   Tetramethylthiuram (TMTD)

a.   Dithiodimorpholine (DTDM)

TMTD also acts as an accelerator as well as a sulfur donor. As a consequence, compounds containing TMTD tend to be cure rate activated, that is, are scorchier and faster in cure rate. These materials are usually used with the objective of improving thermal and oxidative aging resistance. Use of sulfur donors increases the level of mono and disulfidic crosslinks that are reversion resistant and more stable toward oxidative degradation. However, sulfur donors can also be used to reduce the possibility of sulfur bloom (by reducing the level of free sulfur in a formulation) and to modify curing and processing characteristics.

b.   Nonsulfur Crosslinks

The vast majority of rubber products are crosslinked using sulfur. There are, however, special cases or special elastomers where nonsulfur crosslinks are necessary or desirable.

---

* A complete list of the abbreviations used in this report is given in Section VIII.

c. Peroxide Vulcanization

In peroxide vulcanization, direct carbon crosslinks are formed between elastomer molecules as shown in Figure 11.6 (i.e., no molecular bridges as in the case of sulfur cures). The peroxides decompose under vulcanization conditions forming free radicals. The radicals then abstract protons from the polymer backbone–forming polymeric radicals. The polymeric radicals then combine leading to direct crosslink formation. Thus, rubber goods prepares using peroxide vulcanization are said to have a carbon network. Peroxides can be used to crosslink a wide variety of both saturated and unsaturated elastomers, whereas sulfur vulcanization will occur only in unsaturated species.

In general, carbon–carbon bonds from peroxide initiated crosslinks are more stable than the carbon–sulfur–carbon bonds from sulfur vulcanization.[3] Carbon networks are very low in hysteresis and very low in compression set as well. Thus, peroxide initiated cures often give superior aging properties in the rubber products. Carbon networks tend to have lower fatigue, tear, abrasion, and tensile properties compared to sulfur networks.[4] However, peroxide initiated cures generally represent higher cost to the processor and require greater care in storage and processing.

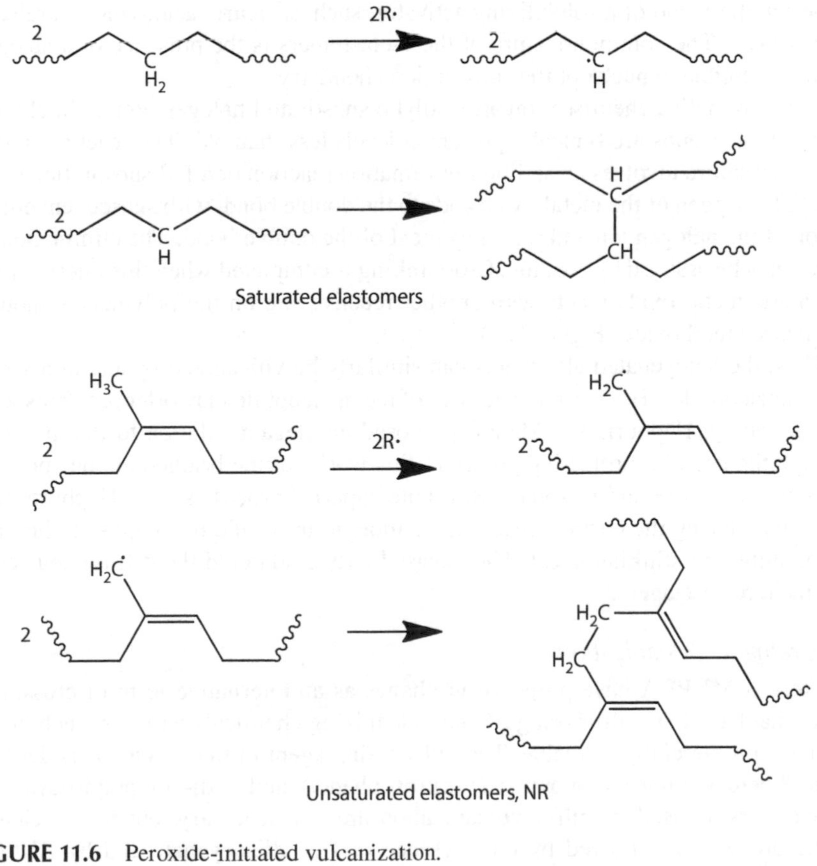

FIGURE 11.6 Peroxide-initiated vulcanization.

A wide variety of organic peroxides are available including products such as benzoyl peroxide and dicumyl peroxide. Proper choice of peroxide class must take into account its stability, activity, intended cure temperature, and effect on processing properties.

Carbon–carbon crosslinks can also be initiated by gamma or x-radiation; these presently find limited commercial application.

d. Resin Vulcanization

Certain difunctional compounds form crosslinks with elastomers by reacting with two polymer molecules to form a bridge. Epoxy resins are used with nitrile, quinone dioximes, and phenolic resins with butyl and dithiols or diamines with flourocarbons. The most important of these is the use of phenolic resins to cure butyl rubber. This cure system is widely used for the bladders used in curing new tires and the curing bags used in the retread industry. The low levels of unsaturation of butyl do require resin cure activation by halogen-containing materials such as $SnCl_2$.

e. Metal Oxide Vulcanization[5]

Metal oxides react with the halogenated elastomers, such as chloroprene, chlorosulfonated polyethylene, and halobutyl rubber. Crosslinking can be effected by the action of zinc oxide, magnesium oxide, litharge alone or in combination, and almost always with the aid of a solubilizing activator, such as stearic acid, zinc stearate, or zinc octoate. The common feature of these elastomers is the presence of a halogen atom susceptible to nucleophilic substitution chemistry.

The most active chemistry involves allylic substituted halogen atoms. In chloroprene, these groups are typically present at levels less than 5%. The reaction in this type of structure involves an addition-elimination reaction or a 1–3 substitution reaction.[6] The oxygen of the metal oxide adds to the double bond, with subsequent elimination of the halogen ion and rearrangement of the double bond. The elimination of the Cl may be assisted by zinc ion. Crosslinking is completed when this intermediate polymeric metal oxide reacts with another reactive site on the polymer of another polymeric metal oxide (Figure 11.7).

Thus, the halogenated elastomers can similarly be vulcanized by any of a series of bis-nucleophiles. However, the activity of the nucleophile may offer problems with process safety. Hepburn and Mahdi[7] proposed an elegant solution to this problem through the use of a protecting group. In this work, amine bridged amides provide good states of cure and in some cases quite superior process safety. Hepburn proposed that during the curing process the amide group would decompose to liberate the diamine crosslinking agent. The released fatty acid could the diffuse and serve as a mold release agent.

f. Urethane Vulcanization

Workers at MRPRA have proposed urethanes as an alternative form of crosslinking to that based on sulfur bridges[8] and vulcanizing chemicals based on such products are commercially available. The vulcanizing agent in these systems is derived from P-benzoquinone monoxime (P-nitrosophenol) and a di- or polyisocyanate. Accelerators as used in sulfur vulcanization are not necessary, but the efficiency of the process is improved by the presence of free diisocyanate and by ZDMC.

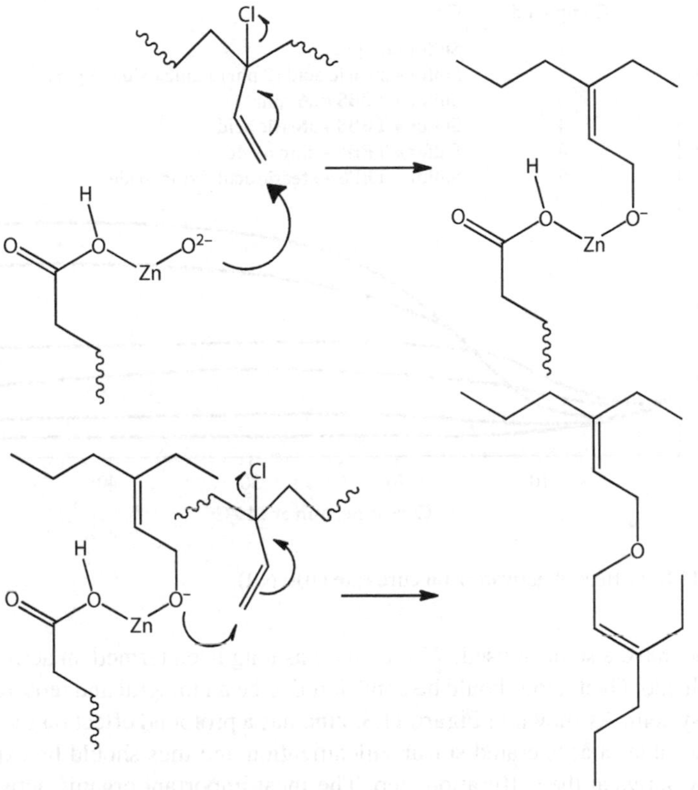

FIGURE 11.7 Metal oxide vulcanization of chloroprene.

The latter catalyzed the reaction between the nitrosophenol and the polymer chain to form pendant groups.

The principal advantage of these systems lies in the high stability of the crosslinks that give very little modulus reversion even on extreme overcure. Problems can occur with their lower scorch, rate of cure and modulus. However, modulus and fatigue life retention on aging is very good. Work in a number of laboratories is aimed at seeking crosslink systems that will be thermally labile at high temperatures but perform elastically at operating temperatures, thus bringing rubber molding closer to plastics technology. One such patent[9] uses an elastomer obtained by reacting a metal salt with a coordinating basic group present in an elastomer containing an electron-donating atom. Co-polymers of BR, SBR and vinylpyridine may be used with zinc, nickel, and cobalt chlorides.

III. ACTIVATORS

Realization of the full potential of most organic accelerators and cure systems requires the use of inorganic and organic activators. Zinc oxide is the most important inorganic activator, but other metallic oxides (particularly, magnesium oxide and

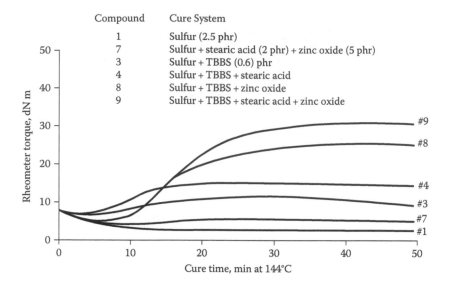

FIGURE 11.8   Effect of activators on cure rate (100 NR).

lead oxide) have also been used. While zinc has long been termed an activator, zinc or other divalent metal ion should be considered to be an integral and required part of the cure system. As shown in Figure 11.8, zinc has a profound effect on the extent of cure achievable in accelerated sulfur vulcanization and thus should be expected to inherently active at the sulfuration step. The most important organic activators are fatty acids, although weak amines, quanidines, ureas, thioureas, amides, polyalcohols, and aminoalcohols are also used.

The large preponderance of rubber compounds today use a combination of zinc oxide and stearic acid as the activating system. Several studies[10–12] have been published on the effects of variations in the concentrations of these activators. In general, the use of the activators zinc oxide and stearic acid improve the rate and efficiency of accelerated sulfur vulcanization. Rheographs obtained on stocks containing various combinations of cure system components are shown in Figure 11.8.

In the absence of an accelerator, the activators, zinc oxide and stearic acid, are ineffective in increasing the number of crosslinks produced (Figure 11.8, compounds 1 and 7). The use of a nonmetal-activated sulfenamide accelerator with sulfur produces a significant increase torque (crosslinks) in a reasonable period of time (compounds 3 and 4). This stock, however, would not be considered to be very well cured by today's standards.

The addition of zinc oxide as the only activator to the accelerated stock produces a dramatic effect and produces a well-cured stock. This demonstrates the critical role of zinc in accelerated sulfur vulcanization. The boost in efficiency suggests that zinc should be considered an integral component of the intermediate responsible for the attachment of sulfur to the rubber in the crosslink reactions. In order for zinc to be used effectively, it must be present in a form that can react with the accelerator system. This means that the zinc must be in a soluble form or a very fine particle size zinc oxide must be used (so that it can be readily solubilized). Most natural rubbers, and some synthetics, contain enough fatty acids to form soluble zinc salts (from added zinc oxide) that interact with the accelerators. Sulfenamide accelerated cures will release free amine that produces a soluble zinc amine complex from the zinc oxide. To assure that sufficient acids are available to solubilize zinc, it is common to add 1–4 phr of stearic acid, or a similar fatty acid. In addition to solubilizing zinc, the fatty acid serves as a plasticizer and/or lubricant to reduce the viscosity of the stock. The use of fatty acid soaps permits full development of crosslinks by the organic accelerator as is shown in compound 9 in Figure 11.8.

Other methods are also used to provide a soluble form of zinc ions. Basic zinc carbonates are more soluble in rubber than fine particle zinc oxide and can, therefore, be used in higher concentrations. Soluble fatty acid zinc salts are used providing both better dispersion and solubility of zinc ions. Common products include zinc stearate and zinc 2-ethylhexoate.

IV. ACCELERATORS

Although many people consider that the development of accelerators began in the early 1900s, the first vulcanization patent[2] issued in the United States described the "combination of said gum with sulfur and white lead to form a triple compound." Whatever the course of Goodyear's experimentation in 1839, his first patent covered an accelerated vulcanization with sulfur. Since that time, many people have studied the use of inorganic and organic compounds as accelerators for sulfur vulcanization.

In the nineteenth century, a number of inorganic compounds, particularly oxides and carbonates, were used as accelerators. These materials did give shorter curing times, but gave little improvement in physical properties. In the early 1900s, the accelerating effect of basic organic compounds was discovered. In 1906, Oenslager[13] found that aniline and other amines accelerated sulfur vulcanization. Since that time, emphasis has been placed on nitrogen and sulfur-containing organic compounds. Important milestones along the way have been the discovery of dithiocarbamates in 1918, 2-mercapthobenzothiazole (MBT) in 1921, and benzothiazole sulfenamides in 1937.

Today, the rubber compounder has available more than one hundred single products of known composition and 37 blends and unspecified materials.[14] Accelerators and accelerator systems are chosen on the basis of their ability to control the following processing/performance properties of rubber compounds:

- Time delay before vulcanization begins (scorch safety).
- Speed of the vulcanization reaction after it is initiated (cure rate).
- Extent of the vulcanization after the vulcanization reaction is complete (state of cure).
- Other factors such as green stock storage stability, fiber or steel adhesion, bloom tendency, cost, etc.

The job of the compounder, therefore, becomes one of selecting and evaluating individual accelerators and/or combinations of accelerators. The proliferation of accelerator types should be viewed as an opportunity since it often gives compounders a chance to custom fit curing systems to their processing/performance needs. This section attempts to categorize and predict performance within and between generic classes of accelerators. Like many reviews, it draws generalizations that may often be violated. The experienced compounder will find numerous instances where performance orders are reversed or otherwise out of order in compounds he has developed. Rather than a definitive list of exact properties, the following is an expectation of what an accelerator response might be, if one has no other data from which to draw conclusions.

A. ACCELERATOR CLASSES

Accelerators may be classified chemically and functionally. The principal chemical classes of accelerators in commercial use today are shown in Table 11.1.

Functionally, these compounds are typically classified as primary or secondary accelerators (including ultra-accelerators or ultras). Those compounds classified as primary accelerators usually provide considerable scorch delay, medium-to-fast cure rates, and good modulus development. Compounds classified as secondary accelerators or ultras usually produce scorchy, very fast curing stocks.

Generally accepted functional classifications of the accelerators are as shown in Figure 11.9.

By proper selection of these accelerators and their combinations, it is possible to vulcanize rubber at almost any desired time and temperature. Of course, the speed of vulcanization is not the same in all polymers. Highly unsaturated elastomers such as natural rubber or butadiene rubber will cure faster with a given vulcanization system than will those polymers containing fewer double bonds such as SBR (85 mol% unsaturation) or EPDM (5–12 mol% unsaturation).

In these polymers, it is common to use higher accelerator levels and less sulfur. However, the relative relationships between accelerators are similar in all of these elastomers, and the comparison between accelerator classes shown in Figure 11.10 is typical.

The development of an activated sulfenamide cure system to meet specific requirements of processing and physical properties requires both a selection and

TABLE 11.1
Accelerator Classes

Class	Response Speed	Acronyms
Aldehyde-amine	Slow	
Guanidines	Medium	DPG, DOTG
Thiazoles	Semi-fast	MBT, MBTS
Sulfenamides	Fast-delayed action	CBS, TBBS, MBS, DCBS
Dithiophosphates	Fast	ZBPD
Thiurams	Very fast	TMTD, TMTM, TETD
Dithiocarbamates	Very fast	ZDMC, ZDBC

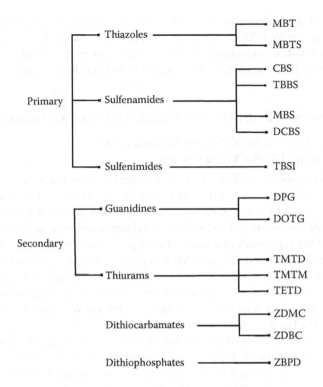

FIGURE 11.9 Primary and secondary accelerators.

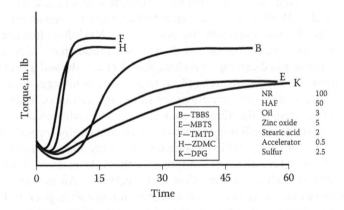

FIGURE 11.10 A comparison of common classes of accelerators.

a refining process. The initial selection of the primary and possibly secondary accelerators to be used is based primarily upon the needs of cure rate, time, and processability balanced by cost. After this decision has been made, a systematic study is required to fit these accelerators to the specific process conditions to be encountered.

To assist in this process, we will first look at a comparison of both primary and secondary accelerators. Then, the effects of primary to secondary ratios and total concentrations will be examined. In each case, the comparison will be based upon Mooney scorch, Rheometer cure characteristics, and tensile modulus.

B. Mechanism of Zinc-Mediated Accelerated Sulfur Vulcanization

1. Historical and General Aspects Related to the Mechanism of Sulfur Vulcanization

Much is known about accelerated sulfur vulcanization of the various diene elastomers. Each elastomer shows differences in various aspects of its vulcanization chemistry. These differences are related to the physical and chemical nature of the elastomer under consideration and to the cure systems employed. Several reviews discuss in detail the early work that has led to the prevailing theories on vulcanization: Chapman and Porter rigorously summarize the chemistry of sulfur vulcanization in natural rubber, and Kresja and Koenig cover sulfur vulcanization in various other elastomers.[15] Most recently, quantitative structure–activity relationship (QSAR) studies have shed more insight as to the nature of the active sulfur–accelerator–zinc complex involved in the vulcanization reaction.[16]

There are many classes of compounds that can serve as accelerators in sulfur vulcanization as shown in Table 11.1. A feature common to vulcanization accelerators is some form of a tautomerizable double bond. In fact, the most active contain the –N=C–S–H functionality. This is the common structural unit found in all of the 2-mercapto-substituted nitrogen heterocyclic accelerators known today. Note that the delayed action precursors, 2-mecaptobenzothiazole disulfide, sulfonamides, and sulfenimides of 2-mercaptobenzothiazole decompose to form 2-mercaptobenzothiazole a structure that contains this N=C–S–H functionality.

By comparing vulcanization activity in accelerators derived from 4-mercaptopyridine and 2-mercaptopyridine, Rostek et al.[17] showed that the position of sulfur "ortho-" to the heteroatom (which in this case is nitrogen) is a structural requirement for activity as an accelerator for sulfur vulcanization. It has been suggested that the function of the nitrogen atom is to act as a hydrogen acceptor during the sulfuration and the crosslinking reactions.[18] This empirically derived mechanism has been used to explain the allylic substitution[19] and concomitant formation of MBT during sulfuration and crosslinking.[20] A typical rubber vulcanizate will contain various components in addition to the sulfur and accelerator. An example of natural rubber vulcanizate prepared using a conventional cure system is given in Table 11.2. As discussed in the preceding section, the rates of vulcanization and states of cure depend not only with the type of accelerator used but also on the amount and type(s) of activator(s) (e.g., stearic acid, zinc oxide, and/or secondary accelerators such as DPG or TMTD, etc.) The time to the onset of cure varies with the class of accelerator used. Some accelerators provide only a relatively short delay before network formation begins. The sulfenamides and sulfenimides are special classes of accelerators that provide for a long delay period before the onset of the network formation.

TABLE 11.2

Composition of a Typical Rubber Vulcanizate

Ingredient	Parts in Formulation
Natural rubber	100
N-330 black	50
Oil	3.0
Stearic acid	2.0
Zinc oxide	5.0
Antidegradant	2.0
Sulfur	2.4
Accelerator (e.g., TBBS)	0.6

Each component of a cure system plays an important role in determining the rate and nature of the vulcanization reaction.

The understanding of the mechanism of vulcanization long remained unclear because of the inherent nature of the problem. During vulcanization, a very small percentage of material reacts with the polymer transforming it into a network of intractable material that is difficult to analyze by traditional methodology. Much of the understanding has been developed through model compound studies, studies of vulcanization reaction kinetics and tracing the fate of the accelerator and sulfur chemicals through extraction and HPLC analysis. Recently, NMR spectroscopic methods have helped to elucidate the nature of the sulfur attachment to the rubber. Most recently, insight has been developed through the use of QSAR studies.

Generally speaking, it is the role of accelerators and cure activators to activate the elemental sulfur and/or the rubber for the crosslinking reaction. Sulfur may be activated by the reaction of the amine with the sulfur molecules generating ammonium polysulfide anions or polysulfidic radical anions. These combine or react to form amine poly(sulfides) or alkyl ammonium polysulfides, which have been proposed as intermediates in vulcanization.[21] Various zinc accelerator complexes have long been postulated as the active sulfurating agents in zinc-containing cure systems.[22] These zinc accelerator complexes having the general structures are shown in Figure 11.11. Such complexes are modified through the action of ligands derived from accelerators (amines from sulfenamides), activators (stearic acid or zinc stearate), or secondary accelerators such as amines, amides, ureas, and guanidines. The complex species of polysulfidic analogues of such structures have been proposed to be involved in the reactions by which sulfur is attached to the rubber and crosslinks are formed.[23]

The zinc accelerator complexes may incorporate additional atoms of sulfur to form zinc accelerator perthiolate type complexes as in Figure 11.11b.[24] Sulfur has also been shown to insert into zinc complexes of dithioacids.[25] The sulfur atoms in the perthiolato zinc complexes are labile and thus readily exchange sulfur atoms. These complexes of labile sulfur have been shown to be effective accelerators.[26] In fact, it was proposed[27] that this type of sulfur insertion reaction may be general in

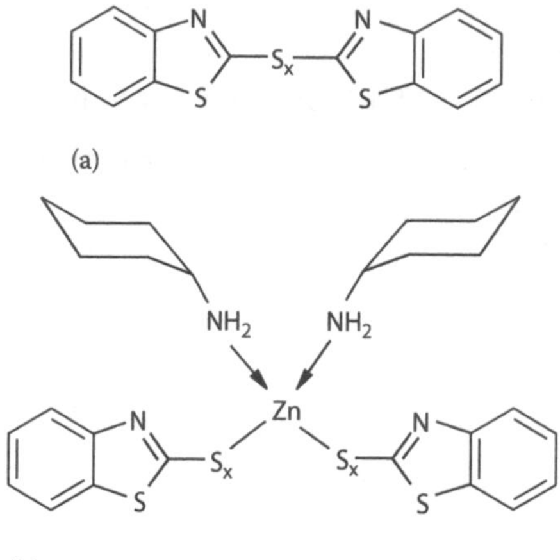

(a)

(b)

FIGURE 11.11   Generalized structures of sulfurating intermediates.

zinc-mediated accelerated sulfur vulcanization. Ultimately, these sulfur exchange and insertion reactions form the bulk of the pre-reactions occurring during delayed action sulfenamide or sulfenimide catalyzed sulfur vulcanization.

Many different mechanisms for sulfur vulcanization have been suggested. Proposed pathways often involve several competitive and/or consecutive reactions, and can involve numerous intermediates. Sometimes, as many as fifteen different chemical intermediates have been proposed.[28] With the large number of competitive reactions and the large number of intermediates, identifying one structure as a critical intermediate appears to be an insurmountable and unrealistic task. In fact, the large numbers of species found through experiment indicate that a complex competitive pathway may provide the best explanation for vulcanization chemistry. Although several intermediates are probably capable of and are likely to cause sulfuration and crosslinking of the rubber, it is likely that one mechanism with a characteristic intermediate dominates the process.

Reaction mechanisms can sometimes be elucidated through the identification of critical chemical intermediates followed by comparison to known reactions. More often, information regarding structure–activity relationships is instrumental in understanding the steps or mechanism of a chemical reaction. These relationships have traditionally correlated empirically derived structural parameters to chemical activity and referred to as QSAR studies.

2.   Historical QSAR Studies

QSARs were born in the first part of the last century. In 1935, Hammett formulated his famous equation in an effort to mathematically relate structural changes to chemical reactivity.[29] Three basic sets of parameters were initially developed. Each set of "σ constants" quantitates the effects of a substituent on a reaction such as the dissociation equilibrium of benzoic acids (σ) or substituted phenols ($\sigma-$) or the rate of solvolysis of cumyl chlorides ($XC_6H_4CCl(CH_3)_2$, $\sigma+$). Since the early days of the Hammett equation, numerous reactivity scales have been generated and large numbers of reactivity constants have been accumulated. Chief among these are the Taft-Hammett σ, and the Taft steric parameters, E_s.

The Hammett relations quantify differences between ground state energies of reactants and transition state energies of active intermediates and are often referred to as linear free energy relationships. Understanding how substituents (or a homologous series of chemical reactants) alter the kinetics of reaction provides direct evidence to identification of chemical nature of transition state complexes and ultimately the mechanism of the chemical reaction under consideration. Thus, defining the "electronic" effects of various compounds or substituents on the kinetics of reaction or understanding influential factors that alter the activation energy of reaction serve to explicitly define the nature of the studied reaction.

While the Hammett Sigma constants account for "resonance" and "inductive" effects in aromatic systems, Taft developed the first generally successful method for numerically explaining the steric effects and inductive polar effects in organic chemistry.[30]

3. Early QSAR Studies of Sulfur Vulcanization

Morita[31] has correlated the inductive effects of a series of sulfenamides and bis-thioformanalides to vulcanization activity. Steric effects were considered negligible (or at least uniform) in this series of substituted phenylthio- and substituted aniline-based mercaptobenzothiazole sulfenamides. Morita showed that pKa values and vulcanization parameters correlated reasonably well to the σ^* constant even though these parameters were developed for conventional organic chemistry (not chemistry involving sulfur and nitrogen).

While the correlations are reasonable for the mercaptobenzothiazole sulfen-amides based on the substituted aniline series used in this example, they are not consistent with the narrow subset of aliphatic amines included in Morita's study. Morita shows, in plots of cure properties versus σ^*, discontinuities that separate aliphatic amines from the substituted phenyl amines. Morita observed two linear relationships having slopes of opposite sign for N-(substituted-phenyl)- and N-alkyl-sulfenamides. Longer scorch delays were observed for electron-withdrawing substituted phenyl compounds and the sterically hindered alkyl substituents. Morita concluded that the more basic amino derivatives generally gave faster acceleration rates and higher crosslink efficiencies and longer scorch delays.

The discontinuity shown in Morita's data suggests that steric factors or electronic (inductive) effects are significantly different in the two amine classes of sulfen-amides. On the other hand, Morita shows that the ^{13}C-NMR plot of the C_2 carbon in the parent sulfenamide versus σ^* are continuous across both classes of amine sulf-enamides. Thus, the factors effecting chemical shifts in the ^{13}C spectra of the parent sulfenamide are different from the factors effecting vulcanization characteristics.

The parameters used in Morita's study have been derived for organic reactions that at most only involve oxygen at the reactive centers or transition states. In the case of sulfur vulcanization, the reactions clearly involve sulfur and carbon and possibly zinc and nitrogen as well. Hence, the relations derived by Morita are surprisingly good considering the differences in chemistry involved. Morita thus showed that the electronic and steric effects of the amine moiety of the derived sulfenamide provide critical influence in controlling the rate of sulfur vulcanization. No insight could be provided toward the role and influence of the heterocyclic portion of the accelerator, mercaptobenzothiazole.

4. Recent QSAR Studies

The previous studies of Morita were based on Hammett constants that had been developed for carbon- and oxygen-centered organic reactions. While sulfur is iso-electronic with oxygen, chemically it is somewhat different, softer, more polarizable and less electronegative, considerably more nucleophilic and less basic than oxygen. Thus, studies utilizing parameters based in sulfur and nitrogen chemistry would be more beneficial in understanding the nature of sulfur vulcanization.

Recently, a detailed QSAR study has provided significant insight into the mechanism of sulfur vulcanization.[32] These studies were based on semi-empirical quantum mechanical calculations describing "proposed" zinc complexes derived from a series of 24 sulfenimides and sulfenamides derived from various amines and sulfur substituted nitrogen heterocycles (see Figure 11.12).

Thus, this study used parameters calculated by first intent to characterize sulfur- and nitrogen-containing structures pertinent to sulfur vulcanization, thereby overcoming the previous shortcomings.

The result of this study clearly showed both the effects of the amine moiety and the heterocyclic thiol on the rate of vulcanization. A model describing the rate of vulcanization was derived employing four terms that accounted for than 96% of the variance in the rates of reaction ($R^2 = 0.9667$). The four parameters were

1. Electron density in the Zn–sulfur bond (electron–electron repulsion)
2. Electron density in the C=N bond (electron–electron repulsion)
3. Interaction parameter for a N–H bond (measure of the quality of interaction of the amine ligand with zinc)
4. Molecular surface area

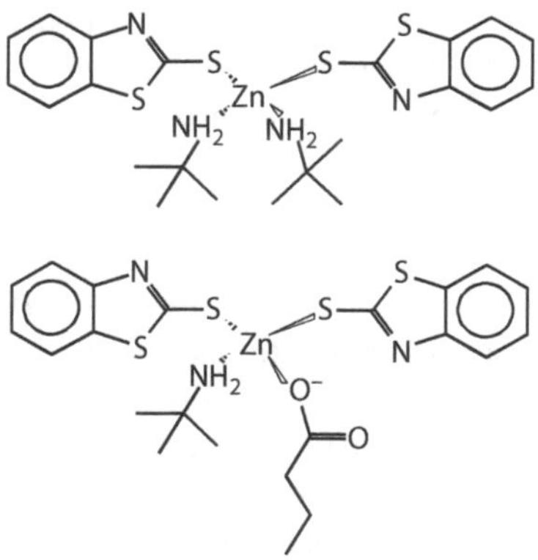

FIGURE 11.12 Sulfenamide and sulfenimide zinc complexes.

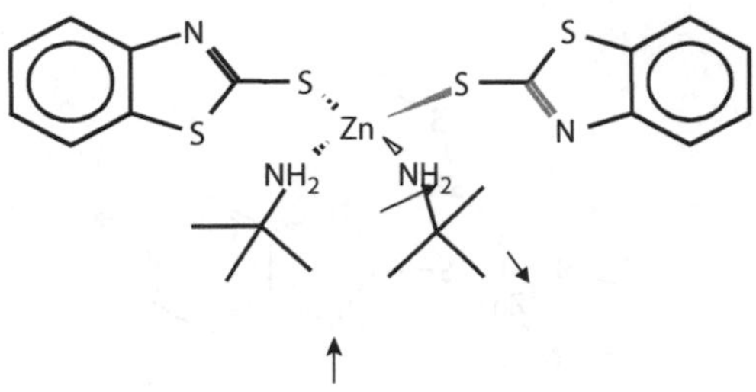

FIGURE 11.13 Arrows indicate the directional "characteristic flow" of electrons favoring faster rates of vulcanization. Note that this flow would naturally facilitate the function of the nitrogen in the heterocycle moiety as a proton acceptor.

Generalized conclusions can thus be drawn from these results. The data support heterocyclic thiols forming strong S–Zn complexes tend to make for slower accelerators. Increasing the electron density in the C=N bond tends to increase the rate of reaction especially when reducing the strength of the S–Zn bond. Improving the quality of the interaction of the amine ligand with the zinc increases the rate of vulcanization. Consequently, structures that favor the general flow of electrons away from the Zn–S bond and into the C=N bend will tend toward faster acceleration (as depicted in Figure 11.13). And finally, since the reaction involves diffusion of metal complexes through a viscous liquid, the rate of reaction is diffusion controlled and thus depends upon the surface area of the complex. Therefore, large accelerator complexes provide for slower reaction kinetics.

This model rationalizes the differences between primary and secondary amine-based sulfenamides. While a more quantitative discussion is given later, the effects are readily understood in qualitative terms. Primary amine-based sulfenamides are typically faster accelerators than those based on secondary amines. In terms of raditional logic, stronger bases would provide for faster reaction kinetics. Thus, neglecting steric effects, secondary amines might be expected to provide for faster vulcanization rates. This discrepancy can now be readily understood as the greater steric nature of the secondary amines reduces the effectiveness of the interaction of the nitrogen to zinc.

The complex as modeled has important significance in understanding the possible structure of a sulfurating intermediate (Figure 11.14). In historically proposed zinc complexes, the heterocyclic thiol was attached to the zinc atom by a chain of sulfur atoms. In the aforementioned structures, accelerator thiolate ions are attached directly to the zinc atom. In the historically proposed structure, it is unlikely that electronic effects derived from the nature of a heterocyclic thiol joined to zinc through a polysulfidic chain (as in structure B, Figure 11.11) would significantly influence the kinetics of sulfur vulcanization. Any number of sulfur atoms in a chain attaching the thiol to zinc should significantly modulate electronic effects of various heterocycles. Thus, for polysulfidic linkages between zinc and the accelerator, the electronic influence on the complex is nonexistent.

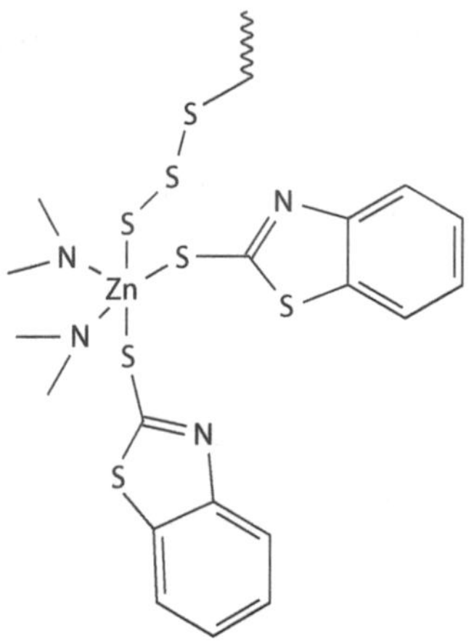

FIGURE 11.14 Proposed structure for the sulfurating intermediate that leads to crosslink formation.

In this model, sulfur (to be added to the polymer chain) is directly attached to zinc at an expanded ligand site (i.e., four coordinate Zn moved to five coordinate zinc, where the fifth coordination site is occupied by the sulfur). This five coordinate structure then interacts with the double bond in the polymer (possibly forming now a hexavalent complex (although elimination of one of the weaker ligands should not be ruled out) and reaction takes place inserting sulfur in the allylic position (Figure 11.15).

Polysulfidic zinc structures are chemically poised for the sulfuration reaction. Zinc hexasulfide complexes have been shown to serve as polysulfidic sulfur donors.[33]

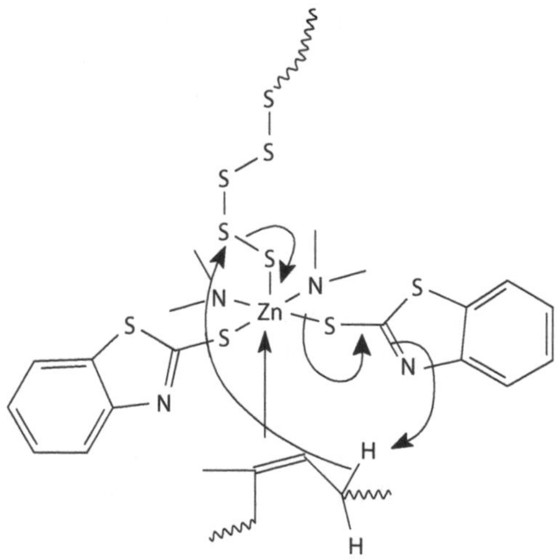

FIGURE 11.15 Crosslink formation.

These complexes are so inherently reactive that when heated to vulcanization temperatures, compounds containing an accelerator (such as a sulfenamide) undergo rapid vulcanization with exceptionally short scorch delay. The resulting network is rich in polysulfidic sulfur crosslinks. Rapid vulcanization is normally achieved through the use of combinations of secondary accelerators with sulfenamides but normally results in networks having short sulfur linkages (primarily mono and disulfidic networks).

5. Molecular Explanations of Various Accelerator Activities

The reactivity of various heterocyclic thiol–based sulfenamides or sulfenimides and the influence of the corresponding amines can now be understood in a more quantitative fashion. So, for example, the accelerators shown in Figure 11.16 can be quantitatively compared. Table 11.3 compares these accelerators having various degrees of activity. For each accelerator, the relative contribution to the maximum rate of vulcanization for the critical structural features is provided along with the overall observed and the predicted maximum rate of vulcanization. In general, as can be seen from the table "sulfenamides are faster than sulfenimides" and "primary amine sulfenamides are faster than secondary amine-based sulfenamides."

The aforementioned accelerators can now be compared using TBBS as a reference point. The steric nature of the dicyclohexylamine is so large that a number of interactions are altered especially the Zn–S bond and the N–Zn bond. As a result, the complex behaves as a rather electron-starved system and the resulting rate is slower than the TBBS system. In addition, the surface area of the complex is so large that this effect alone accounts for a nearly 20% reduction in reactivity of the DCBS accelerator system compared to the TBBS.

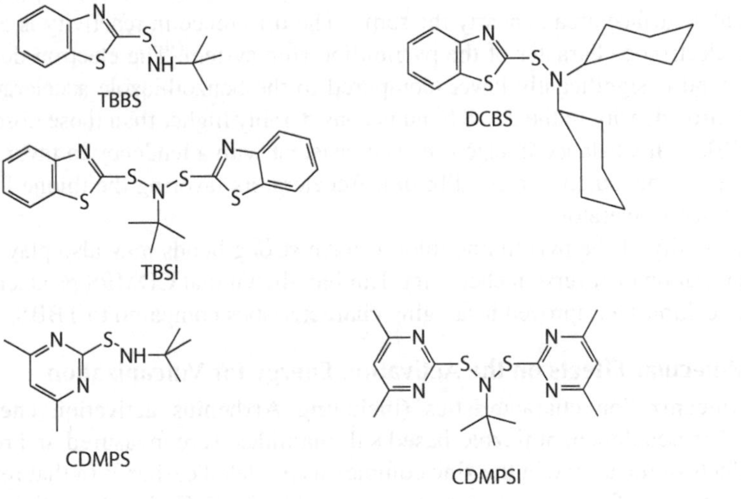

FIGURE 11.16 Structures of accelerators listed in Table 11.3.

TABLE 11.3

Accelerator Type and Rate of Vulcanization[a]

Compound	Intercept	Electron Density Zn–S Bond	Electron Density C–N Bond	Exchange Energy N–H	Molecular Surface Area	Pred.	Obs.
TBBS	−178.98	−1.87	47.22	141.23	−2.46	5.14	5.6
TBSI	−178.98	−3.05	46.63	140.09	−2.53	2.16	2.57
CDMPS	−178.98	−3.26	44.44	141.87	−2.44	1.63	1.1
DCBS	−178.98	−3.60	47.07	138.49	−3.44	2.54	2.6
CDMPSI	−178.98	−3.09	43.93	141.41	−2.50	0.76	1.1

[a] Contributions to the maximum rate of vulcanization from structural features of corresponding zinc complexes.

TBSI, being a sulfenimide, is modeled as a complex having one amine and one acid moiety. The electronic effect of substituting the acid for the amine is to effectively increase electron density in the Zn–S bond. The increase in electron density in the Zn–S bond is a result of reduced steric hindrance allowing for better interaction between the Zn and S atoms and also a result of the inductive effect of the oxygen (oxygen being more electronegative than nitrogen). The inductive effect of the oxygen also reduces the N–Zn interaction as can be seen in the N–H exchange energy. The total result is an accelerator having significantly slower kinetics compared to TBBS.

Finally, CDMPS while having the same amine moiety as TBBS (*t*-butyl amine) has a different heterocyclic thiol (4,6-dimethyl-2-mercaptopyrimidine). In this complex, the N–Zn interaction is similar that observed in the TBBS complex and the molecular surface area is nearly the same. The difference in reactivity is attributed to the electronic character of the pyrimidine ring system. The electron density the C=N bond is significantly lower (compared to the benzothiazole accelerators) and the electron density in the Zn–S bond is considerably higher than those observed for the TBBS. This balance in electrons is consistent with a tendency to favor the thiol tautomer in the tautomeric equilibrium. Accelerators favoring the thione form tend to be faster accelerators.

The ability of the pyrimidine thiol to form strong bonds may also play a role in the maturation or reversion chemistry. Lin has shown that CDMPS produces vulcanizates exhibiting improved heat aging characteristics compared to TBBS.

6. Molecular Effects on the Activation Energy for Vulcanization

The vulcanization characteristics (including Arrhenius activation energy) for seven 2-mercaptobenzothiazole-based sulfenamides were measured and related to the effects of the amine in the zinc complex as modeled earlier.[34] In that report, the maximum rate of vulcanization was correlated to the N–Zn bond length in the zinc complex ($R^2 = 0.987$, df = 13). Other likely amine constants of characterization such as Taft steric constants, pKa, or Hammett Sigma* constants gave poor correlations.

The Arrhenius activation energy correlated well also with the N–Zn bond length ($R^2 = 0.9040$, df = 5). Recent calculations produced a single-parameter model correlating the activation energy for sulfur vulcanization, E_a, with maximum net atomic charge on N having $R^2 = 0.9554$, df = 6. The maximum charge on the nitrogen atom is found on the heterocyclic ring nitrogen. The coefficient for the N charge parameter being negative supports the expectation that the heterocyclic ring nitrogen serves as the hydrogen acceptor in the sulfuration step.[35]

All of these results provide strong support that a complex similar to those shown in figures given earlier are likely to play a strong role in the sulfurization step. These complexes then can be characterized as having a heterocyclic thiol directly bonded to the zinc atom and sulfur attached separately to the zinc as shown in the zinc hexasulfides complexes. Clearly, kinetic effects will be altered in practice as various compounding ingredients can influence the equilibrium and in fact nature of the zinc complex. Practical compounding examples are provided in the next section.

C. PRACTICAL COMPARISON OF PRIMARY ACCELERATORS

The response of an elastomer to a specific accelerator varies with the number and activity of the double bonds present. NR, SBR are typical of the highly unsaturated polymers in use and will be used as examples in this presentation.

1. Natural Rubber

Typical responses of MBTS and the common sulfenamides are compared in NR in Table 11.4 and Figure 11.17. Compared to MBTS, the sulfenamides provide longer scorch delay, faster cure rates, and higher modulus values.

2. SBR

Similar responses in SBR are shown in Table 11.5 and Figure 11.18. The comparison of the thiazole accelerator, MBTS, with the sulfenamides is similar to that found in NR. The differences between sulfenamides are, however, more pronounced than those found in NR.

3. Performance Comparison

At equal concentrations, the sulfenamides can generally be ranked as follows:

a. Scorch Delay
MBTS < CBS ≈ TBBS < MBS < DCBS

b. Cure Rate
CBS ≈ TBBS > MBS > DCBS

c. Modulus Development
TBBS > MBS ≈ CBS > DCBS

TABLE 11.4
Comparison of Primary Accelerators in NR

SMR-5CV	100.0			
N-330 black	50.0			
Sundex 790	3.0			
Zinc oxide	5.0			
Stearic acid	2.0			
TMQ	1.0			
Sulfur	2.4			
MBTS	0.6	—	—	—
CBS	—	0.6	—	—
TBBS	—	—	0.6	—
MBS	—	—	—	0.6
Mooney Scorch at 121°C				
Minimum viscosity	50.0	45.9	45.2	46.9
t_5, min	6.7	12.0	12.5	12.0
Rheometer at 144°C				
Min. torque, dN m	9.5	8.7	8.5	9.2
Max. torque, dN m	34.9	42.7	44.2	42.2
t_2, min	4.9	8.2	9.5	8.0
t_{90}, min	26.3	20.6	23.0	22.8
$t_{90}-t_2$	21.4	12.4	13.5	14.8
Stress–strain: t_{90} cure				
Shore "A" hardness	64.0	67.0	69.0	66.0
100% modulus, MPa	2.0	2.6	3.0	2.5
300% modulus, MPa	10.0	13.3	14.9	13.0
Ult. tensile, MPa	24.6	28.2	29.1	28.4
Ult. elongation, %	570.0	550.0	590.0	555.0

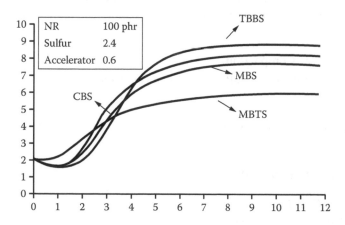

FIGURE 11.17   Primary accelerators—NR 144°C.

TABLE 11.5
Comparison of Primary Accelerators in SBR

SBR 1606	162.0			
Zinc oxide	5.0			
Stearic acid	1.0			
TMQ	2.0			
Sulfur	1.8			
MBTS	1.2	—	—	—
CBS	—	1.2	—	—
TBBS	—	—	1.2	—
MBS	—	—	—	1.2
Mooney scorch at 135°C				
Minimum viscosity	43.2	42.0	42.0	41.3
t_5, min	24.5	34.0	35.7	54.1
Rheometer at 160°C				
Min. torque, dN m	7.6	7.6	7.0	7.7
Max torque, dN m	30.9	35.5	37.2	36.6
t_2, min	5.2	8.0	7.7	10.6
t_{90}, min	37.5	16.7	17.5	21.5
$t_{90}-t_2$	32.3	8.7	9.8	10.9
Stress–strain: t_{90} cure				
Shore "A" hardness	67.0	67.0	68.0	68.0
100% modulus, MPa	1.8	1.9	2.2	2.1
300% modulus, MPa	7.5	9.5	10.3	9.9
Ult. tensile, MPa	20.7	21.6	23.2	21.5
Ult. elongation, %	645.0	570.0	565.0	550.0

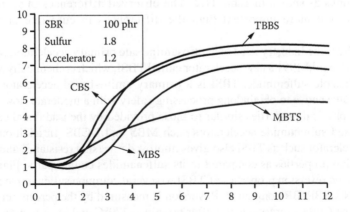

FIGURE 11.18 Primary accelerators—SBR 160°C.

The observed differences in scorch delay are larger and more important than the differences in cure rate or modulus. (These differences are a function of the amine from which the sulfenamide is derived.) Generally, the more basic amines produce sulfenamides that are scorchier and faster curing. Additionally, stearic hindrance will produce slower curing accelerators as in the case of DCBS.

The thiazoles (MBT and MBTS) are normally characterized by less processing safety, slower cure rate and lower modulus at equal use levels than the sulfenamides (CBS, TBBS, and MBS). Nonetheless, the thiazoles have found wide use in rubber products, particularly in mechanical rubber goods because

- In some cases, the longer processing safety of the sulfenamides may not be required and would lead to extended cure times.
- The sulfenamides often cause a slight degree of discoloration of light-colored compounds, thus thiazoles would be preferred in a critical white compound application.
- Reversion tendency of natural rubber is generally less with thiazoles than with sulfenamides, although sulfenamides may often give superior reversion resistance if less sulfur and more accelerators is used.
- A longer history of use of thiazoles, particularly in mechanical goods applications, may make a compounder reluctant to change from a time-proven system.

Sulfenamide accelerators serve to perform two functions. They provide the necessary time period required to mix, process, and shape rubber compounds before vulcanization begins. This portion of the overall vulcanization process is usually referred to as scorch delay. In addition, sulfenamides function as accelerators, in that once the crosslinking process has begun, they speed up this reaction.

Generally, the more basic the amine, the less processing safety and the faster curing the sulfenamide is. In addition, the more steric hindered the amine, the less processing safety but slower curing the sulfenamide as in the case of DCBS. These observations lead to a predicted ranking of some of the common sulfonamides as shown in Table 11.2. The observed differences in scorch delay are larger and more important than the differences in cure rate or modulus (Table 11.6).

To address the industry need of a nitrosamine safe primary accelerator, Monsanto introduced in mid-1991 a new accelerator called TBSI, which is chemically n-t-butyl-2-benzothiazole sulfenimide. TBSI is a primary amine-based accelerator that has provided alternative to obtain long processing safety with a moderate slow cure rate and good physical properties similar to those provided by the traditional secondary amine-based sulfenamide accelerators such MBS and DCBS. In addition, sulfenimide accelerator such as TBSI also gives improved reversion resistance and reduced heat buildup properties as compared to its sulfenamides counterparts. Figure 11.19 compares the reversion properties of TBSI to several common sulfenamide accelerators in a NR/SBR/BR compound. Reversion is measured by the percent retention of Moving Die Curemeter torque unit after 60 min at 170°C and the retention of 100% modulus after overcure at 170°C for 10× rheometer t_{90} min.

TABLE 11.6
Comparison of Sulfenamide Accelerators

Sulfenamide or Sulfenimide	Abbrev.	Mooney Scorch		Rheometer Cure Rate	
		NR t_5, Min. @121°C	SBR t_5, Min. @135°C	NR $t_{90}-t_2$ @150°C	SBR $t_{90}-t_2$ @160°C
	TBBS	42.4	37.3	6.7	12.2
	CBS	32.9	33.2	5.8	11.2
	DCBS	33.7	63.8	12.7	19.0
	MBS	46.5	48.1	8.3	13.0
	TBSI	41.4	42.7	10.0	17.0

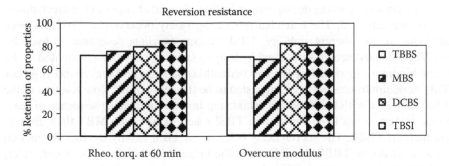

FIGURE 11.19 Reversion resistance.

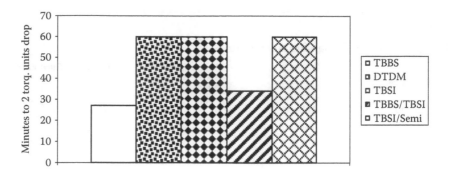

FIGURE 11.20   Reversion-resistant properties of TBSI in a NR/BR (75/25) blend.

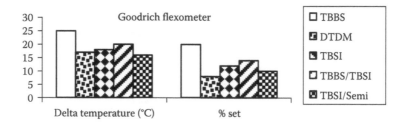

FIGURE 11.21   Goodrich flexometer.

Figure 11.20 gives the reversion-resistant properties of TBSI in a NR/BR (75/25) blend.

Reduced Goodrich heat buildup and permanent set in TBSI accelerated compounds are shown in Figure 11.21.

The improved physical properties observed in vulcanizates accelerated by TBSI suggest significant differences from the conventional sulfenamide cure chemistry. TBSI differs from TBBS (*N-tert*-butyl-2-benzothiazole sulfenamide) in steric effects and mercaptobenzothiazole (MBT) and amine stoichiometric differences. These structural and reactivity differences result in changes in the kinetics of some reactions occurring during vulcanization. Several changes in the mechanistic details are expected. The entended processing safety observed in TBSI has been attributed to its inherent stability. TBSI exhibits excellent resistance to hydrolysis. The extended scorch safety and hydrolysis stability have been attributed to the steric hindrance provided by the two benzothiazole rings and the *tert*-butyl group. This steric hindrance has been suggested to be the cause of the slow reaction of the TBSI molecule with MBT, an important step in the scorch delay sequence of reactions. In fact, it has been shown that TBSI reacts slower with MBT than does its counterpart TBBS. The scorch delay behavior of TBSI is only the first distinction from its analogue TBBS. In addition to the longer scorch delay, slower cure rates, and slower reversion rates, TBSI also provide vulcanizates with lower heat buildup properties.

D. COMPARISON OF SECONDARY ACCELERATORS

There are a large number of secondary accelerators that could be used with each of the sulfenamides, thereby providing a wide range of flexibility. In order to simplify matters, this presentation will examine only the more common secondary accelerators and their effect on TBBS as the primary accelerator. The effects of these materials on the other sulfenamides are similar.

These comparisons have been made in NR, SBR, and NBR using a MBTS/DPG system as a control in each case. Within a given polymer, the sulfur is held at a single concentration. Initial comparisons are made at the same concentration and in the ratio of primary to secondary accelerator. Variations in concentration and in the ratio of primary to secondary accelerator will be discussed later.

1. Natural Rubber

Seven different secondary accelerators were evaluated with TBBS an NR compound and compared with a MBTS/DPG control. The formulations used are shown in Table 11.7.

As is shown in Figure 11.22, all of the activated sulfenamide stocks provide more scorch delay than does the activated thiazole stock.

Of the secondary accelerators, the ZDMC is the scorchiest, and TETD provides the longest scorch delay.

TABLE 11.7
Comparison of Secondary Accelerators in NR[a]

Sulfur	2.5							
MBTS	1.2	—	—	—	—	—	—	—
DPG	0.4	—	—	—	—	—	—	—
TBBS	—	0.6						
TMTD	—	0.4	—	—	—	—	—	—
TMTM	—	—	0.4	—	—	—	—	—
TETD	—	—	—	0.4	—	—	—	—
ZDMC	—	—	—	—	0.4	—	—	—
ZDEC	—	—	—	—	—	0.4	—	—
ZDBC	—	—	—	—	—	—	0.4	—
DOTG	—	—	—	—	—	—	—	0.4
Mooney scorch at 121°C								
t_5, min	7.2	16.8	21.5	23.5	13.7	16.7	20.2	21.7
Rheometer at 143°C								
t_{90}, min	9.2	7.5	9.5	9.8	7.0	7.8	9.3	16.5
t_{90}–t_2	6.2	2.0	2.2	2.5	2.0	2.0	2.6	10.0
Stress–strain: t_{90} cure								
100% modulus, MPa	2.6	3.4	3.5	3.0	3.2	2.9	2.8	2.8

[a] #1RSS-100, FEF Black-40, Circolite® RPO-10, Zinc Oxide-5.0, Stearic Acid-1.5, 6PPD-2.0.

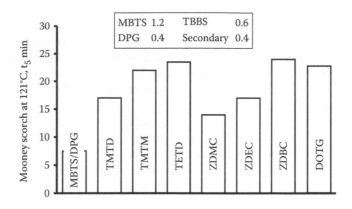

FIGURE 11.22 Comparison of secondary accelerators (100 NR/2.5 sulfur).

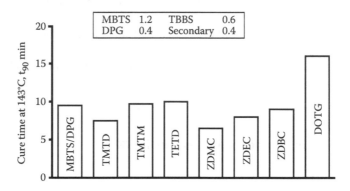

FIGURE 11.23 Comparison of secondary accelerators (100 NR/2.5 sulfur).

Conversely, those stocks containing a dithiocarbamate or thiuram show cure times (see Figure 11.23) at least as short as the activated thiazole control, even though they exhibit much longer scorch delays. Only the use of DOTG as a secondary shows a longer cure time than the control.

Therefore, one can obtain significant improvements in scorch protection with no increase in cure time through the use of an activated sulfenamide.

At a level of accelerator used in this study, all of the activated sulfenamides produced a higher modulus than the activated thiazole (see Figure 11.24). Of course, concentration adjustments can be made to equalize modulus if desired, and such adjustments will be discussed in a later section. The thiurams are known sulfur donors, and therefore, generally require more adjustment to equalize modulus.

2. SBR

These same chemicals have also been evaluated as secondary accelerators in SBR, as shown in Table 11.8.

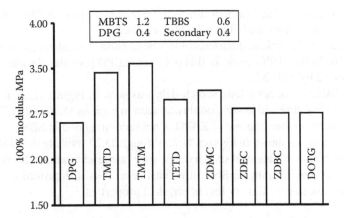

FIGURE 11.24   Comparison of secondary accelerators (100 NR/2.5 sulfur).

TABLE 11.8
Comparison of Secondary Accelerators in SBR[a]

Sulfur	1.8						
MBTS	1.2	—	—	—	—	—	—
DPG	0.4	—	—	—	—	—	—
TBBS	—	0.5					
TMTD	—	0.3	—	—	—	—	—
TMTM	—	—	0.3	—	—	—	—
TETD	—	—	—	0.3	—	—	—
ZDMC	—	—	—	—	0.3	—	—
ZDBC	—	—	—	—	—	0.3	—
ZBPD	—	—	—	—	—	—	0.3
Mooney scorch at 135°C							
t_5, min	10.4	12.3	22.0	14.5	13.2	18.7	24.4
Rheometer at 160°C							
t_{90}, min	9.2	7.4	9.3	8.6	8.7	12.0	21.3
$t_{90}-t_2$	5.9	3.6	3.6	4.2	4.5	6.5	14.5
Stress–strain: t_{90} cure							
100% modulus, MPa	2.0	2.2	2.1	1.9	1.9	1.8	1.6

[a]   SBR 1500-100, N-330-50, Circosol® 4240-10, Zinc Oxide-4.0, Stearic Acid-2.0, 6PPD-2.0.

The responses obtained in SBR are summarized in Figure 11.25 (scorch delay), Figure 11.26 (cure time), and Figure 11.27 (modulus).

Again, all of the activated sulfenamide stocks exhibit greater scorch protection than does the MBTS/DPG stock. In this polymer, ZBPD provides the longest scorch delay, followed by TMTM.

While ZBPD produces a long scorch delay, as seen in Figure 11.25, it also produces a very slow cure and lower modulus, as seen in Figures 11.26 and 11.27, respectively. For these reasons, the use of ZBPD is not recommended in SBR.

The comparisons shown in Figures 11.25 through 11.27 indicate that TMTM provides the better combination of scorch delay, cure rate, and modulus development in the SBR compound. Again, in SBR, it is feasible to obtain improved scorch delay with no increase in cure time or loss of physical properties.

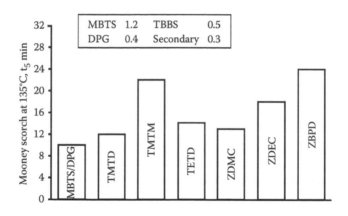

FIGURE 11.25 Comparison of secondary accelerators (100 SBR/1.8 sulfur).

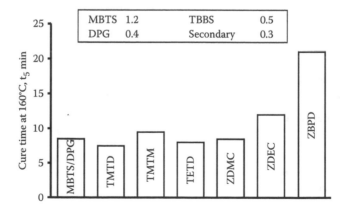

FIGURE 11.26 Comparison of secondary accelerators (100 SBR/1.8 sulfur).

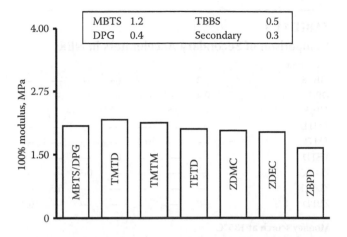

FIGURE 11.27 Comparison of secondary accelerators (100 SBR/1.8 sulfur).

3. Nitrile

Typical responses of these same secondary accelerators, used with TBBS in a black-filled nitrile compound, are shown in Table 11.9.

Again, the responses are compared with the MBTS/DPG control cure system in Figures 11.28 through 11.30. It should be noted that magnesium carbonate–treated sulfur was used to obtain adequate sulfur dispersion.

Figure 11.28 shows the effect of changing from the activated thiazole cure system to an activated sulfenamide. It produces increased processing safety with all secondaries tested in this nitrile stock. Increases in processing safety, depending on the secondary accelerator used, range between 20% and 140%.

As is shown in Figure 11.29, the activated sulfenamide stocks exhibit much shorter cure times than the activated thiazole stock in this polymer.

The improvement in cure time is much greater than that realized in NR or SBR. Even so, the relative relationships between secondary accelerators are similar to those found in NR.

Again, the modulus values realized (see Figure 11.30) are similar to those obtained with the MBTS/DPG system.

The preceding results show that the responses realized with the various secondary accelerators vary significantly from elastomer to elastomer. A logical extension is, therefore, to examine these responses with various fillers. For this reason, the same accelerator combinations studied earlier have been evaluated in natural rubber stocks filled with FEF black, hard clay, and hydrated silica. Concentrations of the vulcanizing agents were held constant, but 2 phr PEG was added to the mineral-filled stocks. The formulation studies are summarized in Table 11.10.

Relative cure time rankings of the secondary accelerators are similar for the three fillers as are the actual cure times. Modulus rankings are also quite similar. However, the actual modulus values produced in this silica-filled stock are lower than those obtained with the other fillers.

TABLE 11.9

Comparison of Secondary Accelerators in NBR[a]

MC sulfur	1.5						
MBTS	1.2	—	—	—	—	—	—
DPG	0.4	—	—	—	—	—	—
TBBS	—	0.5					
TMTD	—	0.3	—	—	—	—	—
TMTM	—	—	0.3	—	—	—	—
TETD	—	—	—	0.3	—	—	—
ZDMC	—	—	—	—	0.3	—	—
ZDBC	—	—	—	—	—	0.3	—
ZBPD	—	—	—	—	—	—	0.3

Mooney scorch at 135°C

t_5, min	2.4	4.7	5.2	5.8	2.9	3.4	3.0

Rheometer at 160°C

t_{90}, min	9.7	3.7	4.5	3.7	2.8	3.5	4.7
$t_{90}-t_2$	8.2	1.7	2.2	1.3	1.3	1.5	3.0

Stress–strain: t_{90} cure

100% modulus, MPa	4.2	4.8	5.2	4.4	4.5	4.1	3.8

[a] Krynac® 34.50-100, N-550 Black-45, N-770 Black-40, DOP-15, Zinc
   Oxide-5.0, Stearic Acid-1.5, Santoflex® 6PPD-2.0.

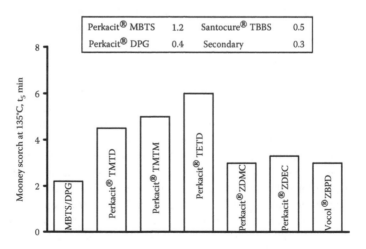

FIGURE 11.28 Comparison of secondary accelerators (100 NBR/1.5 sulfur).

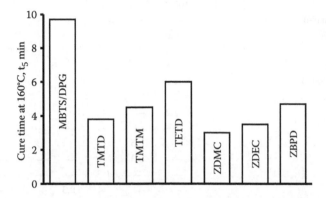

| MBTS | 1.2 | TBBS | 0.5 |
| DPG | 0.4 | Secondary | 0.3 |

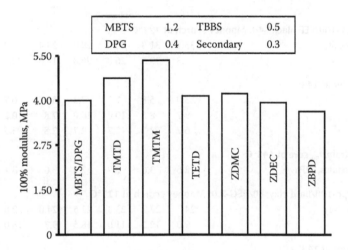

FIGURE 11.29 Comparison of secondary accelerators (100 NBR/1.5 sulfur).

| MBTS | 1.2 | TBBS | 0.5 |
| DPG | 0.4 | Secondary | 0.3 |

FIGURE 11.30 Comparison of secondary accelerators (100 NBR/1.5 sulfur).

The only major difference noted with the change in fillers is that which occurs with TMTM. In the black-filled stock, TMTM produces an excellent combination of long scorch delay, fast cure, and high modulus. With mineral fillers, TMTM does not exhibit an advantage in scorch delay.

Figure 11.31 provides an overall view of the data obtained.

V. RETARDERS

PVI (*N*-cyclohexylthiophthalimide) was the first rubber chemical able to delay the onset of sulfur vulcanization in a predictable manner. PVI is almost the "ideal" retarder, as small additions (0.1–0.4 phr) produce large increases in processing safety (Figure 11.32)

TABLE 11.10
Variation of Secondary Accelerator/Filler NR

NR	100.0						
Filler*(variable)	—	—	—	—	—	—	—
Sundex® 790	10.0						
Zinc oxide	5.0						
Stearic acid	1.5						
Santoflex® 6PPD	2.0						
Sulfur	2.5						
MBTS	1.2	—	—	—	—	—	—
DPG	0.4	—	—	—	—	—	—
TBBS	—	0.5					
TMTD	—	0.3	—	—	—	—	—
TMTM	—	—	0.3	—	—	—	—
TETD	—	—	—	0.3	—	—	—
ZDMC	—	—	—	—	0.3	—	—
ZDBC	—	—	—	—	—	0.3	—
ZBPD	—	—	—	—	—	—	0.3

SMR-5CV-100/FEF black*-40: Mooney scorch at 121°C

Min. viscosity	33.5	31.8	30.0	29.0	29.4	25.7	25.4
t_5, min	7.3	19.5	26.0	26.8	15.3	22.8	21.2

Rheometer at 143°

t_2, min	3.0	5.8	7.3	7.3	5.0	6.7	6.5
t_{90}, min	9.7	8.5	10.0	10.0	7.8	10.2	14.8
$t_{90}-t_2$	6.7	2.7	2.7	2.7	2.8	3.5	8.3

Stress–strain/t_{90} cure at 143°C

100% modulus, MPa	2.6	3.0	2.9	2.8	3.0	3.1	2.1

Pale crepe-100/hard clay*80/PEG-2.0: Mooney scorch at 121°C

Min. viscosity	24.7	23.2	23.3	22.5	24.0	22.0	21.8
t_5, min	9.0	12.2	11.0	18.5	9.7	18.0	12.5

Rheometer at 143°

t_2, min	3.7	4.0	3.8	5.8	3.8	6.2	4.3
t_{90}, min	9.8	6.3	6.2	8.3	6.5	9.3	14.8
$t_{90}-t_2$	6.1	2.3	2.4	2.5	2.7	3.1	10.5

Stress–strain/t_{90} cure at 143°C

100% modulus, MPa	2.9	3.2	3.0	3.1	3.0	2.8	2.5

Pale crepe-100/Hisil 233*-40/PEG-2.0: Mooney scorch at 121°C

Min. viscosity	36.8	40.4	39.4	40.5	40.7	39.4	36.8
t_5, min	12.8	14.7	15.8	22.0	12.2	22.7	14.7

(Continued)

TABLE 11.10 (*Continued*)
Variation of Secondary Accelerator/Filler NR

NR	100.0						
Rheometer at 143°							
t_2, min	4.7	4.7	4.3	6.2	4.2	6.7	5.0
t_{90}, min	12.7	6.7	6.7	9.0	6.9	10.0	16.0
t_{90}–t_2	8.0	2.0	2.4	2.8	2.5	3.3	11.0
Stress–strain/t_{90} cure at 143°C							
100% modulus, MPa	1.6	1.5	1.5	1.7	1.5	1.6	1.2

* It refers to the carbon black grade

Figure 11.33 shows that increases in processing safety are obtained without affecting rate of cure or final cured modulus, at the normal levels used (0.1–0.4 phr). With conventional retarders, reductions in cure rate resulting in increasing cure times have frequently been confused with true retardation of scorch where cure rate is unaffected.

The predictable influence of PVI on processing safety with respect to level of addition and temperature is shown in Figures 11.34 and 11.35.

PVI is highly effective with sulfenamide accelerators as shown in Figure 11.32 where the response of processing safety increasing level of PVI is illustrated for the most commonly used sulfenamides. This linear response enables the exact amount required for a given increase in processing safety to be quickly determined and enables the "calibration" of a compound in processing terms.

Predictability with respect to temperature is demonstrated in Figure 11.35. Unlike other retarders (e.g., *N*-nitrosodiphenylamine), PVI will not decompose over the normal range of processing and curing temperatures. The graph in Figure 11.35 shows two lines representing the relationship between processing safety (Rheometer T2) and processing temperature for a TBBS accelerated SBR compound. The dotted line demonstrates the effect of adding 0.25 phr PVI. This addition produces a parallel shift to the left. Thus, moving from point A at a temperature of 140°C, in a vertical direction by the addition of 0.25 phr PVI will permit a 10°C increase in processing temperature while maintaining the same processing safety. This temperature predictability extends the applications of PVI from a simple retarder to that of a much more versatile additive with which heat input can be considered as a controllable factor in the same way as processing safety.

The linear relationship between PVI level and processing safety shown in Figure 11.34 occurs with a wide range of polymers, accelerators, sulfur levels, filler types and level, and other compounding ingredients. In practically all cases, a straight linear relationship is obtained, the slope and position of which depend on the particular formulation. Although the highest response occurs with sulfenamides, PVI is active with nearly all accelerators for sulfur curable elastomers but normally ineffective with peroxide, resin, or metal oxide curing systems. It is not normally used in latex formulations.

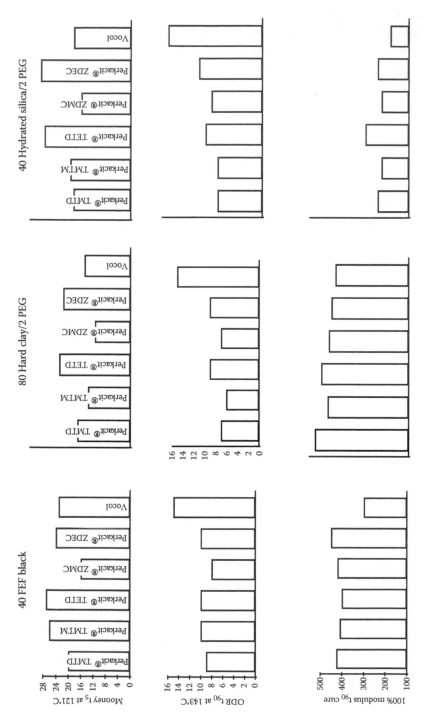

FIGURE 11.31 Secondary accelerators with different fillers (100 NR/0.5 TBBS/0.3 secondary).

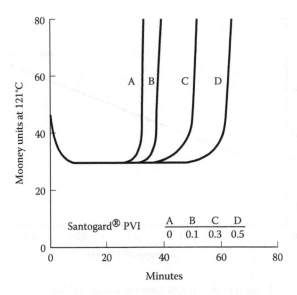

FIGURE 11.32   Effect on Mooney scotch characteristics.

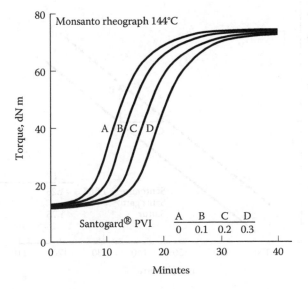

FIGURE 11.33   Effect curing.

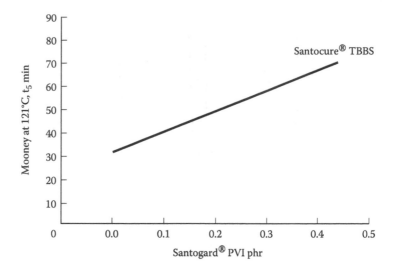

FIGURE 11.34   PVI concentration versus Mooney scorch 100 NR.

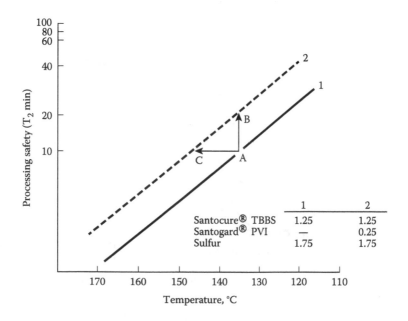

FIGURE 11.35   Effect of Santogard® PVI over a range of processing temperatures 100 SBR.

PVI is most effective with the fastest curing polymers and an approximate order of response is

$$NR > NBR > SBR > EPDM > IIR > CR$$

The response is not only determined by the elastomer but also by the accelerator system. The specialty elastomers show lower response to PVI due to the slower curing nature of the polymer and also to the fact that they are generally cured with accelerators showing a low response to PVI (e.g., thiurams, dithiocarbamates). An exception to this is in butyl formulations where PVI shows the best response with curing system based on dithiocarbamates. PVI has also been used effectively in NBR/PVC, polyacrylic rubber, and sulfur curable polyurethanes. Sulfur level is also very important; the best response is found with conventional levels (1.5–3.0 phr) with a tendency to poorer response as sulfur level increases (ebonites). The response in systems with low sulfur levels is largely dependent on the accelerator.

The addition of PVI produces no deterioration in aging, fatigue, or ozone resistance in compounds cured to optimum. Within normal usage levels (0.1–0.4 phr), it has no effect on modulus, resilience, creep, permanent set, heat buildup, abrasion, oil swelling resistance, etc. PVI is also not known to have any detrimental effects on adhesion of cured rubber to textiles (rayon, nylon, polyester, aramid) or steel (brass- or zinc-coated wire or chemically treated surfaces). It is widely used in skim stocks to maintain high levels of adhesion in steel-belted radial tires.

When levels over 0.4 phr are required, attention must be paid to final cured modulus as this may be reduced slightly with possible effects on compression set, heat buildup, resilience, and creep. If such high levels are required, it is usual to readjust modulus by a small increase in the sulfur level (up to 40% of the level of PVI) or accelerator (up to 20% of the PVI level). A surface bloom may also occur in some cases. PVI will not cause contact or migration staining to painted surfaces but may give slight discoloration in white or light-colored stocks.

VI. CURE SYSTEMS FOR SPECIALTY ELASTOMERS

A. EPDM

Properly compounded EPDM exhibits many desirable vulcanizate properties including resistance to ozone, heat, UV weathering, and chemicals. Because of the attractive combinations of properties, EPDM has gained acceptance in a wide variety of applications. However, the relatively low unsaturation of EPDM requires complex cure systems to achieve the desired properties.

Nearly every conceivable combination of curing ingredients has been evaluated in the various EPDM polymers, and over the years, certain of these have shown particular merit. A brief description of four of these cure systems used in practice follows:

Cure Package 1	**"Low Cost"**	**Sulfur**	**1.5**
		TMTD	1.5
		MBT	0.5

This is one of the earliest cure systems developed for EPDM. It exhibits a medium cure rate and develops satisfactory vulcanizate properties. The primary advantage of this system is its low cost, while a major shortcoming is its severe tendency to bloom.

Cure Package 2	"Triple 8"	Sulfur	2.0
		MBT	1.5
		TDED	0.8
		DPTT	0.8
		TMTD	0.8

This common, nonblooming, cure package has been labeled the "Triple 8" system for obvious reasons. It provides excellent physical properties and very fast cures but tends to be scorchy and is relatively expensive.

Cure Package 3	"Low Set"	Sulfur	0.5
		ZDBD	3.0
		ZDMC	3.0
		DTDM	2.0
		TMTD	3.0

Excellent compression set and good heat aging properties characterize this system. Its drawbacks are a tendency to bloom and very high cost.

Cure Package 4	"General Purpose"	Sulfur	2.0
		MBTS	1.5
		ZDBD	2.5
		TMTD	0.8

This general purpose, nonblooming system offers good performance and is included as another example of a widely used EPDM cure package.

Cure Package 5	"2121 System"	ZBPD	2.0
		TMTD	1.0
		TBBS	2.0
		Sulfur	1.0

An attractive balance of fast cure, good physicals, and good resistance to compression set and heat aging are features of this cure system that has been derived from a complex, statistically designed experiment to optimize the level of each ingredient.

Tables 11.11 through 11.13 summarize the properties obtained with these cure packages when evaluated in three EPDM polymers varying in type and amount of unsaturation. These data confirm the features of each cure system described earlier. The polymers used are as follows.

The advent of the faster curing, more unsaturated EPDMs make it possible to utilize simpler accelerator systems such as the activated thiazoles and activated sulfenamides used in NR, SBR, and the other highly unsaturated polymers. The use of these systems in EPDM is illustrated in Table 11.14.

TABLE 11.11
Low-Saturation EPDM

Masterbatch

Nordel 1070	100.0
N-550 black	100.0
N-774 black	100.0
Paraffinic oil	110.0
TMQ	2.0
Zinc oxide	5.0
Stearic acid	2.0

Cure systems	I	II	III	IV	V
Low cost	X				
Triple 8		X			
Low set			X		
General purpose				X	
"2121"					X

Mooney scorch at 135°C

Minimum viscosity	41.0	49.0	43.0	46.0	41.0
t_5, min	11.4	6.0	17.5	9.5	15.2
t_{35}, min	14.4	8.3	24.8	12.4	19.7

Rheometer at 160°C; 1° Arc

Max. torque, dN m	26.5	33.4	27.7	31.1	25.4
t_2, min	3.5	2.5	4.8	3.0	5.8
t_{90}, min	17.5	17.3	14.5	15.5	18.0

Stress–strain: cure t_{90} at 160°C

Shore "A" hardness	67.0	71.0	69.0	71.0	66.0
100% modulus, MPa	3.3	4.8	3.6	4.1	2.6
Ult. tensile, MPa	11.6	12.8	11.0	11.8	11.0
Ult. elongation, %	320.0	280.0	325.0	295.0	430.0

Stress–strain: after 70 h at 121°C

Shore "A" hardness	72.0	77.0	73.0	77.0	73.0
100% modulus, MPa	5.9	9.4	5.5	9.2	4.8
Ult. tensile, MPa	13.4	13.9	11.5	13.1	12.2
Ult. elongation, %	235.0	160.0	225.0	155.0	280.0
% Retained TE factor[a]	84.0	62.0	72.0	58.0	71.0

Compression set: percent after 22 h at 122°C
Cure $t_{90} + 5'$ at 160°C

% Set	68.0	67.0	40.0	67.0	68.0

[a] Retained TE factor $= \dfrac{\text{Aged Ult. Ten.} \times \text{Aged Ult. Elong}}{\text{Ult. Ten.} \times \text{Ult. Elong}} \times 100$

TABLE 11.12
Medium Saturation EPDM

Masterbatch

Vistalon 5600	100.0
N-774 black	100.0
N-550 black	100.0
Paraffinic oil	110.0
TMQ	2.0
Zinc oxide	5.0
Stearic acid	2.0

Cure systems	I	II	III	IV	V
Low cost	X				
Triple 8		X			
Low set			X		
General purpose				X	
"2121"					X

Mooney scorch at 135°C					
Minimum viscosity	41.0	46.0	38.0	39.0	38.0
t_5, min	7.3	4.2	11.0	7.0	10.5
t_{35}, min	9.8	6.2	17.8	10.0	14.5

Rheometer at 160°C; 1° Arc					
Max. torque, dN m	31.6	35.0	28.3	32.8	31.6
t_2, min	3.2	1.5	3.4	2.5	4.2
t_{90}, min	12.8	9.3	8.0	13.8	12.0

Stress–strain: cure t_{90} at 160°C					
Shore "A" hardness	74.0	76.0	74.0	76.0	74.0
100% modulus, MPa	4.2	4.3	3.1	4.0	4.1
Ult. tensile, MPa	10.4	11.0	9.7	11.1	11.3
Ult. elongation, %	305.0	275.0	375.0	310.0	400.0

Stress–strain: after 70 h. at 121°C					
Shore "A" hardness	78.0	80.0	77.0	79.0	78.0
100% modulus, MPa	6.6	7.2	5.3	6.9	5.2
Ult. tensile, MPa	12.6	12.3	11.1	11.4	12.9
Ult. elongation, %	207.0	175.0	235.0	175.0	280.0
% Retained TE factor[a]	82.0	71.0	72.0	58.0	80.0

Compression set: Percent after 22 h at 122°C					
Cure $t_{90}+5'$ at 160°C					
% Set	67.0	67.0	50.0	65.0	63.0

[a] TE factor is Tensile*Elongation.

TABLE 11.13
High-Saturation EPDM

Masterbatch

Vistalon 6500	100.0
N-774 black	100.0
N-550 black	100.0
Paraffinic oil	110.0
TMQ	2.0
Zinc oxide	5.0
Stearic acid	2.0

Cure systems	I	II	III	IV	V
Low cost	X				
Triple 8		X			
Low set			X		
General purpose				X	
"2121"					X
Mooney scorch at 135°C					
Minimum viscosity	41.0	48.0	44.0	37.0	37.0
t_5, min	9.1	5.3	14.0	12.8	14.5
t_{35}, min	13.5	7.7	29.5	12.8	14.5
Rheometer at 160°C; 1° Arc					
Max. torque, dN m	33.9	39.6	32.8	37.3	31.6
t_2, min	2.8	1.5	3.4	2.5	3.4
t_{90}, min	11.1	8.0	9.0	11.2	9.5
Shore "A" hardness	75.0	77.0	75.0	76.0	76.0
100% modulus, MPa	6.7	8.1	6.1	7.6	5.6
Ult. tensile, MPa	10.3	10.7	9.6	11.2	9.6
Ult. elongation, %	160.0	135.0	165.0	155.0	175.0
Shore "A" hardness	79.0	83.0	78.0	81.0	80.0
100% modulus, MPa	10.3	—	8.7	—	7.8
Ult. tensile, MPa	11.6	11.8	10.2	11.6	10.6
Ult. elongation, %	115.0	85.0	120.0	85.0	140.0
% Retained TE factor[a]	81.0	70.0	77.0	57.0	89.0
Compression set					
Cure $t_{90} + 5'$ at 160°C					
% Set	66.0	62.0	43.0	65.0	67.0

[a] TE factor is Tensile*Elongation.

These data compare one of the faster curing packages (the "Triple 8" system) with these simpler systems. The simple systems offer low-cost, bloom-free stocks and provide good scorch delay and satisfactory physical properties, but they are slower curing. The addition of a second activating accelerator, such as zinc dialkyldithiocarbamate, speeds up the cure with no real change in physical properties.

TABLE 11.14
Thiazole- and Sulfenamide-Curing Systems in Vistalon 5600[a]

CBS	1.2	—	1.2	—	"Triple 8"
MBTS	—	1.5	—	1.5	"Triple 8"
TMTD	0.7	0.8	0.7	0.8	"Triple 8"
ZDEC	—	—	0.7	0.8	"Triple 8"
Sulfur	1.5	1.5	1.5	1.5	"Triple 8"
Mooney scorch at 135°C					
t_5, min	9.2	7.4	7.4	6.3	2.4
Rheometer at 160°C					
t_2, min	3.0	2.7	3.1	2.3	1.1
t_{90}, min	17.7	22.0	13.6	17.0	10.0
Max. torque, dN m	67.8	67.8	67.8	67.8	79.0
Physical properties: Opt. cure at 160°C					
Ult. tensile, MPa	10.1	10.8	10.5	10.9	11.2
100% modulus, MPa	5.7	5.9	5.9	6.4	6.9
Elongation at break, %	220.0	220.0	210.0	200.0	150.0
Compression set: Opt. cure-ASTM-B					
22 h at 100°C, %	54.0	51.0	43.0	48.0	47.0
Compression set: After overcure 1 h at 160°C					
22 h at 100°C, %	27.0	24.0	23.0	22.0	24.0

[a] Polymer 100, FEF Black 200, Oil 120, Zinc Oxide 5, Stearic Acid 1.

A common problem with the widely known EPDM curing packages is the fact that those systems that produce low compression set also exhibit severe bloom. This adverse combination of properties has recently been overcome with the development of the "2828" system (2.0 DTDM/0.8 TMTD/2.0 ZDBC/0.8DPTT) illustrated in Table 11.15. This cure system provides compression set comparable to the low set package discussed earlier; however, no bloom has been observed on stocks cured with this system.

1. Polymer Description

Polymer	3rd Monomer Type	% Unsaturation
Nordel[b] 1070	1,4 hexadiene	2.5
Vistalon[a] 5600	ENB	4.5
Vistalon 6505	ENB	9.5

[a] Nordel is a registered trademark of E.I. duPont de Nemours and Company.
[b] Vistalon is a registered trademark of Exxon Chemical Company.

TABLE 11.15
Curing Systems for "Low Set"

	I	II	III	IV
Vistalon® 3708	100.0			
N-550 black	50.0			
N-762 black	150.0			
Circosol® 4240	120.0			
Stearic acid	1.0			
6PPD	2.0			
Zinc oxide	5.0			
Sulfur	0.5	0.5	—	0.5
DTDM	2.0	1.7	2.0	—
TMTD	3.0	2.5	0.8	1.0
ZDMC	3.0	2.5	—	—
ZDBC	3.0	2.5	2.0	—
DPTT	—	—	0.8	—
CBS	—	—	—	2.0
ZBPD	—	—	—	3.2
Mooney viscometer at 135°C				
Minimum viscosity	20.0	21.4	21.2	20.5
t_5, min	14.2	13.7	16.4	14.2
Rheometer at 160°C				
Min. torque, dN m	4.1	3.7	3.3	3.8
Max. torque, dN m	25.9	24.9	20.9	17.8
t_2, min	5.0	4.8	6.7	5.2
t_{90}, min	11.2	11.2	15.2	11.5
Shore "A" hardness	70.0	69.0	70.0	66.0
100% modulus, MPa	2.9	2.7	2.3	1.9
300% modulus, MPa	6.8	6.4	5.8	4.8
Ult. tensile, MPa	8.8	8.3	7.9	6.5
Ult. elongation, %	550.0	545.0	670.0	640.0
Aged 70 h at 121°C				
Shore "A" hardness	72.0	73.0	70.0	71.0
100% modulus, MPa	3.4	3.2	2.6	2.8
300% modulus, MPa	8.3	7.8	6.5	6.7
Ult. tensile, MPa	9.2	8.9	7.9	8.0
Ult. elongation, %	435.0	425.0	560.0	450.0
% Retained TE factor	82.0	83.0	83.0	87.0
Compression set				
70 h at 121°C				
Cure $t_{90} + 5'$ at 160°C				
% Set	60.0	58.0	57.0	76.0

The advent of the faster curing, more unsaturated EPDMs makes it possible to utilize simpler accelerator systems such as the activated thiazoles and activated sulfenamides used in NR, SBR, and the other highly unsaturated polymers. The use of these systems in EPDM is illustrated in Table 11.14.

These data compare one of the faster curing packages (the "Triple 8" system) with these simpler systems. The simple systems offer low-cost, bloom-free stocks and provide good scorch delay and satisfactory physical properties, but they are slower curing. The addition of a second activating accelerator, such as zinc dialkyldithiocarbamate, speeds up the cure with no real change in physical properties.

A common problem with the widely known EPDM curing packages is the fact that those systems that produce low compression set also exhibit severe bloom. This adverse combination of properties has recently been overcome with the development of the "2828" system (2.0 DTDM/0.8 TMTD/2.0 ZDBC/0.8 DPTT) illustrated in Table 11.15. This cure system provides compression set comparable to the low set package discussed earlier; however, no bloom has been observed on stocks cured with this system.

B. Nitrile Rubber

Nitrile rubber is a general term describing a family of elastomers obtained by the co-polymerization of acrylonitrile and butadiene. Although each polymer's specific properties depend primarily upon its acrylonitrile content, they all exhibit excellent abrasion resistance, heat resistance, low compression set, and high tensile properties when properly compounded. Probably the predominant feature dictating their use is their excellent resistance to petroleum oils.

Cure systems for nitrile rubber are somewhat analogous to those used in SBR except that magnesium carbonate-treated sulfur is usually used to aid in its dispersion into the polymer. Typical cure systems employ approximately 1.5 phr of the treated sulfur with appropriate accelerators to obtain the desired rate and state of cure. Common accelerator combinations include the thiazole/thiuram or sulfenamide/thiuram types. Examples of these sulfur-based cure systems are shown in Table 11.16.

As operating requirements for nitrile rubber become more stringent, improved aging and set resistance become important. These improvements are realized by reducing the amount of sulfur and by using a sulfur donor such as DTDM or TMTD.

Examples of these sulfur donor cure systems are shown in Tables 11.17 and 11.18. The advantages of these systems are shown in improved set resistance and aging while maintaining adequate scorch safety and fast cures. Note also in Table 11.18 that when using equal levels of DTDM/TBBS/TMTD and adjusting only total accelerator concentration, a wide modulus range is achieved while maintaining adequate scorch safety and fast cure rate. Therefore, we have a viable method to control crosslink density in sulfur donor systems.

TABLE 11.16
High Sulfur Nitrile Rubber Systems

MC Treated Sulfur[a]	1.5	1.5	1.5
TMTM	0.4	—	—
MBTS	—	1.5	—
TBBS	—	—	1.2
TMTD	—	—	0.1
Processing and curing properties			
Mooney scorch at 121°C			
t_5, min	6.8	8.1	5.7
Rheometer cure time at 160°C			
t_{90}, min	8.7	15.2	4.7
Physical properties on t_{90} cure			
Hardness	73.0	71.0	75.0
100% modulus, MPa	4.2	3.6	5.0
Ult. tensile, MPa	16.3	16.2	17.3
Ult. elongation, %	380.0	475.0	355.0
Heat aging resistance 70 h at 100°C			
Shore "A" hardness	80.0	78.0	82.0
% Retained TE factor	85.0	68.0	57.0
Compression set: 22 h at 100°C			
% Set	31.0	50.0	55.0

[a] Magnesium carbonate treated sulfur: used to improve dispersion.

C. NEOPRENE

Ethylene thiourea, ETU, has traditionally been the accelerator of choice for attaining maximum physical properties in Neoprene W compounds. However, ETU is now available only in predispersed forms, and the rubber industry is actively looking for a viable replacement.

We have found thiocarbanilide accelerator (A-1) to be effective in neoprene, particularly the Neoprene W types that require additional acceleration beyond that provided by metal oxides alone.

These thiocarbanilide accelerated compounds exhibit good processing safety, fast, level cures with excellent tensile properties, and compression set resistance. The advantages described earlier are demonstrated in Table 11.19 and Figure 11.36. Of particular interest is the very flat plateau obtained with the thiocarbanilide compared with the marching modulus of the other systems.

An unexpected advantage of the thiocarbanilide cure is its dramatic response to PVI for providing longer scorch delay. However, a sacrifice in compression set and

TABLE 11.17
High Sulfur versus Low Sulfur in Nitrile Rubber

MC Treated Sulfur[a]	1.5	0.3
MBTS	1.5	—
TBBS	—	1.0
TMTD	—	1.0
Mooney scorch at 121°C		
t_5, min	8.1	8.1
Rheometer cure time at 160°C		
t_{90}, min	15.2	10.5
Physical properties on t_{90} cure		
Shore "A" hardness	71.0	69.0
100% modulus, MPa	3.6	3.1
Ult. tensile, MPa	16.2	15.1
Ult. elongation, %	475.0	485.0
Heat aging resistance: 70 h at 100°C		
Shore "A" hardness	78.0	74.0
% Retained TE factor	68.0	89.0
Compression set: 22 h at 100°C		
% Set	50.0	24.0

[a] Magnesium carbonate treated sulfur: used to improve dispersion.

modulus is observed as shown in Table 11.20. Also included in this table is A-1/ZBPD-pdr combination that offers very fast, scorchy compounds with excellent compression set resistance. This cure system should be compatible with applications employing continuous vulcanization processes, wherein rapid onset of cure is mandatory.

D. BUTYL RUBBER

Because of its low unsaturation, butyl rubber possesses excellent resistance to weathering, heat and ozone, as well as exhibiting excellent fatigue resistance. Of course, its most predominant attribute is low gas permeability, which makes it the preferred elastomer for interliners, intertubes, bladders, and other air containment parts. The requirements for butyl tubes, for both truck and passenger tires, include good heat resistance and low set upon stretching (or maintaining dimensions after inflation), which can possibly also be related to compression set. Another major problem in most tubes is the weakness of the splice, which results in premature failures due to separation. This appears to be particularly acute in tubes used with steel-belted radial truck tires.

One method to improve splicing behavior is to develop longer scorch times thereby permitting better flow and thus, better knitting at the splice prior to cure. This would have to be accomplished with no loss in other key properties. It is our objective to

TABLE 11.18
Sulfurless Cure Systems for Nitrile Rubber

	1.5	—	—	—
MBTS	1.5	—	—	—
TBBS	—	1.0	1.0	1.0
TMTD	—	1.0	1.0	1.0
DTDM	—	1.0	1.0	2.0
Mooney scorch at 135°C				
t_5, min	8.1	10.7	7.0	7.9
Rheometer cure time at 160°C				
t_{90}, min	15.2	14.7	12.5	13.3
Physical properties on t_{90} cure at 160°C				
Shore "A" hardness	71.0	68.0	71.0	73.0
100% modulus, MPa	3.6	3.1	4.3	5.7
Ult. tensile, MPa	16.2	15.4	16.3	16.8
Ult. elongation, %	475.0	485.0	360.0	290.0
Heat aging resistance: 70 h at 100°C				
Shore "A" hardness	78.0	73.0	75.0	78.0
% Retained TE factor	68.0	87.0	89.0	83.0
Compression set: 22 h at 100°C				
% Set	50.0	22.0	13.0	12.0

compare properties of a Semi-E.V. cure system to those of a conventional sulfur cure system recommended by the polymer manufacturer. Table 11.21 summarizes these formulations and the properties. As observed earlier in the case of nitrile rubber, the DTDM-based cure systems offer significant improvement in heat and compression set resistance as well as improved processing safety, all qualities that contribute to improved product performance.

VII. EFFECTIVE CURATIVE DISPERSION FOR UNIFORM MECHANICAL PROPERTIES

Uniform curative dispersion is important in preparation of rubber articles with uniform mechanical properties. Poor dispersion of curatives resulting from poor mixing will lead to an inhomogeneous network and, thus, significant fluctuations of mechanical properties in the final vulcanizate. Evidence of poor dispersion can be observed by monitoring the vulcanizate final mechanical properties, such as final rheometer modulus, tensile or fatigue properties, or by monitoring the processing characteristics, such as viscosity or curing properties. This section studies the effects of the efficiency of curative mixing on the various properties of simple rubber compounds.

TABLE 11.19
Neoprene W. Curing Systems

Base stock

Neoprene W	100.0
N-990 black	20.0
N-774 black	40.0
Aromatic oil	15.0
TMQ	1.0
Stearic acid	4.0
Stan Mag[a] beads	1.0
Zinc oxide[b]	5.0

Stock	1	2	3
NA-22	0.5	—	—
TMTM	—	1.0	—
DOTG	—	1.0	—
Sulfur	—	0.5	—
A-1	—	—	0.7

Mooney scorch at 135°C

Minimum viscosity	32.8	28.9	30.9
t_5, min	7.7	34.5	9.5

Rheometer at 160°C: MPC dies, ±1° Arc

Max. torque, in. lb.	5.0	4.0	4.1
Max. torque, in. lb.	31.2	29.3	25.6
t_2, min	2.2	5.2	2.3
t_{90}, min	20.8	25.0	5.8

Stress–Strain: Cure t_{90} at 160°C

Hardness	61.0	58.0	57.0
100% modulus, psi	345.0	280.0	270.0
300% modulus, psi	1715.0	1375.0	1320.0
Ult. tensile, psi	2655.0	2440.0	2520.0
Ult. elongation, %	470.0	550.0	540.0

Compression set: percent after 22 h at 100°C

% set	12.0	23.0	10.0

[a] Stan Mag is a Trademark of Harwick Chemical Corp.
[b] Zinc Oxide added to the mill.

A. Accelerator Dispersion: A Review

The effects of MBTS particle sizes on dispersion were chosen because of its high melting point (~174°C). These particles are not likely to melt during mixing. Unmilled MBTS powder (m.p. of 174°C) was sieved to obtain fractions of different

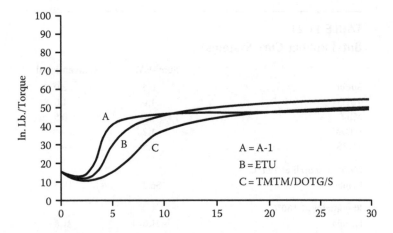

FIGURE 11.36 Comparison of neoprene cure systems.

TABLE 11.20
Variations of A-1 Cure in Neoprene W.

Stock	Control	Faster Cure	Longer Scorch Safety
A-1	0.7	0.7	0.7
ZBPD-pdr	—	0.5	—
PVI	—	—	0.2
Mooney scorch at 121°C			
t_5, min	18.0	7.7	30.0
Rheometer at 160°C			
t_{90}, min	7.4	4.8	10.7
Physical properties at t_{90} cure			
Shore "A" hardness	62.0	63.0	60.0
300% modulus, psi	1950.0	1705.0	1240.0
Ult. tensile, psi	2905.0	2900.0	2700.0
Ult. elongation, %	435.0	475.0	530.0
Compression set after 22 h at 100°C			
% set	17.0	19.0	24.0

TABLE 11.21
Butyl Rubber Cure Systems[a]

	Semi-E.V.	Conventional
Sulfur	0.5	2.0
TMTD	1.0	1.0
MBT	—	0.5
DTDM	1.2	—
TBBS	0.5	—
Mooney scorch at 121°C		
t_5, min	36.2	18.5
Rheometer at 160°C		
t_{90}, min	21.0	21.8
Physical properties		
Shore "A" hardness	68.0	68.0
300% modulus, psi	800.0	1030.0
Ult. tensile, psi	1590.0	1640.0
Ult. elongation, %	600.0	510.0
Compression set 70 h at 121°C		
%, Set	56.0	81.0
Heat age resistance 70 h at 121°C		
% Retained tensile	76.0	57.0

[a] Masterbatch: Butyl 218-100, GPF black 70, paraffinic oil 25, zinc oxide 5.

sizes. These fractions were evaluated in comparison with normal MBTS powder (~150 μm). The formation used in this study is as follows:

Materials	Phr
SBR 1712	137.5
N-330 black	40
Zinc oxide	5
Stearic acid	1
Sulfur	2

Test specimens were cured for 35 min at 153°C.

The data in Figures 11.37 and 11.38 show that accelerator particle greater than 200 μm produced a significant reduction in both tensile and elongation but have only minor effects on rheometer data and modulus value as shown in Figure 11.39.

Other accelerators did not always produce the same results. Figure 11.40 compares MBTS (m.p. of 174°C) to TBBS and MBS (sulfenamides). TBBS (m.p. of 106°C), and MBS (m.p. of 82°C) were chosen to represent medium to low melting point accelerators. Although there is some tendency for tensile properties to fall as

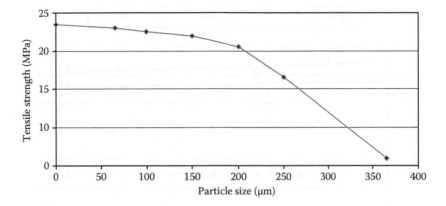

FIGURE 11.37 Effect of MBTS particle size on tensile strength.

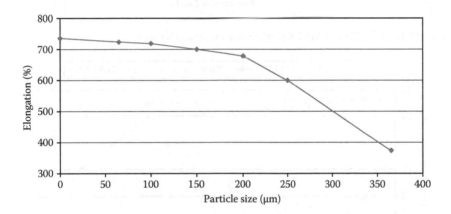

FIGURE 11.38 Effect of MBTS particle size on elongation.

the size of the undispersed particle increases, these data show that for TBBS that melts at a temperature generally below the curing temperature, catastrophic failure does not occur until particles greater than 500 μm are present. With a low melting accelerator such as MBS that melts at typical mixing temperatures, no dispersion effect was detected, indicating that the particles were completely dissolved at an early stage in the compounding process.

The following observations were made from this study and can be summarized in Table 11.22 for accelerator MBTS.

- Validated that large particles are deleterious to good dispersion
- Defined an optimum particle size of ~150 μm for MBTS
- Defined poorer distribution with smaller particles

Good dispersion of curatives, such as accelerators is essential to the preparation of vulcanizates having uniform networks and hence low variation in physical

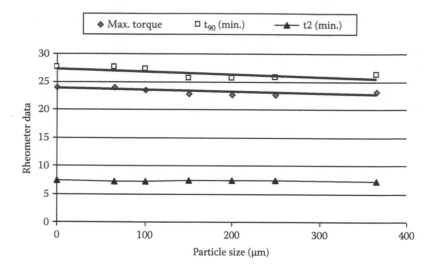

FIGURE 11.39   Effect of MBTS particle size on rheometer data.

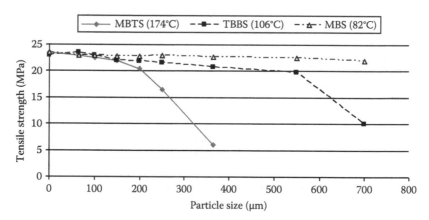

FIGURE 11.40   Effect of melt points on dispersion.

TABLE 11.22
Accelerator MBTS Dispersion

Dispersion	Tensile	Std. Dev.
Homogeneous distribution of large size particles (>200 μm)	L	M–L
Poor distribution of large size particles	L	H
Homogeneous distribution of optimum size particles (~150 μm)	H	L
Poor distribution of optimum size particles	M–H	H
Homogeneous distribution of mixed size particles	M–H	M–L
Poor distribution of mixed size particles	M–L	M–H

Notes: L, low; M, medium; H, high.

properties. Variations in curative dispersion can be simply or qualitatively observed by deformation experiments including stretching or solvent swelling. More quantitative measures of curative dispersion may be assessed using AFM or to a lesser extent modulus measurements such as MDR, RPA, or tensile testing. Finally, a relationship can be shown to exist between visibly undispersed curative and fatigue performance.

VIII. CONCLUSIONS

The chemical linking of a mass of polymer chains into the essence of a single network imparts enhancements in physical properties allowing for applications into highly engineered, durable products. Without vulcanization, the development of the rubber industry would have fallen far short of today's multibillion dollar megalith. As shown in the preceding discussion, vulcanization can play a large role in determining the ultimate performance and applicability of a number of elastomers in highly engineered rubber articles.

Vulcanization systems are chosen on the basis of their ability to control the following processing and performance properties of rubber compounds.

- Time delay before vulcanization begins
- Speed of the vulcanization reaction after it is initiated
- Extent of the vulcanization as measured by modulus, tensile, hardness, etc.
- Type of crosslinking that occurs as it relates to aging and fatigue performance
- Other factors such as green stock storage stability, fiber or steel adhesion, bloom tendency, dispersion, etc.

The job of the compounder, therefore, becomes one of selecting and evaluating individual accelerators and/or combinations of accelerators, vulcanizing agents, etc., that make up the desired vulcanization systems. The proliferation of vulcanization ingredient types should be viewed as an opportunity since it often gives the compounder a chance to custom fit vulcanization systems to the individual processing and performance needs.

ABBREVIATIONS

Abbreviations	Chemical Name
BDITD	Bis(diisopropylthiophosphoryl) disulfide
CBS	N-Cyclohexyl-2-benzothiazolesulfenamide
DBTU	N,N'-Dibutylthiourea
DCBS	N,N-Dicyclohexyl-2-benzothiazolesulfenamide
DETU	N,N'-Diethylthiourea
DOTG	Di-o-tolylguanidine
DPG	Diphenylguanidine
DPTH	Dipentamethylenethiuram hexsulfide
DTDM	Dithiodimorpholine
ETPT	Bis(diethyl thiophosphoryl) trisulfide
ETU	Ethylenethiourea

MBS	2-(Morpholinothio) benzothiazolesulfenamide
MBT	2-Mercaptobenzothiazole
MBTS	Benzothiazyldisulfide
NDPA	N-Nitrosodiphenylamine
PEG	Polyethene glycol
6PPD	N,1,3-Eimethyl-butyl-N-phenyl-p-phenylenediamine
PVI	N-(Cyclohexylthio) phthalimide
TBBS	N-t-Butyl-2-benzothiazolesulfenamide
TDEDC	Tellurium diethyldithiocarbamate
TETD	Tetraethylthiuramdisulfide
TMQ	Polymerized 2,2,4-trimethyl 1,2-dihydroquinoline
TMTD	Tetramethylthiuramdisulfide
TMTM	Tetramethylthiurammonosulfide
TMTU	Trimethylthiourea
ZBDC	Zinc-dibutyldithiocarbamate
ZBPD	Zinc-O-di-n-butylphosphorodithioate
ZDEC	Zinc-diethyldithiocarbamate
ZMBT	Zinc salt of 2-mercaptobenzothiazole
ZMDC	Zinc dimethyldithiocarbamate

REFERENCES

1. Goodyear, Charles; U.S. Patent 3633 (1844).
2. Hancock, Thomas; English Patent 9952 (1843).
3. Nabil, H.; Ismail, H.; Azura, A. R., *Advanced Materials Research* 844, 267–271 (2014); Del Vecchio, R. J.; Ferro, E., Jr.; Micheal, R., *Rubber, Cory, RubberChem 2010, Seventh International Conference Devoted to Rubber Chemicals, Compounding & Mixing*, Vienna, Austria, November 30–December 1, 2010, 14/1–14/6 (2010).
4. Del Vecchio, R. J.; Ferro, E., Jr.; Micheal, R., *Rubber, Cory, RubberChem 2010, Seventh International Conference Devoted to Rubber Chemicals, Compounding & Mixing*, Vienna, Austria, November 30–December 1, 2010, 14/1–14/6 (2010); Gonzalez, H. L.; Rodriguez Diaz, A.; De Benito Gonzalez, J. L.; Fontao Orosa, I.; Marcos Fernandez, A, *Kautschuk Gummi Kunststoffe*, 45(12), 1033–1037 (1992)
5. Ignatz-Hoover, F., *Rubber World*, 220(5), 24 (1999).
6. Hoffman, W., *Vulcanization and Vulcanizing Agents*, Maclaren, London, U.K., 1967.
7. Hepburn, C.; Mahdi, M. S., *Plastics and Rubber Processing and Applications*, 4, 343–348 (1984).
8. Baker, C. S. L.; Barnard, D.; Porter, M., *Rubber Chemical Technology*, 43, 501 (1970).
9. Aquitaine Total Organico, G.B. 1,339,653.
10. Barton, B. C.; Hart, E. J., *Industrial and Engineering Chemistry*, 44, 2444–2448 (1952); Adams, H. E.; Johnson, B. L., *Industrial and Engineering Chemistry*, 45, 1539–1546 (1953); Trivette, C. D., Jr.; Morita, E.; Young, E. J., *Rubber Chemical Technology*, 35, 1370 (1962).
11. Hall, W. E.; Jones, H. C., Presented at a *Meeting of the Rubber Division*, American Chemical Society, Chicago, IL, October 1970.
12. Hall, W. E.; Fox, D. B., Presented at a *Meeting of the Rubber Division*, American Chemical Society, San Francisco, CA, October 5–8, 1976.
13. Oenslager, G., *Industrial and Engineering Chemistry*, 25, 232 (1933).
14. Van Alpen, J., *Rubber Chemicals*, D. Reidel Publishing Co., Boston, MA, 1973.

15. Trivette, C. D., Jr.; Morita, E.; Young, E. J., *Rubber Chemical Technology*, 35(5), 1360–1428 (1962); Chapman, A. V.; Porter, M., in *Natural Rubber Science and Technology*, Roberts, A. D. ed., Oxford University Press, Oxford, U.K., 1988, pp. 511–620; Kresja, M. R.; Koenig, J. L., *Rubber Chemical Technology*, 66, 376 (1993).

16. Ignatz-Hoover, Katritzky et al.

17. Rostek, Charles, J.; Lin, J.-J.; Sikora, D.J.; Katritzky, A.R.; Kuzmierkiewicz, W.; Shobana, N., *Rubber Chemistry and Technology*, 69(2), 180–202 (1996).

18. Wolfe, J. R., Jr., *Rubber Chemistry and* Technology, 41, 1339 (1968); Coran, A. Y., in *Science and Technology of Rubber*, Eirich, F. R. ed., Academic Press Inc., New York, 1978, Chapter 7, p. 301.

19. Skinner, T. D., *Rubber Chemistry and* Technology, 45, 182 (1972).

20. Campbell, R. H.; Wise, R. W., *Rubber Chemistry and Technology*, 37, 635 (1964).

21. Kratz, G. D.; Flower, A. H.; Coolidge, C., *Journal of Indian Engineering and Chemistry*, 12, 317 (1920); Bedford, C. W.; Scott, W., *Journal of Indian Engineering and Chemistry*, 12, 31 (1920); Moore, C. G.; Saville, R. W. *Journal of Chemical Society*, 2082 (1954); Krebs, H., *Rubber Chemical Technology*, 30, 962 (1957); Miligan, B., *Rubber Chemical Technology*, 39, 1115 (1966); Chivers, T., *Nature*, 252, 32 (1974); McCleverty, J. A., in *Sulfur. Its Significance for Chemistry, for the Geo-, Bio-, and Cosmosphere Technology*, Muller, A.; Krebs, B. eds., Elsevier, Amsterdam, the Netherlands, 1984, Vol. 5, pp. 311–329.

22. L. Bateman, C. G.; Moore, M. P.; Saville, M., in *The Chemistry and Physics of Rubber-Like Substances*, Bateman, L. ed., Maclaren, London, U.K., 1963, Chapter 15, p. 449; Barton, B. C.; Hart, E. J., *Journal of Indian Engineering and Chemistry*, 44, 2444 (1952); Coran, A.Y., *Rubber Chemical Technology*, 37, 679 (1964); Coran, A. Y., *Rubber Chemical Technology*, 37, 689 (1964); Coran, A. Y., *Rubber Chemical Technology*, 38, 1 (1965); Milligan, B., *Journal of Chemical Society*, 1, 34 (1966).

23. Higgins, G. M. C.; Saville, B., *Journal of Chemical Society*, 2812 (1963), Coates, E.; Rigg, B.; Saville, B.; Skelton, D., *Journal of Chemical Society*, 5613 (1965); Spacu, G.; Macarovici, C. G., *Bull. Sect. Sci. Acad. Roumaine*, 21, 173 (1938–1939), Lichty, J. G., U.S. Patent 2,129,621 (1938); Tsurugi, J.; Nakabayashi, T., *Journal: Society of Rubber Industry, Japan*, 25, 267 (1952); Gupta, S. K.; Srivastava, T. S., *Journal of Inorganic and Nuclear Chemistry*, 32, 1611 (1970).

24. Bateman, L.; Moore, C. G.; Porter, M.; Saville, M., in *The Chemistry and Physics of Rubber-Like Substances*, Bateman, L. ed., Maclaren, London, U.K., 1963, Chapter 15, p. 449.

25. Fackler, J. P., Jr.; Coucouvanis, D.; Fetchin, J. A.; Seidel, W. C., *Journal of American Chemical Society*, 90, 2784 (1968); Coucouvanis, D.; Fackler, J. P., Jr., *Journal of American Chemical Society*, 89, 1346 (1967).

26. Goda, K.; Tsurgi, J.; Yamamoto, R.; Ohara, M.; Sakuramoto, Y., *Nippon Gomu Kyokaishi*, 46, 63 (1973).

27. Fackler, J. P., Jr.; Coucouvanis, D.; Fetchin, J. A.; Seidel, W. C., *Journal of American Chemical Society*, 90, 2784 (1968); Coucouvanis, D.; Fackler, J. P., Jr., *Journal of American Chemical Society*, 89, 1346 (1967).

28. Chapman, A. V.; Porter, M., in *Natural Rubber Science and Technology*, Roberts, A. D. ed., Oxford University Press, Oxford, U.K., 1988, pp. 511–620.

29. Hammett, L. P., *Chemical Review*, 17, 125 (1935).

30. Taft, R. W., *Journal of American Chemical Society*, 74, 3120 (1952); Taft, R. W., in *Steric Effects in Organic Chemistry*, Newman, M. S. ed., John Wiley & Sons, New York, 1956, p. 556.

31. Morita, E., *Rubber Chemical Technology*, 57, 744 (1984).

32. Ignatz-Hoover, F.; Katritzky, A. R.; Lobanov, V. S.; Karelson, M., *Rubber Chemical Technology* 72, pp. 318–333 (1996).

33. Rauchfuss, T. B.; Maender, O. W.; Ignatz-Hoover, F., U.S. Patent 6,114,469 (2000).

34. Ignatz-Hoover, F.; Kuhls, G., Delayed action accelerated sulfur vulcanization of high diene elastomers; The effects of the amine on vulcanization characteristics of sulfenamide accelerated cure systems, Paper No. 3, *Rubber Division, ACS Spring Meeting*, Chicago, IL, 1990.
35. Ignatz-Hoover, Frederick; unpublished results.

12 Recycling of Rubber

Brendan Rodgers and Bernard D'Cruz

CONTENTS

I.   Introduction 523
II.   Recycling of Rubber 523
      A.   Reduction 525
      B.   Reuse 526
      C.   Recycle 526
      D.   Reclaim 530
      E.   Compounding Application of Recycled Materials 533
III. Summary 534
References 534

I. INTRODUCTION

Not only is there interest in the use of renewable raw materials such as natural rubber and fillers such as calcium carbonate, there are also both environmental and economic reasons to recycle and reclaim scrap rubber. The automotive industry has studied and, in some instances, set targets for recycle content of up to 25% of post-consumer and industrial scrap in their products with no increase in cost or loss in performance. Post-consumer scrap recycling is the reuse of products that have completed their service life. These products can be ground into a fine powder; more recently, attempts are being made to return materials to their original state via devul-canization or other chemical degradation process. Industrial scrap is the waste material generated in the original manufacturing process. In this instance, the goal of recycling is to ensure that all this material is used in the production of high-quality goods. The purpose of this discussion is to provide the rubber technologist with introductory information on how to be compliant with these broad environmental objectives and contain cost while satisfying the end-product design and performance criteria. The discussion will describe the various forms and types of rubber recy-clates available to the compounder and show how they can be incorporated into a rubber compound. The effect of these rubber recyclates on the rubber compound will also be demonstrated in the form of physical and performance data.

II. RECYCLING OF RUBBER

Though achievable levels of recycled materials will most likely be lower than the initial target of 25%, recycling should be a technologist's objective. The first attempt at reusing rubber was through reclaiming. However, in the 1970s and 1980s reclaim

TABLE 12.1
Mesh Size

Mesh Size		Dimension (in.)
10	2.00 mm	0.0787
20	850 μm	0.0331
30	600 μm	0.0234
40	425 μm	0.0165
60	250 μm	0.0098
80	180 μm	0.0070
100	150 μm	0.0059

use declined, due to the growth in the radial tire market. More recently, the use of finely ground rubber (e.g., 20–80 mesh) produced by ambient and cryogenic processes emerged (Table 12.1). This was augmented by the development of wet process grinding of rubber in a water medium to produce very fine particle sizes, that is, 60–200 mesh.

There have been numerous attempts to produce reusable rubber through devulcanization by using some of the following methods and techniques. Ultrasonic, microwave, and bacterial degradation; chemical devulcanization; surface modification; solution swelling in active solvents; and many other methods have been evaluated or are in various stages of development and use. Rubber recyclates include ambient ground rubber (Figure 12.1), cryogenic ground rubber (Figure 12.2), and wet ground rubber, the latter being similar to that produced by the ambient grind process.

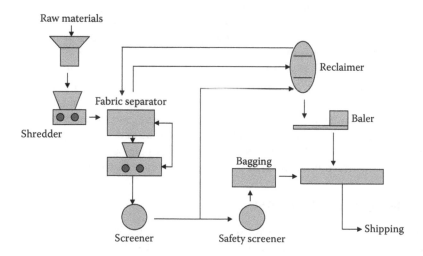

FIGURE 12.1   Simplified schematic of typical ambient grinding and reclaim system.

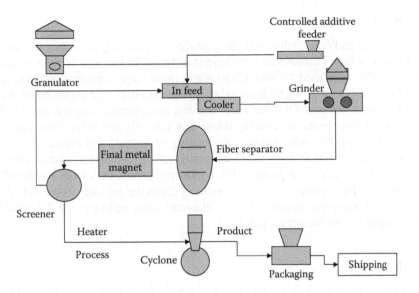

FIGURE 12.2 Schematic of typical cryogenic grinding system.

Four publications worth noting for rubber compounders trying to utilize recycled rubber are as follows:

1. *Rubber Recycling* by Anil Bhowmick (CRC Press) [1]
2. Myhre and MacKillop's review in *Rubber Chemistry and Technology Annual Rubber Reviews*, 2002, of all facets of rubber recycling [2]
3. *Best Practices in Scrap Tires and Rubber Recycling* by Klingensmith and Baranwal, published in 1997, which discusses all aspects of rubber recycling [3]
4. The *Scrap Tire Users Directory*, published yearly by the Recycling Research Institute, which lists all grinders and processors of scrap tire and rubber products [4]

The recycling of rubber products can also be considered, in its entirety, to fall into four basic categories, popularly characterized as reduction, reuse, recycle, and reclaim. Each of these will be considered in turn.

A. REDUCTION

Materials reduction efforts have focused on optimum use of materials, the weight reduction of tires and other engineered products, and gauge optimization of product components. This has largely been facilitated through new manufacturing systems and new product designs. It has received considerable emphasis in recent years due to the efforts at improving tire rolling resistance, work at reducing the contribution of tires to whole vehicle weight, and the positive effect that has on reducing vehicle fuel consumption.

B. REUSE

Reuse of tires and other industrial rubber products has largely been directed toward their use as a fuel. Dismounted tires have thus been put into one of three streams: (1) 60%–65% were used for fuel, (2) tires and other large industrial products have gone into landfills or to stockpiles or other storage facilities, and (3) the balance was recycled into other uses. Of these waste streams, considerable progress over the last 20 years has been made at reducing the size of landfills and other waste tire storage facilities and this will continue. Retreading of aircraft and commercial truck tires is probably the most ideal use of worn products. In the case of aircraft tires, up to four or five retreads are possible. For commercial truck tires from size 9.00R20 to 12.00R24, two retreads are not unusual, or from the highway sizes—11R22.5, 295/75R22.5, the corresponding 24.5 in. diameter sizes, and up to the 315/80R22.5 sizes—up to three retreads are practiced.

C. RECYCLE

The major methods for recycling existing rubber are ambient grinding, cryogenic grinding, and wet grinding. The resulting products are useful for controlling compound cost and improve processing when added to newly compounded rubbers.

In ambient grinding, vulcanized scrap rubber is first reduced to chips on the order of 1–2 in. in size. For some rubber products such as tires, this is normally accomplished by shredding. The shredded rubber is then passed over magnetic, mechanical, and pneumatic separators to remove metals and fibers. These pieces can be reduced in size by further ambient grinding on mills or by freezing them with liquid nitrogen and then grinding them into fine particles. The ambient process uses conventional high-powered mills with close nips that shear the rubber and grind it into small particles. It is common to produce 10–40 mesh material using this method, and the material is the least expensive to produce. The finer the desired particle, the longer the rubber is kept on the mill. Alternatively, multiple grinds can be used to reduce the particle size. The lower practical limit for the process is the production of 40 mesh material (Table 12.1). Any fiber and extraneous material must be removed using an air separator. Steel wires are removed by using a magnetic separator. A flow chart for an ambient grind process including a side stream for reclaiming is shown in Figure 12.1. The process produces a material with an irregular jagged particle shape. In addition, the process generates a significant amount of heat in the rubber during processing. Excess heat can degrade the rubber, and efficient cooling systems are essential.

Cryogenic grinding uses rubber particles of up to 2 in. and freezes them with liquid nitrogen. The frozen pellets are passed into a mill for further grinding. The size of particles typically produced by this method ranges from 60 to 80 mesh. The advantages of this technique are as follows: (1) little heat is generated so there is no thermal degradation of the material such as is found with ambient grinding and (2) finer particles are obtained.

The cryogenic process produces fractured surfaces. The most significant feature of the process is that almost all fiber or steel is liberated from the rubber, resulting in a high yield of usable product. The cost of liquid nitrogen has dropped significantly, and cryogenically ground rubber can now compete on a large scale with ambient ground products. A flow chart for a typical cryogenic process is shown in Figure 12.2.

Many manufacturing organizations wish to incorporate their scrap back into original rubber compound formula. This eliminates scrap disposal, provides better control over cost, and is an environmentally sound business practice. However, several practical problems arise in doing this. First, it may be difficult to accumulate sufficient clean scrap of a given compound or classification type. This is a significant drawback owing to the desire to attain consistent properties and performance of the final compound formulation. A second problem is the need for a recycling organization capable of working with small quantities of a given lot of waste material and keeping it in suitably clean condition. A third is that it is necessary to understand the effects of mesh size and concentration on the rubber properties. The following paragraph on the effects of concentration and mesh size on rubber properties [3,4] is based on the *Cryofine EPDM Handbook* [5].

The effect of variation in recycle content on a styrene butadiene rubber (SBR) based compound is illustrated in Table 12.2. Increasing the amount of 20 mesh

TABLE 12.2
Properties of Ambient Ground Rubber (20 Mesh) SBR 1502 Compounds with 0%, 17%, 33%, and 50 Crumb

	Compound			
	1	2	3	4
Compound ingredient (phr)				
SBR 1502	100.0	100.0	100.0	100.0
N660	90.0	90.0	90.0	90.0
Aromatic oil	50.0	50.0	50.0	50.0
TMQ (polymerized dihydrotrimethylquinoline)	2.0	2.0	2.0	2.0
Stearic acid	1.0	1.0	1.0	1.0
Zinc oxide	5.0	5.0	5.0	5.0
Sulfur	2.0	2.0	2.0	2.0
MBTS	1.0	1.0	1.0	1.0
TMTD	0.5	0.5	0.5	0.5
Crumb (%)	0.0	17.0	33.0	50.0
Property				
Mooney viscosity	40.0	61.0	91.0	111.0
Rheometer maximum torque (MH)	59.0	47.0	33.0	34.0
Tensile strength (MPa)	10.1	7.9	6.0	3.9
Ultimate elongation (%)	330.0	330.0	300.0	270.0

Source: Ball, J., in: Ball, J., ed., *Manual of Reclaimed Rubber*, Rubber Manufacturers Association, Washington, DC, 1956. With permission.

Note: phr, parts per hundred parts of rubber.

TABLE 12.3
Characteristics of Ambient vs. Cryogenically Ground
Whole Tire Recycled Rubber

Physical Property	Ambient Ground	Cryogenic Ground
Specific gravity	Same	Same
Particle shape	Irregular	Fractured
Fiber content	0.5%	Nil
Steel content	0.1%	Nil
Cost	Comparable	Comparable

Source: Ball, J., in: Ball, J., ed., *Manual of Reclaimed Rubber,* Rubber Manufacturers Association, Washington, DC, 1956.

crumb rubber from 0% to 50% results in a drop in tensile strength from 10.1 to 3.9
MPa, an increase in Mooney viscosity from 40.0 to 111.0, and a drop in oscillat-
ing disc rheometer (ODR) rheometer maximum torque (MH) from 59 to 34. The
important observation from these data is that a consistent level of recycle material is essential to ensure uniformity and quality in a product's design specifications. Table 12.3 presents a simplified comparison of the two fundamental types of
ground rubber, which must also be taken into consideration when selecting a material for inclusion in a compound formula. Clearly, cryogenically ground rubber is
preferable to ambient ground material because it is easier to remove fiber or wire.
Also, cryogenic grinding could be more effective in reducing residual fiber size so
as to minimize any potential source of crack initiation. Some typical properties of
an EPDM-based compound with cryogenically ground crumb added are displayed
in Table 12.4. Mesh size of the crumb ranged from 40 to 100, and it was added at
10% and 20% loading. In both cases, loss in fundamental mechanical properties
such as tensile strength was negligible. However, increase in loading as noted in the
data displayed in Table 12.2 did lead to loss in properties.

The data in Table 12.5 were extracted from the *Cryofine Butyl Compounding
Handbook* [6]. They show the effect that an 80 mesh cryogenically ground butyl rubber has on the mechanical and physical properties of a typical halobutyl tire innerliner. The effects of 5%, 10%, and 15% loadings are shown. A 5% level of finely
ground butyl scrap is commonly added to innerliners. Besides reducing compound
cost, the ground butyl provides a path for trapped air to escape from the compound.
The number of tires rejected due to blisters is reduced significantly. This effect of
ground rubber is noted in all elastomers, especially the highly impermeable polymers such as butyl, halobutyl polymers, and fluoroelastomers.

Wet grinding uses a water suspension of rubber particles and a grinding mill.
The material is finely ground to a mesh size of 60–120. These products therefore
find ready use in tire compounds due to their uniformity and minimal contamination. Surface treatment and additives can enhance the mechanical properties
of compounds containing recycled materials. Additives include materials such as

TABLE 12.4
Properties of EPDM Compounds Containing Cryogenic Rubber

	Compound				
	1	2	3	4	5
Basic compound					
EPDM	100.0	100.0	100.0	100.0	100.0
N650	70.0	70.0	70.0	70.0	70.0
N774	130.0	130.0	130.0	130.0	130.0
Paraffin oil	70.0	70.0	70.0	70.0	70.0
Low MW polyethylene	5.0	5.0	5.0	5.0	5.0
TMQ (polymerized dihydrotrimethylquinoline)	1.0	1.0	1.0	1.0	1.0
Stearic acid	1.0	1.0	1.0	1.0	1.0
Zinc oxide	5.0	5.0	5.0	5.0	5.0
Sulfur	1.3	1.3	1.3	1.3	1.3
TBBS	8.0	8.0	8.0	8.0	8.0
TMTD	8.0	8.0	8.0	8.0	8.0
TDEDC (tellurium diethyldithiocarbamate)	8.0	8.0	8.0	8.0	8.0
MBT	1.0	1.0	1.0	1.0	1.0
Mesh	Control	40	60	80	100
Properties at weight 10% loading					
Tensile strength	9.72	8.89	9.86	10.73	9.93
Ultimate elongation	410	330	340	400	380
300% Modulus	8.13	8.4	8.5	8.5	8.4
Hardness (Shore A)	73	70	70	70	71
Properties at 20% loading					
Tensile strength	9.72	8.5	9.4	10.1	9.7
Ultimate elongation	410	320	390	390	390
300% Modulus	8.13	8.4	8.9	8.8	7.9
Hardness (Shore A)	73	72	70	69	68

polyurethane precursors, liquid polymers, oligomers, resin additives, and rubber curatives. In some instances, when the specific chemical composition of the surface treatment is compatible with the materials to be reincorporated, retention of the original mechanical properties of the compound can be achieved. For example, nitrile compounds should be treated with acrylonitrile butadiene copolymers or block copolymers, which have similar solubility parameters [2].

The surface of crumb rubber can be activated by the addition of unsaturated low-molecular-weight elastomers. Latex is added to the crumb rubber in an aqueous dispersion. The water is removed, leaving a coating around the ground rubber. This technique, known as the surface-activated crumb process, has been commercialized by Vredestein Rubber Recycling Company in Europe. There is considerable scope for further development in this area. It is reasonable to state

TABLE 12.5
Properties of Innerline Compounds Loaded with
Cryogenically Ground Butyl Rubber (80 Mesh)

	Compound			
	1	2	3	4
Base compound composition (phr)				
Chlorobutyl rubber (1066)	80.0	80.0	100.0	100.0
Natural rubber (TSR 5)	20.0	20.0	90.0	90.0
N650	65.0	50.0	65.0	50.0
Mineral rubber	4	4	4	4
Phenolic resin	4.00	4.00	4.00	4.00
Naphthenic oil	8.00	8.00	8.00	8.00
Stearic acid	2.00	2.00	2.00	2.00
Zinc oxide	3.00	3.00	3.00	3.00
Sulfur	0.50	0.50	0.50	0.50
MBTS	1.50	1.50	1.50	1.50
Ground butyl rubber loading	0	5	10	15
% Rheometer t90	47.5	46.3	47	46.5
Tensile strength (MPa)	9.7	9.3	8.9	8.8
300% Modulus (MPa)	7.7	7.2	6.9	6.5
Air permeability (Qa)	4.7	4.7	4.5	4.2

Source: Smith, F.G., Reclaimed rubber, in: Babbit, R.O., ed., *The Vanderbilt Rubber Handbook*, RT Vanderbilt, Norwalk, CT, 1978. With permission.

Note: phr, parts per hundred parts of rubber.

that success of the rubber recycling industry will be dependent on developing economically effective means by which the surface of ground recycled rubber is chemically activated so as to enable attainment of the original compound's mechanical properties.

D. RECLAIM

Reclaim of rubber refers to the recovery of original elastomers in a form in which they can be used to replace fresh polymer [1,7,8]. Again, a range of techniques are available at small pilot plant levels to produce such materials:

Ultrasonic devulcanization: Though it has not been achieved commercially, ultrasonic devulcanization continues to be a potential method to allow reclamation of the original polymer. Sulfur–sulfur bonds have lower bond energy than carbon–carbon bonds. Given this, ultrasonic waves can have enough energy to selectively break the sulfur bonds, thereby devulcanizing the compound.

Chemical devulcanization: Chemical devulcanization methods involve mixing rubber peelings in a high-swelling solvent with a catalyst. Heating brings about a significant reduction in cross-link density. Though other chemical techniques are being investigated, any future system will most likely involve catalytic degradation in a solvent at high temperature and pressure.

Thermal devulcanization: This involves the use of microwaves, inducing an increase in temperature with preferential breaking of sulfur–sulfur bonds. Owing to the cost of operating such systems, there has been no successful commercial system, but pilot plant facilities have been in operation.

Chemomechanical and thermomechanical techniques: Such systems have ranged from the simple addition of vulcanization system ingredients to crumb rubber, and polymer surface modification to add functionality to the surface of the particles to treatment at higher temperatures with the intent of activating the surface. No commercially successful systems have been developed, though pilot plant facilities are in operation.

Through the use of reclaiming agents, steam digestion, and/or mechanical shear, it is possible to convert used tires, tubes, bladders, and other rubber articles into a form of rubber that can be incorporated into new rubber compounds. There are many processes available to accomplish this. They include the (1) heater or pan process, (2) dynamic dry digester process, (3) wet digester process, (4) Reclaimotor process, and (5) Banbury process. The purpose of this section is not to discuss the manufacturing methods of reclaim but its benefits and uses. For details on manufacturing, reference should be made to the *RT Vanderbilt Handbook* [8].

Reclaim or recycled rubber was used in significant volume up to the early 1970s in the United States. Then the growth of radial tires, environmental regulations, and the large economy of scale of the new or upgraded styrene butadiene rubber (SBR) and polybutadiene (BR) manufacturing processes resulted in low original rubber prices and a significant contraction of the reclaim rubber industry. In the 1960s, there were estimates that as much as 600 million lb per year of reclaim rubber was used in the United States. The estimate for the year 2002 was 60 million lb, consisting mostly of reclaimed butyl for tire innerliners, reclaimed NR for mats and low-end static applications, and reclaimed silicone for automotive and electrical applications.

Reclaim rubber had several distinct potential advantages. These included lower cost than original rubber, improved rheological characteristics found in manufacture, less shrinkage or better component dimensional stability, and the possible reduction in the need for curing agents in the compound. However, lower compound green strength, durability, and tear strength led to its removal from radial tire compounds. In many regions of the world such as India, China, and some southeastern Asian countries, reclaim is still widely used in footwear, bias tire compounds, mats, automotive parts, and other goods. An estimated 200–300 reclaimers are still operating.

Table 12.6 illustrates the effect of adding reclaim to a radial tire carcass compound [9]. Though caution should be exercised with regard to the impact on cut growth and fatigue resistance, some classical mechanical properties are retained when reclaim

TABLE 12.6
Physical Properties of Radial Tire Casing Compound Containing Wet Ground and Reclaim Rubber

	Control	Substituting 10 Parts GF-80	Substituting 10 Parts Whole Tire Reclaim
Modulus at 300% (MPa)	3	3.7	3.7
Tensile strength (MPa)	14.9	15.1	13.6
Elongation (%)	875	775	740
Hardness (Shore A)	58	58	59
Tear strength (kN m)	20.7	20.9	20.7

Source: *Technical Bulletin*, Rouse Rubber Technical Industries, 1991. With permission.

TABLE 12.7
Reclaim in Automotive Vehicle Mats—Properties

	Compound	
	1	2
Component		
Natural rubber	79.2	
SBR 1502		60.0
Whole tire reclaim	41.6	80.0
Paraffinic oil	9.3	5.0
N550	65.0	45.0
TMQ (polymerized dihydrotrimethylquinoline)	0.6	1.0
Stearic acid	1.4	1.0
Zinc oxide	3.9	5.0
Sulfur	1.9	2.2
MBT	0.4	
MBTS		0.5
TMTM	0.3	
Property		
Tensile strength (MPa)	13.4	12.8
Ultimate elongation (%)	480.0	490.0
Hardness (Shore A)	57.0	65.0
Compression set (%)	19.0	22.0

is added. Static or low performance applications are therefore still preferred. The definition of low performance product applications largely excludes products such as high-performance tires, high-pressure hoses, and conveyor belt covers but does include mats, fenders, and potentially flaps used with tire inner-tubes. Table 12.7 illustrates typical properties achieved when whole tire reclaim is added to either a natural rubber or an SBR automotive mat compound. Basic mechanical properties quoted may be acceptable for this application.

E. COMPOUNDING APPLICATION OF RECYCLED MATERIALS

Recycled material can be added to the original compound formulation in one of two ways. The total elastomer content can be kept at 100 RHC, or the recycle content can be added on top of the original elastomer as a filler. Though at low levels there can be deterioration in the mechanical properties of a compound, there are still a wide variety of successful applications. For example,

1. In cement and concrete, the addition of recycled rubber reduces vibration transmission, improves fracture resistance, and reduces cracking.
2. In fill for sludge treatment plants, crumb rubber is effective at absorbing heavy metals and organic solvents such as benzene, toluene, and other organic solvents.
3. Asphalt containing crumb rubber is more durable, has better resistance to thermal and reflective cracking, reduces the level of noise generated by vehicles traveling on it, and provides a more comfortable ride.
4. Flooring, walkway tiles, and sports surfaces such as running or jogging tracks constitute a growing market for recycled rubber. Such surfaces are very effective. However, the most important technical issue is the removal of all steel from the material. Cryogenic grinding is typically used to produce materials for such applications.

Tables 12.8 and 12.9 list a range of applications for recycled materials in tires and industrial products. Table 12.8 displays tire components that have the potential to contain varying levels of recycled material. Conversely, many components in tires

TABLE 12.8
Use of Reclaim and Recycled Materials in Tires

Component	Passenger Tires	Light Truck Tires	Commercial Tires	Retreads
Treads	Yes	Yes	No	Yes
Subtread	No	No	No	Yes
Casing plies	No	No	No	No
Bead filler	Yes	Yes	No	No
Sidewall	Yes	Yes	No	No
Wedges	Yes	No	No	No
Squeegee/barrier	Yes	Yes	Yes	Yes
Liner	Yes	Yes	No	No

Sources: Klingensmith, W. and Baranwal, K., *Best Practices in Scrap Tires and Rubber Recycling*, ReTAP of the Clean Washington Center, Seattle, WA, 1997; Burlow, F., *Rubber Compounding*, 2nd edn., Marcel Dekker, New York, 1994.

TABLE 12.9

Target Levels for Use of Reclaim and Recycled Materials in Industrial Products

Product	Potential Application	Potential Loading (Wt%)
Belt casing/carcass	No	0
Conveyor belt covers	Yes	3.0
Transmission belts (non-O.E.)	Yes	5.0
Hose covers and inner tubes	Yes	5.0
Low operating pressure tubing	Yes	10.0
Weatherstripping (non-O.E.)	Yes	25.0
Carpet backing	Yes	25.0
Railroad crossings	Yes	50.0

Sources: Klingensmith, W. and Baranwal, K., *Best Practices in Scrap Tires and Rubber Recycling*, ReTAP of the Clean Washington Center, Seattle, WA, 1997; Burlow, F., *Rubber Compounding*, 2nd edn., Marcel Dekker, New York, 1994.

Note: O.E., original equipment.

[a] Rubber blocks laid between rails at highway-railroad crossings and junctions.

cannot contain recycled material owing to potential deterioration in performance. To address this, two requirements may need to be addressed: (1) more finely ground material with better defined particle dimensions and (2) new compounding ingredients to improve factors such as dispersion, fatigue resistance, and tear strength. Table 12.9 similarly shows potential levels for recycled material in industrial products such as belts and hoses. These provide target levels for the materials scientist developing compounds with recycle content.

III.   SUMMARY

The use of recycled rubber will continue to increase throughout the first decades of the twenty-first century. This growth will be driven by regulatory and economic factors rather than technological factors. However, overcapacity in global vehicle and tire production is having a detrimental impact on pricing, which in turn is restricting growth in recycling opportunities. It is anticipated that this will change, and the materials scientist should have the appropriate technologies in place to take advantage of future demand. This discussion has therefore attempted to provide a foundation for the rubber technologist to take advantage of the selection and use of renewable and recycled materials available for the range of products produced by today's rubber industry.

REFERENCES

1. Bhowmick A. *Rubber Recycling*. Boca Raton, FL: CRC Press, 2009.
2. Myhre M, MacKillop DA. Rubber recycling. *Rubber Chem Technol* 2002;72:429–474.
3. Klingensmith W, Baranwal K. *Best Practices in Scrap Tires and Rubber Recycling*. Seattle, WA: ReTAP of the Clean Washington Center, 1997.

4. Sikora M. *Scrap Tire Users Directory*. Washington, DC: Recycling Research Institute, 2001.
5. *Cryofine EPDM Handbook*. Wapakonneta, OH: Midwest Elastomers, 1985.
6. *Cryofine Butyl Handbook*. Wapakonneta, OH: Midwest Elastomers, 1985.
7. Ball J. In: *Manual of Reclaimed Rubber*, Ball J, ed. Washington, DC: Rubber Manufacturers Association, 1956.
8. Smith FG. Reclaimed rubber. In: *The Vanderbilt Rubber Handbook*, Babbit, RO, ed. Norwalk, CT: RT Vanderbilt, 1978.
9. *Technical Bulletin*. Rouse Rubber Technical Industries, 1991.
10. Burlow F. *Rubber Compounding*, 2nd edn. New York: Marcel Dekker, 1994.

13 Industrial Rubber Products

George Burrowes

CONTENTS

I. Introduction 538
II. Hoses... 538
   A. Coolant Hose.. 538
      1. Manufacturing Process... 539
      2. Classification of Hoses and Materials 539
      3. Coolant Hose Materials 541
      4. Ethylene–Propylene Elastomer–Based Coolant Hoses 541
      5. Elastomer Characteristics.. 541
      6. Sulfur Vulcanization 542
      7. Peroxide Vulcanization.. 543
      8. Electrochemical Degradation of Coolant Hoses 544
      9. Silicone Elastomer–Based Coolant Hoses...................................... 546
   B. Fuel Hose... 547
      1. Environmental and Conservation Issues 547
      2. Hose Testing 548
      3. Hose Test Methods 548
      4. Hose Cover Material Development 551
      5. Hose Designs 552
      6. Fuel Hose Classification 553
   C. Hydraulic Hose.. 553
III. V-Belts 556
   A. V-Belt Types.. 557
      1. Materials.. 559
      2. V-Belt Jacket.. 562
      3. Other V-Belt Fabric Components 563
   B. V-Ribbed Belts 563
   C. Materials 563
   D. Belt Performance Criteria 565
IV. Synchronous (Timing) Belts... 567
   A. Materials 568
      1. Oxidative Aging 569
      2. Resin Reinforcement 569
   B. Alkylated Chlorosulfonated Polyethylene (ACSM)............................ 570

C. Short Fiber/Pulp Tooth Reinforcement 571
D. Tensile Member 571
E. Glass 571
F. Aramid 571
G. Belt Performance Criteria 572
V. Conveyor Belts 572
References 574

I. INTRODUCTION

The rubber industry represents an important link in a diverse range of associated manufacturing and service industry. Products find application in such varied applications such as automobiles, medical devices, mining, and many manufacturing systems. The automotive industry in particular has owed much of its success to the quality and performance of tires and associated industrial products such as hoses and belts. High-performance industrial rubber products include automotive hoses, hoses for hydraulic systems, and power transmission belts and conveyor belting. The use of elastomers and rubber compounds in these applications will be reviewed in this chapter.

II. HOSES

The hoses to be reviewed fall into one of three categories: coolant hose, fuel, or hydraulic hose.

A. COOLANT HOSE

Radiator hoses (Figure 13.1) are designed to provide a flexible connection permitting coolant fluid transfer between the engine block and the radiator. These hoses have an inner tube resistant to the coolant fluid (usually ethylene glycol/water mixtures) at the operating temperature, a hydrolysis-resistant textile reinforcement, and are covered by a heat- and ozone-resistant material.

The discussion on radiator hoses also applies in principle to heater hoses (internal diameter normally 19 mm or below), since ethylene glycol/water mixtures are the heating medium for the vehicle interior. However, unlike the radiator type, heater hoses are generally not exposed to continuous movement while the vehicle is in motion. The term "coolant hoses" will be used in this text for information that is pertinent to both radiator and heater hoses.

Automotive bodies and engines are becoming increasingly compact because of aerodynamic styling. At the same time, engines are operating at higher temperatures for improved fuel efficiency; there is an increasing desire for turbocharging, emission control, and power-assist devices. Therefore, the underhood temperatures, including those to which the coolant hoses are exposed, have continued to increase in recent years. The automotive manufacturers' expectation is that the coolant hoses on their engines will perform well over the lifetime of the vehicle. In 1988, a radiator hose life goal of 100,000 miles was quoted [1]. Nowadays, the life goal for these hoses has been extended to 10 years or 150,000 miles [2].

FIGURE 13.1   Radiator hose.

1. Manufacturing Process

In the traditional manufacturing process for coolant hoses, a rubber inner tube material is first extruded, and then passed through a textile knitter, braider, or spiraling equipment to apply reinforcing layer(s) of continuous filament yarn. A rubber cover material is extruded over the reinforced carcass and the unvulcanized hose is cut into predetermined lengths. With the aid of a glycol-based lubricant, the individual hose pieces are placed over the shaped mandrels that hold them in position during vulcanization with high-pressure steam. After that, pieces of vulcanized hose are stripped off the mandrel and trimmed to the required length. Some small internal diameter heater hoses are made on flexible mandrels and vulcanized by continuous processes.

2. Classification of Hoses and Materials

For the automotive industry, the most common performance standard for coolant system hoses is SAE J20, which classifies them according to type of service, for example, SAE 20R3 and SAE 20R4 are normal service heater and radiator hoses, respectively. In addition to outlining a series of other requirements, this standard also defines the physical properties of each "class" of the elastomeric materials to be used in the various hose types [3].

It is common practice in the industry to use compound performance in accelerated aging tests as a predictor of the serviceability of a hose in a vehicle. Some

TABLE 13.1

Material Physical Properties of Main Coolant Hose Types

Hose types used as illustration—SAE 20R3 Heater hose for normal service
SAE 20R4 Radiator hose for normal service

	SAE J20[a] Class D-1	SAE J20[a] Class D-3	SAE J20[a] Class A	GM6250M[b]
Original properties				
Durometer, shore A	55–75	55–75	55–75	60–75
Tensile strength, min, Mpa	7.0	7.0	5.5	7.6
Elongation, min, %	300	300	200	250
Oven age	70 h/125°C	168 h/150°C	70 h/175°C	168 h/165°C
Durometer, points shore A	+15	+15	+10	0–15
Tensile strength change, max, %	−20	−35	−15	−30
Elongation change, max, %	−50	−65	−40	−55
Compression set (ASTM D395 Method B)	125°C	125°C	125°C	150°C
70 h, max, %	75	75	40	60
Coolant immersion (tube only)				
Hours at boiling point	70	168	70	168
Volume change, %	−5 to +20	−5 to +20	0 to +40	+20
Durometer, points shore A	−10 to +10	−10 to +10	−10 to +10	−15 to +15
Tensile strength change, max, %	−20	−20	−30	−15
Elongation change, max, %	−50	−25	−25	−15
Elastomer	EPDM	EPDM	Silicone	EPDM
Vulcanization system	Sulfur	Peroxide	Peroxide	

[a] Property requirements extracted from SAE Standard SAE J20 (October 1997).
[b] Property requirements extracted from General Motors Engineering Standards GM6250M (June 1997).

limited data exist to back this up [4]. Table 13.1 shows the physical property requirements for the three most common classes of hose material: Briefly

1. Class D-1 material requirements are based on oven aging for 70 h at 125°C, with a 125°C compression set; they are usually met by sulfur-vulcanized ethylene–propylene–diene terpolymers (EPDM).
2. Class D-3 material requirements are based on more stringent oven aging, 168 h at 150°C; the same 125°C compression set requirement applies. This material class is usually peroxide-vulcanized EPDM.
3. Class A material requirements are based on oven aging for 70 h at 175°C, with a 125°C compression set; they can be achieved with peroxide vulcanized silicone polymers.

In this context, the compression set test is performed under constant strain conditions for 70 h at the stated temperature [5]. It is a measure of recoverability of the rubber material after aging under 25% compression; low compression set contributes to good coupling retention for a given rubber material. For hose materials of classes D-1 and D-3, with compression set measured at 125°C, the stability of the cross-link is the controlling factor, so a sulfur donor or peroxide cure system is necessary.

Class A materials have the most stringent requirements on aging, compression set, and coolant immersion. Silicone elastomers are usually required for this class.

3. Coolant Hose Materials

Rayon, suitable for 120°C service, has long been used as a cost-effective reinforcing yarn for coolant hoses. However, with increasing underhood temperatures, the more heat-resistant aramids, capable of operating up to 230°C, are used in preference to Rayon for the more demanding coolant hose applications [6]. Though the meta-aramid is significantly more expensive than *para*-aramid, the former is often used for its greater abrasion resistance, essential when yarns get in contact with hoses subjected to high levels of vibration, as well as its greater resistance to hydrolysis and heat.

4. Ethylene–Propylene Elastomer–Based Coolant Hoses

Before the 1960s, natural rubber and styrene–butadiene rubber (SBR) were the base elastomers for the tubes of automotive coolant hoses, with polychloroprene being used whenever an ozone-resistant cover was required. However, with the advent of EPDM technology, ethylene–propylene elastomer compounds rapidly gained widespread acceptance for coolant hoses because of their outstanding resistance to hot coolant fluid exposure and to the dry heat of vehicle engine compartments. Though other elastomers, most notably silicones, find some limited use, EPDM-based coolant hoses are used almost universally by the modern automotive industry. For this reason, most of the following discussion will be devoted to EPDM and its associated compounding issues.

5. Elastomer Characteristics

The following generalizations can be made on the required characteristics of EPDM elastomers for coolant hose compounds [7,8]:

a. *Molecular weight.* Highest molecular weight grades are commonly used since this increases hot green strength and improves tensile strength properties, compression set resistance, and collapse resistance of inner tube during application of reinforcement textile, and improves the capability for filler and oil loading so as to enable cost optimization.

b. *Ethylene content.* Higher ethylene content improves ambient temperature green strength, tensile strength, extrusion rate, and loading capability. High ethylene content can, however, be detrimental to flexibility and set properties at low temperature and may result in nervy extrudates. In practice, coolant hose compounds often contain a blend of high- and low-ethylene EPDM elastomers.

c. *Diene content (unsaturation level).* With sulfur cure systems, increasing levels of termonomer in the EPDM elastomer increase cure rate, tensile modulus and compression set resistance, but reduce scorch safety and in some cases may compromise heat resistance. Ethylidene norbornene (ENB), which gives the fastest cross-linking, is the preferred termonomer for coolant hose EPDM elastomers when compared with dicyclopentadiene (DCPD) or 1,4 hexadiene (1,4 HD). For peroxide curing there is, in principle, no need for diene to be included in the elastomer. However, diene content will improve cure rate and cross-link density.

d. *Molecular weight distribution (MWD).* A broad distribution will improve overall processing characteristics including extrusion smoothness. However, physical properties, especially compression set, may be compromised. The breadth of the MWD can influence cure state and cure rate, broader MWD grades curing to a lower cure state, and slower than narrow grades [4]. A recent development in catalyst technology has resulted in the production of EPDM elastomers with narrow MWD intended to provide good physical properties, along with a high level of chain branching to improve polymer processing [9].

6. Sulfur Vulcanization

Both sulfur and peroxide cure systems find application in coolant hoses. Since the cure system is the most important factor influencing the heat and compression set resistance of a hose, aspects pertinent to coolant hoses will be discussed in detail in the succeeding text.

Several review articles cover the basics of sulfur curing of EPDM elastomers [10–12].

Sulfur-based vulcanizing systems produce excellent stress/strain properties and tear strength in EPDM coolant hoses, and are very cost effective. Low sulfur/sulfur donor systems are preferred for coolant hose compounds as they give a near optimum balance of cure rate, heat resistance, compression set, and mechanical properties. Such cure systems have been reported in the literature [13].

Since EPDM elastomers have far fewer cure sites than diene rubbers, they consequently require higher levels of accelerator to achieve practically useful cure rates.

The heat resistance of a sulfur-cured EPDM compound is improved by the addition of the synergistic combination of zinc salt of mercaptobenzimidazole (ZMB) with poly-trimethyl-dihydroquinoline (TMQ). In the same work, another effective synergistic antidegradant combination was reported, that of nickel dibutyldithiocarbamate (NBC) with diphenylamine–acetone adduct. Further heat aging enhancement was obtained by adding polychloroprene (5 phr), and magnesium oxide and increasing the zinc oxide level [7].

Sulfur vulcanizing agents, with their high polarity, have limited solubility in non-polar EPDM elastomers. When the level of sulfur or accelerator exceeds its solubility in the EPDM, the chemical itself or its reaction products will bloom to the surface of the hose.

To avoid bloom in a hose compound, combinations of several accelerators must be used, each one at a level below its upper solubility limit in the compound. Generally, thiurams and dithiocarbamates have the lowest solubility in EPDM compounds. Bloom of any type on the surface of a coolant hose is unacceptable to automotive customers. Hoses covers must remain black in color with no solid deposit on their surface after being subjected to a 2-week long regimen of cyclic cooling at −30°C and heating at 100°C [14].

7. Peroxide Vulcanization

Several review articles cover the basics of peroxide curing systems of EPDM elastomers [11,12,15]. Comparing bond energies, it is apparent that carbon–carbon cross-links, obtained in EPDM compounds vulcanized with peroxides, have considerably more thermal stability than carbon–sulfur and sulfur–sulfur linkages: [10]

Linkage	Bond Energy (kJ/mol)
C–C	352 (Most thermally stable)
C–S–C	285
C–S–S–C	268
–Sx–	<268 (Least thermally stable)

Peroxide-vulcanized hoses tend to have better resistance to heat and compression set compared to those with sulfur-based systems [4,11].

Dicumyl peroxide and bis(*t*-butylperoxyisopropyl) benzene, on polymer or inert powder binders, are frequently used in EPDM coolant hoses. These peroxides can provide an acceptable balance between scorch safety, cure rate, and required hose properties for a given manufacturing process.

In accordance with the previously discussed trend toward higher underhood service temperatures and increasing warranty periods, some automotive manufacturers have raised the aging requirements on their coolant hose materials in recent years. As an example, General Motors requires tensile strength and elongation changes to be less than 30% and 55%, respectively of original values after 168 h aging at 165°C, with compression set tested at 150°C (Table 13.2). These conditions have tested the upper temperature limits of EPDM-based hose compounds. In addition, there has been some discussion on increasing requirements to the same levels of performance at 175°C. Even though silicone-based coolant hose compounds will meet these demands (Table 13.4, SAE J20 Class A), they are not economical for many applications; therefore, research effort has been put into improving the heat and compression set resistance of EPDM-based compounds for coolant hoses. Studies have concentrated on peroxide-vulcanized compounds as sulfur-based systems are not considered capable of meeting these more stringent levels of heat resistance.

Peroxide-cured compounds based on EPDM elastomers with 2%–3% unsaturation, and high ethylene content (around 70%) with carbon black and silane-treated talc were reported to meet the 175°C target aging requirements. It was also concluded that the addition of magnesium and zinc oxides, together with the antioxidant combination of *p*-dicumyl diphenylamine (DCPA) with zinc 2-mercaptotoluimidazole (ZMTI), all enhanced resistance to heat aging. Though 40%–50% polymer content was preferred, it was claimed that a level of 30%–35% could be formulated for 175°C performance [16,17].

In an extensive study, it was concluded that best aging was obtained with higher molecular weight EPDM elastomers and unsaturation 2% or below. Use of liquid polybutene instead of paraffinic oil, along with the coagent trimethylolpropane trimethacrylate (TMPTMA) and possibly partial replacement of carbon black by silane-treated talc, also improved 175°C aging of the peroxide-cured EPDM test compounds [18].

For optimum performance of a peroxide-cured coolant hose, it is important to select the proper coagent. Peroxide curing coagents improve the efficiency of the cross-linking reaction. At levels above 1 phr, coagents can influence the nature of the resulting network. A difunctional acrylic coagent, supplied with a scorch retarder, is claimed to provide excellent compression set and heat resistance, sufficient to meet the very demanding VW coolant hose specification [19–21].

The metallic coagents, zinc diacrylate (ZDA) and zinc dimethacrylate (ZDMA), both available with added scorch retarder, give higher ultimate elongations for a given modulus than non metallic types. Also, ZDMA will improve tear strength of peroxide EPDM both at ambient and elevated temperatures [22].

There are certain drawbacks specific to the use of peroxides for vulcanizing coolant hoses: [2]

- Based on expensive ingredients (the peroxide itself and the coagent), generally with lower filler loadings and with the potential for higher scrap levels, peroxide-cured coolant hoses are overall more expensive than sulfur-cured versions. To counteract this, one recent proposal has been to blend in a new design EPDM elastomer, one containing vinyl norbornene (VNB), a very efficient termonomer. This is claimed to give equivalent physical properties to an ENB containing EPDM, but requiring a significantly reduced amount of peroxide [23].
- Peroxide-cured EPDM hoses generally have lower tear strength, especially when hot, compared to those with sulfur systems. This is an issue not just for the finished part, but also in processing where high scrap levels may be incurred through tear during unloading from mandrels just after cure. Alternatively, serious design limitations may have to be placed on the angles for peroxide hose at major bends to avoid tear on unloading. A proposal to overcome this has been the use of a new carbon black, which because of its unique morphology and surface modification is claimed to provide significantly improved hot tear strength over that given by standard grades [24]. Also, the coagents ZDMA and two nonmetallic dimethacrylate esters are claimed to provide increased tear resistance at elevated temperatures when used with dicumyl peroxide in EPDM [20,22].
- Metal-forming mandrels used with peroxide-cured EPDM hoses require more frequent cleaning than those used for sulfur-vulcanized hoses. The normally used glycol-based mandrel lubricants react readily with peroxides resulting in the deposit of sticky substances on the mandrel surface.
- Special autoclave purging procedures are needed to minimize the amount of oxygen in contact with the curing hose, since peroxide curing of EPDM under oxygen results in surface stickiness.

8.   Electrochemical Degradation of Coolant Hoses

By the mid-1980s, it was becoming evident that the radiator (predominantly upper) and heater hoses on certain vehicle models and designs were developing longitudinal, generally parallel, microcracks or "striations," which extended from the inside of the inner tube near one or both ends of the hose. The striations were well developing before the compound had reached its expected service life. The tube and cover of the striated hoses remained flexible, so a heat-hardening mechanism did not adequately explain this phenomenon.

Over time, these fluid-exuding striations would become branched into "trees" and tended to grow through the tube to the cover leading to eventual hose cracking, leakage, and even bursting as the yarn became wet and destroyed (Figure 13.2).

Upper radiator hoses were found to exhibit more striations than those on the lower part of the radiator; heater hoses were affected less severely than upper but more so than lower radiator hoses. A reproduction of the automotive hose striations in a lab test led to the identification of the root cause of the failures as being an electrochemical degradation process occurring in the hose, the process being accelerated by high under hood temperatures [25].

The laboratory test used for this investigation (the "Brabolyzer" method) is performed on two pieces of hose or tube (joined by a glass insulator), partially filled with coolant fluid and sealed with stainless steel plugs. A voltage (dc), isolated from the coolant fluid inside the hose, is applied through the end plugs. The entire assembly is placed in an oven set at test temperature, while the specified voltage is applied for a stated time. On completion of the test, a cross section near the negative end of the hose is examined under magnification for the presence of striations.

Other accelerated lab tests have been used to study the striation phenomenon [26,27], two of which have been adopted by the automotive companies and by the Society of Automotive Engineers (SAE) to measure and define resistance to electrochemical degradation [28].

Coolant hose striations form because an electrochemical cell becomes created in the engine cooling system. The metal nipples on the engine and/or radiator form the anode, the coolant mixture (coolant, water, oxygen, ionic stabilizers, and corrosion

FIGURE 13.2 Hose with striations.

inhibitors) being the electrolyte and the carbon in the EPDM hose rubber being the cathode. Thus, a galvanic potential and consequently electrical current may exist at each end of the hose. In the presence of the current, there is a change in the compatibility of the EPDM compounds to the coolant, causing increased fluid absorption by the hose and weakening of the vulcanizate. The effect is accelerated by high underhood temperatures [25–27].

Electrochemical degradation may be minimized or eliminated by reducing the volume loading of carbon black and replacing it in part or totally with inorganic, hydrophobic fillers [26]. Carbon black type, however, may influence resistance to electrochemical degradation; two new carbon blacks are claimed to offer high resistance to degradation even when used at relatively high loadings in hose compounds [29]. Peroxide cures produce compounds with lower conductivity than those with sulfur cures; therefore, peroxide-cured hoses are in general less prone to electrochemical degradation [26,30]. The effect on degradation resistance of various peroxide types has been investigated [31].

In the future, it is likely that vehicles will be converted to 42 V generating systems to accommodate increased demand for electrical current by the new "drive-by-wire" components, for example, electronic steering and braking. EPDM-based hose compounds designed to be resistant to electrochemical degradation have been found in a lab study to be unaffected when the applied potential was increased to 42 V [32].

In another new development, many coolant systems are being factory filled with new "longlife" coolants, still ethylene glycol–based compositions but now containing "organic" acid corrosion inhibitors that largely replace "inorganic" inhibitors (mainly sodium silicate). A lab study has shown that the EPDM hose compounds exposed to coolants with organic inhibitors exhibit reduced electrochemical degradation compared to the same compounds in coolants with inorganic systems [32].

9. Silicone Elastomer–Based Coolant Hoses

As shown in Table 13.1, SAE J20 Class A hose materials are based on significantly more stringent heat aging requirements (70 h at 175°C) compared to those for Classes D-1 and D-3 and a significantly lower allowable compression set. Class A requirements are normally met by silicone-based coolant hose materials.

Under ASTM D2000/SAE J200 classification of elastomers, silicone is shown as Type F (200°C) for heat resistance, while EPDM is Type D (150°C). Silicone vulcanizates do not usually become hard and brittle until the temperature has fallen to about −55°C, though this depends somewhat on hardness. In this regard, they also outperform EPDM vulcanizates.

However, since silicone elastomers are significantly more expensive than EPDM, silicone hoses are only used in situations where extended hose life and reduced service costs will justify a higher purchase price. Silicone radiator hoses are, therefore, used in many turbo-charged engines where the compartment temperatures are elevated, in trucks and buses with high annual mileage and in some emergency and law enforcement vehicles. Silicone heater hoses are sometime used in vehicles where the hose is difficult to access for replacement.

B. Fuel Hose

The modern vehicle's fuel system, of which the hoses are a key element, must not only be capable of storing and delivering fuels to the engine for the expected component lifetime but must also comply with increasingly stringent regulations defining fuel emission levels.

In the earliest vehicles, fuel lines were made of metal tubing, but this fell out of favor due to their inflexibility and capability of transmitting noise. Following the development in 1930s Germany of fuel-resistant acrylonitrile butadiene rubber (NBR) elastomers, formed by the copolymerization of butadiene with acrylonitrile (ACN), flexible rubber hoses rapidly replaced metal tubing in vehicle fuel systems.

Compared to rigid tubing, the flexibility of hoses gives them some important advantages like routability, as well as the capability to isolate noise and vibrations in the engine. Fuel filler neck hoses, for example, must be flexible so that they may absorb shock without rupture in the event of a vehicle crash. Hose materials in direct contact with fuel need to be resistant to the fuel being conveyed, while the whole construction must resist environmental factors for the duration of the vehicle's service life. On the other hand, the fuel itself must not become contaminated by extractables from the hose.

The principal focus of this section is on elastomeric-based hose for fuel line and vapor return lines, but it is also relevant to other hose and tubing in the fuel system, including filler neck and vent hose/tubes.

1. Environmental and Conservation Issues

Automotive fuels, always with some compositional variability, have had their compositions changed still further in the past three decades in response to the series of environmental and conservation initiatives shown in the following text. The materials and constructions of automotive fuel hoses have, consequently, had to accommodate the fuel composition changes.

- Aromatic hydrocarbons, alcohols, ethers, for example, methyl-*t*-butyl ether (MBTE) and other additives present in fuels to compensate for loss in octane number caused by the removal of lead from gasoline.
- Corporate Average Fuel Economy (CAFE) standards adopted in the 1970s resulted in vehicle size/weight reduction and more compact engine compartments with reduced air flow. Coupled with the addition of more underhood heat sources, for example, catalytic converters, this has increased hose and fuel temperatures as well as vapor generation rates within the fuel system.
- Development of fuel injection engines in which recirculated fuel, exposed to air, heat, moisture, and copper ions, forms hydroperoxides (so-called "sour" gasoline) that decompose to form rubber-attacking free radicals.
- Fuel conservation demands leading to the supplementing of gasoline with alcohols to conserve petroleum. In the United States, gasoline is blended with ethanol to give "gasohol."
- Development of biological fuels made from renewable raw materials, for example, "biodiesel" from soybean oil, used either in blends or as replacements for fossil fuels.

- Stringent hydrocarbon emission standards, pioneered by the California Air Resources Board (CARB) and the Environmental Protection Agency (EPA), have been implemented to reduce atmospheric pollution. Under current CARB Low Emission Vehicle (LEV I) and EPA standards, the total allowable vehicle evaporative emission following a diurnal Sealed Housing for Evaporative Determination (SHED) test is only 2 g/day for light-duty vehicles. Starting in the 2004 model year, new CARB LEV II regulations will reduce this number by 75%. These conditions must be met not only when the vehicle is built but also at any time during its defined lifetime. EPA's Tier 2 standards and similar legislation in Europe also mandate significant reductions in evaporative emissions. Fuel permeating through hoses in vehicle fuel systems is a major source of evaporative emissions.
- Legislation passed in the 1990s has mandated the introduction of reformulated gasoline in some areas in order to meet stringent carbon monoxide and ozone standards and to reduce benzene content.

The impact of these issues on the materials and constructions used in fuel hoses has been well reviewed up to 1993 [33].

2. Hose Testing

Commercial fuels are not suitable for material qualification testing as these fuels vary significantly between manufacturers, batches, seasons, and geographical regions. In order to evaluate the effects of fuels on materials and have consistent, comparable test results, it has been found necessary to define worldwide controlled reference fuels that can be used to simulate those used in the real world.

Material performance is determined using the reference fuels, which are designed to exaggerate the effects of fuel on materials and allow testing to be completed in a reasonable time frame with the purpose of predicting hose performance in actual use.

International standards have been published on the compositions (usually expressed as volume% of each component), nomenclature, preparation methods, etc. for recommended reference fuels [34–36]. ASTM Fuel C, iso octane/toluene (50:50), is the reference fuel most often associated with materials testing [34].

3. Hose Test Methods

A range of test methods are of importance and are described as follows.

a.   *Reservoir Method [37]*

One end of the test hose is attached to a metal can acting as a fuel reservoir; the other end has a metal plug. The assembly is positioned such that the hose is always kept full of test fuel. Fuel permeation rate is measured by weighing the assembly at intervals of 24 h, with inversion between weighings to drain and refill the hose with fuel. The rate of fuel permeation is reported as $g/m^2/day$ of exposed tube area.

b.   *Fuel Recirculating Method [38]*

This is a procedure for individual hoses or small assemblies in which the hydrocarbon fluid losses by permeation through component walls and leaks at interfaces are determined as fuel flows through a controlled environment. It is a recirculating system in which liquids that permeate walls and joints are collected by a controlled flow of nitrogen and adsorbed by activated charcoal.

c. Sealed Housing for Evaporative Determination (SHED and Mini-SHED)

This test uses enclosed cells or structures that contain the vehicle, assembly or hose being tested. The environment is controlled and periodically analyzed to determine the quantities of hydrocarbons that are present.

d. Hose Tube Material Development

i. Effect of Heat: Until the mid-1970s, fuel hose constructions were based on a fuel-resistant NBR tube, black and clay loaded, with 32%–34% acrylonitrile (ACN) content. Such constructions were considered to be only capable of 100°C service. By appropriate choice of cure system and use of silica as the main filler, resistance to long-term heat aging at 125°C could be achieved [39,40]. Another approach to obtaining 125°C aging resistance has been through the use of elastomers in which an antioxidant is "bound" to the NBR during the polymerization process [41]. Synergistic combinations of antioxidants, for example, acetone–diphenylamine reaction product with the relatively non-extractable alpha-methylstyrenated diphenylamine (alpha-MSDPA), are now recommended for NBR-based fuel hose tubes to be used in air-aspirated engines with carburetors [42]. Hydrogenated nitrile butadiene rubber (HNBR) and NBR are compatible, and their peroxide-cured blends (50:50) were found to have improved heat resistance compared to NBR alone [43].

ii. Effect of Aromatic Hydrocarbon Content of Gasolines: The aromatic hydrocarbon content of unleaded gasolines can vary from approximately 10% up to 50% depending on producer, season, etc. Aromatic hydrocarbons (toluene, benzene, xylenes) cause more swelling and more adversely affect physical properties of fuel hose tubes than either aliphatic or olefinic hydrocarbons. When vulcanizates based on NBR (34% ACN content) were exposed to ASTM fuels B, D, and C that have toluene contents 30%, 40%, and 50%, respectively [34], swelling and permeation increased with the aromatic content of the fuel. The effect of the aromatic level may be offset to some extent by using NBR-based elastomers with increased ACN content or by blending PVC with NBR [43–45]. Permeability of NBR-based vulcanizates to Fuel C may be reduced by partial replacement of carbon black with platy fillers such as talc. The presence of talc, however, did not significantly affect their swelling in fuel [45]. For test fuels with 29% and 50% aromatic content, the permeation rates through vulcanizates based on vinyliden fluoride–hexafluoropropylene copolymers (FKM) were found to be dramatically lower than for epichlorohydrin–ethylene oxide copolymer (ECO) or NBR (28% ACN) vulcanizates [46].

iii. Effect of Hydroperoxide Containing ("Sour") Gasoline: Government regulations for minimum mileage and pollution control have led to a major increase in the adoption of electronic fuel injection systems. Early failure of rubber fuel hoses in some vehicles fitted with fuel injection was believed to be due to their attack by "sour" or hydroperoxide-containing gasoline. Hydroperoxides are formed in fuel by the combined action of air, moisture, heat, and copper ions. When hydroperoxides decompose, free radicals are formed that can attack some rubbers to impart additional vulcanization (hardening). On the other hand, these free radicals cause other elastomers, for example, those with ether backbones, to undergo "reversion" (softening). Tertiary-butyl hydroperoxide (TBHP) is the chosen hydroperoxide for blending into test fuel compositions to simulate sour gasoline. When immersed in test fuel containing various concentrations of TBHP and catalytic metals (copper, iron), NBR-based vulcanizates, including those with the elastomer-bound antioxidant, became brittle [47]. However, NBR fuel hose tubes that have been compounded for heat resistance were also found to have enhanced resistance to sour gasoline [39]. Fuel hose tubes based on ECO, though they have good resistance to normal fuel, were found to soften drastically on exposure to sour gasoline [47]. An HNBR/fluorinated thermoplastic alloy has been claimed to show promising results for direct and continuous contact with sour gasoline at 60°C in the presence of a copper catalyst [50]. Fluoroelastomers (FKM) are resistant to attack by sour gasoline as evidenced by the high retention of tensile properties and low volume swell of their vulcanizates [47]. After 1000 h exposure at 60°C to sour gasoline, it is claimed that no significant changes in physical properties occurred for the fluorothermoplastic "THV," a terpolymer of tetrafluoroethylene, hexafluoropropylene, and vinylidene fluoride [49].

iv.   Effect of Oxygenates in Fuels: The main oxygenates of interest are ethanol, methanol, and methyl tertiary-butyl ether (MTBE), a commonly used octane booster replacing tetraethyl lead that has been prohibited from most fuels. Stemming from past oil shortages, there was interest in various parts of the world in alternative fuels, especially in supplementing gasoline with alcohols to conserve petroleum. Additionally, environmental benefits accrue from adding alcohols to gasoline. In the United States, the use of "gasohol," gasoline with 10% ethanol, became established. A stringent requirement for hose tube materials has also emerged in that they are expected to be capable of resisting "flex-fuel," that is, methanol blended with gasoline in any proportion. With NBR-based tube compounds, addition of 10%–20% ethanol or methanol to Fuel C was found to markedly increase swelling and permeation compared to Fuel C alone. Methanol had a larger effect than ethanol. Increasing the ACN content of the base NBR reduced the magnitude of the effect, as did blending with PVC [44,45]. Permeation of Fuel C/methanol (85:15) was also decreased by reinforcing NBR vulcanizates with platy fillers [45]. Vulcanizates based on HNBR/fluorinated thermoplastic alloy have been shown to provide improved resistance to flex-fuels containing methanol and, by extension, to those blended with ethanol and MTBE [48]. Compared to NBR vulcanizates, those based on FKM and epichlorohydrin homopolymer were found to have significantly lower volume swell and better retention of physical properties after exposure to methanol-blended fuels. Maximum volume swell for most elastomers occurred with mixtures up to about 25% methanol. Fuel blended with ethanol was shown to be slightly less severe on most of the vulcanizates than blends containing methanol [50,51]. For SAE 30R7 and 30R8 hoses (Table 13.5), Fuel C/ethanol

(90/10) increased permeation rates by 25% and 151%, respectively, compared to Fuel C alone when tested by the reservoir method. For the same hose constructions, Fuel C/methanol (85/15) increased permeation rates by 63% and 342%, respectively. For SAE 30R9 hose (Table 3.1) with FKM inner tube, permeation rates were significantly lower and were not influenced to the same extent by the composition of the blended fuels [54]. Methanol blended 25% and 80% by volume with Fuel C (representing low and high ends of flex-fuel equivalents) deteriorated the physical properties of a tube compound based on FKM with 66% fluorine content. However, FKM elastomers with 68% or greater fluorine content were considerably more resistant to these blends [53]. For veneer construction fuel line, it was found that vulcanizates based on FKM elastomers with 68% fluorine content resisted permeation of methanol-blended fuel significantly better than those based on 66% fluorine grades. FKM elastomers with 68% fluorine content were recommended by these authors as base for the veneer layer in contact with the fuel [54]. The permeation rate of Fuel C/methanol/ethanol (93/5/2) at 40°C through fluorothermoplastics THV, PVDF, and ETFE are claimed to be an order of magnitude lower than for this fuel through FKM or polyamide. A THV terpolymer was also claimed to show no significant changes in physical properties when exposed for 1000 h at 60°C to Fuel C/methanol 50/50 and other alcohol-blended fuels [51]. Oxygenates in diesel fuels are limited to fatty acid esters. In the United States, "biodiesel" contains esters mainly from soybean oil. European RME diesel contains esters from rapeseed oil. RME and biodiesel, as well as low sulfur and regular diesel have higher boiling point ranges than gasoline-based fuels, therefore, have lower levels of evaporative emissions and they are generally not as chemically aggressive as gasoline to elastomeric hoses.

4. Hose Cover Material Development

A fuel hose cover must be able to withstand long-term heat aging, typically at 125°C, as well as having a level of ozone and fuel resistance. Fuel permeation resistance is generally a requirement in order that the cover may act as backup if a small puncture were to occur in a thin tube layer. Other requirements for a cover material are oil and abrasion resistance, as well as good sealing force retention for coupling capability. A fire resistance requirement for the cover is sometimes specified for certain fuel hose constructions.

Polychloroprene compounds capable of resisting air aging at 100°C have been the cover material of choice for fuel hoses until the mid-1970s and still find use today in some nondemanding applications. However, the extraction by the fuel of an antiozonant (NBC) from a polychloroprene cover was shown to lead to premature cover cracking; the use of inherently ozone-resistant cover materials, for example, chlorosulfonated polyethylene (CSM), was then suggested [57]. The terpolymer of epichlorohydrin, ethylene oxide, and allyl-glycidyl ether (GECO) is used as base for fuel hose covers because of its high resistance to heat and to ozone, especially after fuel extraction [58]. A comparison of a series of fuel hose cover compounds concluded that those based on chlorinated polyethylene and CSM provided a good balance of performance and cost. The lower cost NBR/PVC, though fuel and oil resistant, is deficient in heat resistance. GECO had the best combination of fuel, oil, and heat resistance, but was the highest-priced elastomer in the series [57].

5. Hose Designs

The most basic fuel hose consists firstly of an extruded inner tube material that must as much as possible resist the fuel, its permeation, and the extraction of any component by the fuel. The hose is reinforced, depending on the application, by knit, braided, or spiraled yarn (most commonly rayon, polyamide, PET/polyester aramid). On its outside, the hose is covered by a heat- and ozone-resistant material with some other specific requirements. Some applications call for injection-molded–nonreinforced (all-gum) fuel hoses.

As discussed earlier, fluoroelastomers have excellent chemical and permeation resistance to a broad range of fuel types. They are, however, considerably more expensive than other fuel-resistant elastomers. This has resulted in the development of two-layer laminated tubes in which a thin inner layer (veneer), usually based on FKM, contacts the fuel. This inner layer is backed by a second one made from a lower cost fuel-resistant elastomer, for example, ECO [60], NBR, or CSM (Figure 13.3). The thicker backing layer also allows for strike-through (mechanical adhesion) of the reinforcing yarn. By choice of compound ingredients, chemical adhesion is achievable between an FKM veneer and an NBR-based backing layer [59].

In addition to FKM elastomers, thermoplastics are sometimes used as the veneer layer. Application of a thin veneer of Nylon 11 to the surface of an NBR compound was found to dramatically reduce permeability to ethanol and methanol blends with Fuel C [60]. Polyamide veneer layers provide a low-cost alternative to FKM, but can cause coupling retention and noise transmission problems.

An alternative construction also uses a laminated tube structure—an innermost elastomeric tube layer with a permeation-resistant "barrier" material between it and another rubber tie gum layer (usually of the same material as the inner tube)

FIGURE 13.3   Veneer hose design.

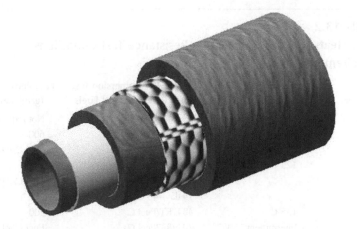

FIGURE 13.4 Barrier hose design.

(Figure 13.4). The tube and tie layers are typically based on NBR or ECO elastomers. Polyamide could be used as the permeation-resistant barrier material; however, some fluorothermoplastics are preferred because their methanol/Fuel C permeation rate was found to be only about 1:10 that for Nylon 12 and Nylon 12, 12 [61].

Of the fluorothermoplastics, THV terpolymers are the most flexible of the melt processable types, resistant to a range of fuels, bondable to FKM-, ECO-, and NBR-based materials and claimed to act as a very effective barrier against hydrocarbon emissions [49,62]. Reinforcing yarn plies, along with a heat- and ozone-resistant elastomeric cover, complete the constructions [63].

6. Fuel Hose Classification

SAE Standard J30 classifies a wide range of automotive fuel hoses. In Table 13.2, portions of the standard that pertain to heat, chemical, and fuel permeation resistance are reproduced. Moving down the table from standard hose design to sophisticated veneer and barrier constructions illustrates the evolution of these products in order to meet increasingly stringent demands.

C. HYDRAULIC HOSE

Hydraulic hose is designed to accommodate the fluid type, fluid-operating temperature, operating pressure, service temperature and environmental conditions. Hoses are used when pipes or tubes cannot be used, usually to provide flexibility for machine operation or maintenance such as with excavators. Acrylonitrile butadiene nitrile rubber (NBR), because of its oil resistance is used for the tube compound, and the tube is reinforced with one to six layers of braided or spiraled brass coated and intermediate rubber layers. The exterior is designed principally for abrasion resistance and can be compounded with polychloroprene, chlorinated polyethylene, nitrile, or rubberized fabric. The bend radius of hydraulic hose is designed for the equipment, to minimize hose failures that can result in equipment downtime. Hydraulic hoses have steel fittings or couplings mounted at the ends. The weakest

TABLE 13.2

Hose Heat, Fuel and Permeation Resistance Test Conditions/ Requirements

SAE Spec	Upper Service Temperature	ASTM D471 Immersion Test ASTM Reference Test Fuels	Fuel Permeation (g/m²/day[b])
30R2 and 30R3	100°C	48 h/RT/Fuel B	Not required
30R6	100°C	48 h/RT/Fuel C	600
		70 h/RT/Fuel G	Fuel C/RT[a]
30R7	125°C	48 h/RT/Fuel C	550
		70 h/RT/Fuel G	Fuel C/RT[a]
		14 days/40°C/Sour Gas#1	
30R8	135°C	48 h/RT/Fuel C	200
	Intermittent 150°C	70 h/RT/Fuel G	Fuel C/RT[a]
30R9	135°C	48 h/RT/Fuel C	15
	Intermittent 150°C	70 h/RT/Fuel G	Fuel C/RT[a]
		14 days/40°C/Sour Gas#1	
30R11	100°C T1	48 h /RT/Fuel C	100,50,25
	125°C T2	1000 h/40°C/Fuel I	Fuel I/40°C[b]
30R12	100°C T1	48 h/RT/Fuel C	100,50,25
	125°C T2	168 h/RT/Fuels I and K	Fuel I/40°C[b]
	135°C T3	168 h/RT/Sour Gas#2 and 3	
	150°C T4		

Source: SAE Standard J30, October 2001.

Notes: Fuel B, toluene/iso octane (30/70); Fuel G, Fuel D/ethanol (85/15); Sour Gas#1, Fuel B/ TBHP; Fuel C, toluene/iso octane (50/50); Fuel I, Fuel C/methanol (85/15); Sour Gas#2: Fuel I/TBHP/copper ion; Fuel D, toluene/iso octane (40/60); Fuel K, Fuel C/methanol (15/85); Sour Gas#3, Fuel K/TBHP/copper ion.

[a] Reservoir method.

[b] SAE J1737.

part of the high-pressure hose is the connection of the hose to the fitting due to the crimping operation. Couplings are typically installed by distributors and not by the hose manufacturer, because the hoses are cut to the required length to suit the specific equipment on which it will be installed. Hose service lives are in the order of 5–7 years. One of the governing standards to which hydraulic hose is described and specified is the SAE specification SAE J517. The hoses that the specification describes are summarized in Table 13.3.

Steel wire–reinforced hydraulic hoses are required when there are high operating pressures. Steel wire reinforced hydraulic hoses can be built by either braiding or spiraling brass–coated steel cord over a tube extruded on a mandrel. Braiding utilizes equipment that can continuously lay sets of wire over one another to form long lengths of reinforced hose. Due to heavier gauges, spiraling steel wire is more difficult than braiding and is only used to reinforce the highest-pressure

TABLE 13.3
Hydraulic Hose Constructions

SAE Type	Tube	Reinforcement	Cover	Working Pressure	Temperature Range
100R1AT	Oil Resistant Synthetic Rubber	1 Braid Steel Wire	Oil & Weather Resistant Rubber	Medium	-40 to +100C
100R2AT	Oil Resistant Synthetic Rubber	2 Braids Steel Wire	Oil & Weather Resistant Rubber	High	-40 to +100C
100R3	Oil Resistant Synthetic Rubber	2 Braids Fiber	Oil & Weather Resistant Rubber	Low	-40 to +100C
100R4	Oil Resistant Synthetic Rubber	Textile Plies/Braids; Spiral Body Wire	Oil & Weather Resistant Rubber		-40 to +100C
100R5	Oil Resistant Synthetic Rubber	1 Wire Braid	Textile Braid		-40 to +100C
100R6	Oil Resistant Synthetic Rubber	Fiber Layers	Oil & Weather Resistant Rubber	Low	-40 to +100C
100R7	Fluid Resistant Thermoplastic	Synthetic Fiber	Fluid & Weather Resistant Thermoplastic	Medium	-40 to +93C
100R8	Fluid Resistant Thermoplastic	Synthetic Fiber	Fluid & Weather Resistant Thermoplastic	High	-40 to +93C
100R12	Oil Resistant Synthetic Rubber	4 Spirals Steel Wire	Oil & Weather Resistant Rubber	Very High	-40 to +121C
100R13	Oil Resistant Synthetic Rubber	Multiple Spirals Heavy Steel Wire	Oil & Weather Resistant Rubber	Very High	-40 to +121C
100R14	PTFE	1 Braid 303XX Stainless Steel Wire	None		-54 to +204C
100R15	Oil Resistant Synthetic Rubber	Multiple Spirals Heavy Steel Wire	Oil & Weather Resistant Rubber	Very High	-40 to +121C
100R16	Oil Resistant Synthetic Rubber	1 or 2 Braids Compact Steel Wire	Oil & Weather Resistant Rubber	High	-40 to +100C
100R17	Oil Resistant Synthetic Rubber	1 or 2 Braids Compact Steel Wire	Oil & Weather Resistant Rubber	High	-40 to +100C
100R18	Fluid Resistant Thermoplastic	Synthetic Fiber	Fluid & Weather Resistant Thermoplastic	High	-40 to +93C
100R19	Oil Resistant Synthetic Rubber	1 or 2 Braids Compact Steel Wire	Oil & Weather Resistant Rubber	High	-40 to +100C

FIGURE 13.5   Hydraulic hoses–braided reinforcement.

FIGURE 13.6   Hydraulic hoses–spiral reinforcement.

hydraulic hoses, that is, those having burst pressures greater than 6000 psi. Steel wire spirals are also more rigid than braided plies. The wire is brass-coated to promote for adhesion to rubber. The adhesion mechanism via formation of copper sulfide dendrites is similar to that found in steel wire–reinforced tire belts and plies. Generic hose constructions are shown in Figures 13.5 and 13.6.

III.   V-BELTS

From prehistoric times, it was known that mechanical power could be transmitted by the friction between some type of "belt" and the "pulley" in which it was travelling. First, this took the form of an open-ended strap wrapped around a pole which it rotated in alternating directions, for example, bow drill. The ends of the strap were then joined to form an endless loop able to transmit rotary motion between two shafts. Flat pulley drives using spliced leather flat belts evolved in early-nineteenth century England with the Industrial Revolution in which hand tools were replaced by power-driven machines concentrated in factories. There were problems maintaining belt tensions, keeping the belts on badly aligned drives, and the space requirements for the drives were excessive. Later on, some drives used multiple spliced rope drives wedged in deep grooves rather than on flat pulleys. The next evolution, in the 1890s, was the plying up and cutting of leather and textiles into "V"-shaped belts to run in similarly shaped grooves.

From its beginning, the automobile industry, paralleling the industrial situation, used leather flat belts on two-pulley fan drives to cool their engines. By 1916, engines of over 100 hp were in use, requiring larger cooling fans. With the addition of electrical accessories, a generator had to be added to the fan belt.

Shortly afterward, leather flat belts were replaced by fully molded endless rubber V-belts with cotton cord fabric for strength and a cover of woven cotton fabric rubberized (with natural rubber–based materials) to increase the coefficient of friction for improved power transmission. The "V"-shaped belt cross-section was a technological breakthrough of its day. A V-shaped pulley groove with a belt "wedged" into it produces more belt/pulley friction than a flat belt at the same tension—the pressure at the pulley wall is thereby magnified.

Today, V-belts are used in a wide variety of automotive, industrial, agricultural, and domestic applications where power is transmitted from a driving pulley connected to the power source to one or more driven parts of the engine/equipment.

A. V-Belt Types

Belts may be classified into two types, "synchronous" for drives that require synchronization or timing (Section IV) and "non synchronous" where synchronization is not required. Examples of the latter are the following:

- Fabric-wrapped V-belts, used mainly on industrial drives (Figure 13.7).
- Raw edge V-belts, in which the wrapping fabric is removed (Figure 13.8). The inner surface or base of the V-belt is sometimes notched to increase its flexibility around small pulleys and increase airflow for cooler running. They are used on automotive drives, though are being replaced by V-ribbed belts in many instances. However, as they are more flexible than wrapped belts, raw edge belts are being increasingly used in many industrial applications.
- "Variable speed" belts, used on continuously variable transmissions (CVTs) in which precise continuous control of pulley speed ratios is required. They are much wider than ordinary V-belts. Notches are either molded or cut into

FIGURE 13.7 Fabric wrapped V-belt.

FIGURE 13.8   Raw edge V-belt.

FIGURE 13.9   Variable speed belt.

the base of the belt after vulcanization. Used on agricultural, industrial, and automotive drives (most notably for snowmobiles) where high-impact loads must be withstood and high loads must be transmitted (Figure 13.9).

- Banded V-belts, in which belts are held together side by side in a single unit by being vulcanized to a tie band (of rubberized fabric) laid across the top. Instead of a series of individual V-belts on a drive, a banded belt may be used to distribute power more evenly, reduce vibration, and prevent belt turnover. Used on agricultural, textile, and heavy industrial equipment (Figure 13.10).
- Double-V (hexagonal) back-to-back V-belts with a central cord line (Figure 13.11). They can transfer power from either side in drives where the belt passes in a zigzag pattern around a number of pulleys. Used in agricultural and some industrial applications.

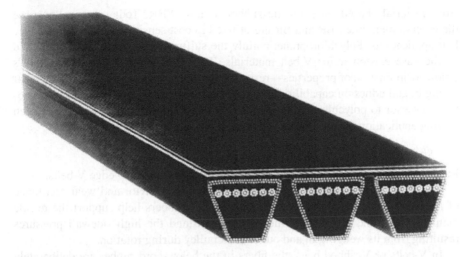

FIGURE 13.10 Banded V-belts.

FIGURE 13.11 Hexagonal belt.

1. Materials

V-belts are made up of several components (Figures 13.7 and 13.8). They are as follows:

- A rubber-based material, usually containing short fibers.
- Load-carrying textile tensile member. These are grouped in a plane close to the top of the belt section. The cross-sectional view of the tensile members is known as the "cord line."
- An adhesion rubber encapsulating the tensile member to increase adhesion to other belt components.
- Conventional fabric-covered belts are surrounded by a rubberized woven cotton/synthetic envelope, one or more plies, for protection against wear and environmental hazards and control of coefficient of friction.
- Raw edge belts do not have an envelope but include a rubberized cotton/synthetic backing fabric.

The various belt components are discussed in more detail in the succeeding text.

a. Base Elastomer

Natural rubber was the earliest elastomer of choice, but due to shortages of this commodity during World War II, V-belts from the mid-1940s on were being produced from materials based on synthetic rubber, mainly SBR. Today, general-purpose diene elastomers like SBR and BR are utilized in cost-sensitive, lower performance belt applications. Polychloroprene, mainly the sulfur-modified grades, is also used as the base elastomer for V-belt materials. This is because of polychloroprene's unique combination of properties—resistance to flex fatigue, wear, and oil, high tear strength, and adhesion capability to other belt components. In recent years, elastomers superior to polychloroprene in heat resistance and other properties have been finding application in some high-performance V-ribbed and timing belt areas.

b. Short Fiber Reinforcement

Rubbers reinforced with fiber are used in both wrapped and raw edge V-belts.

The fibers in belt sidewalls modify the frictional behavior and wear resistance of raw edge belts. More importantly, fiber-loaded rubbers help support the tensile member cords and also enable the belt to withstand the high sidewall pressures resulting from its wedging in and out of each pulley during rotation.

In V-belts or V-ribbed belts, the fibers in the below-cord rubber are deliberately oriented so that they lie perpendicular to the cords. This allows development of a high low-strain modulus in the belt transverse direction, which will ensure that during belt rotation, the cords remain in a horizontal plane with the load evenly distributed over all of them. This is critical if a good belt fatigue life is to be obtained. The high modulus in the transverse direction enables the belt to withstand the compressive sidewall forces. Along with this, a low modulus is developed perpendicular to the fiber direction, which enables the belt to be very flexible in its longitudinal direction. Fibers, therefore, provide anisotropic reinforcement of the below-cord rubber material.

In a calendered sheet of short fiber-reinforced rubber, the fibers will be oriented in the direction of rubber flow (referred to as the "machine direction"). To achieve the fiber orientation required in the belt, the fibers must be perpendicular to the cording direction on the build machine. Therefore, the calendered sheet is cut at a right angle, turned 90° and joined before it is supplied to the build operation.

Commonly used fibers are cotton, polyester (PET), unregenerated cellulose, and aramids. Cut fiber lengths are typically 1–10 mm; the fiber content of some belt rubber materials may vary up to about 20% by weight. Aspect ratio (length/diameter ratio) is an important property in determining the reinforcing capability of a particular cut fiber [64].

To be effective in a belt, the fibers need to be very evenly dispersed in the rubber matrix since fiber clumps act as stress raisers resulting in premature belt failure. The mixing procedure is critical. In addition, a high level of adhesion must be obtained between fiber and elastomer. The adhesion may be purely mechanical, for untreated fiber, or chemical in nature, where the fiber carries an adhesive treatment.

Fibrillated para-aramid fiber ("pulp"), produced in an elastomer masterbatch form to facilitate its dispersion in rubber, has been shown in a lab study to produce higher low-strain modulus, higher anisotropy, and better dynamic properties than reinforcing with other short fibers [65]. Dipped resorcinol–formaldehyde latex (RFL)–dipped chopped aramid fibers and a different pulp masterbatch have been claimed, based on lab data, to be effective anisotropic reinforcements for belts [66]. The pulp often finds application in belts requiring increased lateral stiffness for high-load carrying capability.

The design considerations associated with the use in belt applications of para-aramid pulp, as well as chopped fiber with and without adhesive treatment, have been reviewed [67].

c. Tensile Member/Cord

During the 1940s, rayon, with its higher modulus and lower stretch, replaced cotton as the tensile cord of choice for V-belts. The introduction of PET cord in the 1950s offered further improvement in reducing belt stretch. Today, PET, offering the best price/strength ratio of all reinforcing materials, has become the dominant material for tension carrying cords used in V-belts [68]. The constructions and processing of PET cords for V-belts have been reviewed in detail [69,70].

For high-performance V-belts and V-ribbed belts, the following properties are required of the PET cord:

- Resistance to fatigue
- Excellent dimensional stability
- Adhesion to the rubber components
- High modulus
- Minimized heat shrinkage
- Sufficient tack (adherence to other components) for the build process
- Unaffected by moisture.

In the 1980s, a special high modulus low shrinkage (HMLS) PET filament yarn was developed for the V-belt market. Belt cords made from this yarn are claimed to have high dimensional stability, low elongation, low creep, and high dynamic integrity [71].

Yarns are twisted and then plied into cords of the required linear density suitable for the particular belt application. The twist factor influences cord properties, for example, higher number of turns per unit length giving lower modulus but better resistance to bending fatigue.

For polyester PET and other cords, adhesion to the elastomer matrix is obtained by dipping the cord in a resorcinol-formaldehyde latex RFL dip with vinylpyridine or an elastomer latex. The RF component promotes adhesion to the fiber, while the dried latex rubber (L) co-vulcanizes with the rubber matrix. The resulting modulus, creep, and heat shrinkage force are all dependent on the "tension" applied to the cord during the heat treatment stage following the dip, the cure "time," and "temperature" ("3T") used in the treating unit.

For raw edge belts, it is additionally required that the cords be stiff to prevent the filaments from fraying during the belt cutting operation. For PET cord, this is achieved by including a predip based on an isocyanate in solvent (usually toluene), which must penetrate into the center of the cord. As well as reacting with the RFL applied at the next stage, the isocyanate bonds the filaments together, cross-linking to form a stiff network.

A coating of rubber cement is often applied over the RFL layer to improve cord tack for the build operation and to prevent dip deterioration on storage.

The production of stiff PET cord for raw edge belts can cause converters to violate the Clean Air Act guidelines if they do not have solvent recovery or incineration systems. In a recent development, it was found that surrounding the PET filament core with a sheath of polybutylene terephthalate (PBT) produces a cord that will self-stiffen. The PBT sheath melts in the treatment process, flows into the spaces between filaments, and bonds them together. When the cord cools below the PBT melting point, it stiffens. The isocyanate treatment step, with its solvent emission, is thus removed from the cord treating process [72].

d. Aramid

Because of their very high modulus, low growth, and thermal stability, *para*-aramid cords are used in some heavy duty belts, for example, for agricultural machinery, which must withstand high shock loads, and in variable speed belts that transmit high loads.

To obtain adequate adhesion between *para*-aramid and most rubbers, treatment with RFL alone is not sufficient. A two-part dip system must be used in which an isocyanate- or epoxy-type predip is added ahead of an RFL dip. Isocyanates in toluene will penetrate the cord, improving cuttability and fray resistance but decreasing tensile strength; aqueous solutions of epoxies have poor penetration but generally have less effect on tensile strength. The epoxy pre dip may be applied to the aramid during the production of the yarn, which is then twisted into the cord and RFL treated, or aramid cord may be dipped first and then have the RFL applied [73].

e. Adhesion Rubber

The level of adhesion between cord and rubber has an important influence on belt life. The adhesion rubber material must be fairly rigid, with high tear resistance, capable of encapsulating and bonding to the cord and adjacent belt components during vulcanization.

f. Fabrics

The constructions and processing of fabric components of V-belts have been reviewed in detail [69,70].

2. V-Belt Jacket

When applied to the surface of a fabric, rubber improves its friction and wear characteristics in contact with a pulley. As well as protecting the internal components of the belt, the rubberized fabric envelope provides the V-belt with an abrasion-resistant surface with the required friction characteristics (Figure 13.7). The base fabrics are usually cotton or cotton-rich blends with synthetics, plain weave in square construction. Generally, only one layer of fabric jackets a belt, but in some applications two or more layers may be required. The fabric is spread or calendered with, generally, a polychloroprene-based material having the required wear, friction, adhesion, tack, and antistatic characteristics. If the yarns of the treated fabric were laid parallel and perpendicular to the length of the belt, flexing of the belt over the pulley would be difficult. To counteract this, the treated jacket fabric is applied to the belt, not straight, but in the bias direction. This is achieved by cutting the fabric (in a 45° direction) into rhombus-shaped pieces and then reassembling them by joining the selvedges with overlap joints. Alternatively, "wide angle" bias fabrics (warp/weft intersect at about 120° instead of 90°), made by special production methods [74], are also used for V-belt jackets.

After being slit to the required width, the jacket fabric is wrapped evenly and tightly around the circumference of the unvulcanized "skived" belt (V-shaped cross-section) in the "flipping" operation. The belt is then vulcanized in a ring shaped or linear mold.

3. Other V-Belt Fabric Components

Instead of using fiber-reinforced rubber, or in addition to it, some raw edge V-belts may include layers of nylon or PET cord fabric for increased transverse rigidity to support the cord line. Rubberized cotton fabrics, or cotton synthetic blends may be used as the top and bottom fabrics. Wide angle bias fabrics, as described earlier, may be used for belt components when high bending pliability is required.

B. V-RIBBED BELTS

V-ribbed belts, which are becoming increasingly important, have been the subject of intense research activity in recent years.

Basically a flat belt, V-ribbed belts have a series of longitudinal ribs, of full or truncated V-profile, on the driving face that mate with corresponding grooves in the pulley rim and also have the capability to drive off the flat backing face. They offer the power transmission capability of a V-belt together with the flexibility of a flat belt.

Though first patented in 1952 and used initially in some industrial applications, it was the use of a single V-ribbed belt to drive the front-end accessories in the Ford Mustang V-8 engine of the late 1970s that established these "serpentine" belt drives in the automotive industry. V-ribbed belts, which are relatively thin, perform better than conventional V-belts on drives with very small pulleys (e.g., on alternators), high speeds and high pulley speed ratios. Modern automotive drives may use one V-ribbed belt where three or more conventional V-belts would have been needed in the past, allowing size and weight reductions in the engine compartment. This is why virtually all automobiles produced today in North America and Japan use V-ribbed belts on their multi-pulley accessory drives [75].

V-ribbed belts are also used extensively in the industrial sector in applications as diverse as blowers, centrifugal pumps, compressors, fans, generators, machine tools, washing machines, clothes dryers, and fitness equipment, utilizing their smooth running capability and the other advantages outlined earlier (Figure 13.12).

C. MATERIALS

V-ribbed belts have the following components:

- A tensile member, most commonly PET cord. Fiberglass and aramid cords may be used in certain specialized applications. Low-modulus/high-twist polyamide cords, which impart "elastic" character to belts, are used in some

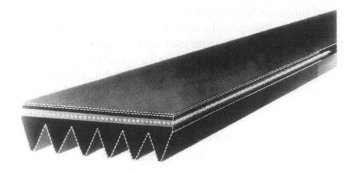

FIGURE 13.12   V-ribbed belt.

washing machine and dryer applications to eliminate the need for tension-
ing mechanisms.
- A rubber layer that encapsulates and adheres to the tensile member.
- Rib material, usually rubber reinforced by short fibers oriented in the belt
  transverse direction. In the manufacturing process, to obtain the orientation required in the ribs, the fibers must be perpendicular to the cording
  direction on the build machine. As for V-belts, this involves cutting the
  calendered sheet at a right angle, turning it 90° and joining. Either a grinding process after cure or mold curing is used to produce the rib profiles.
  In certain constructions, the ribs are covered with a flock or a facing fabric.
- A backing material, either rubber or rubberized fabric.

Polychloroprene, particularly the sulfur-modified grades, was the original elastomer of choice for V-ribbed belt materials. Today, polychloroprene-based materials are still commonly found in V-ribbed belts for industrial applications, since the
elastomer offers a unique blend of properties as discussed earlier. However, in the
automotive sector, the drive for improved fuel efficiency has led to hotter operating
engines and more aerodynamic vehicle styling that in turn has restricted the size of
the engine compartment and decreased air circulation. Engine compartment temperatures have increased, now often averaging 125°C–135°C with peaks up to 150°C.
At the same time, vehicle manufacturers' warranty periods are being extended on all
components, including the belts. This has meant that polychloroprene-based materials, having only moderate heat resistance, are no longer suitable for much of today's
automotive V-ribbed belt service. Improvements to the heat resistance and other performance aspects of polychloroprene-based materials have been claimed by the use
of elastomer modification, elastomer blending, and compound modification [76–78].
However, despite this effort, non-polychloroprene V-ribbed belts have become firmly
established in the automotive industry in the recent times.

Ethylene–alpha-olefin elastomers, with particular emphasis on EPDM, are increasingly being used as the base for automotive V-ribbed belt materials. Since this type of
elastomer is more cost effective and offers broader operating temperature ranges than
polychloroprene or its other alternatives, EPDM-based V-ribbed belts have been the
subject of intense research and development activity in the industry.

Products based on both sulfur-cured and peroxide-cured EPDM belt materials have been developed and are being produced.

Even though the saturated backbone of EPDM provides good heat resistance, it is claimed that certain antioxidant packages will further enhance it. For example, the combination alpha-MSDPA with ZMTI is suggested for peroxide-cured EPDM belt compounds [42].

Peroxide-coagent cure systems for EPDM elastomers, while providing improved heat aging over sulfur systems, can also be formulated to have the required balance of properties for automotive V-ribbed belts. Dynamic properties and fatigue resistance of peroxide-cured EPDM compounds were found to be significantly enhanced by using ZDA or ZDMA. On peroxide curing, the ZDA becomes incorporated into the cross-link network. When subjected to stress, bonds to the zinc in the network will undergo cleavage and will reform. In another benefit to belt applications, ZDA addition is claimed to enhance rubber/textile adhesion for peroxide-cured EPDM [19].

The aforementioned benefits of using ZDMA as coagent, including improved pilling resistance with proper elastomer selection, are claimed to have been realized for peroxide-cured EPDM-based V-ribbed belts [79].

CSM was once proposed as a base elastomer for automotive V-ribbed belts as its saturated backbone would offer improved heat and ozone resistance over polychloroprene. However, CSM-based materials tend to have high hysteresis and limited low-temperature capabilities. To address these concerns, pendant alkyl groups were later introduced onto the backbone, reducing chlorine content and crystallinity. The result was a family of alkylated chlorosulfonated polyethylene (ACSM) elastomers covering a range of chlorine content and distribution, alkyl type and amount, as well as molecular weight and MWD. ACSM elastomers retained the chlorosulfonyl group of CSM, so they could be cross-linked by sulfur cure systems, while peroxide curing would be possible at the hydrocarbon backbone.

ACSM elastomers are claimed to be suitable for applications requiring dynamic flex and an operating temperature range of −40°C to 135°C with peaks up to 150°C. These properties, along with their claimed high abrasion resistance, made certain ACSM elastomers candidates for V-ribbed belts [80].

D. BELT PERFORMANCE CRITERIA

The effective life of a V-ribbed belt may be defined as the amount of time for which it will transmit the required torque to a required efficiency [81]. This is not necessarily the same as the time for which the belt is capable of operating.

In addition to vehicle testing, a number of belt testing methods have been devised by the automotive industry and belt manufacturers to assess how V-ribbed belts will perform in service. On a series of lab bench test configurations, belts are exposed to high, low, and cycling temperatures, fluid contamination, multi-pulley flexing, abrasive wear, misalignment, and pulley slip conditions. The details of the tests and belt acceptability vary slightly from customer to customer; however, a standard set of tests and performance criteria has been agreed upon for V-ribbed belts [82].

In general, a belt is considered inadequate for service if any of the following situations occur during the lab tests:

1. Cord fracture.
2. Fraying or cord separation.
3. Delamination of layers.
4. Loss of rib ("chunking").
5. Cracking across belt cross section.
6. Excessive rib cracking.
7. Excessive belt slip (in dry or wet conditions).
8. Excessive weight loss due to abrasion.
9. Pilling in excess of a permissible level (visual rating scale). Under certain conditions, tacky wear debris (or "pills") can accumulate along belt ribs, be retained on the belt, and affect its performance on the drive.
10. Unacceptable belt noise levels during operation. In recent years, vehicle manufacturers have put much effort into reducing engine noise. In doing so, they have exerted pressure on the belt suppliers to eliminate noise produced by the belt. In response to this, one notable development has been to change the design of the V-ribbed belt to cogged instead of continuous ribs. Parallel, equally spaced grooves are molded across the longitudinal ribs at an angle other than perpendicular, so that transverse grooves in adjacent ribs are offset from each other (Figure 13.13). This design feature has been found to significantly reduce belt-operating noise [83].

Studies on the performance and failure mechanisms suggest that an improved V-ribbed belt should have low bending stiffness, high coefficient of friction belt/pulley, high transverse stiffness (transverse oriented fibers), and a wear-resistant rib material with good fatigue resistance, especially at the rib top [81].

FIGURE 13.13   Gatorback® V-ribbed belt.

IV. SYNCHRONOUS (TIMING) BELTS

Synchronous belts, the most highly engineered member of the belt family, are used on drives (both automotive and non automotive) that require synchronization or timing. This type of belt differs from the others discussed earlier in having a toothed inner surface, the teeth meshing with corresponding axial pulley grooves to give non slip engagement, resulting in a fixed rpm ratio between driver and driven pulleys (Figure 13.14).

The first timing belt patent, issued in 1946, was for the use of a belt to synchronize movement in a sewing machine. Further developments have occurred in other industrial applications. However, in 1965, Pontiac introduced its new six-cylinder overhead cam engine with a synchronous belt drive. This marked the beginning of what is today the major application for synchronous belts, namely to drive the valve camshaft from the crankshaft on overhead camshaft (OHC) engines.

Starting from the first belt-driven overhead cams in the early 1960s, synchronous belts have grown in importance with the increasing use of small-sized four-cylinder OHC engines in passenger cars, particularly in Europe and Japan. However, larger displacement engines (above 2 L) generally tend to use timing chains. In diesel engines, synchronous belts are also used to drive diesel injection pumps and accessories. Since belt failure on an "interfering" engine design can result in serious damage to the cylinder head, automobile manufacturers' goal is a maintenance-free belt to last the entire life of the engine.

In the non automotive area, synchronous belts are used instead of chains in many applications such as compressors, machine tools, business machines, and fitness equipment, as well as in textile, wood, and paper processing machinery.

An increasingly important application for synchronous belts is on positioning devices, for example, in manufacturing automation equipment, where the belt's primary function is not power transmission but rather to transmit complex motions and varying loads with high accuracy. Motion control performance is claimed to be predictable from belt stiffness and damping characteristics [84].

Modern drive systems based on belts are generally quieter, more efficient, lighter weight, and lubrication free compared to those based on the alternative of metal chains or gears. The original belts had trapezoidal teeth, but closely spaced curvilinear tooth profiles with greater load carrying capability were introduced into the automotive market in the 1980s.

A more recent development has centered on obtaining a significant reduction in belt tooth engagement noise and a major improvement in power capacity through the use of a helical offset tooth pattern, like a "herring bone" gear (Figures 13.15), meshing with a corresponding pulley [85].

FIGURE 13.14 Conventional synchronous belt.

FIGURE 13.15   Helical offset tooth synchronous.

A.   MATERIALS

Synchronous belts have the following components:

- Tensile member.
- High tenacity facing fabric, usually of woven polyamide, adhered to the surface of the teeth to provide wear resistance and reinforce the teeth. The fibers in one direction are crimped giving the fabric a one-way stretch. This enables the fabric to be pushed by the flow of the compound into the mold cavities to form the tooth surface.
- Although some high performance belts are composed of polyurethane, this section will focus on the rubber material forming the belt teeth and backing; the requirements for this material are discussed in the succeeding text.

Originally, automotive synchronous belts were made from polychloroprene-based materials, taking advantage of the flex fatigue and tear resistance of that elastomer as well as its adhesion capability.

In the 1980s and 1990s, changes in vehicle engines and design resulted in significantly increased average engine compartment temperatures (from 80°C–90°C up to 125°C–135°C with peaks to 150°C). At the same time, belt warranty periods were being extended. As a result, polychloroprene, with heat resistance only up to about 100°C, became inadequate as the base for many automotive original equipment synchronous belts.

Today, polychloroprene-based belts only account for a minor portion of the automotive market, mostly for the aftermarket. Polychloroprene-based synchronous belts do, however, continue to be used in a number of non automotive industrial applications in which a high level of heat resistance is not a major requirement.

Following their introduction in Japan in the mid-1980s, and then by BMW in Europe (1987), synchronous belts based on hydrogenated nitrile butadiene HNBR elastomers have, despite their significantly higher cost, largely replaced polychloroprene-based belts in the automotive original equipment sector.

HNBR elastomers are made by catalytic hydrogen of the carbon–carbon double bonds in an NBR-based elastomer to form more thermally and oxidatively stable single bonds.

Partially hydrogenated HNBR elastomers (4%–8% double bonds remaining unconverted), which are curable using sulfur-based systems, are the base for many automotive synchronous belt tooth compounds.

With residual double bond content approaching zero, fully hydrogenated grades, which must be vulcanized with peroxides, are the base for some high-performance belts as discussed in the succeeding text.

In a study on thermo-oxidative aging, Arrhenius plots of time/temperature to "failure" show that sulfur-vulcanized HNBR elastomers achieve a service life that corresponds to long-term exposure of about 1000 h at 130°C. In the same study, the "1000 h" temperature for peroxide-vulcanized HNBR was reported as 150°C. On the other hand, that for polychloroprene was only about 110°C [86].

Compared to polychloroprene, HNBR-based vulcanizates are claimed to have improved heat resistance, as discussed, as well as improved ozone resistance and low-temperature performance, better resistance to flex crack initiation after aging, and better retention of properties over a wide temperature range [56].

1. Oxidative Aging

Under thermo-oxidative conditions, HNBR-based materials undergo cross-linking reactions leading to embrittlement and failure.

For peroxide-cured HNBR materials, it is generally recommended to include the synergistic antioxidant combination of alpha-MSDPA with ZMTI [42].

A proprietary heat stabilizer technology for HNBR, effective with both peroxide and sulfur donor cures, has been claimed to significantly delay the onset of the aging mechanism more so that it can be obtained through adding conventional antidegradants. By use of this technology, it is claimed that the effective service temperature range for HNBR materials can be extended by about 10°C. Alternatively, predicted service life at the current service temperature can be expected to increase by a factor of 2 [87].

2. Resin Reinforcement

In response to the demand for even further improvements in HNBR synchronous belts, like increased durability, narrower belt width, and increased heat resistance, a new generation of enhanced performance belt materials was developed. These new HNBR-based materials take advantage of the improved heat resistance of the fully hydrogenated grades, in combination with boosting reinforcement (load carrying capability), yet maintain other characteristics essential to the performance of the belt.

The preferred means of increasing HNBR elastomer reinforcement is by the use of resin coagents, for example ZDA or ZDMA, selected because of their proven reinforcing capability in other elastomers.

One approach to obtaining this additional reinforcement has been through the addition of preformed acrylate salts of zinc, particularly ZDA, to peroxide-cured HNBR compounds. Compared to a sulfur-cured test compound, the HNBR with peroxide/ZDA had significantly higher modulus values and heat resistance, while other properties were comparable [88].

The alternative approach to obtaining enhanced reinforcement and heat resistance is through HNBR prepared by the in situ reaction of methacrylic acid with zinc oxide to give HNBR/ZDMA composites [89].

When an HNBR/ZDMA composite is cured with peroxide, a continuous interpenetrating network of poly-ZDMA forms to reinforce the elastomer. Compared to HNBR with preformed ZDA, a compound based on an HNBR/ZDMA composite is claimed to have better heat resistance, more stable dynamic modulus over a given temperature range, lower dynamic heat build-up, and better abrasion resistance. These materials also have the capability for very high tensile strength [90].

B. ALKYLATED CHLOROSULFONATED POLYETHYLENE (ACSM)

In the 1980s, because of their heat resistance advantage over polychloroprene and their cost effectiveness compared to HNBR, CSM elastomers were evaluated in automotive synchronous belts. This direction was not widely adopted because of the hysteresis and low-temperature limitations of CSM.

The ACSM family of elastomers, developed to address the shortcomings of CSM, was proposed as a more cost-effective candidate than HNBR to replace polychloroprene in synchronous belts.

ACSM has a completely saturated backbone that gives the required heat and ozone resistance. Control of the elastomer's chlorine content, the amount and type of pendant alkyl groups, and the molecular architecture is claimed to result in the necessary mechanical and dynamic performance for high-performance synchronous belts [91].

On a power absorption test at 149°C, the mean belt lives of ACSM- and HNBR-based synchronous belts have been reported as 401 and 370 h, respectively. On the same test, polychloroprene belt life, extrapolated from data at lower temperatures, was 44 h. On a stop/start cold flex test at −30°C, ACSM-based belts were claimed to have significantly better belt life than those based on polychloroprene or HNBR, even after the belts were aged for 48 h at 150°C [80].

In a study simulating belt service conditions, a comparison was made of synchronous belt model compounds based on polychloroprene, CSM, ACSM, and HNBR (sulfur cured). Based on data from a lab dynamic test, it was reported that the ACSM compound generated less internal heat than the other materials (except polychloroprene) and had least change in loss modulus due to aging. These data claimed to indicate that an ACSM belt would have longer belt life at high temperatures [92].

NBC (typically at 1 phr) is a recommended heat stabilizer for sulfur-cured ACSM. For maximized heat resistance, it is recommended that 1 phr metaphenylene bismaleimide (MBM) should replace sulfur and NBC level be increased to 3 phr to give heat aging performance claimed to be similar to that of a peroxide-cured vulcanizate [78].

C. SHORT FIBER/PULP TOOTH REINFORCEMENT

Aramid fibers or pulp are sometimes present in the tooth rubber where they are oriented in the belt longitudinal direction. The fibers increase modulus at the base of the tooth, thereby increasing its resistance to being sheared off during service, which is the most common failure mode of synchronous belts. In addition, when present at the tooth surface, short fibers will improve wear resistance and reduce belt noise [67].

D. TENSILE MEMBER

The cord used in synchronous belts must exhibit virtually no growth over the belt's operating load range in order to maintain its pitch (distance between adjacent teeth along the pitch line or center of the cord) and mesh properly with the pulley grooves.

Cords are applied in alternating "S" and "Z" configurations to prevent the belts tracking on pulleys. When building a sleeve of belts, cording tension is controlled within tight limits as it determines belt pitch.

E. GLASS

Fiberglass cord, with its high tensile strength, low extensibility, low hysteresis, and high dimensional stability, is the predominant choice for synchronous belt tensile member. Fiberglass cord is made up of fine glass filaments coated in rubber latex, twisted and then plied together to form a cord. The rubber latex coating on the filaments is cured by an RFL system to give adhesion between the cord and belt. Glass cord is degraded by water, so adhesion systems are designed where possible to protect the cord from moisture. It has been reported that "high-strength" glass types are increasingly being used for automotive synchronous belts [93]. Two advantages are reported:

1. For the same tensile properties, cord diameter can be reduced with high-strength glass, giving less hysteresis/lowered heat build up in the belt.
2. For the same cord diameter, high-strength glass increases the load-bearing capability of the belt.

F. ARAMID

p-Aramid cords are used in some industrial applications with high-load rating and requiring resistance to shock loads.

G. BELT PERFORMANCE CRITERIA

For automotive synchronous belts, the minimum acceptable belt performance on vehicle testing and lab bench dynamic tests is generally required to obtain customer approval. Likewise, the belt and its materials must meet a set of minimum acceptable physical property requirements (e.g., belt tensile strength, tooth hardness, adhesion to cord and fabric, resistance to high and low temperatures, etc.). Although test procedures can be specific to certain vehicle manufacturers, global industry standard methods are also used [94,95]. When operating, the teeth of the synchronous belt engage with the pulley grooves, the torque being transmitted through friction and loading of the tooth. On automotive drives, if belts fail before their scheduled replacement, they may do so by several failure modes: [81]

1. Cord failure, where a length of cord comes loose from the side of a belt; controlled by the adhesion levels within the belt.
2. Fabric separation, where a layer of fabric and teeth becomes separated from the cord layer; controlled by the adhesion levels within the belt.
3. Excessive tooth wear, where teeth are worn down to the extent that they cannot support the load; controlled by the wear resistance of the fabric.
4. Back cracking, where transverse cracks on the back can ultimately lead to belt failure; controlled by the fatigue resistance of the rubber material.
5. Tooth root cracking, the most common failure mode for automotive belts, can cause the tooth to be sheared off completely, or at least made too weak to support the load, making the belt unable to transmit power. For the root cracking failure mode, a large distortion of the tooth during meshing results in reduced belt life. Tooth distortion decreases with increasing tooth stiffness. Also, a rigid tooth will reduce fabric fatigue failure by minimizing fabric deformation. Since the belt operates in a dynamic environment, the dynamic storage modulus of the tooth compound should be considered; it must remain as high as possible over the operating temperature range and during the entire life of the belt if early failure by tooth root cracking is to be avoided.

V. CONVEYOR BELTS

Large-scale conveyor belt systems are highly complex with typical systems consisting of six components, the conveyor belt, drive system, support assemblies, tightening system, feeding system, and discharge system. The most common form of conveyor belt have a core reinforcement. For fabric belts, (Figure 13.16) this can be either PET or hybrid PET nylon composites. Steel cord–reinforced belts (Figure 13.17) use brass-coated wire or cables, galvanized steel cables, or brass wire mesh. The reinforcement is covered with a layer of rubber whose formulations are similar in composition to those found in wire- and fabric-coated compounds in tires.

Steel cable–reinforced belts typically use galvanized wire for corrosion resistance. The wires are laid down in a creel room and then moved to a compactor unit

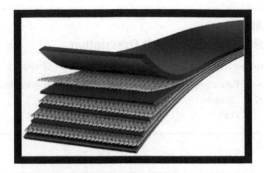

FIGURE 13.16 Fabric reinforced belt.

FIGURE 13.17 Steel cable reinforced belt.

where the belt's components are laid down. The wires are assembled in alternating "S" and "Z" twists to help reduce belt warping.

Industry descriptions for the two classes of belts considered here are illustrated in Tables 13.4 and 13.5. Fabric belts may start with a two-letter designation that defines the fabric type such as PET or nylon, or newer hybrid cords. Steel-reinforced belts begin with the designation ST. The term "carcass," is defined as the reinforcement layers, friction or cushion layers, and the friction and transition components between the reinforcements and the cover layers.

The cover compounds found in typical conveyor belts are again similar in nature to those found in large commercial truck tires. A premium, high-performance, conveyor belt will typically have a cover compound based on natural rubber such as RSS#2 (number 2 smoked sheet), with highly reinforcing carbon black. Lower performance belts, which may not require the level of tear strength required in premium belts, might use an emulsion SBR such as IISRP 1500 in place of natural rubber. Belts put into service in food-processing facilities have covers based on EPDM or food grade emulsion SBR. In those environments where the belt may be exposed to oils, a polychloroprene (CR)-, nitrile (NBR)-based compound, or NBR/PVC blend can be used. Mining belts require resistance to flames and burning and such cover compounds will contain flame retardants such as antimony trioxide and chlorinated paraffins.

TABLE 13.4
Fabric-Reinforced Belt Description

Nomenclature		EP400/4
PN	Fabric type	Polyester/nylon
400	Strength	kN/m
4	Plies	4 ply belt

TABLE 13.5
Steel Cord-Reinforced Belt Description

Nomenclature		ST3500/48–3/4 * 1/4
ST	Steel cord	
3500	Breaking strength	kN/m
48	Belt width	Inches
3/4	Top cover gauge	Inches
1/4	Lower or pulley cover gauge	Inches

The power consumption within a conveyor system can be reduced by minimizing the energy lost as the loaded belt is moved over the supporting pulleys. This is achieved by minimizing the belt bottom cover's rolling resistance by compounding approaches similar to those used in tire treads.

REFERENCES

1. E.J. Duda, What do automotive engineers want from vendors? *Elastomerics*, 18, September 1988.
2. G. Vroomen, EPDM coolant hoses based on peroxide vulcanization, *IRC2000 Rubber Conference*, Helsinki, Finland, June 2000.
3. SAE Standard J20, Coolant system hoses, October 1997.
4. R.C. Keller, Meeting coolant hose compound requirements of the future, SAE Technical Paper 870190, February 1987.
5. ASTM D395-14 Standard Test Methods for Rubber Property—Compression Set 1 Method B. 2014.
6. A.A. Schapp and MWGM. Peters, Twaron para-aramid fiber in knitted hoses; Influence of the yarn properties on the hose performance, Paper 50, *Meeting of Rubber Division*, American Chemical Society, Cleveland, OH, October 1997.
7. R.C. Keller, Advances in ethylene-propylene elastomer compounding for hose applications, *Rubber World*, 33, May 1983.
8. G. Vroomen, J. Noordermeer, and M. Wilms, Automotive coolant hose technology, Paper 42, *Meeting of Rubber Division*, American Chemical Society, Nashville, TN, October 1998.
9. M.J. Dees, New Keltan® EPDM grades for the automotive industry, Paper 104, *Meeting of Rubber Division*, American Chemical Society, Orlando, FL, September 1999.

10. S. Koch and A.G. Bayer (eds.), *Bayer Manual for the Rubber Industry*, 2nd edn., Bayer, Leverkusen, Germany, 1993.

11. Selecting a Cure System, ND-310.1 (R1), 1995, technical literature Dupont-Dow Elastomers.

12. M.A. Fisler, Compounding Nordel® for Compression Set Resistance, ND-530.1, Dupont-Dow Elastomers.

13. R. Ohm and R.T. Vanderbilt (eds.), *The Vanderbilt Rubber Handbook*, 13th edn., Norwalk, CT, pp. 140, 1990.

14. General Motors Engineering Standards, GM6250M, "Coolant Hose" (June 1997), section 7.11.

15. P.R. Dluzneski, Peroxide vulcanization of elastomers, *Rubber Chemistry and Technology*, 74(3), 451–492, July 2001.

16. J.R. Dunn, Compounding for modern automotive requirements, Paper 37, *Meeting of Rubber Division*, American Chemical Society, Cincinnati, OH, October 1988.

17. J.R. Dunn, D. Keller, and J. Patterson, EPDM compounding for high temperature aging requirements in automotive hose, Paper 78, *Meeting of Rubber Division*, American Chemical Society, Cleveland, OH, October 1987.

18. M.T. Gallagher, EPDM Extrusions Resistance to 175°C, Paper 85, *Meeting of Rubber Division*, American Chemical Society, Orlando, FL, October 1993.

19. R. Costin and W. Nagel, Selecting acrylic type coagents for hose and belt applications, Paper 53, *Meeting of Rubber Division*, American Chemical Society, Chicago, IL, April 1999.

20. "A New Coagent for Improved EPDM Properties for Radiator Hose Applications", Sartomer Application Bulletin 4114, November 2001.

21. Volkswagen Specification TL52361. Coolant Hose.

22. R. Costin and W. Nagel, Selecting acrylic coagents for hose applications, *IRC2000 Rubber Conference*, Helsinki, Finland, June 2000.

23. M.C. Bulawa and P.S. Ravishankar, Reduction of peroxide usage in EPDM coolant hose via polymer redesign, Paper 33, *Meeting of Rubber Division*, American Chemical Society, Chicago, IL, April 1999.

24. S. Monthey and M. Lucchi, A new carbon black for peroxide cured EPDM coolant hose, Paper 34, *Meeting of Rubber Division*, American Chemical Society, Orlando, FL, September 1999.

25. H. Schneider, H. Tucker, and E.T. Seo, Electrochemical degradation of coolant hoses, technical paper No. 73 presented at the 141st ACS Rubber Division Meeting, Louisville, KY (May, 1992).

26. R.C. Keller, Performance studies of ethylene-propylene rubber automotive coolant hoses, SAE Technical Paper 900576, February 1990.

27. G. Vroomen and H. Verhoef, Electrochemical degradation of EPDM cooling water hoses, *Kautschuk Gummi Kunststoffe*, 48(10), 749–753, 1995.

28. SAE Standard J1684, *Test Method for Evaluating the Electrochemical Resistance of Coolant System Hoses and Materials*, January 1994.

29. M. Lucchi, New cabot carbon blacks for improved electrochemical degradation resistance, Paper 9, *IRC2000 Rubber Conference*, Helsinki, Finland, June 2000.

30. R.C. Keller and D.A. White, Electrochemical influence on carbon black filled ethylene-propylene elastomers, *Rubber World*, 20–24, June 1992.

31. M. Bennett and L. Nijhof, The effect of peroxide curatives on the electrochemical degradation behavior of hose compounds, Paper 10, *IRC2000 Rubber Conference*, Helsinki, Finland, June 2000.

32. M.C. Bulawa and R.C. Keller, Impact of 42 volt electrical systems on the electrochemical resistance of peroxide cured automotive coolant hoses, Paper 117, *Meeting of Rubber Division*, American Chemical Society, Cincinnati, OH, October 2000.

33. R. Vara and J.R. Dunn, Developments in fuel hoses to meet changing environmental needs, Paper 13, *Meeting of Rubber Division*, American Chemical Society, Orlando, FL, October 1993.
34. ASTM D471-98, *Test Method for Rubber Property—Effect of Liquids*, Table 13.3.
35. DIN Standard 51 604.
36. SAE Standard J1681, (R.) *Gasoline, Alcohol and Diesel Fuel Surrogates for Materials Testing*, SAE International, Warrendale, PA, January 2000.
37. SAE Standard J30, *Fuel and Oil Hoses*, October 2001.
38. SAE Standard J1737, *Test Procedure to Determine the Hydrocarbon Losses from Fuel Tubes, Hoses, Fittings and Fuel Line Assemblies by Recirculation*, August 1997.
39. J.R. Dunn, H.A. Pfisterer, and J.J. Ridland, NBR vulcanizates resistant to high temperature and sour gasoline, *Rubber Chemistry and Technology*, 52, 331–352, 1979.
40. J.R. Dunn, Heat and fuel resistance of NBR and NBR/PVC blends, *Plastics and Rubber Processing and Applications*, 2(2), 161–168, 1982.
41. J. Horvath, High performance nitrile rubber automotive fuel hose, *Rubber Chemistry and Technology*, 52, 883–894, 1979.
42. D.P. Sinha, T. Jablonowski, B. Ohm, and V. Vanis, Compounding for hose and belt applications, Presented at a *Meeting of the All India Rubber Industries Association*, Mumbai, India, April 2001.
43. R.F. Karg, C.L. Hill, K. Dosch, and B. Johnson, Ultra-high ACN polymer in fuel system application, SAE Technical Paper 900196, February 1990.
44. H.A. Pfisterer and J.R. Dunn, New factors affecting the performance of automotive fuel hose, *Rubber Chemistry and Technology*, 53, 357–367, 1980.
45. J.R. Dunn and R.G. Vara, Fuel resistance and fuel permeability of NBR and NBR blends, *Elastomerics*, 29–40, May 1986.
46. J.D. MacLachlan, Automotive fuel permeation resistance—A comparison of elastomeric materials, SAE Technical Paper 790657, June 1979.
47. A. Nersasian, Effect of sour gasoline on fuel hose rubber materials, SAE Technical Paper 790659, June 1979.
48. G.J. Griffin and M.T. Gallagher, New HNBR polymers for fuel contact applications, Paper 9, *Meeting of Rubber Division*, American Chemical Society, Orlando, FL, September 1999.
49. D.E. Hull, J.R. Balzer, and P.F. Tuckner, Unique fluoroplastic for low permeation fuel system applications, SAE Technical Paper 950363, February 1995.
50. I.A. Abu-Isa, Effects of mixtures of gasoline with methanol and with ethanol on automotive elastomers, SAE Technical Paper 800786, June 1980.
51. I.A. Abu-Isa, Effects of methanol/gasoline mixtures on elastomers, SAE Technical Paper 840411, February 1984.
52. G.F. Baltz, Effects of alcohol extended fuels on the rate of fuel hose permeation, SAE Technical Paper 880709, February 1988.
53. J.R. Balzer and A.M. Sohlo, Effects of long-term flex-fuel exposure on fluorocarbon elastomers, SAE Technical Paper 900118, February 1990.
54. J.R. Balzer and A. Edmonson, Effects of methanol blend fuels on fuel hose permeation, SAE Technical Paper 910106, February 1991.
55. M.G. Wyzgoski, Migration of antiozonant from neoprene hose covers, SAE Technical Paper 790658, June 1979.
56. K. Hashimoto, A. Maeda, K. Hosoya, and Y. Todani, Specialty elastomers for automotive applications, *Rubber Chemistry and Technology*, 70, 449–519, 1998.
57. C. Hooker and R. Vara, A comparison of chlorinated and chlorosulfonated polyethylene elastomers with other Materials for Automotive Fuel Hose Covers", Paper 128, *Meeting of Rubber Division*, American Chemical Society, Cincinnati, OH, October 2000.

58. C. Cable and C. Smith, Epichlorohydrin in fuel hose, Paper 8, *Meeting of Rubber Division*, American Chemical Society, Louisville, KY, October 1996.
59. M. Sugimoto, T. Okumoto, T. Kurosaki, M. Ichikawa, and K. Terashima, Direct adhesion between modified nitrile rubbers and fluorocarbon rubber through vulcanization, Paper 16B16, IRC 1985, Kyoto, Japan, 1985.
60. J.R. Dunn and H.A. Pfisterer, Resistance of NBR-based fuel hose tube to fuel-alcohol blends, SAE Technical Paper 800856, June 1980.
61. D.R. Goldsberry, S.E. Chillous, and R.R. Will, Fluoropolymer resins: Permeation of automotive fuels, SAE Technical Paper 910104, February 1991.
62. D. Duchesne, D. Hull, and A. Molnar, THV fluorothermoplastics in automotive fuel management systems, SAE Technical Paper 1999-01-0379.
63. US Patent 5,639,528; US Patent 6,261,657.
64. J.W. Rodgers, Using different fibers in V-belt compounds, *Rubber and Plastics News*, 28–31, June 22, 1981.
65. L.L. Outzs, The use of aramid fiber pulp/neoprene masterbatch to develop unique engineering and dynamic properties in rubber, Paper 72, *Meeting of Rubber Division*, American Chemical Society, Washington, DC, October 1990.
66. J.F. van der Pol, Short para-aramid fiber reinforcement, *Rubber World*, 210, 32–37, June 1994.
67. J.L. Aherne, New developments in p-aramid short fiber reinforcement of rubbers, *IRC 2001*, Birmingham, U.K., June 2001.
68. H.W. Stanhope, V-belt reinforcement—Polyester, the most popular fiber, Paper 82, *Meeting of Rubber Division*, American Chemical Society, May 1984.
69. W.K. Donaldson, Power transmission belts, Chapter 9, in: *Textile Reinforcement of Elastomers*, eds. WC. Wake and DB. Wootton, Essex, U.K., Applied Science Publishers Ltd., 1982.
70. M. Fukuda, T. Shioyama, and Y. Mikami, V-belt and fan belt manufacturing technology, Chapter 17, in: *Rubber Products Manufacturing Technology*, eds. AK. Bhowmick, MM. Hall, and HA. Benarey, New York, Marcel Dekker, Inc., 1994.
71. A. Roetgers, Schrumpfarme Polyester-Hochmodulgarne fur den Einsatz in Antriebselementen, *Kautschuk Gummi Kunststoffe*, 44(12), 1137–1141, 1991.
72. E. Gebauer and D. Gajewski, A new polyester bicomponent fiber that allows a solvent free treating for power transmission belts, Paper 29, *Meeting of Rubber Division*, American Chemical Society, Nashville, TN, September 1998.
73. H. Janssen, Aramid fibers and new adhesion systems to elastomers, Paper 16, *Rubber Bonding'98*, Frankfurt, Germany, December 1998.
74. US Patent 3,784,427.
75. J. Miller, Serpentine belts take over auto market, *Rubber and Plastics News*, 21–22, June 17, 1996.
76. R. Musch, R. Schubart, and A. Sumner, Heat resistant curing system for halogen-containing polymers, Paper 6, *Meeting of Rubber Division*, American Chemical Society, Louisville, KY, October 1996.
77. P. Arjunan, Technological compatibilization of dissimilar elastomer blends: Part 1. Neoprene and ethylene-co-propylene rubber blends for power transmission belt application, Paper 52, *Meeting of Rubber Division*, American Chemical Society, Nashville, TN, September 1998.
78. R.F. Ohm, P.A. Callais, and L.H. Palys, Cure systems and antidegradant packages for hose and belt polymers, Paper 52, *Meeting of Rubber Division*, American Chemical Society, Chicago, IL, April 1999.
79. U.S. Patent 5,610,217.
80. J.G. Pillow and R.E. Ennis, ACSM—A new generation of chlorinated elastomer for high temperature dynamic applications, Paper 19, *Meeting of Rubber Division*, American Chemical Society, Detroit, MI, October 1989.

81. K.W. Dalgarno, Short review—Power transmission belt performance and failure, *Rubber Chemistry and Technology*, 71, 619–636, July 1998.
82. SAE Standard J2432, January 2000.
83. U.S. Patent 5,382,198.
84. M.N. Tranquilla, Predicting behavior of synchronous belts, *Machine Design*, 68(14), 64–69, August 8, 1996.
85. M.J.W. Gregg, Synchronous drive belts using helical offset teeth, Paper 55, *Meeting of Rubber Division*, American Chemical Society, Chicago, IL, April 1999.
86. *Hydrogenated Nitrile Rubber, Properties and Recent Trends in Application Development*, Bayer Rubber Business Group, 1999.
87. E.C. Campomizzi, H. Bender, and W. von Hellens, Effect of a novel heat stabilizer additive on the heat resistance of HNBR, Paper 32, *Meeting of Rubber Division*, American Chemical Society, Orlando, FL, September 1999.
88. T.A. Brown, Compounding for maximum heat resistance and load bearing capacity in HNBR synchronous belt compounds, Paper 80, *Meeting of Rubber Division*, American Chemical Society, Detroit, MI, October 1991.
89. U.S. Patent 4,918,144.
90. M.E. Wood and W.G. Bradford, Synchronous and serpentine belt compounds: The new benchmark in dynamic performance, Paper 67, *Meeting of Rubber Division*, American Chemical Society, Orlando, FL, October 1993.
91. J-M Tixhon, New rubber family for automotive use, *Plastics and Rubber Weekly*, 7, December 21, 1991.
92. F.K. Jones, L.L. Outzs, and G. Liolios, The effect of heat generation on the performance of several elastomers for automotive use, Paper 58, *Meeting of Rubber Division*, American Chemical Society, Las Vegas, NV, May 1990.
93. C.A. Hamand, R. Knowles, and C.A. Stevens, *The Benefits and Opportunities of New High Strength Glass Cord*, Rubber Technology International, pp. 66–68, 1999.
94. ISO Standard 10917, Synchronous Belt Drives—Automotive Belts and Pulleys—Fatigue Test, 1995.
95. ISO Standard 12046, Synchronous Belt Drives—Automotive Belts—Determination of Physical Properties, 1995.

14 Tire Technology

Brendan Rodgers and Bernard D'Cruz

CONTENTS

I.   Introduction 579
II. Types of Tires 580
    A. Basic Tire Design 581
    B. Tire Nomenclature and Dimensions.. 582
    C. Tread Design... 584
III. Tire Materials 586
    A. Reinforcements... 586
    B. Tire Compounds 588
    C. Compound Development 590
        1. Information Technology 590
        2. New Compound Development.. 590
    D. Compound Mechanical Properties... 591
IV. Tire Testing... 593
V. Trends in Tire Technology.. 596
VI. Summary 597
References.. 598

I. INTRODUCTION

Tires are essential for effective operation of most forms of transportation, ranging from passenger automobiles to heavy-duty trucks, farming vehicles, and aircraft. The pneumatic tire is not only the most complex composite product in mass production but also the most complex component on a vehicle. It consists of a variety of materials such as natural rubber (NR), synthetic elastomers such as styrene butadiene rubber (SBR) copolymers, and a wide variety of chemicals; textiles such as nylon and polyester; and a range of brass- and bronze-coated steel wires. Over the last 30 years, tires have undergone a major evolution. In the 1950s, tires were predominantly of a bias ply construction and required the use of an inner tube for air retention. Bias or diagonal ply tires describe a tire construction where the ply cords extend to the beads and where the ply cords are laid at an angle between 36° for passenger tires and 42° for heavy-duty large truck tires (Figure 14.1). Automobile tires lasted for between 15,000 and 25,000 miles, after which they were replaced. Though the use of bias tires in many markets has now been virtually eliminated, they still have a very large market share in Asia wherein some severe applications can outperform radial constructions.

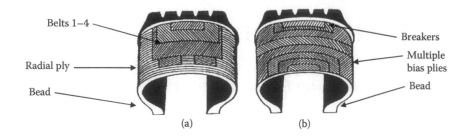

FIGURE 14.1   Tire constructions. (a) Radial construction and (b) bias construction.

In the 1960s, the radial tire emerged with significant improvements in mileage, fuel economy, and safety. Radial tires have a structure where the ply cords extend to the beads and are laid at a 90° angle to the centerline of the tread. The casing is stabilized by a ridged belt system consisting of brass-coated steel cord. This was followed by the introduction of the tubeless tire, which used an airtight membrane inside the tire, and an airtight seal between the tire and the rim was formed through compression fitting of the tire bead on an inclined steel wheel. The tubeless tire owes its success to the development and application of halogenated butyl rubbers, which allow very little air permeation.

"Ultralow" aspect ratio tires are now emerging in which the sidewalls are considerably shorter, which allows improvements in vehicle stability, rolling resistance, and, in the case of commercial trucks, cargo volume through lowering of the trailer platform height. It also provides a basis for the development of "run-flat" technology, which allows tires to operate at near-zero inflation pressure.

II. TYPES OF TIRES

Tires fall into essentially eight categories based on the mission profile for the product:

1. Racing and sports vehicles and recreation uses that could also encompass bicycle tires
2. Passenger car vehicles
3. Light truck and sports utility vehicles (SUVs), whose gross vehicle weight (GVW) does not exceed 7250 kg
4. Commercial trucks
5. Farm vehicles including implements such as plows
6. Earthmoving equipment, also described as "off the road" or OTR, whose GVW can approach 300 tons
7. Aircraft
8. Nonpneumatic specialty tires for applications such as forklift trucks

All tires, whether they are designed for something as simple as a bicycle or for mounting on large commercial aircraft, must meet a fundamental set of performance functions. They must

1. Be durable and safe through the expected service life of the product
2. Provide load-carrying capacity
3. Provide cushioning and damping
4. Generate minimum noise and vibration
5. Transmit driving cornering, steering, and braking torque
6. Resist abrasion
7. Have low rolling resistance

A. Basic Tire Design

Figure 14.2 illustrates the components of a typical commercial truck tire used on vehicles that have a GVW of up to 40 tons [1]. The major components of the tire can be described as follows:

Tread: The tread is the component of the tire that is in contact with the road and provides wear resistance, good traction characteristics, and fuel economy and service-related damage.

Tread base: The component underneath the tread that is designed to ensure good adhesion between the tread and the tire casing and dissipate heat from the tread. The shoulder region, which in larger tires is coextruded as a unique component, also assists in heat dissipation.

Sidewall: The sidewall provides long-term weathering protection and casing durability. It also protects the tire from impact and curb scuffing. It is compounded to resist fatigue and flex-related cracking.

Bead region: The bead or lower region of the tire consists of the following components: (1) bead ring, which locks the tire onto the wheel or rim; (2) apex or bead filler; (3) chipper wire, which protects the ply from rim damage; and (4) the chafer, which is the abrasion-resistant rubber component in contact with the rim.

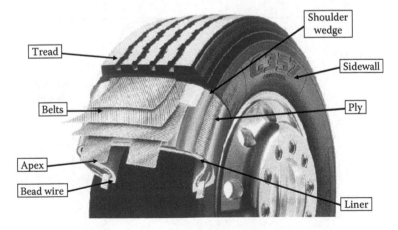

FIGURE 14.2 Tire cross section displaying major components and reinforcements and mounted on an aluminum rim.

Plies: Ply cords consist of steel or fabric, extend from the bead on one side of the tire to the opposite side, and serve as the primary reinforcing member on the tire. They provide the strength to the tire that enables it to retain air at high pressures.

Belts: Belts are layers of steel wire and sometimes fabric that form a hoop under the tread. They restrict deformation of the casing, provide rigidity to the tread region, and provide a stable foundation for the tread components. This allows improved wear performance, vehicle stability, handling performance, damage resistance, and protection of the ply cords.

Shoulder wedge: A stiff elastomeric component located at the edge of the belt assembly and designed to provide stability to the shoulder region of the tire, provide a uniform footprint, and help dissipate heat.

Innerliner: Membrane consisting of compounded low-permeability rubber such as bromobutyl rubber (BIIR) or chlorobutyl rubber (CIIR) and whose function is to retain compressed air inside the tire. The liner typically reaches from the bead on one side of the tire to the opposite side, thereby providing a seal.

B. TIRE NOMENCLATURE AND DIMENSIONS

Figure 14.3 illustrates the dimensions used to describe a tire mounted on a rim. The section width is the width of the tire from sidewall to sidewall. The section height is the distance from the bottom of the bead region of the tire to the top of the tread. These two parameters are important in describing the tire size seen on the sidewall. Three basic tire size designations are used:

1. Conventional-size tires used on flat base rims normally for tube-type tires
2. Conventional-size tires used on 15° rims for tubeless tires
3. Metric sizes used on 15° rims for tubeless tires

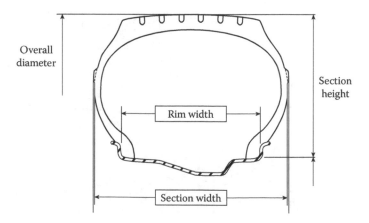

FIGURE 14.3   Dimensions of a tire mounted on a 15° rim.

The letter R in the size description designates "radial," D or (–) is "bias," and ML indicates that the tire is for mining and logging applications. The letter P denotes passenger automobile tires, and LT describes a tire for a light truck or an SUV. For example, the tire size designation LT235/75R15 denotes

LT   Light truck
235 Approximate section width in millimeters
75   Aspect ratio or the ratio of section width to section height
     Conventionally sized tires have an aspect ratio of 80
R   Radial construction
15   Nominal rim diameter in inches

Other sidewall markings include construction and manufacturing information, speed, and load-carrying capability. The speed symbol indicates the maximum speed at which a tire can carry a load corresponding to its load index. For example, a tire with a speed rating of S can operate at speeds up to 180 km/h; T, to 190 km/h; and V to 240 km/h. The speed rating symbol can be found as part of the size description or part of the load index symbol. Tires with the marking ZR such as a 195/50ZR15 are for speeds over 240 km/h (Table 14.1).

The load index (Table 14.2) is a numerical code associated with the maximum load a tire can carry at the speed indicated by its speed symbol. Load index tables are published by the tire industry associations and the individual tire manufacturers. For illustration, a typical load index of 149/145 J defines a tire that can carry a load of 3250 kg when single mounted or 2900 kg when dual mounted, at a speed rating (J) of 100 km/h. Such conditions might be found for tires mounted on city delivery trucks.

Reference should be made to the Tire and Rim Association Inc. *Year Book* or the European Tyre and Rim Technical Organization (ETRTO) *Standards Manual* for further information on tire nomenclature and rim specifications for a given tire size [2–5].

TABLE 14.1

Tire Speed Ratings

Speed Symbol	Maximum Speed (km/h)	Typical Application
D	50	Agricultural vehicles
L	120	Commercial trucks
M	130	Commercial trucks
S	180	Passenger cars and light trucks
T	190	Passenger cars
H	210	Luxury passenger cars
V	240	High-performance cars
Z	Above 270	Superhigh-performance cars

TABLE 14.2
Examples of ETRTO Load Index Numerical Code
for Tire Load-Carrying Capability

Load Index	Load-Carrying Capability (kg)
144	2800
145	2900
146	3000
149	3250
152	3550
155	3875
160	4500

C. TREAD DESIGN

The tread pattern influences the ability of the tire to transmit steering, braking, and driving forces while operating on a broad range of highway and off-road surfaces. The design of a tread pattern involves the introduction of grooves into a smooth tread face such as illustrated in Figure 14.4 [6]. This enables improvement in tire road grip and the dissipation of water to minimize hydroplaning and improves

Highway steer (rib)

Highway drive (rib–lug)
combination with inboard lugs

Highway trailer tire

Highway drive (rib–lug)
combination with outboard lugs

On-/off-road and
off-road design

FIGURE 14.4   Tread patterns typically found on commercial trucks.

tire–vehicle steering qualities. The tread ribs can be further divided into blocks or elements to facilitate improvement in traction qualities. This becomes important in high-torque applications such as the drive axle position in heavy-duty trucks or in high-performance passenger tires. The tread elements are arranged to give optimum depth, tread width, and net-to-gross contact area. For a truck tire, tread patterns can be classified into five basic categories:

1. Highway rib.
2. Highway rib–lug combination, where the rib can be either inboard or outboard in the shoulder area. The lugs will be correspondingly opposite to the ribs, outboard or inboard.
3. Highway lug design (no rib configuration present).
4. On-/off-highway.
5. Off-highway or off-road.

Rib designs with design elements principally in the circumferential direction are the most common type of tread pattern and show overall good service for a broad range of conditions. On commercial trucks, they are used nearly exclusively on the steer axle and trailer axle positions because of their lateral traction and uniform wearing characteristics. Rib–lug (semilug) combinations tend to find use on all-season tires, which require a balance of wear, traction, and wet skid resistance. On heavy-duty truck drive axles, where forward traction is a prime requirement and where fast wear can occur as a result of torque-induced slip, a highway lug design is required. For off-highway service conditions, the tread pattern assumes a staggered joint lateral circumferential direction for both lateral and forward directions. Grooves tend to be larger and deeper, with the rib walls angled to prevent stone retention.

The ratio of net road or pavement contact area, excluding voids, to gross tread surface area decreases as wet traction becomes more important. Table 14.3 illustrates the net-to-gross percentage for typical commercial truck tire tread patterns illustrated in Figure 14.4 [1,6]. Similarly, automobile tire tread designs can be placed in some broad categories such as for high-mileage tire tread patterns, all-season or broad-market tires, traction, high performance, and traction–high performance. There are some simple guidelines that the design engineer is aware of such as a

TABLE 14.3
Tread Pattern Contact Area with Nature of Service

Type of Service	% Net Contact to Gross Tire Footprint
Highway steer and trailer tires	73–82
Highway drive axle tire	65–75
On-/off-road (mixed service)	60–70
Off-highway	55–65

TABLE 14.4
Automobile Tire Tread Pattern Classes

Category	Design	Application	Net to Gross Pavement Contact Area
1	Central solid rib, outer rib block configuration	High mileage	High
2	Block–rib	All season Broad market Baseline	Medium-high
3	All block	Traction	Medium
4	Central groove	Traction, high performance	Medium-low
5	Directional Asymmetric Symmetrical	High performance	Low-medium

minimum structural rigidity and aspect ratio of the tread blocks to prevent irregular wear and curvature at the base of tread elements to prevent cyclic high strains and consequent fatigue cracking. Thus, though tread pattern designs are very diverse with very few published guidelines on pattern optimization, Table 14.4 shows one possible schematic on how tire tread design is tuned for a given performance level or mission profile [7].

III. TIRE MATERIALS

A tire is a textile/steel cord/rubber composite where the steel and fabric cords reinforce the rubber compounds and are the primary load-bearing structure within the tire. Effective tire engineering requires knowledge of the materials to be used in the tire structure.

A. REINFORCEMENTS

Tire reinforcing materials include steel wire, which is typically used for the belts in radial passenger and truck tires; polyester cords, which are typically used in the ply of passenger car tires; and nylon, which may be used also in the belt and ply. Other materials used for reinforcement include rayon, cotton, fiberglass, and aramid [8].

Steel cords used in tires are typically brass-coated. A steel rod is first drawn down to a diameter of 1.2 mm, then the brass plating is applied. The filaments are then drawn down to their final diameter, which can range from 0.15 to 0.45 mm, depending in the application for which the cord is needed. The filaments are then wound into the desired cord construction. Steel cords are typically used in the belts of nearly all radial tires, the plies in radial truck medium tires, and the plies of off-road truck tires.

Reinforcement science and technology uses a specialized terminology with which rubber technologists must be familiar. Some definitions pertaining to both fabrics and steel cord are presented here for ease of reference:

Fiber: Fibers are one of the three classes of polymers, the other two being elastomers and plastics. Fibers consist of linear macromolecules oriented along the length of the fiber axis.

Filament: The smallest continuous component in a textile or steel cord composing the strand or cord.

Yarn: Filaments assembled so as to form a continuous strand.

Cord: A structure composed of two or more strands used as plied yarns or as the end product.

Warp: Cords that run lengthwise in a tire or other rubber product such as a conveyor belt.

Weft: Cords in a fabric running crosswise to the direction of the warp.

Pick: The light threads that are placed at right angles to the warp cords; also referred to as the *pick*.

Rivit: Distance between cords in a fabric. A fabric with a low EPI number (cord ends per inch width of fabric) has high rivit. Low rivit describes a greater number of cord ends per inch of fabric.

Twist: The number of turns per unit length in a cord or yarn. Twist can be in one of two directions.

S twist: S twist spirals are around the central axes of the cord so as to conform to the slope of the central portion of the letter S, that is, clockwise.

Z twist: Z twist describes a cord that conforms to the slope of the central portion of the letter Z, where the cord runs counterclockwise around the central axis of the cord.

Denier: The weight per unit length of cord expressed in g/9,000 m.

Decitex: Similar to denier but describes the weight of cord in g/10,000 m.

Tenacity: Cord strength.

Strength: May also be defined by the term tenacity. It is a measure of the tensile strength of the cord and can be expressed in g/denier.

LASE: Load applied to a cord for a specified elongation (load at specified elongation).

Brass weight: Typically 3.65 g/kg of cable. Brass thickness is typically 0.3 μm.

Construction: Textile cords are defined by strength (decitex), the number of yarns in the ply, and the number of plies in the cord. For example, a cord described as 1400/1/3 has a decitex of 1400, only one yarn in the ply, with three plies composing the final cord. Typically the number of fibers and filaments is not defined.

A number of general rules govern the design of a steel cord for use in tires. For example, if the cord is used in the ply of a steel-reinforced heavy-duty truck tire, it will undergo a greater amount of flexing than in a lighter vehicle and will require resistance to fatigue. If the application is belts, stiffness and rigidity become major design parameters. The thicker a cord is, the stiffer it will be. Thinner cords tend to show better fatigue resistance.

B. TIRE COMPOUNDS

Compounding ingredients fall into five general categories:

1. Polymers, such as NR, SBR, and polybutadiene.
2. Fillers such as carbon black, silica, clays, and calcium carbonate, though the latter two are used more in rubber products for industrial applications.
3. Protectant systems consisting of antioxidants, antiozonants, and waxes.
4. Vulcanization system.
5. Special-purpose materials such as processing aids and resins. The materials scientist designing a rubber formulation has a range of objectives and restrictions within which to operate. Product performance objectives define the initial selection of materials. These materials must not raise environmental concerns, must be processable in production plants, and must be cost effective.

NR is classified into essentially three categories: (1) technically specified rubbers, for which contaminant levels such as ash, nitrogen-containing chemicals, and viscosity are defined; (2) visually inspected rubbers such as smoked sheets and crepe, which are shipped in 115 kg bales; and (3) specialty rubbers such as epoxidized NR and powdered rubber. NR use in radial tires has increased over that used in bias tires because of its greater green strength (strength before vulcanization), tear strength, tire component-to-component adhesion, lower heat generation (hysteresis) under dynamically loaded conditions, and lower tire rolling resistance.

Classification of synthetic rubbers is governed by the International Institute of Synthetic Rubber Producers. For SBR, polyisoprene, and polybutadiene, a series of numbers have been assigned that define the basic properties of the polymer. For example, the IISRP 1500 series describes the range of commercially available, cold polymerized emulsion SBR. Series 1200–1249 define non-oil-extended polybutadienes, and 2200–2249 are polyisoprene and copolymers of isoprene [9].

A series of empirical guidelines can be used in designing a tread compound polymer for a set of tire performance requirements: [8]

1. There is a near linear drop in abrasion resistance or tread wear as polymer glass transition temperature (T_g) increases.
2. As T_g increases, wet grip or traction improves.
3. Inclusion of styrene or an increase in the styrene level increases tire wet traction and decreases tire tread wear performance.

Selection of the appropriate grade of carbon black described in ASTM D1765 leads to significant improvements in performance [10]. For example, the larger surface area carbon blacks are required for tread wear. However, with an increase in surface area and a corresponding improvement in tread wear, rolling resistance of the tire can increase, with adverse effects on fuel economy. Reduction in carbon black loading will, however, lower the rolling resistance.

Addition of silica to a compound will improve tear strength, reduce heat buildup, and improve compound adhesion in multicomponent products such as tires. Two fundamental properties of silica influence their use in rubber compounds: ultimate particle size and the extent of surface hydration.

Unsaturated elastomers are susceptible to oxidation. Atmospheric ozone will also readily degrade elastomers. For example, tire sidewalls that are subjected to a high degree of flexing will show crazing and cracking after a certain service life, which is clearly visible in most tires after about 6 years of operation. To protect an elastomer from oxidation or ozonolysis, three categories of materials are added to the elastomeric formulation: waxes, antioxidants, and antiozonants.

Antioxidants are mainly amine or phenolic derivatives such as 1,2-dihydro-2,2,4-trimethylquinoline (TMQ) and function by reacting with oxides or broken polymer chain ends caused by reaction with oxygen. They therefore prevent the propagation of an oxidative process, enabling the retention of the product's physical properties.

Ozone protection is obtained through use of two materials; waxes and, typically, *para*-phenylenediamines. A number of empirical guidelines can be used to develop an antidegradant system for a compound in a tire. Short-term static protection is achieved by the use of paraffinic waxes. Microcrystalline waxes provide long-term ozone protection while the finished product is in storage. A critical level of wax bloom is required to form a protective film for static ozone protection. Optimized blends of waxes, *para*-phenylenediamines such as 6-PPD, and antioxidants such as TMQ provide long-term tire protection under both static and dynamic applications and over an extended temperature range.

The vulcanization system in a typical tire compound consists of three components: the activation system, typically zinc oxide and stearic acid; the vulcanizing agent, typically sulfur; and accelerators. In an idealized tire compound, zinc oxide and stearic acid react to form zinc stearate at tire vulcanization temperatures. The zinc stearate then reacts with the accelerators in the compound to form a "sulfurating complex," which with sulfur creates the required cross-link network. Accelerators can be classified according to their reaction rate and the nature of the cross-links they create. Thiurams give predominantly monosulfidic sulfur crosslinks. Sulfenamides, when used with high sulfur concentrations, build predominantly polysulfidic cross-links. The structure of the cross-link network can therefore be designed to meet a specific set of service requirements.

A broad range of other materials are also used in a tire compound. For example, titanium dioxide is used for tire white sidewalls, and processing oils are used to assist in the mixing, extrusion, and calendering of compounds in the factory. Plasticizers such as dioctylphthalate are also used as processing aids. Chemical peptizers such as pentachlorothiophenol are added to the compound during the initial mixing to reduce energy consumption in mixing and provide a more uniformly mixed compound. Resins include extending or processing resins, tackifying resins that help in handling of components during the tire assembly operation, and curing resins that are used to improve properties such as tensile strength.

A comprehensive review of materials used in rubber compounding was recently compiled by Klingensmith [11], to which further reference is recommended.

C. COMPOUND DEVELOPMENT

Compound formulation development and reformulation provides a rapid form of new product development, provides a means to rapidly meet new regulatory requirements, responds to competitive concerns, improves existing products, and enables new product development. The sources of information for new compound development center on a number of areas, raw materials supplier, scientific publications, universities and research institutes, and internal company development activities. The techniques available to the compound development scientist center on several tools:

1. Information Technology

Information technology (IT) systems centered on the deployment of knowledge management systems and tools for experimental designs are basic to an efficient operation of a compound development team.

The functions provided include the following:

1. Information such as approved formulations
2. Vendor-supplied data
3. Knowledge records, that is, reports
4. Experimental data storage and easy retrieval that would include formulations and associated compound properties such as vulcanization kinetics and rheological properties, classic mechanical properties, and dynamic and hysteretic properties

2. New Compound Development

Formulation development to meet a new performance requirement can be conducted at various levels:

1. The most elementary is screening of a series of formulations based on the experience of the scientist. This may involve incremental changes in one or more selected components in a formula. Alternatively, it may involve substitution of one material for another.
2. More sophisticated tools using "designed experiments" can be used. These essentially fall into two categories: simple factorial designs where two or more components in a formulation are varied in an incremental manner and full multiple regressions where three or more components in a formulation are changed in defined increments, data collected, multiple regression equations computed, graphical representation of data computed, and optimized formulations calculated with the desired mechanical properties.
3. Computational techniques based on neural networks and genetic algorithms are now being used. This enables boundaries to be established within which

designed experiment may be developed to tune a specific formulation. Such techniques when developed enable many more components in a formulation to be considered without the experimenter being overcome with excessive amounts of data.

4. *Predictive modeling*: Many tire companies have developed proprietary models that enable estimation of how a rubber compound formulation will perform in a product such as a tire. There are a number of elementary relationships available to the researcher such as the effect of tangent delta on tire traction and influence of compound rebound on rolling resistance. Basic computational tools can be readily assembled to calculate the effect of changing the hysteretic properties of several compounds in a tire simultaneously and estimating the resulting rolling resistance.

On completion of the laboratory development phase, adequate testing is essential to verify that the product will meet the performance expectations and predicted performance parameters to which it has been designed.

Formulations are available in several industry publications such as the *Natural Rubber Formulary and Property Index* published by the Malaysian Rubber Producers Research Association [12,13]. Typical examples of compound formulations cited frequently in the technical literature are tabulated for general reference purposes (Tables 14.5 and 14.6). Further optimization can be conducted on these formulations should a specific set of mechanical properties be required to meet the product mission profile, product manufacturing environment, or comply with regulatory constraints. A further point to be noted in the context of this discussion is the importance of defining optimum compound mixing temperatures, internal mixer compound dwell time, and required final compound viscosity. Compound viscosity is important in ensuring quality component extrusions, which is a function of throughput, extrudate temperature, adherence to contour or gauge control, and appearance that may be adversely affected by bloom of any compound constituents.

D. COMPOUND MECHANICAL PROPERTIES

Compounds are designed to meet a defined set of properties. These properties fall into four fundamental categories: (1) processing, (2) classic mechanical properties such as tensile strength, (3) dynamical properties such as fatigue resistance and dynamic stiffness, and (4) component-specific properties such as gas permeability of innerliner compounds. The dynamic mechanical properties of compounds are essentially described by the compound storage modulus (E' in tension, G' in shear) and the loss modulus (E'' and G''). From these two measurements, a hysteresis coefficient, termed the tangent delta, or tan delta, can be calculated from the ratio of E''/E', or G''/G' (Table 14.7). From the two terms, storage modulus and loss modulus, a series of additional parameters can be estimated and includes a complex modulus (E^*, G^*), elastic compliance modulus (D' in tension, J'' in shear), loss compliance (D'', J''), and complex compliance (D^*, J^*).

The storage modulus, loss modulus, and tan delta terms can be used as predictive tools for tire compound performance. The stiffness of a rubber compound is

TABLE 14.5
Model Tread Compounds

Model truck tire tread compound, example 1

Natural rubber	50.00
Polybutadiene	25.00
SBR	25.00
Carbon black (N220)	65.00
Peptizer	0.25
Paraffin wax	1.00
Microcrystalline wax	2.00
Paraffinic oil	10.00
Polymerized dihydrotrimethylquinoline (TMQ)	1.00
7PPD	2.50
Stearic acid	2.00
Zinc oxide	5.00
TBBS	1.25
Sulfur	1.00
DPG	0.30
Retarder (if required)	0.25

Model truck tire tread compound, example 2

Natural rubber	100.00
Carbon black (N220)	50.00
Peptizer	0.25
Paraffin wax	1.00
Microcrystalline wax	2.00
Paraffinic oil	3.00
Polymerized dihydrotrimethylquinoline (TMQ)	1.00
Stearic acid	2.00
Zinc oxide	5.00
TBBS	1.00
Sulfur	1.00
DPG	0.25
Retarder (if required)	0.20

a function of many factors, one of which is the T_g of the polymer. The higher the T_g, the higher the storage modulus and also the loss modulus. The ratio of these terms allowing the calculation of the tangent delta will then provide a prediction of the tire tread compound traction performance. From the work of Nordseik and others, a tangent delta temperature profile can be used to identify the temperature at which the testing for tangent delta will occur [14,15]. For example, for a given set of tread compounds, the one with the highest tangent delta at 0°C would be expected to demonstrate the best wet traction performance, at 20°C, the best ice performance, and at +60°C, the lowest rolling resistance (Figure 14.5).

TABLE 14.6

Tire Sidewall and Casing Compounds (phr)

Model truck tire tread compound

Natural rubber	60.00
Polybutadiene	40.00
Carbon black (N330)	48.00
Peptizer	0.15
Paraffin wax	1.00
Microcrystalline wax	2.00
Paraffinic oil	3.00
Polymerized dihydrotrimethylquinoline (TMQ)	1.50
7PPD	3.50
Stearic acid	2.00
Zinc oxide	3.00
TBBS	0.95
Sulfur	1.25
Retarder (if required)	0.15

Model tire casing ply compound (phr)

Natural rubber	65.00
Polybutadiene	35.00
Carbon black (N660)	65.00
Peptizer	0.25
Paraffin wax	1.00
Microcrystalline wax	1.00
Paraffinic oil	8.00
Polymerized dihydrotrimethylquinoline (TMQ)	1.00
N-1,3-Dimethylbutyl-N-phenyl-p-phenylenediamine (6PPD)	2.50
Stearic acid	2.00
Zinc oxide	3.00
DCBS	0.90
Sulfur	4.50
Retarder (if required)	0.25

IV. TIRE TESTING

Tire testing occurs in three stages: (1) initial laboratory testing, (2) general proving ground testing for which tires are mounted on vehicles that are then tested on specially prepared roads, and (3) commercial fleet tests [8]. Laboratory testing includes tire uniformity tests, speed rating determination, rolling resistance, durability assessment, and basic handling characteristics such as the cornering coefficient.

Proving grounds consist of high-speed test tracks, gravel roads, wet and dry skid pads, and tethered tracks for the testing of farm tractor tires. Testing is conducted under defined conditions such as certain loads, inflations, and speeds and

TABLE 14.7

Dynamic Mechanical Properties

Property	Description	Measurement	Temperature for Measurement	Tire Performance Parameter	Improvement
E′, G′	Storage modulus	Tension, shear	T_g and 23°C	Irregular wear, cornering coeff. tread wear	Increase E′, G′
E″, G″	Loss modulus	Tension, shear	0°C, 60°C	Rolling resistance, wet traction	Decrease at 60°C for RR, increase at 0°C for traction
Tangent delta	E″/E′, G″/G′	Tension, shear	0°C, 60°C, 80°C	Rolling resistance, wet traction, high speed	Decrease at 60°C for RR, increase at 0°C for traction
T_g	Glass transition temperature		−80°C to 10°C	Glass transition temperature, controls hysteresis	—

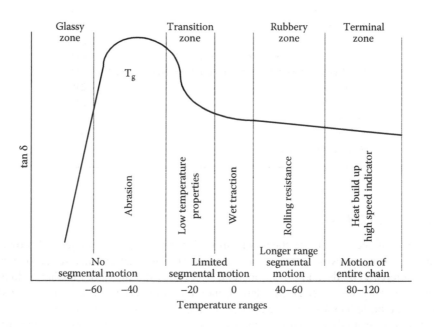

FIGURE 14.5 Tangent delta curve for tread compound property prediction. (From Ford, T.L. and Charles, F.S., *Heavy Duty Truck Tire Engineering, 34th Buckendale Lecture*, SAE, Warrendale, PA, SP729, 1988.)

the use of specific vehicle types. The vehicles are specified to meet a set of defined conditions such as wheel alignment, vehicle horsepower, wheelbase, and axle configuration.

Finally, commercial testing enables the tire engineer to obtain an assessment of how a design will perform under a broad range of service conditions that will be experienced by the end consumer. The testing protocol tires undergo results in a broad range of products to meet the needs of both the vehicle manufacturers and the end users for optimum performance under a variety of service conditions.

It is appropriate to present definitions of important tire performance parameters and indicate what compound parameters influence these properties (1, 6):

Cornering coefficient: The lateral force divided by the vertical load at a defined slip angle. Stiffer tread compounds tend to improve the cornering coefficient.

Conicity: The tendency of a tire to pull a vehicle to one side or another. It is caused by off-centered or misplaced components during the tire-building process.

Balance: The weight distribution around the circumference of a tire. Poor balance is due to components having irregular dimensions. It can also be affected by irregular component splice widths and poor application of component parts during tire building.

Force variation: Periodic variation in normal vertical force of a loaded free-rolling tire that repeats with every revolution.

Harmonic: Sinusoidal periodic or rhythmic force variations around the tire. One phase is described as the first harmonic. When two phases are noted, the second is described as a second harmonic. First harmonic lateral force variation is typically due to a tread splice. A radial harmonic may be due to irregular placement of the belt lay up.

Lateral force: Side force that is exerted by a tire as it rotates under a load.

Lateral force variation: Variation or change in force from one side of the tire to the other as it rotates under a load. It may cause the tire to wobble and is due to irregular tire component dimensions. Lateral force variation is a summation of the lateral first, second, third harmonics, etc.

Lateral force coefficient: The lateral force divided by the vertical load.

Lateral runout: Difference between the maximum and minimum measurements parallel to the spin axis at the widest point of each tire sidewall when the tire is mounted on a wheel.

Radial force: Force acting on a tire perpendicular to the centerline of rotation or the direction of the axle. It is caused by heavy tire component splices and will increase radial force. Soft spots in the tire such as that due to stretched ply cords cause a decrease in radial force.

Radial force variation: A summation of the radial first, second, third harmonics, etc. It is the change in radial force as the tire is rotated. Radial force variation will cause the vehicle to have a rough ride (as if on a poorly surfaced road).

Runout: The differential between the maximum and minimum lateral or radial forces.

Radial runout: Difference between the maximum and minimum measurements on the tread surface and in a plane perpendicular to the spin axis while the tire is mounted on a wheel. A measure of the out-of-roundness of the tire. Also termed centerline runout.

Rolling resistance: Resistance of a tire to rolling. It has a direct impact on vehicle fuel economy and is influenced most by compound hysteretic properties.

Self-aligning torque: The stabilizing reaction to slip angle that helps the tire and vehicle to return to neutral conditions at the completion of a maneuver.

Slip angle: Angle between the vehicle's direction of travel and the direction in which the front wheels are pointing.

Uniformity: Measure of the tire's ability to run smoothly and vibration-free; sometimes measured as tire balance, radial force variation, or lateral force variation.

Speed rating: Alphabetic rating that defines the design speed capability of the tire. The letter is incorporated into the size description of the tire. For example, a 195/75SR14 has a speed rating of "S." Tables of speed ratings and corresponding alphabetic designations are published by the Tire and Rim Association (2).

The laboratory test to which the industry tests tires is the Federal Motor Vehicle Safety Standard, FMVSS139. This consists of several individual tests.

V. TRENDS IN TIRE TECHNOLOGY

The tire industry has continued to make some very significant improvements in the design, quality, and performance of its products [5,7].

There are three factors driving the changes in the tire industry: government regulations, vehicle manufacturer's needs for ongoing improvements in performance and quality, and the tire companies responding to consumer trends (Table 14.8). In the automobile tire segment, there will be continued increase in both tire and rim diameters, this in many instances necessitating redesign of the vehicle corners to compensate for change in tire spring rate and damping properties (Table 14.9). Decrease in sidewall height and development of "ultralow-profile" tire constructions, which can also aid in improving fuel economy, has led to improved vehicle handling qualities. Ongoing efforts to further reduce rolling resistance, decrease noise generation, and improve durability and traction under a wide variety of conditions will always continue, this being achieved in some instances by increase in tread width. Increased air retention capability to meet extended vehicle service intervals, maintain fuel economy, and help prevent decrease in ultimate casing life is continuing. Run-flat capability, though important, has only captured a very small percentage of the automobile tire market.

As shown in Table 14.9, increase in the numbers of tire sizes will continue, leading to increases in manufacturing complexity, but offset by the overall growth of the automotive industry.

TABLE 14.8
Tire Trends

Source	Tire Property	Reason
Government regulations	Durability	Safety
	Traction	Safety
	Noise	Environmental
	Rolling resistance	Environmental
	Wear	Consumer protection
Vehicle manufacturers	Uniformity	Driver comfort
	Load-carrying capability	Increasing vehicle weights
	Low-profile constructions	Improve tire–vehicle handling
	Inflation pressure retention	Zero maintenance between service
Tire manufacturers	Size and design proliferation	Increase number of vehicle platforms
	Manufacturing automation	Lack of skilled labor, uniformity
	Novel materials	Environmental

TABLE 14.9
Trends in Tire Sizes

Year	1951	1975	2003
Sizes (Tire and Rim Association)	21	91	254
Average section width (mm)	155	235	245
Lowest aspect ratio	96	50	30
Rim sizes (in.)	12–15	12–15	12–22
Maximum rated load (kg)	735	1000	1320

VI. SUMMARY

This chapter has reviewed some of the many applications for which elastomeric materials and rubber compounds are used. The role of the modern materials scientist in the tire and rubber industry is to develop new products and improve the quality and characteristics of existing products. The key parameters governing this include the following:

1. *Performance*—The product must meet end user requirements.
2. *Quality*—The product must be durable and uniform and have good appearance.
3. *Environmental factors*—Not only must the product, manufacturing processes, and materials composition be in compliance with existing regulatory requirements but also the modern materials scientist must be aware of any future liabilities that might need to be accounted for.
4. *Cost*—The materials scientist must consider additional factors such as compound uniformity, consistency, manufacturing throughput, raw materials cost, and conversion costs.

In meeting these objectives, rubber compounding has developed from an intuitive art to a quantitative science necessitating knowledge of advanced chemistry, physics, and mathematics, and structural mechanics.

REFERENCES

1. Kovac FJ, Rodgers MB. Tire engineering. In: Mark JE, Erman B, Eirich FR, eds. *Science and Technology of Rubber*. San Diego, CA: Academic Press, 1994, pp. 675–718.
2. The Tire and Rim Association Inc. *Year Book*. Copley, OH, 2013.
3. European Tyre and Rim Technical Organization. *Standards Manual*. Brussels, Belgium, 2013.
4. Clark SK. *Mechanics of Pneumatic Tires*. Washington, DC: U.S. Department of Transportation, 1982.
5. Gent AN, Walter JD. *The Pneumatic Tire*. Washington, DC: National Highway Traffic Safety Administration, U.S. Department of Transportation, 2005.
6. Ford TL, Charles FS. *Heavy Duty Truck Tire Engineering. 34th Buckendale Lecture*. Warrendale, PA: SAE, SP729, 1988.
7. Rodgers MB, Tracey D, Waddell W. Tire applications of elastomers, 1. Treads. Paper H. In: Presented at a *Meeting of the American Chemical Society Rubber Division,* Grand Rapids, MI, 2004.
8. Rodgers MB, Waddell WH. The science of rubber compounding. In: Mark JE, Erman B, Eirich FR, eds. *Science and Technology of Rubber*. San Diego, CA: Academic Press, 2004.
9. International Institute of Synthetic Rubber Producers. *The Synthetic Rubber Manual*, 11th edn. Houston, TX, 1989.
10. ASTM D1765. *Standard Classification for Carbon Blacks Used in Rubber Products*, 2012.
11. Klingensmith W. Rubber compounding. In: Kroschwitz JI, Howe-Grant M, eds. *Kirk-Othmer Encyclopedia of Chemical Technology*, 4th edn., New York: John Wiley, Vol. 21. 1997, pp. 481–561.
12. Malaysian Rubber Producer's Research Association. *The Natural Rubber Formulary and Property Index*. Imprint of Luton, England, 1984.
13. Duda EJ. What do automotive engineers want from vendors? *Elastomerics* 120: 18, September 1988.
14. Nordsiek NH. The "integral rubber" concept—An approach to an ideal tire tread rubber. *Kautsch Gummi Kunst* 38: 178–185, 1985.
15. Gatti LF, Huffcut RJ. Applying dynamic mechanical properties for tire compound development. In: *International Tire Conference*, Akron, OH, September 1996.

Index

A

Accelerator
  CDMPS, 484
  classes
    activated sulfenamide cure system, 474
    highly unsaturated elastomers, 474
    primary/secondary, 474–475
    principal chemical, 474
    specific process conditions, 475
  description, 465
  dispersion, MBTS particle size, 514, 519
    elongation, 516–517
    rheometer data, 516, 518
    tensile strength, 516–517
  inorganic compounds, 473
  rubber compounder, 473
  structures, 483
  TBSI, 484
  type, 483–484
  zinc-mediated accelerated sulfur
      vulcanization
    Arrhenius activation energy, 484–485
    historical and general aspects, 476–478
    QSAR studies, 478–483
Acrylic elastomers (ACMs), 93–94
Acrylonitrile-butadiene rubber (NBR)
      derivatives, 94–95
Activators
  cure rate, 472
  description, 465
  zinc oxide, 471–472
Adhesives, 10, 129–130, 268
Aliphatic TPUs, 154–155
Alkylated chlorosulfonated polyethylene
      (ACSM), 570–571
Alkylated melamine-formaldehyde resin,
      407–408
Ambient vs. cryogenically ground whole tire
      recycled rubber, 528
Anionic polymerization and anionic polymers
  chain transfer, 58
  commercial anionic polymers and processes,
      59–60
  initiation
    aggregation of organolithium
        species, 46
    functional initiators, 49
    head group loss, 49
    lithium amide initiators, 47

  polymerization of butadiene, 49–50
  solubility and effectiveness, 48
  tin tetrachloride, 47
  tributyltin lithium, 49
  other types, 58–59
  propagation
    kinetic factors, 50
    mechanism, 50
    microstructure formation, 51
    monomer/initiator ratio, 50–51
    polar modifiers effect, 52
    potassium butoxide/lithium ratio, 52–53
  termination
    agents, 57
    alkoxysilanes, 56
    amine-and tin-containing electrophiles,
        53
    coupling of solution SBR, 54–55
    lithium polymerization, 53–54
    loss tangent vs. temperature for tin-carbon
        bonds, 56
    Mooney viscosity, 55
    tin-containing compounds, 54
    tin content position effect, 54–55
    tin tetrachloride, 56
    tributyltin-capped low-molecular-weight
        polybutadiene, 54
Antioxidants
  antiozonants
    NR/BR and SBR compounds, 443–448
    solubility, 448–450
  arylalkyl-PPDs, 442–443
  charge density, 441–442
  commercially available, 420
  compounding evaluation of waxes,
      451–453
  cure systems, 421
  dialkyl-PPDs, 443–444
  experiments
    aromatic amines (see Aromatic amine
        antioxidants)
    phenolic antioxidants, 428–430
  mechanisms
    aromatic amine, 425–427
    hydroperoxide-decomposing antioxidants,
        427–428
    phenolic, 423–424
  oxidative degradation, 421–423
  petroleum waxes, 451–452
  protective systems, 458

scavenger theory, 420
TAPDT, 443, 445
vulcanization system
    compounding evaluation, 455–457
    peroxide cure system, 454–455
    sulfur cure system, 454–455
Antiozonants
    chemical reactivity
        arylalkyl-PPDs, 442–443
        charge density, 441–442
        dialkyl-PPDs, 443–444
        TAPDT, 443, 445
    commercially available, 441–442
    NR/BR and SBR compounds, 443–445
        dynamic migration, 446–447
        measurements, 447–448
        static migration, 444–446
    solubility
        ozone crack resistance, 450
        SBR, NR /BR, and NR compounds,
            448–449
        water and acid rain, 448–450
Antistatic plasticizers, 363
Aromatic amine antioxidants
    experiments
        bead filler, 436–437
        hydroperoxide-decomposing antioxidants,
            437–441
        tire casings and internal components,
            431–436
    mechanisms, 425–427
Aromatic process oils, 371
Asphalt, 353
Atomic force microscopy (AFM), 263–264
Automotive hoses, 129–132

B

Bamboo fiber-filled NR, 28
Barrier hose design, 552
Bicycle tires, 322–323
Biotechnology filler, 28
Bitumen, 353
Boots and bellows, 303–304
Brominated isobutylene-*co-para*-methylstyrene
        (BIMSM)
    structure, 116
    synthesis and manufacture, 113–114
Brunauer-Emmett-Teller (BET) adsorption
        method, 256
Bulk polymerization, 91
Butadiene polymerization, 67–70
n-t-Butyl-2-benzothiazolesulfenimide, *see* TBSI
Butyl rubber
    cure systems, 513, 516
    DTDM-based cure systems, 513
    fatigue resistance, 512

product performance, 513
splicing behavior, 512
structure, 114
synthesis and manufacture, 110–112

C

Calcium carbonate, 28
Carbon black, 277
    classification and various grades
        ASTM grades, 232, 234–235
        carcass/semireinforcing grades, 236–237
        nitrogen surface area, 232–233
        N100 series, 233
        N200 series, 233
        N300 series, 233
        thermal grades, 237
    definitions, 210–211
    in-process modification
        carbon black-silica dual-phase filler,
            244–245
        chemical addition, 243
        metal addition, 243–244
        wide aggregate size distribution blacks,
            244
    manufacturing process, 211–212
        drying, 216–217
        filtration/separation, 215–216
        furnace process, 212, 214
        pelletizing, 216
        reactor conditions *vs.* properties, 215
        schematic, 212–213
        thermal process, 214–215
    postprocess modification
        gas adsorption, 240
        oxidation, 240–241
        plasma treatment, 242
        polymer grafting, 242–243
        reaction with diazonium salts, 241–242
    quality control
        impurities, 223–224
        in-rubber tests, 224–225
        pellet properties, 222–223
        specific surface area, 217–220
        structure, 220–221
        tint strength, 221–222
    rubber properties
        cured, 228–232
        mixing and dispersion, 225–228
        uncured, 228
    surface chemistry, 238–239
Chemical devulcanization, 531
Chemomechanical devulcanization, 531
Chloroprene, metal oxide vulcanization of,
        470–471
Chlorosulfonated polyethylene, 98–99
Clays, 28

Cobalt catalysts, 70
Cohesive energy density, 380
Commercial emulsion polymers and process, 43–46
Commercial truck tire, 581
Commercial Ziegler-Natta polymers and processes, 75
Compatible resins
    aromatic process oils, 390
    base formulations, 389, 391
    composition of tread compound, 389, 391
    crack propagation, 389, 391
    liquid hydrocarbon resins, 391, 393
    mechanical dynamic properties, 389, 392
    styrene butadiene tread compound, 389–390
    tear propagation performance, 390
    tensile strength and elongation, 389
Conveyor belts
    fabric reinforced belt, 572–574
    power transmission belts, 301
    steel cable reinforced belt, 572–574
Coolant hoses
    classification, 539–541
    diagram, 538–539
    elastomer characteristics, 541–542
    electrochemical degradation, 544–546
    ethylene-propylene elastomer-based, 541
    manufacturing process, 539
    materials, 541
    peroxide vulcanization, 543–544
    silicone elastomer-based, 546
    sulfur vulcanization, 542
Copolymers, 353
CORTERRAT, 156
Coumarone-indene resins, 383–384
Coumarone resins, 351–352
Crepe rubber
    estate brown, 10–11
    flat bark, 11–12
    pale latex, 9–10
    pure smoked blanket, 11–12
    thick blanket, 10–11
    thin brown, 10–11
Cured rubber properties
    accelerated sulfur vulcanization systems, 228
    vs. carbon black loading, 229–230
    vs. carbon black nitrogen surface area, 229
    carbon black structure, 228
    vs. carbon black structure, 229–230
    low-strain properties, 231
    N234 and N330, 231–232
    N234-filled SBR and unfilled D3191, 229, 231
    physical properties, 228
    strain properties, 231
    viscous modulus, 230
Cure rate, 464
Cure time, 464

D

Deproteinized NR, 21
1,2-Dihydro-2,2,4-trimethylquinoline (TMQ), 426
Dispersing agents, 355
Dithiodimorpholine, 468
Drying, carbon black, 216–217
Dynamic parts, 130, 133
Dynamic vulcanization
    definition, 174
    development, 174–176
    interphase structure, 185
    melt viscosity control, 195–196
    morphology control, 192–195
    plastic phase crystallinity
        brittle material, 186
        hydrodynamic volume, 186
        iPP morphology and mechanical properties, 187, 189–191
        melt viscosity, 186
        plane strain condition, 187–188
        plane stress condition, 187–188
        plastic crystal melting point, 185
        principal tensile stress, 187
    rubber and plastic compatibility
        bending, 179
        compatibilization, 178
        diblock polymer, 178
        emulsification, 178
        formulations and properties, 180–181
        impact strength, 179
        interfacial adhesion, 177
        mechanical property improvement, 178
        melt morphology, 179
        phase images, 181
        polymer interfacial tension, 177
        stiffness, 179
        transformation of thermoplastic olefins, 182–185
    rubber vulcanization, 190–192

E

Earth mover and off-the-road tires, 317–318
Ebonite, 22
Elastomeric copolyesters and copolyamides, 155–157
Elastomers
    butyl rubber, 512–513, 516
    EPDM, 503–510
    neoprene, 511–512, 514–515
    nitrile rubber, 510–513
Emulsion-filler masterbatches, 43

Emulsion polymerization and polymers, 91
   commercial emulsion polymers and process,
     43–46
   emulsion-filler masterbatches, 43
   functional emulsion polymers, 42
   oil-extended emulsion polymers, 42–43
   polymerization, 39–41
Engine mounts, belts, and air springs, 303
EPDM. *see* Ethylene-propylene-diene terpolymer
   (EPDM)
Epoxidized NR, 21
Estate brown crepes, 10–11
Ethylene-acrylic (EAM) elastomers, 99–100
Ethylene-propylene copolymers, 101–103
Ethylene-propylene-diene terpolymer (EPDM),
   503, 509
   cure systems for low set, 509–510
   and EPM, 101–102
   ExxonMobil Chemical Vistalon™, 102–103
   high-saturation, 504, 507
   low-saturation, 504–505
   medium saturation, 504, 506
   nonblooming system, 504
   thiazole-and sulfenamide-curing systems,
     504, 508
   Triple 8 system, 504
Ethylene-propylene-diene terpolymers, 101–103
Ethylene-propylene elastomer-based coolant
   hoses, 541
Ethylene thiourea (ETU), 511

F

Fatty acid, 24
   amides, 346
   esters, 345
Fatty alcohols, 346
Filtration/separation, carbon black, 215–216
Finish/surface modifier, 268
Flame retardant ester plasticizers, 363
Flat bark crepes, 11–12
Floor coverings, 303–304
Fluorocarbon elastomers, 104–105
Forming, 336
Free volume theory, 360–361
Fuel hose
   classification, 553–554
   environmental and conservation issues,
     547–548
   fuel recirculating method, 548–549
   hose cover material development, 551–552
   hose designs, 552–553
   hose tube material development, 549–551
   reservoir method, 548
   sealed housing for evaporative determination,
     549
   testing, 548

Functional emulsion polymers, 42
Furnace process, carbon black, 212, 214

G

Gas-phase polymerization, 75
Gel theory, 360
General-purpose elastomers
   anionic polymerization and anionic
     polymers
     chain transfer, 58
     commercial anionic polymers and
       processes, 59–60
     initiation, 46–50
     other types, 58–59
     propagation, 50–53
     termination, 53–58
   emulsion polymerization and emulsion
     polymers
     commercial emulsion polymers and
       process, 43–46
     emulsion-filler masterbatches, 43
     functional emulsion polymers, 42
     oil-extended emulsion polymers, 42–43
     polymerization, 39–41
   SBRs in tire compounds
     anionic polymerization, 61
     characterization and stress relaxation,
       64–65
     microstructural and macrostructural
       differences, 60
     RMS radius *versus* molar mass of
       solution, 64, 66
     styrene and vinyl effect, 63–65
     styrene level effect, 63
     traction and wear, 61
     wear resistance, 61–62
     wet skid resistance, 61
   structure-property relationships
     glass transition temperature, 35–37
     laboratory testing methods, 34–35
     molecular weight and molecular weight
       distribution, 37–38
     sequence distribution in solution
       SBR, 38
   Ziegler-Natta polymerization and Ziegler-
     Natta polymers, 66–67
     cis-polybutadiene, 70–73
     commercial Ziegler-Natta polymers and
       processes, 75
     gas-phase polymerization, 75
     mechanism of butadiene polymerization,
       67–70
     syndiotactic polybutadiene, 73–75
Golf balls, 305
Granular chemicals, 363–364
Guayule, 21–22

H

Halogenated butyl rubber
    structure, 114–115
    synthesis and manufacture, 111, 113
Hammett equation, 478–479
Hammett Sigma constants, 479
Heat-resistant vulcanizates, 363
High-saturation EPDM, 504, 507
High sulfur nitrile rubber systems, 510–511
Homogenizing agents
    asphalt and bitumen, 353
    copolymers, 353
    coumarone resins, 351–352
    difficult-to-blend elastomers, 351
    fillers, 351
    lignin, 353–354
    petroleum resins, 352–353
    phenolic resins, 353–354
    processing with, 354–355
    raw materials, 351
    rosins, 353–354
    solubility parameters, 351–352
    terpene resins, 353
Hydraulic hose
    acrylonitrile butadiene nitrile rubber, 553
    braided reinforcement, 556
    service lives, 554
    specification, 554–555
    spiral reinforcement, 556
    steel wire-reinforced, 554
Hydrocarbon resins
    acidity and basicity, 388
    color, 388
    compatible resins performance, 389–393
    coumarone-indene resins, 383–384
    glass transition temperature/softening point,
        385–386
    incompatible resins performance, 391–394
    molecular weight and molecular weight
        distribution, 387
    performance, 388–389
    petroleum-based hydrocarbon resins,
        384–385
    processing at high temperature, 388–389
    residual unsaturation, 387–388
    solubility and compatibility, 387
    viscosity, 386–387
Hydroperoxide-decomposing antioxidants
    distearyl thiopropionate, 428
    nonstaining polymer stabilization systems, 437
    organic phosphite esters and sulfide, 427
    oxidation, 427
    stabilization system
        cis-BR, 437–438
        SBR, 439–441
    tris(nonylphenol)phosphite, 428

I

Imino melamine dimers, 408–409
Incompatible resins
    complex modulus, 392–394
    dynamic stiffness, 393
    mixing and processing temperature, 391
    softening point, 392
    static stiffness, 393
Industrial rubber products
    boots and bellows, 303–304
    conveyor belts
        fabric reinforced belt, 572–574
        power transmission belts, 301
        steel cable reinforced belt, 572–574
    coolant hose
        classification, 539–541
        diagram, 538–539
        elastomer characteristics, 541–542
        electrochemical degradation, 544–546
        ethylene-propylene elastomer-
            based, 541
        manufacturing process, 539
        materials, 541
        peroxide vulcanization, 543–544
        silicone elastomer-based, 546
        sulfur vulcanization, 542
    engine mounts, belts, and air springs, 303
    floor coverings, 303–304
    fuel hose
        classification, 553–554
        environmental and conservation issues,
            547–548
        fuel recirculating method, 548–549
        hose cover material development,
            551–552
        hose designs, 552–553
        hose tube material development,
            549–551
        reservoir method, 548
        sealed housing for evaporative
            determination, 549
        testing, 548
    golf balls, 305
    hydraulic hose
        acrylonitrile butadiene nitrile
            rubber, 553
        braided reinforcement, 556
        service lives, 554
        specification, 554–555
        spiral reinforcement, 556
        steel wire-reinforced, 554
    rice hulling rollers
        DIN abrasion, 300, 302
        NBR, 300, 302
        nonstaining properties, 299
        SBR, 300–301

seals, cables, profiles, and hoses
    calcined clays, 297, 299
    chlorosulfonated polyethylene rubber,
        299–300
    fuel sealing compound, 297, 299
    in-rubber data, 295, 297
    oil seals, 296, 298
    peroxide-curing system, 297
    pre-reacted silicas, 296
    stress-strain curves, 296, 299
    washing machine sealing compounds,
        295, 298
soft rollers, 300–301, 303
synchronous belts, 567–568
    alkylated chlorosulfonated polyethylene,
        570–571
    aramid, 571
    glass, 571
    materials, 568–570
    performance criteria, 572
    short fiber/pulp tooth reinforcement, 571
    tensile member, 571
V-belts, 556–557
    banded, 557, 559
    fabric components, 563
    fabric-wrapped, 557–558
    hexagonal, 557, 559
    jackets, 558, 562–563
    materials, 558–562, 564–565
    performance criteria, 565–566
    raw edge, 557–558
    variable speed, 557–558
    V-ribbed belts, 563–564
Infrared (IR) spectroscopy, 262–263
Infuse™, 158
Innerline compounds, properties of, 528, 530
In-process carbon black modification
    carbon black-silica dual-phase filler,
        244–245
    chemical addition, 243
    metal addition, 243–244
    wide aggregate size distribution
        blacks, 244
In-rubber tests, 224–225
In situ reinforcing resin networks
    cure system, 395
    dihydroxybenzoazine, 394–395
    hexamethoxymethylmelamine, 393
    hexamethylenetetramine, 393
    methylene donors, 396
    phenol-and resorcinol-based novolak resins,
        395
    resorcinol, 394
    steric factors and chemical reactivities, 395
    thermoset resin network, 393
International Institute of Synthetic Rubber
        Producers (IISRP), 43

Isobutylene-based elastomers
    applications
        automotive hoses, 129–132
        dynamic parts, 130, 133
        pharmaceuticals, 130, 134
        tire black sidewall, 125–126
        tire curing bladders and envelopes,
            128–129
        tire innerliner, 121–125
        tire treads, 126–128
        tire white sidewall and cover strip,
            126–127
    chemical properties
        solubility, 118
        stability, 119
        vulcanization, 119–121
    physical properties
        dynamic damping, 118
        permeability, 116–118
    structure
        brominated isobutylene-co-para-
            methylstyrene, 116
        butyl rubber, 114
        halogenated butyl rubber, 114–115
        polyisobutylene, 114
        star-branched butyl rubber, 115–116
    synthesis and manufacture
        brominated isobutylene co-para-
            methylstyrene, 113–114
        butyl rubber, 110–112
        halobutyl rubbers, 111, 113
        star-branched butyl rubbers, 113

K

Koresin®, 358

L

Lignin, 353–354
Liquid low-molecular-weight rubber, 20
Low-saturation EPDM, 504–505
Lubricants
    discussion, 344–346
    influence, 350
    processing, 348–350
    properties and mode of action, 346–349
Lubricity theory, 360

M

Mastication, 338–341
Mechanistic theory of plasticisation, 361
Medium saturation EPDM, 504, 506
Melt viscosity control, 195–196
Metal oxide vulcanization, 470–471

Metal soaps, 345
Methyl methacrylate grafting, 20
Microdispersion, 227
Mixing, 335–336
Motorcycle tires, 321

N

Naphthenic process oils, 371
Natural rubber (NR)
  chemistry
    ammonia, 6
    biosynthesis, 3–4
    cis-and trans-isomers, 2–3
    definitions, 5
    forms of rubber, 5
    gutta-percha, 3
    molecular-weight distribution, 2
    tapped latex, 3
  naturally occurring materials, 27–29
  production, 6–7
  products and grades
    crepe rubber, 9–12
    schematic of production
      process, 7–8
    sheet rubber, 8–9
    technical classification, 12–13
    TSR, 12–15
  quality
    consistency and uniformity, 23
    contamination, 23–24
    fatty acids, 24
    packaging, 23
  special-purpose, 20–22
  tires, 25–27
  viscosity and viscosity stabilization
    baling temperature, 19
    bound rubber, 19
    crystallization, 20
    cure characteristics, 17
    field methods, 19
    hardening phenomenon, 18
    hydroxylamine neutral sulfate, 18
    molecular-weight distribution, 17
    Mooney viscometer, 15–17
    natural antioxidants, 17–18
    peptizers, 18
    Wallace plasticity test, 18
Neodymium catalysts, 71–73
Neoprene, 511–512, 514
Network maturation, 464
Nickel catalysts, 70–71
Nitrile rubber
  low vs. high sulfur, 510, 512
  sulfur-based cure systems, 510–511
  sulfurless cure systems, 510, 513
NMR spectroscopy, 261–262

Nonmetal-activated sulfenamide accelerator with
    sulfur, 472
Nonsulfur crosslinks, 468
Novolak resins, 397–399
NR. see Natural rubber (NR)

O

Oil absorption number, 258–259
Oil-extended emulsion polymers, 42–43
Oil-extended NR, 20–21
Olefin block copolymers (OBCs), 106–107,
    157–159
One-component melamine resins, 407
Organosilicones, 346

P

Pale latex crepes, 9–10
Paraffinic process oils, 371
Passenger car tire treads
  abrasion behavior, 307, 309
  abrasion resistance, 312, 314–315
  compound and vulcanizate data,
    307–308
  dispersion behavior, 307
  dynamic modulus, 312–313
  filler-filler network, 310, 312
  green tire tread formulation, 308, 310
  hysteresis loss, 306
  improved tread wear, 307, 309
  loss factor, 312–313
  mixing procedure, 308, 311
  Mooney viscosity, 310–311
  RPA measurement, 312, 314
  typical analytical properties, 308–309
  typical green tire tread formulation, 306
  wet grip performance, 305
Pelletizing, carbon black, 216
Peroxide cure system, 454–455
Peroxide vulcanization, 469–470, 543–544
Petroleum resins, 352–353, 384–385
Petroleum waxes, 451–452
Pharmaceuticals, 130, 134
Phenolic antioxidants
  experiments
    nonstaining and nondiscoloring
      properties, 428
    stabilization system, 429–430
    white sidewall compound, 429–430
  mechanisms, 423–424
Phenolic resins, 353–354
  in situ reinforcing resin networks, 393–396
  novolak resins, 397–399
  polymerized cashew nutshell oil, 400–403
  resole-type, 399–402
Phosphazenes, 105

Physical and chemical peptizers
  mastication
    common peptizing agents, 340–341
    dispersion, 338
    mechanical breakdown process, 339
    polyisoprene, 338–339
    rubber, 338–339
    rubber-reaction sequence, 338–339
    viscosity reduction *vs.* temperature, 338,
      340
    zinc soaps, 340
  peptizing agents
    influence of, 342–343
    processing with, 341–342
Plasticized poly(vinyl chloride) (PVC),
    142–143
Plasticizers
  compatibility, 361–362
  functions, 359–360
  processing, 363
  selection, 361–363
  theory, 360–361
Poly(α-methylstyrene) (PαMS), 168
Polychloroprene, 95–97
Poly(2,6-dimethyl-1,4-phenylene oxide) (PPO),
    168–169
Polyethers, 105–106
Polyisobutylene, 114
Polymer grafting, 242–243
Polymerization, 39–41
Polymerized cashew nutshell oil
  acid catalyzed polymerization, 401
  Cardolite NC 360®, 401
  components, 401–402
  preparation and performance, 401, 403
  resorcinol, 400
Polyterpene resins, 383
Post-consumer scrap recycling, 523
Postprocess carbon black modification
  gas adsorption, 240
  oxidation, 240–241
  plasma treatment, 242
  polymer grafting, 242–243
  reaction with diazonium salts, 241–242
Powder carbon black, 217
Pre-vulcanization inhibitor (PVI), 465
Primary accelerators, 474–475
  cure rate, 485
  modulus development, 485
  natural rubber, 485–486
  reversion resistance, 488–489
  SBR, 485, 487
  scorch delay, 485
  sulfenamide accelerators, 488
  thiazoles, 488
Primer, 267–268

Process additives for silica-loaded tread
    compounds
  ester process additive, 367
  low-rolling-resistance tires, 366
  mixing and processing behavior,
    368–369
  silica-to-silane coupling reaction, 367
  STRUKTOL HT 207, 369
  viscosity, 369
  zinc–potassium soap, 367
Process oils
  characterization
    analytical tests, 371–372
    aniline point, 373
    aromatic content, 374
    color, 373
    environmental considerations, 374
    evaporative loss, 374
    flash point, 373
    pour point, 373–374
    properties, 371–372
    refractive index, 373
    relative density, 372
    viscosity, 372–373
    viscosity-gravity constant, 374
  compatibility, 374–375
  discussion, 369–370
  manufacturing, 370–371
  trends, 375–376
Pure smoked blanket crepe, 11–12
PVI (*N*-cyclohexylthiophthalimide)
  dithiocarbamates, 503
  Mooney scotch characteristics, effect on,
    497, 501
  processing safety, 499
  PVI concentration *vs.* Mooney scorch 100NR,
    499, 502
  Santogard® PVI, effect of, 499, 502
  skim stocks, 503

Q

Q rubbers. *see* Silicone rubber
QSAR studies
  characteristic flow of electrons, 481
  crosslink formation, 482
  dissociation equilibrium, 478
  Hammett relations, 479
  parameters, 480
  polysulfidic zinc structures, 482
  primary amine-based sulfenamides, 481
  secondary accelerators, 483
  sulfenamide and sulfenimide zinc complexes,
    480
  sulfurating intermediate structure, 481–482
  sulfur vulcanization, 479–480

Quality control, carbon black
    impurities, 223–224
    in-rubber tests, 224–225
    pellet properties, 222–223
    specific surface area, 217–220
    structure, 220–221
    tint strength, 221–222
Quantitative structure-activity relationship
        (QSAR), 483
    crosslink formation, 482
    Hammett equation, 478–479
    historical studies, 478–479
    sulfenamide and sulfenimide zinc complexes,
        480
    sulfurating intermediate structure,
        481–482
    of sulfur vulcanization, 479–480

R

Radial tire casing compound, properties of,
        531–532
Radiator hoses. *see* Coolant hoses
Rate of cure, 464
Raw materials handling, 334
Raw rubber, 463
Recycled materials
    industrial products, 533–534
    tires, 533
Recycling of rubber
    ambient grinding, 526
    ambient ground rubber SBR 1502
        compounds, 527
    cryogenic grinding, 524–527
    EPDM compounds, properties of,
        528–529
    mesh size, 524
    reclaim rubber
        advantages, 531
        automotive vehicle mats, 532–533
        chemical devulcanization, 531
        chemomechanical and thermomechanical
            devulcanization, 531
        radial tire casing compound, properties
            of, 531–532
        thermal devulcanization, 531
        ultrasonic devulcanization, 530
    reclaim system, 524
    recycled material application, 533–534
    reduction, 525
    reuse of tires, 526
    surface-activated crumb process, 529–530
    wet grinding, 528–529
Resins
    cohesive energy density, 380
    compatible, 381

cured rubber, 415
    hydrocarbon
        acidity and basicity, 388
        color, 388
        compatible resins performance, 389–393
        coumarone-indene resins, 383–384
        glass transition temperature/softening
            point, 385–386
        incompatible resins performance, 391–394
        molecular weight and molecular weight
            distribution, 387
        performance, 388–389
        petroleum-based hydrocarbon resins,
            384–385
        processing at high temperature, 388–389
        residual unsaturation, 387–388
        solubility and compatibility, 387
        viscosity, 386–387
    phenolic
        in situ reinforcing resin networks,
            393–396
        novolak resins, 397–399
        polymerized cashew nutshell oil,
            400–403
        resole-type, 399–402
    plasticizers, 380
    rosin and terpene-based, 382–383
    rubber reinforcements, 401, 403–407
    softener, 380
    solubility parameter, 380
    synthetic hydrocarbon, 382
    3D solubility parameter approach, 381
    trends, 412–415
    vulcanization, 470
    weight-average molecular weights, 379
    wire adhesion
        alkylated melamine-formaldehyde resin,
            407–408
        ASTM D 2229-93a, 408, 411
        ASTM de Mattia crack growth, 408, 412
        cured properties, 408, 411
        imino melamine dimers, 408–409
        vulcanized melamine formaldehyde MF
            resin, 408, 410
        wire coat formulation, 408, 410
Resole-type phenolic resins
    butyl rubbers, 400, 402
    cure systems, 400–401
    free formaldehyde, 399
    heat reactive, 400
    molar ratio, 399
    preparation, 399–400
Retarders
    description, 465
    Mh/Rmax, 466
    Ml/Rmin, 466

PVI
  dithiocarbamates, 503
  Mooney scotch characteristics, effect on, 497, 501
  processing safety, 499
  PVI concentration *vs.* Mooney scorch 100NR, 499, 502
  Santogard® PVI, effect of, 499, 502
  skim stocks, 503
  t25, 466
  t_{90}, 466
  ts2, 466
Ribbed smoked sheet (RSS) rubber, 5
Rice bran oil, 29
Rice hulling rollers
  DIN abrasion, 300, 302
  NBR, 300, 302
  nonstaining properties, 299
  SBR, 300–301
Ring-opened polymers, 106
Rosins
  pine trees, 353
  terpene-based resins, 382–383
  unsaturated acids, 353–354
Rubber properties
  cured
    accelerated sulfur vulcanization systems, 228
    vs. carbon black loading, 229–230
    vs. carbon black nitrogen surface area, 229
    carbon black structure, 228
    vs. carbon black structure, 229–230
    low-strain properties, 231
    N234 and N330, 231–232
    N234-filled SBR and unfilled D3191, 229, 231
    physical properties, 228
    strain properties, 231
    viscous modulus, 230
  mixing and dispersion
    ASTM standard test method, 227
    carbon black-filled rubber compound, 226
    deagglomeration, 226
    filler–filler network, 227
    microdispersion, 227
    resistivity, 228
    shear forces, 225
    specific surface area, 225
    uncured, 228
Rubber reinforcements, 401, 403–407
  general considerations, 277–278
  in-rubber performance
    green compound, 283–285
    vulcanization, 285–287
  peroxide-cured EPM compound, 287, 289
  sulfur-cured S-SBR/BR compound, 286, 288

  types of interactions
    CB-CB and CB-polymer interactions, 279
    silica-silane-rubber interaction, 281–283
    silica-silica and silica-rubber interaction, 279–281
Rubber vulcanization, 190–192

S

SBR. *see* Styrene-butadiene rubber (SBR)
Scanning electron microscopy (SEM), 261
Scorch, 463
Seals, cables, profiles, and hoses
  calcined clays, 297, 299
  chlorosulfonated polyethylene rubber, 299–300
  fuel sealing compound, 297, 299
  in-rubber data, 295, 297
  oil seals, 296, 298
  peroxide-curing system, 297
  pre-reacted silicas, 296
  stress-strain curves, 296, 299
  washing machine sealing compounds, 295, 298
Secondary accelerators, 474–475
  different fillers, 497, 500
  natural rubber, 491–492
  nitrile, 495–497
  SBR, 492–494
  variation, 497–499
Sheet rubber, 8–9
Shoe soles, 292–296
Silanes. *see also* Silica
  characterization
    methods of analysis, 271–273
    silane-rubber coupling, 274–275
    silica-silane coupling, 273–274
  definitions, 267–268
  essentials
    general structure, 269
    history, 269
    types and applications, 269–270
  function, 268
  overview of reactions, 268–269
  process and technology, 270–271
  product overview and applications, 275–276
  and silica
    carbon black, 277
    improved processing, 290–291
    reduced ethanol emission, 290
    rubber reinforcement, 277–289
    tire label and silane coupling efficiency, 287–290
    white fillers, 277
Silica
  bicycle tires, 322–323

characterization, 255–256
  AFM, 263–264
  chemical bulk analyses, 264–265
  chemical reactions of silanol groups, 261
  hexadecyl-trimethyl-ammonium bromide
    surface area, 256–257
  intra-aggregate and interaggregate
    structure, 257–258
  IR spectroscopy, 262–263
  microscopic methods, 259–260
  NMR spectroscopy, 261–262
  oil absorption number, 258–259
  PaSD, 261
  pore volume and pore size distribution, 257
  SAS, 264
  Sears determination, 256
  SEM, 261
  specific surface, 256
  TEM, 260–261
  thermoanalytical method, 263
  void volume, 259
commercial products, 266–267
earth mover and off-the-road tires, 317–318
essentials, 255
general considerations and basic information,
  253–255
history, applications, 291–292
industrial rubber goods
  boots and bellows, 303–304
  conveyor belts and power transmission
    belts, 301
  engine mounts, belts, and air springs, 303
  floor coverings, 303–304
  golf balls, 305
  rice hulling rollers, 299–302
  seals, cables, profiles, and hoses, 295–300
  soft rollers, 300–301, 303
motorcycle tires, 321
passenger car tire treads
  abrasion behavior, 307, 309
  abrasion resistance, 312, 314–315
  compound and vulcanizate data, 307–308
  dispersion behavior, 307
  dynamic modulus, 312–313
  filler-filler network, 310, 312
  green tire tread formulation, 308, 310
  hysteresis loss, 306
  improved tread wear, 307, 309
  loss factor, 312–313
  mixing procedure, 308, 311
  Mooney viscosity, 310–311
  RPA measurement, 312, 314
  typical analytical properties, 308–309
  typical green tire tread formulation, 306
  wet grip performance, 305
process and technology
  dispersibility and surface activity, 266

  production process, 265
  product properties, 265–266
  typical process, 266
shoe soles, 292–296
solid tires, 318–319
tire body, 319–321
truck tires, 316–317
winter tire treads, 313–316
Silicone elastomer-based coolant hose, 546
Silicone rubber, 102, 104
Small-angle scattering (SAS), 264
Softener, 380
Soft rollers, 300–301, 303
Solid tires, 318–319
Solution polymerization, 91
Solvent-resistant specialty elastomers
  acrylic elastomers, 93–94, 97–98
  acrylonitrile-butadiene rubber, 92–93
  chlorosulfonated polyethylene, 98–99
  ethylene-acrylic elastomers, 99–100
  generic compound formulation, 93–94
  NBR derivatives, 94–95
  polychloroprene, 95–97
Special-purpose elastomers
  attributes, 84
  generic nomenclature, 84–85
  international abbreviations, 85–86
  need, 85–89
  olefin block copolymers, 106–107
  solvent-resistant
    acrylic elastomers, 93–94, 97–98
    acrylonitrile-butadiene rubber, 92–93
    chlorosulfonated polyethylene, 98–99
    ethylene-acrylic elastomers, 99–100
    generic compound formulation, 93–94
    NBR derivatives, 94–95
    polychloroprene, 95–97
  synthesis and manufacturing process
    anionic polymerization, 90
    bulk polymerization, 91
    cationic polymerization, 90
    chain shuttling polymerization, 92
    emulsion polymerization, 91
    free radical polymerization, 89–90
    ionic polymerization, 90
    solution polymerization, 91
    suspension polymerization, 91–92
  temperature-resistant elastomers, 100
    ethylene-propylene copolymers and
      ethylene-propylene-diene terpolymers,
      101–103
    fluorocarbon elastomers, 104–105
    phosphazenes, 105
    polyethers, 105–106
    ring-opened polymers, 106
    silicone rubber, 102, 104
vulcanization, 89

Standard Indonesian Rubber (SIR), 13
Standard Thai Rubber (STR), 13
Star-branched butyl rubbers (SBBs)
    structure, 115–116
    synthesis and manufacture, 113
Starch, 28
State of cure, 464
Steel wire-reinforced hydraulic hoses, 554
STRUKTOL® W 33 Flakes, 355
Styrene-butadiene rubber (SBR)
    anionic polymerization, 61
    characterization and stress relaxation,
        64–65
    microstructural and macrostructural
        differences, 60
    RMS radius *vs.* molar mass of solution,
        64, 66
    styrene and vinyl effect, 63–65
    styrene level effect, 63
    traction and wear, 61
    wear resistance, 61–62
    wet skid resistance, 61
Styrenic block copolymers
    compounded materials, 164–165
    cross-link density, 160
    discrete plastic phase morphology, 161
    effect of paraffinic oil, 163–164
    frozen-in morphology, 162
    high order-disorder transition temperature, 162
    Kraton G1651, 160
    morphology, 165–166
    phase incompatibility, 161
    phase separation, 160
    polymer microstructure and morphology, 161
    selected new developments, 171–173
    spaghetti and meatball morphology, 160
    upper service temperature improvement,
        167–170
Styroflex 2G66, 172
Sulfenamide accelerators, 488
Sulfenamide-curing systems, in Vistalon 5600,
    504, 508
Sulfur cure system, 454–455
Sulfur preparations, 364–365
Sulfur vulcanization, 542
Superior processing rubber, 21
Surface-activated crumb process, 529–530
Suspension polymerization, 91–92
Synchronous belts, 567–568
    alkylated chlorosulfonated polyethylene,
        570–571
    aramid, 571
    glass, 571
    materials, 568–570
    performance criteria, 572
    short fiber/pulp tooth reinforcement, 571
    tensile member, 571

Syndiotactic polybutadiene, 73–75
Synthetic polyisoprene, 22

T

Tackifiers
    definition and manufacturing importance,
        355–356
    processing with, 359
    theories of autohesion and tack,
        356–359
Talc, 28
TBSI, 484, 488, 490
Technically specified rubber (TSR), 12–15
Temperature-resistant elastomers, 100
    ethylene-propylene copolymers and
        ethylene-propylene-diene
        terpolymers, 101–103
    fluorocarbon elastomers, 104–105
    phosphazenes, 105
    polyethers, 105–106
    ring-opened polymers, 106
    silicone rubber, 102, 104
Terpene resins, 353
Tetramethylthiuram disulfide (TMTD)
    dithiodimorpholine, 468
    metal oxide vulcanization, 470–471
    nonsulfur crosslinks, 468
    peroxide vulcanization, 469–470
    resin vulcanization, 470
    urethane vulcanization, 470–471
Thermal devulcanization, 531
Thermal process, carbon black, 214–215
Thermomechanical devulcanization, 531
Thermoplastic elastomers (TPE)
    classification, 141–143
    definition, 141
    olefinic block copolymers, 157–159
    segmented block copolymer
        cross-link density, 146
        elastomeric copolyesters and
            copolyamides, 155–157
        hard-phase content, 145–146
        polymer microstructure and
            morphology, 145
        products of commerce, 142–143
        property comparison, 143–144
        thermoplastic polyurethanes, (*see*
            Thermoplastic polyurethanes
            (TPU))
    styrenic block copolymers, 160–162
        compounded materials, 164–165
        effect of paraffinic oil, 163–164
        morphology, 165–166
        selected new developments, 171–173
        upper service temperature improvement,
            167–170

thermoplastic vulcanizates, 173–174
  dynamic vulcanization (*see* Dynamic
    vulcanization)
  hardness control, 199
  processability, 199
  rationalization of PP/EPDM elastic
    recovery, 197–199
Thermoplastic NR, 21
Thermoplastic olefins (TPOs), 143
Thermoplastic polyurethanes (TPU)
  addition step-growth polymerization, 146
  aliphatic, 154–155
  elastomeric copolyesters and copolyamides,
    155–157
  morphology and microstructure, 148–149
  phase mixing, 148
  phase separation, 147
  polyether-based, 146
  soft and hard segment, 147
  thermal characteristics
    aliphatic and aromatic diamines, 153
    annealing, 152
    annealing temperature, 149
    aromatic diol chain extenders, 153
    crystallization, 150
    elastic recovery, 152
    endotherm, 149
    hydrocarbon diols, 153
    Koberstein schematic model, 150–151
    melt memory, 151
    microphase separation transition, 150
    molecular heterogeneity, 152
    solubility, 151
    Spandex, 154
    thermogravimetric analysis, 150
Thermoplastic vulcanizates, 173–174
  dynamic vulcanization
    definition, 174
    development, 174–176
    interphase structure, 185
    melt viscosity control, 195–196
    morphology control, 192–195
    plastic phase crystallinity, 185–191
    rubber and plastic compatibility
      bending, 179
      compatibilization, 178
      diblock polymer, 178
      emulsification, 178
      formulations and properties,
        180–181
      impact strength, 179
      interfacial adhesion, 177
      mechanical property improvement,
        178
      melt morphology, 179
      phase images, 181
      polymer interfacial tension, 177
    stiffness, 179
    transformation of thermoplastic
      olefins, 182–185
    rubber vulcanization, 190–192
  hardness control, 199
  processability, 199
  rationalization of PP/EPDM elastic recovery,
    197–199
Thiazole-curing systems, in Vistalon 5600, 504,
  508
Thick blanket crepes, 10–11
Thin brown crepes, 10–11
Timing belts. *see* Synchronous belts
Tire black sidewall, 125–126
Tire body, 319–321
Tire casings and internal components
  alkylperoxyl radicals, 435
  antifatigue mechanism, 432, 435
  AO 445, 431, 434
  BLE, 431, 434
  fatigue phenomenon, 431
  fatigue protection, 436
  heat aging, 436
  mechanism of antioxidant action, 436
  nitroxyl radical formation, 432, 434–435
  polymerized1,2-dihydro-2,2,4-
    trimethylquinoline, 431, 433
  strength and durability, 431
  tire ply compound, 431–432
Tire curing bladders and envelopes, 128–129
Tire innerliner, 121–125
Tire label and silane coupling efficiency, 287–290
Tire technology
  bias/diagonal ply, 579–580
  constructions, 579–580
  materials
    compound development, 590–591
    compound mechanical properties,
      591–594
    compounds, 588–590
    reinforcements, 586–588
  parameters, 597–598
  testing, 593, 596
  trends, 596–597
  types, 580–581
    basic tire design, 581–582
    tire nomenclature and dimensions,
      582–584
    tread design, 584–586
  ultralow aspect ratio, 580
Tire treads, 126–128
Tire white sidewall and cover strip, 126–127
Titanium catalysts, 70
TMTD. *see* Tetramethylthiuram disulfide
  (TMTD)
TPE. *see* Thermoplastic elastomers (TPE)
TPU. *see* Thermoplastic polyurethanes (TPU)

Transmission electron microscopy (TEM),
    260–261
Truck tires, 316–317
TSR 5, 14
TSR 10, 14
TSR 20, 15
TSR 50, 15
TSR CV, 13–14
TSR L, 14

U

Ultralow aspect ratio tires, 580
Ultrasonic devulcanization, 530
Uncured rubber properties, 228
Uniform curative dispersion, 513
Urethane vulcanization, 470–471

V

V-belts, 556–557
    banded, 557, 559
    fabric components, 563
    fabric-wrapped, 557–558
    hexagonal, 557, 559
    jackets, 558, 562–563
    materials, 558–562, 564–565
    performance criteria, 565–566
    raw edge, 557–558
    variable speed, 557–558
    V-ribbed belts, 563–564
Veneer hose design, 552
Vulcanization, 89, 336–338
    compounding evaluation, 455–457
    description, 462–463
    effects, 467
    peroxide cure system, 454–455
    sulfur cure system, 454–455
Vulcanized melamine formaldehyde MF resin,
    408, 410

Vulcanized rubber, 463
Vulcanizing agents, 465
    sulfur vulcanization, 466–468
    tetramethylthiuram disulfide
        dithiodimorpholine, 468
        metal oxide vulcanization, 470–471
        nonsulfur crosslinks, 468
        peroxide vulcanization, 469–470
        resin vulcanization, 470
        urethane vulcanization, 470–471

W

White fillers, 277
Wide Aggregate size distribution blacks, 244
Winter tire treads, 313–316

X

X-ray diffraction, 264
Xylene-formaldehyde resins, 358–359

Z

Ziegler-Natta polymerization and Ziegler-Natta
        polymers, 66–67
    cis-polybutadiene
        cobalt catalysts, 70
        neodymium catalysts, 71–73
        nickel catalysts, 70–71
        titanium catalysts, 70
    commercial Ziegler-Natta polymers and
        processes, 75
    gas-phase polymerization, 75
    mechanism of butadiene polymerization,
        67–70
    syndiotactic polybutadiene, 73–75
Zinc-free rubber-processing additives,
        366–367